SP
Pvt. Ltd.

Current Concepts in
CROP PROTECTION

Susanta Banik

Department of Plant Pathology
School of Agricultural Sciences and Rural Development
Nagaland University, Medziphema-797106
Nagaland, India

2013

Studium Press (India) Pvt. Ltd.

Current Concepts in
CROP PROTECTION

ISBN: 978-93-80012-42-1

Published by:

Studium Press (India) Pvt. Ltd.
4735/22, 2nd Floor, Prakash Deep Building
(Near Delhi Medical Association),
Ansari Road, Darya Ganj, New Delhi-110 002
Tel.: 91-11-43240200-15 (15 lines), Fax: 91-11-43240215;
E-mail: studiumpress@gmail.com

Printed at:

List of Contributors

Achintya Pramanik: Bidhan Chandra Krishi Viswavidyalaya, Mohanpur Nadia, West Bengal, India, *E-mail*: achintyapramanik@gmail.com

Ashish Yadav: Defence Institute of High Altitude Research (DIHAR), C/o 56 APO, Leh-Ladakh, India

A. K. Mandal: Scientist (Plant Pathology), Directorate of Seed Research Kushmaur, P.O. Kaithauli, Dist. Mau (UP) 275 101, India, *E-mail*: asitmandal.iari@gmail.com

Biswarup Saha: College of Fisheries, Central Agricultural University, Lem bucherra, Tripura West - 799 210, India, *E-mail*: biswarup.ext@gmail.com

C. Sudhakar: Scientist (Agronomy), ARS, Acharya N.G. Ranga Agricultural University, Tandur, Rangareddy (Dt.)- 501 141, Andhra Pradesh, India, *E-mail*: chouratsudhakar@yahoo.com

C.V. Sameer Kumar: Sr. Scientist (Breeding), ARS, Acharya N.G. Ranga Agricultural University, Tandur, Rangareddy (Dt.)- 501 141, Andhra Pradesh, India, *E-mail*: cv_sameerkumar@yahoo.com

Chandan Kapoor: Scientist, Plant breeding, ICAR Research Complex Sikkim Centre, Tadong, Gangtok-737 102, India, *E-mail*: chandann aarm@gmail.com

D.M.J.B. Senanayake: Research Officer, Field Crops Research and Development Institute, Mahaillu ppalama, Sri Lanka, *E-mail*: jsenanayake@gmail.com

Gulzar S. Sanghera: Sr. Scientist (PBG), SKUAST-K, Mountain Research Centre for Field Crops, Khudwani, Anantnag, 192 102, J&K, India, *E-mail*: g_singh72@rediffmail.com

Jaydeep Halder: Economic Botanist-VII, Bethuadahari, Nadia, West Bengal, India. *E-mail*: jaydeep.halder@gmail.com, *Present address*: Crop Protection Division Indian Institute of Vegetable Research Varanasi, Uttar Pradesh-221305, India

M. Azadvar: Department of Plant Protection Research, Agricultural Research Center of Jiroft and Kahnouj, Jiroft, Iran, *E-mail*: m_azadehvar@yahoo.com

M. Suresh: Scientist (Pathology), Agricultural Research StationAcharya N.G. Ranga Agricultural University, Tandur, Rangareddy (Dt.)- 501 141, Andhra Pradesh, India, *E-mail*: sureshiari@yahoo.co.in

Malkhan Singh Gurjar: Scientist (Plant Pathology), Central Potato Research Station (CPRI), Shillong (Meghalaya) - 793 009, India, *E-mail*: malkhan_iari@yahoo.com

Mousa Najafinia: Department of Plant Protection, Agricultural Research Center of Jiroft and Kahnouj, Jiroft, Iran, *E-mail*: mnajafinia10@yahoo.com

Murlidhar Sadawarti: Scientist (Seed Technology), Central Potato Research Station, Gwalior (M.P.), India, *E-mail*: murlidhar_ex@rediffmail.com

Naresh M. Meshram: Scientist, Division of Entomology, Indian Agriculture Research Institute, New Delh - 110 012, India, *E-mail*: nmmeshram@gmail.com

Pranab Dutta: Department of Plant Pathology, Assam Agricultural University, Jorhat, Assam-785013, India, *E-mail*: pranabd1974@indiatimes.com

Prashant P. Jambhulkar: Assistant Professor (Plant Pathology), ARS, Borwat Farm, MPUAT (Udaipur) Banswara, Rajasthan - 327 001, India, *E-mail*: prashant_pj@rediffmail.com

Prem Lal Kashyap: Scientist (Plant Pathology), National Bureau of Agriculturally Important Microorganisms (NBAIM), Kusmaur, Post Bag No. 6, Mau Nath Bhanjan, Uttar Pradesh - 275 101, India, *E-mail*: plkashyap@gmail.com

R.B. Srivastava: Defence Institute of High Altitude Research (DIHAR), C/o 56 APO, Leh-Ladakh, India

R. Gopi: Scientist, Plant Pathology, ICAR Research Complex for NEH Region, Sikkim Centre, Tadong, Gangtok - 737 102, India, *E-mail*: ramaraj.muthu.gopi2@gmail.com

S.C. Dubey: Principal Scientist, Division of Plant Pathology, Indian Agricultural Research Institute, New Delhi - 110 012, India, *E-mail*: scdube2002@yahoo.co.in

S. Ramesh Babu: Assistant Prof, Dept of Entomology, Agricultural Research Station, MPUAT, Banswara, Rajasthan - 327 001, India, *E-mail*: babuento@yahoo.co.in

Somen Acharya: Defence Institute of High Altitude Research (DIHAR), C/o 56 APO, Leh-Ladakh, India, *E-mail*: someniari@gmail.com

Sudheer Kumar: Senior Scientist (Plant Pathology), National Bureau of Agriculturally Important Microorganisms (NBAIM),Kusmaur, Post Bag No. 6, Mau Nath Bhanjan, Uttar Pradesh - 275 101, India, *E-mail*: sudheer.nbaim@gmail.com

Susanta Banik: Department of Plant Pathology, School of Agricultural Sciences and Rural Development, Nagaland University, Medziphema, Nagaland - 797 106, India, *E-mail*: susant20@rediffmail.com

Tushar K. Dutta: Scientist (Plant Nematology), Room No-327, Division of Nematology, Indian Agricultural Research Institute Pusa Road, New Delhi - 110 012, India, *E-mail*: nemaiari@gmail.com

V.K. Baranwal: Principal Scientist, Division of Plant Pathology, IARI, New Delhi - 110 012, India, *E-mail*: vbaranwal2001@yahoo.com

Preface

Quite often crop protection poses a bigger challenge than crop production. At this point of time when world population crossed seven billion mark, saving the crop from losses due to various pests on farm and storage conditions could be a pragmatic decision as the cultivable land area is limited and shrinking. Strategies of crop protection are ever-evolving with the needs of various ages, societies and preferences. Crop production without adequate and timely crop protection is a fruitless exercise. Biotic and abiotic factors interrelated with the crop life determine the success of crop production. Technological breakthrough, ecological considerations and traditional wisdom have been mainly responsible for generating and shaping the idea of innovative crop protection measures.

Available crop protection measures range from cheap, easy-to-follow cultural practices which sometimes do not involve any money to highly sophisticated, cost-intensive research products of biotechnology and nanotechnology.

Though chemical measure of crop protection is the most popular choice owing to its rapid visible effect, whatever measures are being contemplated on these days, are mostly revolving around the idea of reducing hazardous synthetic chemicals and shifting towards green chemistry. Newer chemical molecules adhering to the stringent safety norms are only being developed and promoted. Crop protection here faces a big challenge since development of safe pesticidal compounds is not an overnight work.

Reduction of pesticidal load to the environment has also been tried with site specific crop protection in precision farming. Computer-controlled, need-based and targeted delivery of specific amount of plant protection chemicals in parts of the affected crop field holds the promise of future crop protection measures.

Change in the climatic factors all over the globe though at minor degrees are now quite prominent and influencing the flora-fauna interrelationships in all ecosystems especially the ones concerned with food production. Extinction or relative abundance of species of both beneficial and harmful ones misbalancing the equilibrium through which present day crops evolved and such phenomenon is throwing the our valuable crops to unknown and immeasurable threatening situation.

Development in the field of biotechnology which happened at an unprecedented rate brought us new hope of doing away with all evils of pesticides and seemed to be panacea of all problems of agricultural production and protection. Till date, success stories of biotechnology in relation to crop protection mainly pertain to management of weeds, a handful of insects and a few plant viruses. A lot more remains to be done as the development is skewed in terms of geographical coverage and target pests managed.

Nanotechnology has made inrush in agricultural field and crop protection products are being redesigned and redefined though with several safety precautions. In this highly networked global village, information and communication technology has hastened the process of dissemination of crop protection technologies at the doorstep of farming communities. Software packages on crop protection problems, pest identification, distribution, crop loss assessment, prediction and forecasting and suggested measures of overcoming the pests are available through various government and non-government agencies.

Besides these, inter-country trade on agriculture raised the concern on easy movement of agricultural pests from one geographical area to another. Such concern necessitated the imposition of legal barriers which are under constant revision. Several check posts or quarantine stations established at strategic locations of a country or region are playing instrumental role in preventing the inadvertent introduction of dreaded pests.

Cultivation of high value crops in controlled environment is being encouraged in many countries. Such protected culture is not immune to crop pests; rather it is challenged by a different set of crop protection problems.

When all the plant protection measures researched and developed in the present age by the plant protectionist are being followed as standard measures commercially across countries and regions, some time-tested indigenous ideas of plant protection are gaining importance and have been the focus of renewed interest in effective, sustainable and chemical free crop protection at small scale.

Nature's bounty on crop protection has also been widely explored in terms of botanicals, beneficial insects, fungi, bacteria and viruses. Several such bio-products find prominent place in crop protection in organic as well as sustainable agriculture.

In this context, the present book entitled *'Current Concepts in Crop Protection'* attempts to deal with the above mentioned issues in the form of fifteen various chapters contributed on invitation by the young and active crop protectionists of repute. The chapters in the book highlight the present crop protection scenario targeting the needs of students, educationists, researchers engaged in crop protection in particular and agriculturists in general.

I express my deepest gratitude to the contributors whose cooperative efforts have made the book possible. The materials, opinions provided in the chapters are exclusively of the respective contributors who also certified in writing that their contributions do not violate any copyright issues. There is vast amount of information published in scientific journals and books on various aspects of crop protection all of which would have not been possible to accommodate in the chapters of this book. In this connection I would like to acknowledge the contributions of those authors whose works have been presented in the chapters but might have not been acknowledged.

I would like to extend my thankfulness to the publisher Studium Press for accepting the book plan and taking keen interest in bringing out the publication.

I sincerely hope that the book will be able to contribute to better understanding of the present crop protection measures by all concerned.

Susanta Banik
May 2012

About the Author

Dr. Susanta Banik has studied B.Sc (Ag) from BCKV, Nadia; M.Sc (Ag) Plant Pathology from ANGRAU, Hyderabad and Ph.D in Plant Pathology from IARI, New Delhi. He enjoyed several scholarships (BCKV Merit scholarship) and fellowships (ICAR JRF, CSIR-JRF) during the study period. He has authored two books, three book chapters, several popular articles and several research articles in reputed journals. He has been conferred 'KCMehta and Manoranjan Mitra Award' by Indian Phytopathological Society. He is life member of some reputed national and international journals. He is also a member of editorial board in an international journal and reviewer of some national journals. He is at present actively engaged in research projects and teaching various courses of plant pathology at UG and PG level as well as guiding M.Sc and Ph.D students at various capacities.

Contents

1

Impact of Climate Change on Plant Pathogens and Pests

SOMEN ACHARYA[1], ASHISH YADAV[1], A.K. MANDAL[2] AND R.B. SRIVASTAVA[3]

ABSTRACT

Despite increasing knowledge of the predicted impacts of climate change, potential threats from diseases and pests to agriculture remain uncertain because some climatic factors may favour pathogens and insects while others may inhibit a few insects and pathogens. Climate change is going to influence the ecology of insects, pests and pathogens, with possible implications for crop protection and pesticide use. Climate change could affect disease by either decreasing or increasing the encounter rate between pathogen and host by changing rates of the two species. Temperature, rainfall, humidity, radiation, storm or dew and elevated CO_2 can affect the growth and spread of fungi and bacteria. Most studies indicate that insect pest activity will increase under climate change scenario, leading to greater risk of crop losses. Increased temperature causes migration of species northwards and into higher latitudes, while in the tropics higher temperatures might adversely affect specific pest species. The predominance of evidence indicates that there will be an overall increase in the number of outbreaks of a wider variety of insects and pathogens. The chapter discusses the impact of climate change on diseases, pests and their natural enemies; fate and behavior of pesticides and the possible crop protection strategies under the climate change scenario.

Key words: Climate change, Pests, Plant diseases, Natural enemies, Pesticides.

[1] Defence Institute of High Altitude Research, (DIHAR), DRDO, C/o 56 APO, Leh–Ladakh, India.
[2] Directorate of Seed Research, Kushmaur, Maunath Bhanjan, U.P. - 275101, India.
[3] Defence Institute of High Altitude Research (DIHAR), DRDO, C/o 56 APO, Leh–Ladakh, India.
Corresponding author: E-mail: someniari@gmail.com

1. INTRODUCTION

Agriculture provides human society with food, fiber and energy and for many people in developing countries, it is the main source of income. Pests and plant pathogens are among the most important limiting factors affecting agriculture and food security. Pest and disease outbreaks occur with the change in climatic conditions such as temperature, rainfall, humidity, soil moisture etc. and biotic factors such as host condition and natural enemy abundance. A change in climatic conditions can cause a pest or disease to expand its normal range into a new environment, extending agricultural losses and affecting natural plant communities (Rosenzweig *et al.*, 2001). There are several examples of crop damage caused by the sudden spread of agricultural pests and diseases. Several historical evidences described the impact of plant diseases and pests on the food security. One of the most cited events is the Irish famine due to a series of catastrophic epidemics of potato late blight caused by *Phytophthora infestans* during 1845–47, resulting undernourishment, loss of million lives and eventual overseas migration (Ristaino, 2002). In 1989, the glassy–winged sharpshooter and its associated crop diseases had destroyed the majority of grape vines in the southern California. Other large scale events were the Bengal famine in 1942 due to rice brown spot (caused by *Helminthosporium oryzae*, synonymous = *Drechslera oryzae*) epidemics; the 1960 famine in China due to a wheat stripe rust (*Puccinia striiformis*) outbreak and the recent epidemics of potato late blight caused by a new race of *Phytophthora infestans* (Chakraborty *et al.*, 2000; Rosenzweig *et al.*, 2001).

Agriculture will face significant challenges in the 21st century, largely due to the need of increasing global food supply under the declining availability of soil and water resources and increasing threats from climate change. The climate change effects on agriculture will differ across the world. Determining how climate change will affect agriculture is complex; varieties of effects are likely to occur. Figure 1 shows various economic, environmental and social impacts of change in disease/pest pressure in agriculture due to climate change.

Changes in temperature as well as changes in precipitation patterns and the increase in CO_2 levels projected to accompany climate change will have important effects on global agriculture. While there is clear evidence that climate change is altering the distribution of plant diseases and pests, the full effects are difficult to predict. Changes in temperature, rainfall, humidity and atmospheric gases affect the life cycle of the plant, pest/pathogen, vectors and their dispersal, by changing the rate of infection or infestation altering the interactions between pests, their natural enemies and their hosts. There may also be a change in cropping patterns under climate change scenario leading to novel plant pathogen or pest interactions (Parker and Gilbert, 2004) through the movement of a new pest/pathogen into an existing crop which then infects a previously unaffected indigenous (native and non-native) host; the introduction of a new host through changes in cropping practices which an indigenous pest/pathogen infects; and the introduction of both new host and new pest/pathogen which interact with each other. Conversely, climate change may restrict the prevalence of a pathogen/pest or a crop, subsequently restricting the occurrence of disease or pest infestation.

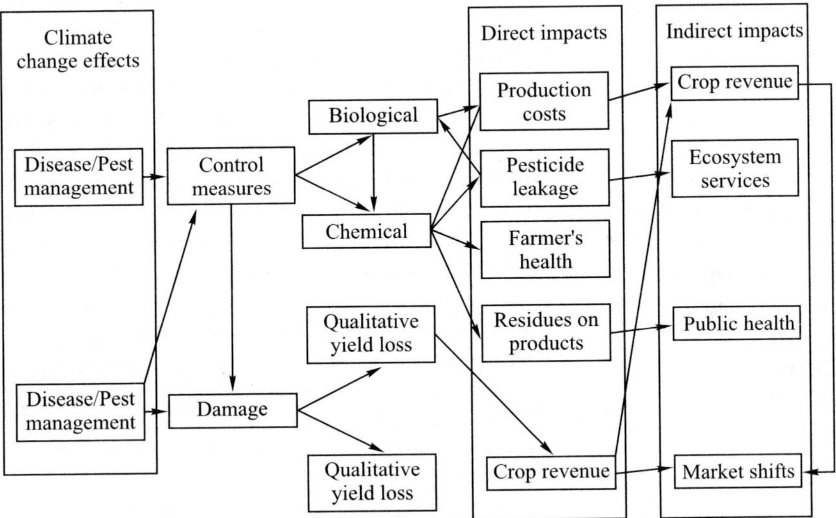

Fig. 1: Economic, environmental and social impacts of change on disease/pest pressure due to climate change

2. CLIMATE CHANGE AND DISEASES IN AGRICULTURE

The effects of climate change on plant systems are often evaluated without due consideration of plant diseases and pests (Gregory *et al.*, 2009). One specific element that seems to have received little attention so far is that plant diseases may also be seen as markers of climate change (Garrett *et al.*, 2011). Temperature, rainfall, humidity, radiation, storm or dew can affect the growth and spread of fungi and bacteria (Patterson *et al.*, 1999). Other important factors influencing plant diseases are air pollution, particularly ozone and UV–B radiation (Manning and von Tiedemann, 1995) as well as nutrient (especially nitrogen) availability. Shift in optimal temperatures for crops, plant pathogens and biological control agents may have beneficial or negative impacts depending on regional situations and crops (Olfert and Weiss, 2006). Table 1 shows some examples of disease outbreaks due to change in rainfall, moisture, temperature, air currents, drought etc.

Table 1: Effects of different climatic factors on some disease outbreaks in agriculture (Rosenzweig *et al.,* 2000)

Event		*Effects: Disease damage to crops*
Floods and heavy rains	•	Increased moisture benefits epidemics and prevalence of leaf fungal pathogens.
	—	Rice leaf blight caused great famine in Bengal (1942), 2 million people died.
	—	Wheat stripe rust outbreak in major production regions of China contributed to the 1960s famine.
	—	In the U.S. Midwest (1993) fungal epidemics in corn, soybean, alfalfa and wheat.
	—	In the U.S. Great Plains (1993) mycotoxin produced by wheat scab (*Fusarium* sp.) reached a record high.
	—	Humid summers drove epidemics of gray leaf spot of maize (Iowa and Illinois, 1996).
	•	Water—induced soil transport increases dissemination of soil–borne pathogens to non-infected areas.
	—	Outbreaks of soybean sudden death syndrome in the north central U.S. states (1993).
	•	Continuous soil saturation causes long-term problems related to rot development and increase damage by diseases.
	—	In maize, crazy top and common smut.
Drought	•	Water stress diminishes plant vigour and alters carbon-to-nitrogen ratios, lowering plant resistance to diseases. Attack by fungal pathogens of stems and roots are favoured by weakened plant conditions.
Storms and air currents	•	Air currents provide large scale transportation for disease causing agents (*e.g.*, spores of fungi) or insects from overwintering areas to attacking areas.
	—	The spread of the stem rust fungus that overwinters in Mexico and Texas is always favoured by moist southern air currents.
	—	The southern leaf blight of corn spread from Mississippi to the Midwest by air currents of a tropical storm in the Gulf of Mexico during 1970.
Higher temperature	•	Warm winters increase overwintering populations of all pests.
	•	Data reported for the European wheat scab and wheat rust.
	•	Increased population of aphids that carry the soybean mosaic virus.

2.1. Effect of Rising Temperatures on Pathogens and Diseases

Temperature has potential impacts on plant disease through both crop plant and the pathogen. Research has shown that host plants such as wheat and oats become more susceptible to rust diseases with increased temperature; but some forage crops show increased lignification at higher temperatures which enhances host resistance to pathogens (Coakley *et al.*, 1999). Many mathematical models that have been useful for forecasting plant disease epidemics are based on increase in pathogen growth and infection within specified temperature ranges. Generally, fungi that cause plant disease grow best in moderate temperature ranges. Temperate climate zones that include seasons with cold average temperatures are likely to experience longer periods of temperatures suitable for pathogen growth and reproduction due to global warming. For example, predictive models for potato and tomato late blight (caused by *Phytophthora infestans*) show that the fungus infects and reproduces most successfully during periods of high moisture that occur when temperatures are between 7.2°C and 26.8°C (Wallin and Waggoner, 1950). Earlier onset of warm temperatures could result in an earlier threat from late blight with the potential for more severe epidemics and increases in the number of fungicide applications needed for control. Using datasets spanning about 69 years, Hannukkala *et al.* (2007) has linked early occurrence and increased frequency of potato late blight epidemics in Finland to the effects of climate change and a lack of crop rotation. In rice, some resistance genes, such as *Xa7* effective against some bacterial leaf blight (*Xanthomonas oryzae* pv. oryzae) races, are more effective at high temperatures than low, whereas other resistance genes are less effective (Webb *et al.*, 2010).

A simulation modeled rice leaf blast epidemics in Japan, China, Thailand, South Korea and Philippines under increasing temperature and ultraviolet B (UV-B) radiation. Elevated CO_2 was not considered in that model. The simulation showed that in the cooler regions of Japan and northern China a temperature increase might lead to more severe blast epidemics, while in humid tropics and warm humid tropics this risk might decrease. The researchers concluded that in these regions blast development is inhibited by high temperatures. UV-B radiation will enhance the severity of blast, but more in cooler, than in warmer regions (Luo *et al.*, 1995).

2.2. Rising CO_2 Levels and Plant Pathogens and Diseases

Increased CO_2 levels can impact both the host and the pathogen in various ways. Researchers have shown that higher growth rates of leaves and stems observed for plants grown under elevated CO_2 concentrations may result in more canopy size and density resulting in a greater biomass with a much higher microclimate relative humidity. This is likely to promote plant diseases such as rusts, powdery mildews, leaf spots and blights (Manning and von Tiedemann, 1995). Research on rice leaf blast and rice sheath blight in the temperate climes of Japan with Free Air CO_2 Enrichment (FACE) technology showed that elevated CO_2 increased the potential risks for infection from leaf blast and epidemics of sheath blight. Plants grown under elevated CO_2 concentration has lower leaf silicon content, which may have contributed to the increased susceptibility to rice blast under elevated CO_2 concentrations. One hypothesis for sheath blight infection is that the higher number of tillers observed under elevated CO_2 concentrations may have increased the chance for fungal sclerotia to adhere to the leaf sheath at the water surface (Kobayashi *et al.*, 2006). Lower plant decomposition rates observed in high CO_2 situations could increase the crop residue on which disease organisms can overwinter, resulting in higher inoculum levels at the beginning of the growing season, and earlier and faster disease epidemics. Delay in pathogens establishment and infection, and a potential increase in pathogen's reproductive rate were observed for the fungus *Colletotrichum gloeosporioides* and *Maravalia cryptostegiae* or for *Erysiphe graminis* on barley under elevated CO_2 concentration. However, increased CO_2 can result in changes in host morphology, physiology and composition that can increase host resistance to pathogens (Coakley *et al.*, 1999) *e.g.*, resistance of pasture legume *Stylosanthes scabra* to anthracnose, caused by *C. gloeosporioides* (Pangga *et al.*, 2004).

Host and its physiology may affect the host–pathogen interaction in increased CO_2 levels. In a long term study Mcelrone *et al.* (2005), in controlled environment, observed reduction in disease incidence caused by the fungi *Phyllosticta minima* on leaves of *Acer rubrum*, grown in high atmospheric CO_2. Reduction in damage was not related to the pathogen germination or infection rates, but to a lower stomatal opening of the host leaves, which showed altered chemistry and lower nutritive value due to reduced N content and higher C/N ratio (Mcelrone *et al.*, 2005).

Astonishing results were also observed in an experiment on oat (*Avena sativa*) grown under elevated CO_2 (700 ppm) and infected by *Barley yellow dwarf virus* (BYDV). Root mass of virus infected plants increased by 37–60% with CO_2 enrichment, but was largely unaffected in healthy plants. CO_2 enrichment increased photosynthesis and water use efficiency by 34 and 93% in healthy plants and by 48 and 174% in infected plants respectively. Increased photosynthesis and higher water use efficiency due to reduced stomatal conductance were considered to be responsible for better performance of virus infected plants. Potentially this could increase the virus reservoir due to improved winter survival of infected plants (Malmstrom and Field, 1997). The beneficial effects of elevated atmospheric CO_2 levels are also expected to counterbalance losses due to pathogens attacking plant organs other than leaves. In a hydroponic experiment, on the effect of elevated CO_2 (700 ppm) on the tomato root pathogen *Phytophthora parasitica*, results showed increased plant biomass by 30%, thereby counterbalancing the losses caused by the pathogen (Jwa and Walling, 2001). In one similar experiment, wheat grown at elevated atmospheric CO_2 (700 ppm) and under different fertilization and water regimes, the host water content, the plant N content and the infection rate with powdery mildew were investigated. In all fertilization regimes, the mean % leaf area infected with mildew was significantly reduced under elevated atmospheric CO_2 compared to ambient CO_2 (Neumeister, 2010). In soyabean elevated CO_2 increased Septoria brown spot but reduced downy mildew (Eastburn *et al.*, 2010).

2.3. Changes in Moisture Affect Pathogens and Diseases

Like temperature, moisture can impact both host plants and pathogens in various ways. Occurrence of some pathogens such as late blight of potato, downy mildew of grapes, apple scab, fire blight of apple and several vegetable root pathogens are often correlated with the amount and distribution of rainfall in a year and more likely to infect plants with increased moisture. Forecast models for these diseases are based on leaf wetness, relative humidity and precipitation measurements. Other pathogens like the powdery mildew tend to thrive in conditions with lower (but not low) moisture conditions. Spores of most of them are inhibited in presence of free water on leaf surface and do not germinate if relative humidity is high *e.g.*, *Erysiphe graminis* in wheat and *Erysiphe cichoracearum* in cucurbits. For most of the species severe infection occurs when relative humidity is much lower (about 50–70%). More frequent and extreme rainfall events that are predicted by some climate change models could result in more and longer periods with favorable pathogen environments. Host crops with canopy size limited by lack of moisture might no longer be so limited and may produce canopies that hold moisture in the form of leaf wetness (presence of thin film of water on the leaf surface) or high canopy relative humidity for longer periods resulting increasing risk from pathogen infection (Coakley *et al.*, 1999). Some climate change models predict higher atmospheric water vapour concentrations with increased temperature. This would also favour pathogen and disease development. Long-term experiment on wheat at Rothamsted in the UK spanning over 160 years, applying biotechnological tools, demonstrated link between fluctuations in two wheat pathogens *Phaeosphaeria nodorum* and *Mycosphaerella graminicola* and changes in spring rainfall, summer temperature and SO_2 emission (Shaw *et al.*, 2008). The change of rainfall patterns such as more late-season rains will result in more severe Ascochyta blight infection on chickpea pods and Fusarium head blight on wheat.

Changes in moisture affect plant physiology and host–disease interactions, particularly during the early process of host infection, as well as the survival of pathogen's resting propagules or infective spores. Long term experimental data on Indian chickpea showed that highest incidence of Ascochyta blight (caused by *Ascochyta rabiei*) is mainly dependent on atmospheric moisture and temperature. Highest incidence of the disease is expected during a late crop risky period of a few weeks with more than 50% afternoon relative humidity and mean temperature around 20°C which is optimal for germination of pycnidiospores (Jhorar *et al.*, 1997).

In India, pearl millet downy mildew (*Sclerospora graminicola*) is widespread and highly destructive, causing yield losses upto 30%. Other downy mildew species like *Plasmopara halstedii* and *Peronospora parasitica* cause severe losses on some cultivars of sunflower and mustard respectively. Onion downy mildew (*Peronospora destructor*) in Himachal Pradesh resulted upto 75% yield loss in the period of 1988–89 (Thakur and Mathur, 2002). These host dependent obligate biotrophs survive during dry seasons as thick walled, long lived oospores and spread through infective conidia or sporangia, which release infective zoospores during suitable wet season. Changes in hydrologic cycles due to alterations in monsoon regimes, affecting occurrence and length of the host plants' surface wetness (essentially important for initiating infection and rapid spread inside the host tissue) and relative humidity (required for spore production), affect spreading and prevalence of the disease.

2.4. Effect of Ozone (O_3) on Plant Diseases

Ozone (O_3) is a secondary pollutant that is increasing downwind of major metropolitan areas around the world. It is a highly phytotoxic pollutant that decreases carbon assimilation of O_3-sensitive plants through direct effects on photosynthesis, leaf area and leaf area duration (Karnosky *et al.*, 2007). The effect of O_3 on plant diseases may be strongly affected by factors like plant stage and physiology, available soil nutrients and concentrations of other atmospheric gases. Some reports have shown that there is a significant enhancement for diseases with increasing O_3 concentration such as leaf spot (*Septoria nodorum*), powdery mildew (*Erysiphe graminis*) and spot blotch (*Bipolaris sorokiniana*) on wheat flag leaves (Von Tiedemann *et al.*, 1991), tan spot fungus (*Pyrenophora tritici–repentis*) on wheat. On the contrary, occurrence of mycorrhizal and non-mycorrhizal root-infecting fungi, powdery mildew (*Sphaerotheca fuliginea*) on cucumber, spot blotch on barley and fescue and wheat stem (*Puccinia graminis tritici*) and leaf rust (*Puccinia recondita*) on wheat have been reported to decrease with O_3. Experiments in controlled conditions showed moderate ozone injury on *Sphaerotheca fuliginea*-inoculated cucumbers.

Powdery mildew development was found severe on plants exposed to 50 ppb as O_3 exposures stimulated the fungal conidial germination. But at higher concentration about 100 to 200 ppb, there was a significant decline in fungus colonization. Same results were found in case of bottle gourd powdery mildew (Khan and Khan, 1998). In soybeans, elevated CO_2 alone or in combination with O_3 significantly reduced downy mildew (*Peronospora manshurica*) disease severity by 39–66% across a 3-year study. In contrast, elevated CO_2 alone or in combination with O_3 significantly increased brown spot (*Septoria glycines*) severity, but the increase was small in magnitude (Eastburn *et al.*, 2010). But one study has shown no correlation between O_3 and pathogen occurrence in wheat (Chakraborty *et al.*, 2008).

In conclusion, global climate change will affect plant diseases in concert with other global change phenomena (Garrett *et al.*, 2006), but so far there are no indications if disease pressure increases or not. Solid regional models considering agricultural practices including precipitation patterns along with consideration of elevated CO_2 levels and increased temperature are needed in this regard.

3. CLIMATE CHANGE AND PEST ACTIVITY

Most studies indicate that insect pest activity, the second major cause of crop damage, will increase under climate change, leading to greater risk of crop losses. Climate change has been altering the underlying agro-ecosystems through elevated temperatures and CO_2 levels, leading to changes in pest activity and population levels (IPCC, 1996). Insects are poikilothermic (cold-blooded) organisms; the temperature of their bodies is approximately the same as that of the environment. Therefore, temperature is probably the single most important environmental factor influencing insect behaviour, distribution, development, survival and reproduction. Some researchers believe that the effect of temperature on insects largely overwhelms the effects of other environmental factors (Bale *et al.*, 2002). It has been estimated that with a 2°C temperature increase in temperate regions, insects might experience one to five additional life cycles per season (Yamamura and Kiritani, 1998). Other researchers have found that effects of moisture and CO_2 on insects can be potentially important considerations in a global climate change scenario (Hunter, 2001; Hamilton *et al.*, 2005). The host range of many insects will change or expand and new combinations of pests may emerge as natural ecosystems respond to shifts in temperature and precipitation profiles. The effect of climate on pests may add to the effect of other factors such as the overuse of pesticides and the loss of biodiversity, which already contribute to plant pest outbreaks.

3.1. Impact of Temperature on Geographic Distribution and Population Dynamics of Insect Pests

Climate change may result in changes in geographical distribution of species and population growth rates, increases in the number of generations, extension of the growing season, changes in crop-pest synchrony, changes in interspecific interactions and increased risk of invasion by migrant pests (Porter *et al.*, 1991). Low temperatures are often more important than high temperatures in determining geographical distribution of insect pests (Hill, 1987). Higher temperatures will increase rates of development and the number of pests surviving the winter temperatures. Geographic distributions of crops, pests and predators are expected to shift, with pests extending to higher latitudes (Elphinstone and Toth, 2008) and sudden outbreaks of insect pests can wipe out certain crop species and also encourage the invasion by exotic species (Kannan and James, 2009). Spatial shifts in distribution of crops under changing climatic conditions will also influence the distribution of insect pests in a geographical region (Parry and Carter, 1989). Warmer conditions may be expected to promote the poleward extension of the range of species currently limited by low temperature or the altitude at which they can survive. For example, northern distribution of cotton pink bollworm is limited by cold winter temperatures. In USA, it has been restricted from spreading to the San Joaquin Valley in California because of heavy frosts common in this area. Climate change may enable this pest to invade the valley and populations of the corn earworm (*Heliothis zea*) in the North America might move to higher latitudes/altitudes, leading to greater damage in maize and other crops (EPA, 1989). For all the insect species, higher temperatures, below the species' upper threshold limit will result in faster development and rapid increase in pest populations as the reproductive maturity period will be reduced considerably. In addition to the direct effects of temperature changes on development rates, improvement in nutritional content of some crops due to elevated CO_2 levels may result in dramatic increases in growth of some insect species (White, 1984) while the growth of certain insect pests may be adversely affected (Maffei *et al.*, 2007).

Pests may cause damage for longer periods within a year. Warmer temperatures in the spring and winter will enable certain pests to become active sooner in the season and persist longer. For example, increased atmospheric temperature will lead to earlier infestation by corn earworm (*Heliothis zea*) in North America (EPA, 1989) and *Helicoverpa armigera* in north India (Sharma, 2010) resulting in increased crop loss. An increase of 3°C in mean daily temperature would cause the carrot fly (*Delia radicum*) to become active a month earlier than at present (Collier *et al.*, 1991) and temperature increases of 5 to 10°C would result in completion of four generations each year, which will require adoption of new pest control strategies. An increase of 2°C will reduce the generation turnover of the bird cherry aphid, *Rhopalosiphum padi* by varying levels, depending on the changes in mean temperature (Morgan, 1996). Rising temperatures may help in availability of new niches for insect pests. An increase of 1 and 3°C in temperature will cause northward shifts in the potential distribution of the European corn borer, (*Ostrinia nubilalis*) up to 1,220 km, with an additional generation in nearly all regions where it is currently known to occur (Porter *et al.*, 1991). Musolin (2007) observed that in Japan, warmer climate led to the northward migration of the green stinkbug (*Nezara viridula*) a major agricultural pest damaging soybean, rice, cotton and many other crops. Table 2 shows the possible responses on heteroptera species to slight and substantial temperature increase due to climate change (Musolin, 2007).

Table 2: Expected responses of heteroptera species and communities under two scenarios of further climate change (Musolin, 2007)

Categories of responses	***Slight temperature increase (<2°C)***	***Substantial temperature increase (>2°C)***
Distribution range	Likely to shift in some species, especially those capable of long-distance flights and associated with ornamental plants and/or urban habitats	Likely to shift in many species
Abundance	Likely to increase in multivoltine species with flexible life cycles	Likely to change, depending on the community response
Phenology	Slight to moderate advance of early-season events	Substantial advance of early-season and some delay of late-season events
Voltinism (the number of generations of an organism per year)	An additional generation in some multivoltine (more than two generations per year) species with flexible life cycles	One or more additional generation(s) in some multivoltine and univoltine (one generation per year) species (with facultative diapause)
Physiology and behaviour	Slight/undetectable changes	Evident/detectable changes (*e.g.*, in parameters of photoperiodic responses)
Community structure	Similar to currently observed	Increased species richness; substantial changes in structure

Overwintering of insect pests will increase as a result of climate change, producing larger spring populations as a base to build-up in numbers in the following season. These may be vulnerable to parasitoids and predators if the latter also overwinter more readily. Diamond back moth, *Plutella xylostella* overwintered in Alberta in 1994 (Dosdall, 1994), and if overwintering becomes common, the status of this insect as a pest in North America will increase dramatically. Several studies suggest that increased overwintering survival of aphid populations will result in more aphid outbreaks and increased virus infection further into the growing season (Zhou *et al.*, 1995). The promotion of aphid vectors and the consequent transmission of plant viruses in crop plants are particularly concerning. A rise in temperature would contribute to an increase in number of generations for Russian wheat aphid (a major pest of wheat and barley in dryland areas) and several virus-transmitting aphid species. There may also be increased dispersal of air-borne insect species in response to atmospheric disturbances. Many insects such as *Helicoverpa* sp. are migratory, and therefore, may be well adapted to exploit new opportunities by moving rapidly into new areas as a result of climate change (Sharma, 2005).

Global rising temperature may be responsible for species substitution, like recent decline in abundance of diamond back moth (*Plutella xylostella*) on *Brassica* plants in Japan, substituted by *Helicoverpa armigera* and *Trichoplusia* sp. in general, global warming was considered to favour natural enemies by increasing their number of generations more than in corresponding hosts (Kiritani, 2006). Resistance genes will lose their effectiveness under high temperature regimes; for instance, resistance genes in wheat for Hessian fly are particularly vulnerable (El Bouhssini *et al.*, 1999).

3.2. Impact of Changing CO_2 and Precipitation on Insects

Some researchers have found that beside rising temperature, rising CO_2 can potentially have important effects on insect pest problems. It seems that insects can detect CO_2 sources such as plants and elevated levels might affect the insect's CO_2 sensing system (Guerenstein and Hildebrand, 2008). Population densities of chewing insects will remain unaffected or decrease, but will not increase while sap sucker (phloem feeder) population densities might increase probably due to reduced nutrient quality of C_3 plants (Whittaker, 1999; DeLucia *et al.*, 2008). However, the results from experiments with aphids (phloem feeders) feeding on plants grown under elevated CO_2, and/or at elevated temperature have not shown consistent results (Holopainen, 2002). Increased aphid infestations as well as reduced infestation in response to elevated CO_2 have been observed in various experiments. Recently, FACE technology experiment is being carried out all over the world to see the effect of elevated levels of CO_2 on crop damage by insect pests and diseases. In this technology atmosphere has been created with CO_2 and O_2 concentrations similar to what climate change models predict for the middle of the 21st century. In one FACE study, soybeans grown in elevated CO_2 atmosphere had 57% more damage from insects (primarily Japanese beetle, potato leafhopper, western corn rootworm and Mexican bean beetle) than those grown in today's atmosphere. It is thought that measured increases in the levels of simple sugars in the soybean leaves may have stimulated the additional insect feeding (Hamilton *et al.*, 2005).

A Chinese research team has conducted several studies on *Helicoverpa armigera* and CO_2. Chen *et al.* (2005) reared larvae of *Helicoverpa armigera* on milky grains of spring wheat grown in ambient CO_2 concentrations, at 550 ppm and at 750 ppm. The results show that the larvae developed quite similarly under all CO_2 concentrations, even though the larvae under elevated CO_2 consumed much more than those under ambient CO_2. Quite interesting is the fact that the adult moth raised under elevated CO_2 lived longer, but laid significantly less eggs. The researcher suggests that net damage of cotton bollworm on wheat will be less under elevated atmospheric CO_2 levels because increased consumption is offset by slower development and reduced fertility (Wu *et al.*, 2006). Chen *et al.* (2007) also carried out a three generation experiment on consumption, growth and performance of *Helicoverpa armigera* feeding on transgenic Bt–cotton versus conventional cotton grown under elevated CO_2 (750 ppm) versus ambient CO_2 (375 ppm). The results suggest that on the one hand damage caused by the cotton bollworm might be higher under elevated CO_2 conditions, regardless of the cotton variety. On the other hand population abundance might be lower under elevated CO_2 compared to that under ambient CO_2 probably due to nutritional changes under elevated CO_2 (*e.g.*, individual compensatory feeding on host plants). In South America's southern areas, a major grain basket in the world new threats are likely to occur on wheat such as head blast (due to the fungus *Magnaporthe grisea*) that may induce grain yield losses over 50% in warm (25–28°C) and humid environments under climate change. In Brazil, Prestes *et al.* (2007) observed head infection ranging from 10% to 86% among cultivars and breeding lines of wheat.

Very little research has been carried out to investigate effects of precipitation on insects than temperature. Some insects are sensitive to precipitation and are killed or removed from crops by heavy rains like onion thrips (Reiners and Petzoldt, 2005). For some insects that overwinter in soil, such as the cranberry fruit worm and other cranberry insect pests, flooding the soil has been used as a control measure (Vincent *et al.*, 2003). More frequent and intense precipitation events forecasted with climate change would reduce incidence of these insects. Other insects such as pea aphids are not tolerant of drought (Mcvean and Dixon, 2001).

3.3. Impact of Climate Change on Soil Nematodes and Spider Mites

Soil nematodes are very small (0.3–5.0 mm long) worm–like organisms which occur in millions per square meter soil. Nematodes feed on a wide range of soil organisms (bacteria, fungi, slug eggs, insect larvae etc.) as well as plant roots and also work as natural enemies of insect larvae and slugs. Herbivorous nematodes can cause crop losses especially in root crops like potatoes and beets grown mostly in Europe, but also in soybean grown in Asia (CPC, 2007).

Movement of soil nematode is connected with the continuity of soil water films. Their activities are largely controlled by soil physical and biological conditions (Yeates and Bongers, 1999). Scientific research on climate change and its impact on herbivorous nematodes on agricultural crops in arable soils are very limited. However, based upon their environmental requirements some assumptions are given. Severe droughts resulting in a reduction of soil water will most likely negatively affect soil nematodes. Higher atmospheric temperatures will probably have little effect, since thermal conductivity of soils is low. Similar to other organisms which feed on plants, increased CO_2 levels are believed to have an impact on herbivorous nematodes (Ayres *et al.*, 2008). Regarding impact of climate change on nematode activity, basically, all kinds of results were determined: increase, decrease and no change of nematodes populations (Sticht *et al.*, 2009). Apparently, a prediction of how climate change will affect herbivorous soil nematodes and thus yields cannot be made. There are some evidence that population dynamics may change but so far no trend is clear.

Spider mites are not insects. They belong to the class of Arachnida and some of them are among the most important plant pests worldwide. They feed on leaves of over 150 plant species including field crops, vegetables and fruits. A common member of the group is red spider mite or two–spotted spider mite (*Tetranychus urticae*) which is common in tropical and warm temperate zones and in glasshouses in temperate zones (Neumeister, 2010). In one study two-spotted spider mites were raised on common beans grown at 600 and 700 ppm CO_2. A significant decrease in the number of the offspring in the first and second generations (34% and 49%) respectively was observed compared to ambient CO_2 (Joutei *et al.*, 2000). But in a similar experiment on clover grown at different CO_2 levels (395–748 ppm) showed a quite opposite effect: under elevated CO_2 spider mite reproduction increased significantly compared to lower CO_2 (Heagle *et al.*, 2002). They observed slight temperature differences could cause significantly different reproduction rates.

It seems that current knowledge does not allow a generalization regarding the impact of climate change on herbivorous insects, especially not for the tropics. Even the trend of a northward shift of insects must not coercively translate into a pest problem-ecosystem is not that simple and human influence is quite strong. Basically, it would be necessary to investigate at least over three trophic levels with several generations of plants, herbivores, predators/parasites, under elevated temperature and elevated CO_2.

3.4. Impact of Climate Change on Natural Enemies

The term natural enemy describes naturally occurring organisms which reduce weed, pest or disease pressure in a specific ecosystem. Depending on the agro-ecosystem the types of natural enemies can vary. In orchards, higher animals such as birds (Mols and Visser, 2002), bats and shrews (Soricidae) might considerably reduce pests. In palm oil plantations in Malaysia, owls effectively control rats. In rice paddy fields frogs, toads and fishes might play a role in controlling certain pests. Even fungi such as grape powdery mildew have a mite as a natural enemy, which suppresses powdery mildew density (Norton *et al.,* 2000). In general, each common pest has numerous natural enemies. Crop Protection Compendium (CPC) shows that for the cotton bollworm (*Helicoverpa armigera*) there are about 216 known natural enemies.

Natural enemies can be divided roughly into three functional groups: pathogens, parasites/parasitoids and predators. Considering the tremendous economic importance of natural enemies in agro-ecosystems, surprisingly little research has been done on the effects of climate change on them. Ecologists argue that the tritrophic interactions between plants, herbivorous insects, and their natural enemies (predators, parasitoids and pathogens) result from a long co-evolutionary process specific to a particular environment and relatively stable climatic conditions (Hance *et al.,* 2007). Abrupt environmental changes as induced by current climatic change and elevated CO_2 may influence the biology of each component of a system differently, provoking a destabilization in their population dynamics that may lead to the extinction of part of the system (Fajer, 1989).

3.4.1. *Pathogens*

Fungi, bacteria, microsporidia and viruses can successfully affect rodents, insect pest, mites and plant pathogens. They are widely used in biological control (Roy *et al.,* 2009), with the bacterium *Bacillus thuringiensis* and the fungus *Beauveria bassiana* being prominent examples.

Effects of climate change on the efficiency of pathogens depend on the environment they live in. In general fungi and bacteria benefit from warm and moist environments, therefore mild and wetter winters as predicted in temperate zones will benefit them, especially those living in the soil (*e.g., Beauveria bassiana*). Since many larvae or pupae of pests also overwinter (pass through or wait out the winter season) in soils, fungi and bacteria might affect them more strongly. Gutierrez *et al.* (2008) found out that during the normally wet Northern California winter, the fungal pathogen (*Pandora neoaphidis*) causes catastrophic mortality to pea aphid (*Acyrthosiphon pisum*), but during hot dry periods, the impact of the pathogen declines.

Most entomopathogenic fungi have optimal growth temperatures between 25 and 35°C. *Beauveria bassiana* grows at a wide temperature range (from 8 to 35°C) with a maximum thermal threshold for growth at 37°C (Fernandes *et al.*, 2008). Higher temperatures, low humidity as well as direct exposure to UV radiation reduces efficiency of pathogens. However, each pathogen responds to temperatures differently and behavior of the host in response to temperature is also important. Some pathogens, which always live in the host body might not be affected directly by climatic changes, they basically follow the development of their hosts. Effects of higher temperature on the impact of the microsporidia *Nosema lymantriae* on the gipsy moth (*Lymatria dispar*) clearly showed a much higher and earlier mortality of gipsy moth larvae at higher temperatures (Neumeister, 2010). Pathogens, especially viruses, become more deadly if the vector/host is weakened, therefore environmental stress such as high or low temperature might lead to higher mortality. Considering that herbivorous pests are potentially weakened by the lower nutritional quality of C_3 plants grown under elevated CO_2, it could be assumed that mortality of pests feeding on C_3 crops increases when infected with pathogens. However, more research work is needed in this aspect.

3.4.2. *Parasitoids*

Parasitoids are organisms which need to live part of their life in or on another organism *i.e.,* host. Some parasitoids paralyze or kill their host quickly, while others need to develop with their host. It was suggested that among all natural enemies parasitoids have the strongest impacts on herbivore species (Hawkins *et al.,* 1997). In biological pest control parasitoids, especially wasps of the genus *Trichogramma*, are widely used, and some estimates suggest that 10% to 20% of all insects may be parasitoid wasps (Pennacchio and Strand, 2006).

Parasitoids which live on crop pests belong to the third trophic level. Thus they are indirectly or directly affected by any changes of the first (plant) and second level (herbivore). If a herbivore reproduces less, because of low nutritional value, less potential hosts are available for the parasitoid. If the host changes its seasonal appearance or behavior due to climatic changes the parasitoid might not be able to locate the host. Finally parasitoids might be adversely affected, if the host dies too early due to additional environmental stress. However, in temperate zones milder winters might enhance survival of parasitoids. Legrand *et al.* (2004) have shown that parasitoids of cereal aphids are active in winter and this winter activity can considerably reduce spring aphid populations. No experiments have been conducted to investigate changes of all three trophic levels together (plant–herbivore–parasitoid) under climate change (elevated CO_2 and temperature). Bezemer *et al.* (1998) conducted an experiment involving several plant species, aphids and parasitoids under elevated CO_2 (+200 ppm of ambient concentrations) and showed that elevated CO_2 did not influence parasitism. Elevated temperature (+2°C of ambient temperature) increased parasitism by about 300% on average, but due to high variation between the replicates no significance could be detected. Mathematical model developed by Hoover and Newman (2004) predicts responses of grasses, cereal aphids and parasitoids to combined effects of elevated CO_2 and temperature. Results suggest that aphid and parasitoid populations will develop more similar to current ambient conditions than expected from the individual effects of CO_2 or temperature increases. Stireman *et al.* (2005) looked at frequency of parasitism in 15 Lepidoptera (Butterfly) rearing programs from a broad spectrum of climatic regimes and locations, from the region between southern Canada and central Brazil. They conducted a statistical analysis and found that the precipitation variability seems to be a key factor influencing parasitism. A higher variability led to a decrease in parasitism. These findings basically support the theory that interaction, which evolved due to stable conditions, is weakened when frequent changes occur.

3.4.3. *Predators*

Predators are basically all organisms which prey on/hunt pest organisms. The range stretches from predatory nematodes, to spiders, to eagles. Predators not only reduce pest population by feeding on them, their simple presence causes pests to cease feeding, to forage at less favorable sites, and to drop off host plants altogether in an escape response. The resulting effect is usually a slowing of prey population growth, which delays the outbreak phase. However, dropping from a plant or field crop floor may result in mortality as well due to desiccation and predation by generalist predators (Nelson *et al.,* 2004). Like parasitoids, predators which prey on crop pests belong to the third trophic level. Thus they are indirectly or directly affected by any changes of the first (plant) and second level (herbivore). Atmospheric CO_2 levels may affect the performance of natural enemies and/or susceptibility of prey via a variety of indirect effects. Some of these impacts, which potentially make prey more susceptible to their enemies, include:

- Herbivores that feed on poor host plants under elevated CO_2 conditions often spend more time in the more vulnerable, early stages of development, and thus may suffer greater mortality from natural enemies;

- Herbivores may be physically weakened while feeding on poor hosts under elevated CO_2 conditions, and are thus less able to defend themselves against predators and parasitoids; and enriched CO_2 may alter enemy–avoidance behaviour; some aphids, for example, show reduced responses to alarm pheromones under elevated CO_2, potentially making them more susceptible to enemy attack (Awmack *et al.,* 1997). Such effects would increase the susceptibility of herbivores to natural enemies, reducing herbivore population size under elevated CO_2 conditions (Coll and Hughes, 2008).

4. FATE AND BEHAVIOR OF PESTICIDES UNDER CLIMATE CHANGE SCENARIO

The use of agrochemicals will likely increase under climate change conditions as crop diseases and pests become more prevalent and as a result, loading of such chemicals *viz.*, pesticides in the environment will also increase. Amounts of pesticide applied to food items will therefore increase. The tolerance of crops to pesticides may reduce under higher levels of ozone. In such scenario, increased pesticide applications may start to reduce crop yield and this may lead to the discovery and manufacture of new active ingredients. The rate of atmospheric degradation of airborne pesticides may increase due to higher concentrations of tropospheric ozone. Changes in seasonal rainfall may lead to changes in the spatial and temporal distribution of wet deposition of airborne pesticides and their degradation products. The transport of particle-associated pesticides into water bodies is likely to increase significantly and there will be a moderate increase in the transport of hydrophilic and volatile pesticides.

Climate change is also likely to affect fate processes of different pesticides that determine their persistence in an environmental compartment. For pesticides, biodegradation, transformation and volatilization are expected to increase (Table 3), whereas sequestration of sorptive contaminants might decrease because soil organic carbon is predicted to decrease (Smith *et al.,* 2005).

Higher temperatures will give rise to increased volatilization and faster degradation of pesticide residues in the soil and surface waters. Increased winter rainfall will lead to wetter soils and so pesticide-rich water will move more rapidly from the surface through the soil matrix. More by-pass flow due to more intense rainfall events will increase the likelihood of rapid pesticide movement to surface waters and vertically to below the soil layer. Surface runoff may become more significant leading to increased erosion of pesticide-rich soil particles from fields to surface waters. More frequent and intense storm events will affect groundwater recharge. Periodic flushing into aquifers of pesticides sorbed onto colloids and sediments may be enhanced. High groundwater levels may occur more frequently due to increased winter rainfall. Periodic high groundwater levels can intercept pesticides and other diffused agricultural pollutants in the unsaturated zone and soil zone, reducing the time for pesticide degradation and leading to seasonal increases in the concentration of pesticides in groundwater.

Table 3: Impact of climate change on fate processes for pesticides applied in agriculture (Boxall *et al.*, 2010)

Fate of pesticides	***Impact of climate change***
Hydrolysis	Not sensitive
Photolysis	Increases as UV radiation increases in summer
Biodegradation/transformation	Higher temperatures increase rate
	Wetter winters increase rate
	Drier summers decrease rate
Sequestration	Lower for contaminants that sorb to soil organic matter
	Might be affected by drier soils
	Small temperature effect
Volatilization	Increases with increasing temperature
Bioconcentration	Increases with increasing temperature
Biomagnification	Not sensitive
Dilution	Increases in periods of high rainfall
	Decreases in prolonged dry periods

5. CONCLUSIONS

The precise impacts of climate change on insects and pathogens are somewhat uncertain because some climate changes may favour pathogens and insects while others may inhibit a few insects and pathogens. The prevalence of evidence indicates that there will be an overall increase in the number of outbreaks of a wider variety of insects and pathogens.

The possible mitigation strategies include:

- Careful monitoring of disease and pest damage, including associated pathogen and insect populations is very much necessary. This would be the first step in a strategy to understand and deal with the effect of global climatic changes as they occur. The acquired knowledge would provide tools to enhance resistance/tolerance to biotic stresses, which would lead to improved Integrated Pest Management (IPM) and sustained crop productivity.

- Specific predictions will be difficult because of the complex multi-trophic relationships between many microbes including pathogens, their vectors and their respective host plants. Nevertheless, such mechanistic understandings under likely multi-factorial climate change scenarios need to be obtained through development of linked experimentation and prediction modeling for insect pests and disease outbreaks.
- One key issue in modeling is the level of data resolution achieved and its effects on forecasting. The asymmetric increase of minimum and maximum temperature and the non-linearity in growth and development responses of plant pathogens require increased resolution (in terms of day or hours) when modeling the effects of climate change (Scherm and Van Bruggen, 1994). Modeling plant diseases in a global change scenario includes the integration of a global climate model (GCM) with a number of sub-models accounting for plant diseases, growth and losses (Scherm *et al.*, 1998).
- Risks maps (maps generated through modeling of plant scale conditions from regional scale data, used to represent disease likelihood) interacting with a GCM may also produce maps under estimated climate change with different scenarios. It is very useful in predicting invasion of areas previously free of pathogens or disease epidemics within a region (Seem *et al.*, 1998).
- A further modern tool to predict climate change on diseases and pests is the probability distribution map (PDM). Their use allows identification of potential risks related to distribution of pests, vectors or plant diseases. PDMs are calculated on the basis of combined use of local climate records, local records of pest or disease occurrence and subsequent statistical analysis. The model shows the areas susceptible of colonization or of endemism for the given organism (Morales and Jones, 2004). The model predicts possibility of early identification of areas susceptible of invasion by pest or disease in a different climatic scenario; possibility to anticipate the insurgence of epidemics for secondary pests already present in the areas with sub optimal conditions for their life cycle.
- Diversity to give more options and build spatial and temporal heterogeneity into the cropping system will enhance resilience to both abiotic and biotic stress challenges.
- For dryland areas, conservation agriculture like new higher value crop rotations with high yielding drought and pest resistant varieties that include oilseeds, medicinal and pharmaceutical crop options; shorter duration and earlier sowing varieties to increase the options for changing climates; better weed management; potential of

combining supplemental irrigation and water harvesting are some of the options to mitigate climate change.

- Particular attention must be paid at research programmes aiming at the management of insect pests and diseases of agricultural crops through the introduction of robust resistance genes and biochemical response mechanisms in commonly used varieties. It needs to be evaluated about how long the resulting plant genetic pool will remain useful, on a scale of decades or even years. Protection of natural biodiversity and plant genetic resources have a more practical and immediate justification.
- Also it appears wise to increase the investments in research on plant genetics and on the adaptive potentialities of traditional as well as less used varieties and accessions from germplasm collection and to test the application of integrated management tools or of new biological control agents in sustainable agriculture.
- The cooperation between farmers, either at the regional and the national level, may provide a first low cost network for monitoring, which may receive a further advantage by the diffusion of information technology and communications tools. Evaluating the effects of the forecasted changes on the traditional control methods adopted by the farmers and on pesticide use also appears as a useful research field. Emphasis should be given on more efficient use of natural resources. Finally, quarantine efforts must be improved, with particular efforts required in less developed agriculture systems and eradication or management strategies in order to provide a first barrier for target pests or disease spreading especially if affecting strategic crops of global importance.

REFERENCES

Awmack, C.S., Woodcock, C.M. and Harrington, R. (1997). Climate change may increase vulnerability of aphids to natural enemies. *Ecological Entomology,* **22:** 366–68.

Ayres, E., Wall, D.H., Simmons, B.L., Field, C.B., Milchunas, D.G., Morgan, J.A. and Roy, J. (2008). Below ground nematode herbivores are resistant to elevated atmospheric CO_2 concentrations in grassland ecosystems, *Soil Biology and Biochemistry,* **40:** 978–85.

Bale, J.S., Masters, G.J., Hodkinson, I.D., Awmack, C., Bezemer, T.M., Brown, V.K., Butterfield, J., Buse, A., Coulson, J.C., Farrar, J., Good, J.*E.G.*, Harrington, R., Hartley, S., Jones, T.H., Lindroth, R.L., Press, M.C., Symrnioudis, I., Watt, A.D. and Whittaker, J.B. (2002). Herbivory in global climate change research: direct effects of rising temperatures on insect herbivores. *Global Change Biology,* **8:** 1–16.

Bezemer, T.M., Jones, T.H. and Knight, K.J. (1998). Long-term effects of elevated CO_2 and temperature on populations of the peach potato aphid *Myzus persicae* and its parasitoid *Aphidius matricariae*. *Oecologia,* **116:** 128–35.

Boxall, A., Hardy, A., Beulke, S., Boucard, T., Burgin, L., Falloon, P., Haygarth, P., Hutchinson, T., Kovats, S., Leonardi, G., Levy, L., Nichols, G., Parsons, S., Potts, L., Stone, D., Topp, E., Turley, D., Walsh, K., Wellington, E. and Williams, R. (2010). Impacts of climate change on indirect human exposure to pathogens and chemicals from agriculture. *Ciencia and Saude Coletiva,* **15(3):** 743–56.

Chakraborty, S., Luck, J., Hollaway, G., Freeman, A., Norton, R., Garrett, K.A., Percy, K., Hopkins, A., Davis, C. and Karnosky, D.F. (2008). Impacts of global change on diseases of agricultural crops and forest trees. *Perspectives in Agriculture, Veterinary Science, Nutrition and Natural Resources,* **3(54):** 1–15.

Chakraborty, S., Tiedemann, A.V. and Teng, P.S. (2000). Climate change: potential impact on plant diseases. *Environmental Pollution,* **108:** 317–26.

Chen, F., Wu, G., Parajulee, M.N. and Ge, F. (2007). Long-term impacts of elevated carbon dioxide and transgenic *Bt* cotton on performance and feeding of three generations of cotton bollworm. *Entomologia Experimentalis et Applicata,* **124:** 27–35.

Chen, F.J., Wu, G., Lu, J. and Ge, F. (2005). Effects of elevated CO_2 on the foraging behaviour of cotton bollworm. *Helicoverpa armigera*, *Insect Science,* **12:** 359–65.

Coakley, S.M., Scherm, H. and Chakraborty, S. (1999). Climate change and disease management. *Ann. Rev. Phyto.,* **37:** 399–426.

Coll, M. and Hughes, L. (2008). Effects of elevated CO_2 on an insect omnivore: A test for nutritional effects mediated by host plants and prey. *Agriculture, Ecosystems and Environment,* **123:** 271–79.

Collier, R.H., Finch, S., Phelps, K., and Thompson, A.R. (1991). Possible impact of global warming on cabbage root fly (*Delia radicum*) activity in the UK. *Annals of Applied Biology,* **118:** 261–71.

CPC. (2007). Crop Protection Compendium. CAB International. http://www.cabi.org/compendia/cpc/accessed via German national licence, http://www.nationallizenzen.de/

DeLucia, E.H., Casteel, C.L., Nabity, P.D. and O'Neill, B.F. (2008). Insects take a bigger bite out of plants in a warmer, higher carbon dioxide world. *PNAS,* **105(6):** 1781–82.

Dosdall, L.M. (1994). Evidence for successful overwintering of diamondback moth, *Plutella xylostella* (L.) (Lepidoptera: Plutellidae), in Alberta. *Canadian Entomologist,* **126:** 183–85.

Eastburn, D.M., Degennaro, M.M., DeLucia, E.H., Dermody, O. and McElrone, A.J. (2010). Elevated atmospheric carbon dioxide and ozone alter soybean diseases at SoyFACE. *Global Change Biology,* **16:** 320–30.

El Bouhssini, M., Hatchett, J.H. and Wilde, G.E. (1999). Hessian fly (Diptera: Cecidomyiidae) larval survival as affected by wheat resistance alleles, temperature and larval density. *J. Agric. Urban Entomology,* **16:** 245–54.

Elphinstone, J. and Toth, I.K. (2008). *Erwinia chrysanthemi* (Dikeya spp.)—the facts. Oxford, U.K: Potato Council.

Environment Protection Agency (EPA). (1989). The potential effects of global climate change on the United States. Vol 2: National Studies. Review of the Report to Congress. US Environmental Protection Agency, Washington DC, p. 261.

Fajer, E.D. (1989). How enriched carbon dioxide environments may alter biotic systems even in the absence of climatic changes. *Conservation Biology,* **3(3):** 318–20.

Fernandes, E.K.K., Rangel, D.E.N., Moraes, A.M.L., Bittencourt, V.R.E.P. and Roberts, D.W. (2008). *Cold activity of Beauveria* and *Metarhizium*, and thermotolerance of *Beauveria*. *Journal of Invertebrate Pathology,* **98:** 69–70.

Garrett, K.A., Dendy, S.P., Frank, E.E., Rouse, M.N. and Travers, S.E. (2006). Climate change effects on plant disease: genomes to ecosystems. *Annual Review of Phytopathology,* **44:** 489–509.

Garrett, K.A., Forbes, G.A., Savary, S., Skelsey, P., Sparks, A.H., Van Bruggen, A.H.C., Valdivia, C., Willocquet, L., Djurle, A., Duveiller, E., Eckersten, H., Pande, S., Vera Cruz, C. and Yuen, J. (2011). Complexity in climate change impacts: a framework for analysis of effects mediated by plant disease. *Plant Pathology,* **60(1):** 15–30.

Gregory, P., Johnson, S.N., Newton, A.C. and Ingram, J.S.I. (2009). Integrating pests and pathogens into the climate change/food security debate. *J. Exp. Bot.,* **10:** 2827–38.

Guerenstein, P.G. and Hildebrand, J.G. (2008). Roles and effects of environmental carbon dioxide in insect life. *Annual Review of Entomology,* **53:** 161–78.

Gutierrez, A.P., Ponti, L., d'Oultremont, T. and Ellis, C.K. (2008). Climate change effects on poikilotherm tritrophic interactions. *Climatic Change,* **87:** 167–92.

Hamilton, J.G., Dermody, O., Aldea, M., Zangerl, A.R., Rogers, A., Berenbaum, M.R. and Delucia, E. (2005). Anthropogenic changes in tropospheric composition increase susceptibility of soybean to insect herbivory. *Envirn. Entomol.,* **34(2):** 479–85.

Hance, T., Van Baaren, J., Vernon, P. and Guy, B.G. (2007). Impact of extreme temperatures on parasitoids in a climate change perspective. *Annual Review of Entomology,* **52:** 107–26.

Hannukkala, A.O., Kaukoranta, T., Lehtinen, A. and Rahkonen, A. (2007). Late blight epidemics on potato in Finland, 1933–2002; increased and earlier occurrence of epidemics associated with climate change and lack of rotation. *Plant Pathology,* **56:** 167–76.

Hawkins, B.A., Cornell, H.V. and Hochberg, M.E. (1997). Predators, parasitoids and pathogens as mortality agents in phytophagous insect populations. *Ecology,* **78(7):** 2145–52.

Heagle, A.S., Burns, J.C., Fisher, D.S. and Miller, J.E. (2002). Effects of carbon dioxide enrichment on leaf chemistry and reproduction by two-spotted spider mites (Acari: Tetrachynidae) on white clover. *Environmental Entomology,* **31:** 594–601.

Hill, D.S. (1987). Agricultural insects pests of temperate regions and their control. Cambridge, UK: Cambridge University Press. p. 659.

Holopainen, J.K. (2002). Aphid response to elevated ozone and CO_2, *Entomologia Experimentalis et Applicata,* **104:** 137–42.

Hoover, J.K. and Newman, J.A. (2004). Tritrophic interactions in the context of climate change: a model of grasses, cereal aphids and their parasitoids. *Global Change Biology,* **10:** 1197–1208.

Hunter, M.D. (2001). Effects of elevated atmospheric carbon dioxide on insect-plant interactions. *Ag. Forest. Entomol.* **3:** 153–59.

IPCC. (1996). Climate Change 1995: Impacts, Adaptations, and Mitigation of Climate Change: Scientific–Technical Analysis. W.G.II, 2nd Assessment Report. Cambridge: Cambridge U. Press.

Jhorar, O.P., Mathauda, S.S., Singh, G., Butler, D.R. and Mavi, H.S. (1997). Relationships between climate variables and Ascochyta blight of chickpea in Punjab, India. *Agricultural and Forest Meteorology,* **87:** 171–77.

Joutei, A.B., Roy, J., Van Impe, G. and Lebrun, P. (2000). Effect of elevated CO_2 on the demography of a leaf sucking mite feeding on bean. *Oecologia,* **123(1):** 75–81.

Jwa, N.S. and Walling, L.L. (2001). Influence of elevated CO_2 concentration on disease development in tomato. *New Phytologist,* **149:** 509–18.

Kannan, R. and James, D.A. (2009). Effects of climate change on global diversity: a review of key literature. *Tropical Ecology,* **50:** 31–39.

Karnosky, D.F., Skelly, J.M., Percy, K.E. and Chappelka, A.H. (2007). Perspectives regarding 50 years of research on effects of tropospheric ozone air pollution on US forests. *Environmental Pollution,* **147:** 489–506.

Khan, M.R. and Khan, M.W. (1998). Interactive effects of ozone and powdery mildew (*Sphaerotheca fuliginea*) on bottle gourd (*Lagenaria siceraria*), *Agriculture, Ecosystem and Environment,* **70:** 109–18.

Kiritani, K. (2006). Predicting impacts of global warming on population dynamics and distribution of arthropods in Japan. *Population Ecology,* **48:** 5–12.

Kobayashi, T., Ishiguro, K., Nakajima, T., Kim, H.Y., Okada, M. and Kobayashi, K. (2006). Effects of elevated atmospheric CO_2 concentration on the infection of rice blast and sheath blight. *Phytopathology,* **96:** 425–31.

Legrand, M.A., Colinet, H., Vernon, P. and Hance, T. (2004). Autumn, winter and spring dynamics of aphid Sitobion avenae and parasitoid *Aphidius rhopalosiphi* interactions. *Annals of Applied Biology,* **145:** 139–44.

Luo, Y., TeBeest, D.O., Teng, P.S. and Fabellar, N.G. (1995). Simulation studies on risk analysis of rice blast epidemics associated with global climate in several Asian countries. *J. of Biogeography,* **22:** 673–78.

Maffei, M.E., Mithofer, A. and Boland, W. (2007). Insects feeding on plants: Rapid signals and responses proceeding induction of phytochemical release. *Phytochemistry,* **68:** 2946–59.

Malmstrom, C.M. and Field, C.B. 1997. Virus—induced differences in the response of oat plants to elevated carbon dioxide. *Plant, Cell and Environment,* **20:** 178–88.

Manning, W.J. and von Tiedemann, A. (1995). Climate Change: Potential effects of increased atmospheric carbon dioxide (CO_2) and Ozone and ultraviolet–B (UV–B). *Environmental Pollution,* **88:** 219–45.

Mcelrone, A.J., Reid, C.D., Hoye, K.A., Hart, E. and Jackson, R.B. (2005). Elevated CO_2 reduces disease incidence and severity of a red maple fungal pathogen via changes in host physiology and leaf chemistry. *Global Change Biology,* **11:** 1828–36.

Mcvean, R. and Dixon, A.F.G. (2001). The effect of plant drought-stress on populations of the pea aphid *Acyrthosiphon pisum*. *Ecol. Entomol.,* **26:** 440–43.

Mols, C.M.M. and Visser, M.E. (2002). Great tits can reduce caterpillar damage in apple orchards. *Journal of Applied Ecology,* **39:** 888–99.

Morales, F.J. and Jones, P.G. (2004). The ecology and epidemiology of white fly-transmitted viruses in Latin America. *Virus Research,* **100:** 57–65.

Morgan, D. (1996). Temperature changes and insect pests: a simulation study. *Aspects of Applied Biology,* **45:** 277–83.

Musolin, D. (2007). Insects in a warmer world: ecological, physiological and life-history responses of true bugs (*Heteroptera*) to climate change. *Global Change Biology,* **13:** 1565–85.

Nelson, E.H., Matthews, C.E. and Rosenheim, J.A. (2004). Predators reduce prey population growth by inducing changes in prey behavior. *Ecology,* **85(7):** 1853–58.

Neumeister, L. (2010). Climate Change and Crop Protection: Anything can happen. PAN Asia and the Pacific Publication.

Norton, A.P., English–Loeb, G., Gadoury, D.G. and Seem, R.C. (2000). Mycophagous mites and foliar pathogens: leaf domatia mediate tritrophic interactions in grapes. *Ecology,* **81:** 490–99.

Olfert, O. and Weiss, R.M. (2006). Impact of climate change on potential distribution and relative abundance of *Oulema melanopus*, *Meligethes viridescens* and *Ceutorhynchus obstrictus* in Canada, *Agriculture Ecosystem and Environment,* **113:** 295–301.

Pangga, I.B., Chakraborty, S. and Yates, D. (2004). Canopy size and induced resistance in *Stylosanthes scabra* determine anthracnose severity at high CO_2, *Phytopathology,* **94:** 221–27.

Parker, I.M. and Gilbert, G.S. (2004). The evolutionary ecology of novel plant–pathogen interactions. *Annual Review of Ecology, Evolution and Systematics,* **35:** 675–700.

Parry, M.L. and Carter, T.R. (1989). An assessment of the effects of climatic change on agriculture. *Climatic Change,* **15:** 95–116.

Patterson, D.T., Westbrook, J.K., Joyce, R.J.V. and Rogasik, J. (1999). Weeds, insects and diseases. *Climatic Change,* **43:** 711–27.

Pennacchio, F. and Strand, M.R. (2006). Evolution of developmental strategies in parasitic hymenoptera. *Annual Review of Entomology,* **51:** 233–58.

Porter, J.H., Parry, M.L., and Carter, T.R. (1991). The potential effects of climate change on agricultural insect pests. *Agricultural and Forest Meteorology,* **57:** 221–40.

Prestes, A.M., Arendt P.F., Fernandes, J.M.C. and Scheeren, P.L. (2007). Resistance to *Magnaporthe grisea* among Brazilian wheat genotypes. *In*: Buck, H.T., Nisi, J.E. and Salomón, N. (*Eds*) Wheat Production in Stressed Environments. Springer, Dordrecht, the Netherlands, pp. 119–23.

Reiners, S. and Petzoldt, C. (2005). Integrated crop and pest management guidelines for commercial vegetable production. Cornell Cooperative Extension publication #124VG, http://www.nysaes.cornell.edu/recommends/

Ristaino, J.B. (2002). Tracking historic migrations of the Irish potato famine pathogen. *Phytophthora infestans, Microbes and Infection,* **4:** 1369–77.

Rosenzweig, C., Iglesias, A., Yang, X.B., Epstein, P.R. and Chivian, E. (2000). Climate Change and U.S. Agriculture: The Impacts of Warming and Extreme Weather Events on Productivity, Plant Diseases, and Pests. http://www.chge.med.harvard.edu/publications/documents/agricultureclimate.pdf

Rosenzweig, C., Iglesias, A., Yang, X.B., Epstein, P.R. and Chivian, E. (2001). Climate change and extreme weather events. Implications for food production, plant diseases and pests. *Global Change and Human Health,* **2:** 90–104.

Roy, H., Hails, R.S., Hesketh, H., Roy, D.B. and Pell, J.K. (2009). Beyond biological control: non-pest insects and their pathogens in a changing world. *Insect Conservation and Diversity,* **2:** 65–72.

Scherm, H. and Van Bruggen, A.H.C. (1994). Global warming and non-linear growth: how importance are changes in average temperature? *Phytopathology,* **84:** 1380–84.

Scherm, H., Newton, A.C. and Harrington, R. (1998). Working on a global scale for global change impact assessment in plant pathology. Proceedings 7th International Congress of Plant Pathology, Edinburgh, Scotland, UK.

Seem, R.C., Magarey, R.D., Zack, J.W. and Russo, J.M. (1998). Whether forecasting at the whole plant level: implications for estimating disease risk with global climate models. Proceedings 7th International Congress of Plant Pathology, Edinburgh, Scotland, UK.

Sharma, H.C. (*Ed.*). (2005). *Heliothis/Helicoverpa* Management: Emerging Trends and Strategies for Future Research. New Delhi, India: Oxford and IBH, and Science Publishers, USA. p. 469.

Sharma, H.C. (2010). Effect of climate change on IPM in grain legumes. *In*: 5th International Food Legumes Research Conference (IFLRC V), and the 7th European Conference on Grain Legumes (AEP VII), pp. 26–30.

Shaw, M.W., Bearchell, S.J., Fitt, B.D.L. and Fraaije, B.A. (2008). Long term relationships between environment and abundance in wheat of *Phaeosphaeria nodorum* and *Mycosphaerella graminicola*, *New Phytologist*, **177:** 229–38.

Smith, J.U., Smith, P., Wattenbach, M., Zaehle, S., Hiederer, R., Jones, R.J.A., Montanarella, L., Rounsevell, M.D.A., Reginster, I. and Ewert, F. (2005). Projected changes in mineral soil carbon of European croplands and grasslands, 1990–2080. *Global Change Biology,* **11:** 2141–52.

Sticht, C., Schrader, S., Giesemann, A. and Weigel, H.J. (2009). Sensitivity of nematode feeding types in arable soil to free air CO_2 enrichment (FACE) is crop specific. *Pedobiologia,* **52:** 337–49.

Stireman, J.O., Dyer, L.A., Janzen, D.H., Singer, M.S., Lill, J.T., Marquis, R.J., Ricklets, R.E., Gentry, G.L., Hallwachs, W., Coley, P.D., Barone, J.A., Greeney, H.F., Connahs, H., Barbosa, P., Morais, H.C. and Diniz, I.R. (2005). Climatic unpredictability and parasitism of caterpillars: implications of global warming. *Proceedings of the National Academy of Science,* **102:** 17384–87.

Thakur, R.P. and Mathur, K. (2002). Downy mildews of India. *Crop Protection,* **21:** 333–45.

Vincent, C., Hallman, G., Panneton, B. and Fleurat–Lessardu, F. (2003). Management of agricultural insects with physical control methods. *Annual Review of Entomology,* **48:** 261–81.

Von Tiedemann, A., Weigel, H. and Jager, H.J. (1991). Effects of open top chamber fumigations with ozone on 3 fungal leaf diseases of wheat and the mycoflora of the phyllosphere. *Environmental Pollution,* **72:** 205–24.

Wallin, J.R. and Waggoner, P.E. (1950). The influence of climate on the development and spread of *Phytophthora infestans* in artificially inoculated potato plots. *Plant Dis. Reptr. Suppl.* **190:** 19–33.

Webb, K.M., Ona, I., Bai, J., Garrett, K.A., Mew, T., Vera Cruz, C.M. and Leach, J.E. (2010). A benefit of high temperature: increased effectiveness of a rice bacterial blight disease resistance gene. *New Phytol.,* **185:** 568–76.

White, T.C.R. (1984). The abundance of invertebrate herbivores in relation to the availability of nitrogen in stressed food plants. *Oecologia,* **63:** 90–105.

Whittaker, J.B. (1999). Impacts and responses at population level of herbivorous insects to elevated CO_2. *European Journal of Entomology,* **96:** 149–56.

Wu, G., Chen, F.J. and Ge, F. (2006). Response of multiple generations of cotton bollworm *Helicoverpa armigera* Hübner, feeding on spring wheat, to elevated CO_2. *Journal of Applied Entomology,* **130(1):** 2–9.

Yamamura, K. and Kiritani, K. (1998). A simple method to estimate the potential increase in the number of generations under global warming in temperate zones. *Appl. Ent. and Zool.,* **33:** 289–98.

Yeates, G.W. and Bongers, T. (1999). Nematode diversity in agroecosystems. *Agriculture, Ecosystems and Environment,* **74:** 113–35.

Zhou, X.L., Harrington, R., Woiwod, I.P., Perry, J.N., Bale, J.S. and Clark, S.J. (1995). Effects of temperature on aphid phenology. *Global Change Biology,* **1**: 303–13.

2

Crop Protection with an Emerging Approach: Nanotechnology

SUSANTA BANIK[1] AND JAYDEEP HALDER[2]

ABSTRACT

Methods of crop protection are ever-evolving to suit specific purpose and to apply the fruits of scientific and technological breakthrough in other areas. Change in pathogen-host dynamics owing to intensive and extensive farming practices has only pressurized the crop protection researchers to look for novel approaches. Nanotechnological approach has been gradually emerging as far as crop protection is concerned. Though nanotechnological application in crop protection is mainly at its infancy, research and development are vigorously being pursued to harness its full potential. Much has been envisaged with respect to the points of merit of nanotechnology over other methods in crop protection though it is understood that all existing methods of crop protection have limitations. Any new method obtained by technological breakthrough or other innovative practices have always supplemented the widely practiced ones. In this chapter, an attempt has been made to present the areas of crop protection where nanotechnology has made its footprints.

Key words: Nanotechnology, Pathogen detection, Nanoformulations, Crop protection, Environment.

1. INTRODUCTION

With the ever increasing population in the world, the demand for food grain and other essential agricultural commodities is also increasing. Scientists

[1] Department of Plant Pathology, School of Agricultural Sciences and Rural Development, Nagaland University, Medziphema - 797 106, Nagaland, India.

[2] Sugarcane Research Station, Bethuadahari, Nadia, West Bengal, India.

Corresponding author: E-mail: susant20@rediffmail.com

are contemplating on bringing newer technologies in their constant endeavour to meet the demand and supply gap. One of the limitations in obtaining higher agricultural production is crop loss due to biotic factors like various pests for example insects, plant pathogens, weeds etc. It has been estimated that of the 36.5% average of total crop losses annually worldwide, 14.1% are caused by plant diseases, 10.2% by insects, and 12.2% by weeds. Losses due to plant diseases alone amount to $220 billion worldwide (Agrios, 2005). Though the present tools and technology available to suppress the pest population is quite adequate, it is heavily dependent on synthetic inorganic chemicals, the so called pesticides. Avoiding the use of pesticide because of its ill effects on environment is the prime target in any management practice being evolved these days. Restricting the chemical option in management practices creates a vacuum which can be filled up only by newer technologies like nanotechnology.

The term nanotechnology, the buzzword of present day science owes its origin from the Greek word 'nano' literally meaning dwarf. When it is expressed in terms of dimension one nanometer equals to one billionth of a meter (1 nm = 10^{-9} m). The subject nanotechnology deals with manufacturing, study and manipulation of matter at nano-scale (or atomic scale) in the size range of 1–100 nm which may be called as nanoparticles (Rajan, 2004). The study of nanoparticles dates back to 1831 when Michael Faraday investigated gold colloids. It was more than 125 years later in 1959, that the potential benefits of fabricating matter at nano-level were visualized by Noble Laureate Richard Feynman. One Japanese researcher Norio Taniguchi finally engineered materials at nanometer scale in 1974 and coined the term 'nanotechnology'.

Any material when miniaturized at nanometer scale (less than 100 nm) exhibits new properties that are entirely different from its bulk counterpart due to small size and high surface to volume ratio. It is shown in terms of higher plasticity at high temperature; higher hardness, breaking strength and toughness at low temperature; higher chemical reactivity and surface energy; and high mobility in the body of an organism including cellular entry (Rajan, 2004).

Earlier, biotechnology had brought the prospect of a timely relief. Therefore, merging together the molecular biology and nanotechnology gave rise to a new subject called nanobiotechnology, a term coined by Lynn W. Jelinski of Cornell University.

The progress so far, which has been made in the broad area of plant protection due to the advent of nanotechnology in respect of detection and management of plants pathogens, studying host plant morphology in response to pathogen attack, 'smart' drug delivery in plant, nano-formulations for insect management etc. have been reviewed here.

2. DETECTION OF PLANT PATHOGENS

Most of the advanced techniques of today for detection of plant pathogens are either nucleic acid or protein based. PCR technology that allows multiplication of trace amount of target nucleic acid represents the ultimate in terms of sensitivity but has numerous drawbacks in terms of complexity, sensitivity to contamination, cost, lack of portability and major challenges with respect to multiplexing.

Protein-based diagnostics are largely dominated by enzyme-linked immunosorbent assay (ELISA) (PM detection limits) that relies on fluorophore labelling and is extraordinarily general and not equivalent to PCR. Drawbacks of molecular fluorophores include susceptibility to photobleaching, broad absorption and emission bands, and a reliance on relatively expensive equipment to probe their presence in an assay (Rosi and Mirkin, 2005).

In this backdrop, though the techniques have not been perfected yet and things are in budding stage in terms of plant pathogen detection, researchers are looking forward to harness the benefits of nanomaterials by addressing all possible drawbacks as found in currently available diagnostic tools. As said earlier nanoparticles are different from their bulk counterparts. Metals reduced to nanosize (1–100 nm) achieve certain properties which make them suitable for development as diagnostic probe of biological markers (Sharon *et al.*, 2010). These properties are large aspect ratio (surface to volume ratio), chemically alterable physical properties, change in the chemical and physical properties with respect to size and shape, strong affinity to target (particularly of gold nanoparticles to proteins), structural sturdiness in spite of atomic granularity and enhanced or delayed particles aggregation depending on the type of the surface modification, enhanced photoemission, high electrical and heat conductivity and improved surface catalytic activity (Liu, 2006; Garg *et al.*, 2008; McNeil, 2005; Rosi and Mirkin, 2005; Shrestha *et al.*, 2007).

Usually in a typical ELISA technique for detection of pathogens in general and plant pathogens in particular enzyme-conjugated antibody is used with a suitable substrate to be added later on for colour development. Presence or absence and intensity of colour are regarded as signals for qualitative and quantitative assessment of the target pathogen. Instead of enzyme, fluorescent silica nanoprobes were conjugated with the secondary antibody of goat anti-rabbit IgG in an experiment (Yao *et al.*, 2009) for detection of a bacterial plant pathogen *Xanthomonas axonopodis* pv. *vesicatoria* (bacterial spot on solanaceous plants). Basically, an organic dye tris–2, 2′–bipyridyl dichlororuthenium (II) hexahydrate (Rubpy) was incorporated into the core of circular silica nanoparticles with average diameter of 50 ± 4.2 nm. Thus silica nanoparticles became fluorescent which was photostable. Such fluorescent silica nanoprobes have potential to be used for rapid diagnosis of plant diseases caused by other pathogens also.

Use of micromechanical cantilever arrays for detection of fungal spore (*Aspergillus niger* and *Saccharomyces cerevisiae*) was demonstrated by Nugaeva *et al.* (2005). Proteins like concanavalin A, fibronectin or immunoglobulin G were surface grafted on micro-fabricated uncoated as well as gold-coated silicon cantilevers. These proteins were found to have different affinities to bind to the molecular structures present on fungal cell surface. Spore immobilization and germination of the test fungi led to shift in resonance frequency which was measured by dynamically operated cantilever arrays. This took a few hours in contrast to several days in conventional techniques. The finding that shift was proportional to the mass of single fungal spore can be used for quantitative estimation. The biosensors detected the target fungi in the range of 10^3–10^6 cfu ml^{-1} in the investigation made by Nugaeva *et al.* (2005).

Lateral flow immunoassays are currently used for qualitative, semiquantitative detection and to some extent quantitative monitoring in the fields. Lateral flow assay involves two sources of antibody (polyclonal or monoclonal), one of which is immobilised onto a nitrocellulose–based membrane, using a sophisticated reagent dispenser, and the other is sensitised onto blue-dyed latex or red-nanogold particles. On-site detection using lateral flow assays have been developed for *Rhizoctonia solani* (Thornton *et al.*, 2004), *Trichoderma hamatum* (Thornton, 2008), *Phytophthora ramorum* (Lane *et al.*, 2006) and *Peronospora destructor* (Kennedy and Wakeham, 2008).

2.1. BioMEMS (Bio Micro-Electro-Mechanical Systems)

The emphasis of detection technologies has been moved to BioMEMS/sensor technology because this provides equally reliable results in a fraction of time of the conventional methods. BioMEMS are devices or systems, constructed using techniques inspired by micro/nanoscale fabrication, that are used for performing identification, immobilization, growth, separation, purification and manipulation of single or multiple cells, biomolecules, toxins and other chemical/biological species (Bashir 2004; Kua *et al.*, 2005). In general, the use of micro/nanoscale detection technologies has the following potential advantages:

1. Higher sensitivity
2. Less requirement for reagent
3. Less detection time
4. Higher portability

2.2. Quantum Dots (QDs)

"QDs are few nm in diameter, roughly spherical (some QDs have rod like structures), fluorescent, crystalline particles of semiconductors whose excitons are confined in all the three spatial dimensions" (Sharon *et al.*, 2010). QDs have emerged as important tool for detection of a specific biological marker in medical field with extreme accuracy as they are photostable, optically sensitive, very stable light emitters and can be used as labelling. They are also easy to handle and can be traced easily with ordinary equipment. They have been used in cell labelling, cell tracking, in vivo imaging and DNA detection (Sharon *et al.*, 2010).

2.3. Nanoscale Biosensor/Nanosensors

Benefits of nanosensors are numerous. They promise to do away with the problems faced by others sensors of present day use. The nanosensors which would be small and portable would provide rapid response and real-time processing with accurate, quantitative, reliable, reproducible, robust, specific and stable results.

Detection of infection in non-symptomatic plant followed by targeted delivery of treatment would be an essential component for precision farming.

2.4. Utilization of Carbon Nano Material as a Sensor

Carbon nanomaterials have been developed to act as electrode for electrochemical analysis (Sharon and Sharon, 2008). They have the potential to be developed as electro chemical sensor to detect pesticide residue in plants.

Though there has been no patent filed so far exclusively for diagnosis for plant disease through nanotechnology method, the methods developed for diagnosis of animal diseases can be applicable to plants as well (Kalpana Sastry *et al.*, 2010).

Researchers at Texas AgriLife Research are planning to use nanotechnology for detecting plant disease at an early stage so that tons of food is protected from the possible outbreak. At present the detection technique takes days to find the plant disease and researchers are focusing to find a short and possible detection system that can give results within a few hours. For the present project, researchers will make a kit and will evaluate and test the kit in real agricultural fields. The futuristic kit has been named PADLOC (Pathogen Detection Lab-On-a-Chip).

3. MANAGEMENT OF PLANT PATHOGENS USING NANOSCIENCE AND TECHNOLOGY

Sharon *et al.* (2010) are of the view that plant pathologists made a late start in harnessing the benefits of nanomaterials for their use. Nevertheless, there have been some exciting results obtained especially in plant disease management aspects concerning fungi, bacteria and flowering plant parasite. Prospect of nanoformulations in controlling flowering plant parasite has been reviewed recently by Perez-de-Luque and Rubiales (2009).

3.1. Nanoparticles for Control of Various Plant Diseases

3.1.1. *Nanosized Silver*

Silver (Ag) is known to have antimicrobial activity both in ionic or nanoparticle forms. The powerful antimicrobial effect of silver especially in unicellular microorganisms is believed to be brought about by enzyme inactivation (Kim *et al.*, 1998). Other heavy metals like Cu and Zn are inferior to Ag in antimicrobial actions. In ionic state Ag is highly microbicidal (Kim *et al.*, 1998; O'Neill *et al.*, 2003; Thomas and McCubin, 2003). However, silver is not stable in ionic state which is amenable to be oxidized or reduced quickly. Moreover, metallic Ag is not a good antimicrobial. These disadvantages of silver in ionic and metallic state have helped in shifting the focus to nano-sized silver particles.

The most studied and utilized nanoparticle is nano-silver whose antimicrobial effect has been tested against many disease causing pathogens of animals and plants. Silver is also an excellent plant growth stimulator (Sharon *et al.*, 2010). Antifungal effect of nano-silver colloids (average diameter of 1.5 nm) was studied against the powdery mildew pathogen of rose. Powdery mildew of rose caused by *Sphaerotheca pannosa* var *rosae* was effectively controlled (95% disease reduction) by use of nano silver colloids at concentration of 10 ppm. 'Silver is now an accepted agrochemical replacement' and maximum number of patents are filed for 'nano-silver for preservation and treatment of diseases in agriculture field' (Sharon *et al.*, 2010). There is concern that nano-silver is classified and its use be regulated as a pesticide (Anderson, 2009).

Application of silver in management of plant diseases has been tested by Jo *et al.* (2009) with reference to two fungal pathogens of cereals *viz.*, *Bipolaris sorokiniana* (spot blotch of wheat) and *Magnaporthe grisea* (rice blast). *In vitro* assays indicated that silver both in ionic and nanoparticle forms inhibited colony growth of both the pathogens but *M. grisea* was comparatively more sensitive to silver application. When tested *in vivo* with perennial ryegrass (*Lolium perenne*) silver ions and nanoparticles brought significant reduction in disease severity when applied 3 hours prior to pathogen inoculation. Silver was not effective when applied 24 hours after pathogen inoculation. Therefore, silver which had preventive role was found to inhibit colony formation of spores of the fungi tested.

In another study, silver nano particles synthesized extracellularly by *Alternaria alternata* were found to cause significant enhancement in the antifungal action of the triazole fungicide fluconazole against *Candida albicans*, *Phoma glomerata* and *Trichoderma* sp. (Gajbhiye *et al.*, 2009). However, no significant enhancement was observed with respect to the fungi *Phoma herbarum* and *Fusarium semitectum*.

3.1.2. *Nanosized Silica-Silver*

Silica is well known to enhance stress resistance to plants including plant diseases (Brecht *et al.*, 2003; Ma *et al.*, 2001) through promotion of plant physiological activity and growth (Garver *et al.*, 1998; Kanto *et al.*, 2004) but it has no direct antimicrobial effect. On the other hand silver is known to have excellent antimicrobial effect as discussed earlier. Thus a new composition of nano silica-silver was developed (Park *et al.*, 2006) to combat plant disease problems.

Nanosized silica-silver (Si–Ag) particles were produced and tested by Park *et al.* (2006) against oomycetous pathogen (*Pythium ultimum*), fungal pathogens (*Magnaporthe grisea*, *Colletotrichum gloeosporioides*, *Botrytis cinerea*, *Rhizoctonia solani*) and bacterial species (*Bacillus subtilis*, *Azotobacter chroococeum*, *Rhizobium tropici*, *Pseudomonas syringae*, and *Xanthomonas campestris* pv. *vesicatoria*) including plant pathogenic ones. *In vitro* test showed higher effectiveness of silica-silver nano particles towards fungi at the dose of 10 ppm causing 100% inhibition of vegetative growth (Table 1). It was found that smaller size of silver nanoparticles was more effective against fungi. Most of the bacteria tested were inhibited completely with only 100 ppm of silica-silver nanoparticles.

Table 1: *In vitro* antimicrobial activity test of nanosized silica-silver

Microorganisms	***Percent growth inhibition in presence of different concentration of nanosized silica-silver***			
	0.3 ppm	***3.0 ppm***	***10 ppm***	***100 ppm***
Pythium ultimum	15.5	66.7	100	100
Magnaporthe grisea	1.4	27.0	100	100
Colletotrichum gloeosporioides	11.8	21.6	100	100
Botrytis cinerea	2.6	82.7	100	100
Rhizoctonia solani	54.8	94.8	100	100
Bacillus subtilis	0	0	50	100
Azotobacter chroococcum	0	0	0	100
Rhizobium tropici	0	0	0	100
Pseudomonas syringae	0	0	0	100
Xanthomonas campestris pv. *vesicatoria*	0	0	0	100

(*Source*: Park *et al.*, 2006).

Germination of spores of *Botrytis cinerea* was completely inhibited when incubated with silica-silver nanoparticles at concentration of 3 ppm. Inhibition of mycelial growth of the fungus is also reported in presence of silica-silver nanoparticles.

When nanosized silica-silver particles were applied under field condition to control powdery mildew diseases of cucurbits, 100% control was achieved after 3 weeks (Park *et al.*, 2006). These nanoparticles were found to be phytotoxic only at a very high dose of 3200 ppm when tested in cucumber and pansy plants.

Park *et al.* (2006) reported that smaller sized nano particles of 1–5 nm pass through a protoplasmic membrane and silica was well absorbed by fungi. When the nano sized silica silver is absorbed in to the fungal cell, silver nanoparticles function to increase the disinfecting activity, and silica (that induces dynamic resistance to disease) acts as a physical barrier to pathogenic fungi (Kim *et al.*, 2002).

When the nanosized silica silver attempted against the bacteria it is shown that it inhibits the growth and development of both Gram-positive and Gram-negative bacteria, in which the inhibitory effect on the growth and development of Gram-positive bacteria is higher than that on the growth and development of Gram negative bacteria.

3.1.3. *Mesoporous Silica Nanoparticles*

These are silica (SiO_2) nanoparticles with regularly arranged pores which increase the surface area of the nanoparticles. Mesosporous silica was invented to be used as molecular sieves, now their use has been expanded in nanotechnology. Targeted delivery of chemicals and DNA can be made by mesoporous silica nanoparticles (Wang *et al.*, 2002). It offers the possibility of genetic manipulation of plants at ease for a desired quality. Moreover, delivery of chemicals at targeted site in plant will reduce chemical use; improve efficiency of used chemical and reduce the chemical residue problem to the minimum while achieving pest control at its highest level thus mitigating the fear of chemical use to a large extent.

3.1.4. *Nano-Copper*

Nano-copper was reported to be highly effective in controlling bacterial diseases *viz.* bacterial blight of rice (*Xanthomonas oryzae pv.oryzae*) and leaf spot of mung (*X. campestris pv. phaseoli*) (Gogoi *et al.*, 2009).

3.1.5. Nano-Iron

Movement and behaviour of nanoparticles and their curative affect is being studied more extensively involving humans. Similar study to deliver the nanoparticles in the targeted site of a diseased plant has been done by Corredor *et al.* (2009).

They applied iron nanoparticles coated with carbon to pumpkin plants for treating specific plant part that is infected. The study was mainly aimed at analyzing the penetration and movement of applied nanoparticles into plant cells and the ways to keep them in specific plant parts applied with nanoparticles. They found that the nanoparticles that penetrated inside the living pumpkin cells did no damage to cells. Localization of applied iron nanoparticles only to diseased plant part would be achieved by use of magnet.

3.1.6. *Carbon Nanotubes*

Carbon nanotubes have shown growth enhancing effect on tomato when grown in soil containing carbon nanotubes (Khodakovsky *et al.*, 2000). It is believed that carbon nanotubes entered the germinating tomato seeds thus facilitating water uptake and plant growth. This observation if proved true, will pave way for using carbon nanotubes as delivery vehicle for plant protection chemicals to kill internally seed borne pests which are at the moment not so easy to control through seed treatments. It is an added advantage that carbon nanotubes are not harmful but growth promoting at least to tomato crop as reported.

3.2. Nanotechnology-Based Products for Plant Disease Management

(i) ***Fungicides*:** Manufacturers are developing nanoformulations of existing fungicidal compounds by reducing the size of active ingredients to nanoscale and also by nanoencapsulating them. Syngenta have developed fungicide formulation containing nanoparticles for example Banner MAXX Fungicide (a.i. propiconazole), Apron MAXX (a.i. fludioxonil) RFC for seed treatments. Similarly, Primo MAXX has been developed as plant growth regulator but it helps the plant in withstanding abiotic as well as biotic stresses including plant pathogens (Gogoi *et al.*, 2009).

(ii) **Other Products**

(a) ***Nano-5*:** It is a marketed product and is projected as natural mucilage organic solution to control several plant pathogens and pests as well besides improving crop yield. Life system in any organism is controlled at molecular level whose dimension ranges between 15–20 nm. Nanomolecules in the product can easily permeate cells of pathogenic microorganisms and kill them quickly. One needs to completely cover the infected area of plant for best result since Nano-5 is a contact solution. Nano-5 solution becomes stable 6 hours after mixing with water. Nano-molecules present in the solution enlarge to their original size after 8 hours of making aqueous solution. A few examples of plant diseases and pests (insect and mites) against which Nano-5 was found effective (at dilution of 1:500) are given below (Tables 2 and 4).

Table 2: Management of plant diseases with Nano-5

Plant pathogens	*Mode of application*	*Killing time*
1. Gray Mold, blast, Fusarium wilt, early blight	Spray **Nano-5** onto the surface of leaves once every 3 days	1–2 hrs
2. Late blight, Phytophthora diseases, southern blight, white root rot, blister blight of tea, rust	Apply to the roots twice	1–2 hrs
3. Sclerotinia rot, ergot, powdery mildew, Fusarium root rot, downy mildew	Spray onto the surface of leaves once every 5–7 days	Stops infection within 1–2 hrs
4. Bakanae disease, white rust, leaf blight, soft rot	Apply to the roots twice	Stops infection within 1–2 hrs
5. Bacterial wilt, leaf spot, rot, brown leaf spot, black rot, canker	Apply to the roots twice	Stops infection within 1–2 hrs
6. Mosaic, ringspot, transitory yellowing, tristeza virus, exocortis viroid	Spray nano–5 onto the surface of leaves and apply to the roots once every 3 days	7 days perfect control
7. Stem and bulb nematode, cyst nematode, spiral nematode	Apply to the roots until very wet	1–2 hrs

Source: http://www.unofortune.com.tw/index.htm, site visited on 09-04-2010.

(b) ***Systemic acquired resistance-inducer***: A product of nanotechnology research in agriculture with the name of Nano-Gro has been launched (Agro Nanotechnology Corp., Florida, http://www.agronano.com). It is available in the form of pellets of the size of a rain drop which is highly concentrated and required to be dissolved in water to make working solution. It is claimed that the product of one kg is sufficient for treating seeds of 3,333 ha. It is a cheap and convenient way of producing stronger, bigger and healthier plant. The product is a plant growth regulator and immunity enhancer. It is neither a source of nutrient for plant nor does it contain any hormone or proteins nor does it bring about any change in genetic structure of plant. When applied at right dose, it acts as a messenger and communicates with the plant with a message that the plant is in imminent danger. Plants immediately respond to the message and initiate its survival mechanisms like acceleration of growth, augmenting protein and sugar reserve, increase in fertility and reproduction which translate into higher yield and better quality product. Because the stress message given by Nano-Gro is false and there is no real stress in nature, plants yield more than expected.

- Plants treated with Nano-Gro show an average yield increase of 20% with maximum of 50% in case of grain yield of sunflower; increase in protein and sugar content by about 10% and plants can fight various diseases. The product is certified to be an organic one and harmless to plants and soil. Such a product would bring miracle which is sought after for not only managing plant diseases but increase crop yield in organic way (Agro Nanotechnology Corp., Florida, http://www.agronano.com).

(c) ***Nano green*:** It is a product prepared by mixing several bio-based chemicals was reported to eliminate blast disease (*Magnaporthe grisea*) from infected rice plant. The test was conducted in University of Georgia and the product was found to outperform any other pesticide or fungicides currently in use in agriculture (Gogoi *et al.*, 2009).

3.3. Nanoformulations of Chemicals for Plant Disease Management

The agrochemicals in present day use are generally highly toxic and not easily biodegradable which pose the problem of residual toxicity and subsequent fall out in ecosystem. The problem lies primarily with kind of formulations of agrochemicals including those for plant disease management. Only a little percentage of applied chemicals actually works against the target pathogen and the rest is wasted and finds ways into soil, water bodies and atmosphere. If chemical management is indispensable, can the formulation be made in such a way that only required amount of chemical is applied which can kill the target with high efficiency with minimum damage to the environment. Can the chemical be delivered only to affected plant part? Nanotechnology may be an answer to such problems. Toxicity of nanosized antimicrobial compounds is higher and this will minimize the quantity of chemical to be applied.

Formulation like nano-emulsion exhibits greater stability and increased coating of leaves because of reduced surface tension which enhances plant uptake of the sprayed chemical. Active ingredients in the form of nanoemulsions do not require organic solvents for solubilization and equally importantly there is no precipitation or creaming thereby extra caution for continuous stirring is not needed to be taken up. Nano-emulsions can be developed for hydrophobic or poorly water–soluble anti-fungal active ingredients. Such nano-emulsions have already been developed for certain insecticides like pyrethroids (Gogoi *et al.*, 2009).

Nanotechnology also offers encapsulation of pesticidal compound with nano-particles thereby achieving greater control over release of active ingredients in circumstances when mortality of target pest is maximum. In other word it gives a switching mechanism to the active ingredient which is released (after decoating of nanoparticle layer) in response to changes in anyone or more of the following such as pressure, heat, pH, water, sound frequency, magnetic field etc.

Besides this, nano-encapsulation protects the chemical compound from various causes of degradation like evaporation, hydrolysis, photolysis, microbial degradation etc.

4. 'SMART' DELIVERY IN PLANT

An interesting and fascinating area of nanoparticles is 'smart' or targeted drug delivery in the biological system. This is being vigorously pursued in cancer treatment (Kukowska–Latallo *et al.*, 2005). Use of nanoparticles in 'smart' delivery was postulated as 'magic bullets' more than a century ago by P. Ehrlich (Himmelweit, 1960). In order to develop smart treatment–delivery system in plant, Gonzalez–Melendi *et al.* (2008) worked with *Cucurbita pepo*, which were treated with carbon–coated Fe nanoparticles *in vitro*. The magnetic core consisting of iron nanoparticles allow themselves to be guided to a place of interest in the body (affected part) of an organism using small magnets that create a magnetic field. The carbon coating provides biocompatibility and acts as a surface for adsorption where various types of molecules of interest (drug/DNA/chemical/enzyme) can be adsorbed. Moreover, the carbon shell helps in good visualization of the nanoparticles under microscopes. Torney *et al.* (2007) successfully delivered DNA and chemicals in plant cells through mesosporous silica nanoparticles. The work of Gonzalez–Melendi *et al.* (2008) is probably the first to report the penetration and transport of nanoparticles inside whole plant. They showed how the magnetic nanoparticles can penetrate into the plant, travel through the vascular system and how they can be concentrated in desired plant part by creating a magnetic field.

These results indicate the future possibility which is imminent and potential of nanoparticles in delivery of substances inhibitory to various plant pathogens in the targeted site and at the same time offers the prospect of reduction in the use of inorganic chemicals that are currently being used in plant disease management.

4.1. Use of Nanofabrication Techniques for Host Pathogen Interaction Studies

Nanofabrication techniques are used in creating artificial plant parts such as stomata and xylem vessel which are then used to study the infection process and behaviour of pathogens inside host plant for example *Uromyces appendiculatus* (fungus causing rust disease of bean), *Colletotrichum graminicola* (fungus causing anthracnose in corn) and *Xylella fastidiosa* (xylem limited bacterium causing Pierce's disease of grapevine) (Meng *et al.*, 2005). Such study would help breeder to look for specific stomatal characters to prevent entry inside host through stomata or leaf surface characters to prevent appressorium formation before penetration of fungi or vascular characters to obstruct the movement of vascular pathogens like bacteria, fungi etc. In other words, it would help formation of proper breeding strategy to screen for or to develop disease resistant crop plants.

5. INSECT–PEST MANAGEMENT

Encapsulation, development of slow release formulations of pesticides and enhancing the pesticidal activity of the existing plant protection compounds are being vigorously pursued these days (Table 3). Pesticidal compounds

Table 3: Nanotechnology based agrochemicals under development

Type of product	*Product name and manufacturer*	*Nature of product*	*Purpose*
Herbicide	Tamil Nadu Agricultural University (India) and Technologico de Monterry (Mexico)	Nano-formulated	Designed to attack the coating of weeds, destroy soil seed banks and prevent weed germination
Super combined fertilizer and pesticide	Pakistan US Science and Technology Cooperative Program	Nano-clay capsule contains growth stimulants and bio-control agents	Because it can be designed for slow release of active ingredients, treatment requires only one application over the life of the crop
Pesticides, including herbicides	Australian Commonwealth Scientific and Industrial Research Organization	Nano-encapsulated active ingredients	Very small size of nanocapsules increases their potency and may enable targeted release of active ingredients

Source: Gogoi *et al.* (2009).

of 100–250 nm size get dissolved in water more effectively than existing ones and show enhanced activity. Formulations based on suspensions of nanoscale particles (nanoemulsions) containing uniform suspensions of pesticidal or herbicidal nanoparticles in the range of 200–400 nm can have multiple applications for preventative measures, treatment or preservation of the harvested product (Gogoi *et al.*, 2009).

6. NANOINSECTICIDES IN INSECT PEST MANAGEMENT

Fifty per cent nano-formulated imidacloprid product in combination with photo catalysts SDS/Ag.TiO_2 when evaluated against adults of *Martianus dermestoids* (Tenebrionidae: Coleoptera), a store grain pest, resulted highest mortality at 142 hour with LC_{50} value 9.86 mg/l (95% confidence intervals 6.4–21.1 mg/l) than the normal imidacloprid formulation (LC_{50} = 13.45 mg/l) (Guan *et al.*, 2008).

Sason *et al.* (2007) evaluated that the toxicity of nano-sized novaluron (an insect growth regulator) SC formulation as compared to standard commercial EC and SC formulation against Egyptian cotton leaf worm, *Spodoptera littoralis* and observed that nano sized novaluron formulation was about 2.5 folds more potent than the commercial standard SC formulation. The LC_{50} and LC_{90} (at 95% F.L) for novaluron SC (commercial) and novaluron SC (nanosized) were 3.6 (2.9–4.4), 17 (14–22) and 0.3 (0.2–0.5), 2 (1–7), respectively. They also concluded that nanosized SC formulation is able to penetrate the digestive tract and reach the biochemical site faster than the standard SC formulation.

Coconut mite (*Dermanyssus gallinae*) is controlled by nano silica without showing any adverse effect on plant physiology. Application of surface functionalized nanosilica killed the coleopteran store grain pests like *Sitophilus oryzae* and *Tribolium castaneum* by means of physicosorption of lipid from insect cuticle (Gogoi *et al.*, 2009).

Wan (2005) studied the effect of action of mixture of two nano particles with two insecticides to phytophagous mite (*Epitrimerus pyri*). The active ingredients Cypermethrin and alpha Terhienyl mixed with nano-particles of zinc oxide and copper oxide showed synergistic effects to the tested mite.

Presently the mode of action and biosafety of different nanopesticides like fungicides (elemental sulphur and hexaconazole), insecticides (acephate and fly ash) and fumigants in nano and nano-encapsulated forms are being studied by Indian Statistical Institute, Kolkata and Indian Agricultural Research Institute, New Delhi under the National Agricultural Innovation Projects, India.

Immunity in beneficial insects can be developed using nanosilica. Grasserie disease in *Bombyx mori* caused by nuclear polyhedrosis virus

(NPV) could be controlled with the help of lipophilic amorphous silica nanoparticles (LASN)–live BmNPV conjugate as drug. The International Agency for Research on Cancer (IARC) has rated amorphous silica dust as non–carcinogenic and USDA also considers amorphous silica as safe for anthropogenic use (Gogoi *et al*., 2009).

Table 4: Management of insect and mite using Nano-5

Pest (insect and mite)	*Mode of application*	*Killing time*
Red mite, thrips, aphids, scale, plant bug	Spray Nano-5 onto the surface and onto the backs of leaves one time per day for three days	30 minutes
Fruit fly, silver whitefly	Only spray Nano-5 onto the surface of leaves	30 minutes
Banana skipper, leaf hopper, brown plant hopper, leaf roller	Spray Nano-5 onto the surface of leaves twice	12 sec
Diamond back moth	Spray Nano-5 onto the surface of leaves twice	3–5 sec
Common cutworm	Spray Nano-5 onto the surface and onto the backs of leaves plus apply to the roots twice	5–12 sec
Phylloxera	Spray Nano-5 onto the surface and onto the backs of leaves twice	15 sec
Cabbage looper, Psyllid nymph	Spray Nano-5 onto the surface of leaves twice	7 sec
Leaf miner	Spray Nano-5 onto the surface of leaves twice	12 minutes

Source: http://www.unofortune.com.tw/index.htm, site visited on 09-04-2010.

7. ENVIRONMENTAL FATE OF NANOMATERIALS

As the research and development of nano-products are increasing more products containing nanoparticles are developed and thereby greater scope for environmental exposure. These particles may be released in the environment during the process of manufacture, formulation and uses.

7.1. Fate of Nanomaterials in Air

Fate of nanomaterials in the air depends on several factors *viz*., initial dimension of the nanoparticles, and chemical characteristics like the length of time that the particles remain suspended in air, their interaction with other airborne particles and the total distance that they travel prior to deposition (EPA, 2007).

The rate of diffusion is inversely proportional to particle diameter, while the rate of gravitational settling is proportional to particle diameter (Aitken *et al.*, 2004). According to Bidleman (1988) a particle can be classified by size and behavior into three general groups: small particles (diameters <80 nm) are described as being in the agglomeration mode; large particles (>2000 nm) described as being in the coarse mode and are subject to gravitational settling and intermediate-sized particles (>80 nm and <2000 nm) are described as being the accumulation mode and can remain suspended in air for the longest time, and can be removed from air via dry or wet deposition. Many nanosized particles are reported to be photoactive (Colvin, 2003), but their susceptibility to photodegradation in the atmosphere has not been studied. However, no studies are currently available that examine the interaction of nanosized adsorbants and chemicals sorbed to them, and how this interaction might influence their respective atmospheric chemistries (EPA, 2007).

7.2. Fate of Nanomaterials in Soil

Nanomaterials released to soil can be strongly sorbed to soil due to their high surface areas and therefore may be immobile. The strength of the sorption of any intentionally produced nanoparticles to soil will be dependent on its particle size, chemistry, applied particle surface treatment, and the conditions under which it is applied etc. (EPA, 2007).

7.3. Fate of Nanomaterials in Water

Fate of nanopesticides in aquatic environments is controlled by many factors such as aqueous solubility/dispersibility, interactions between the nanomaterials and natural and anthropogenic chemical system, biological and abiotic processes (EPA, 2007), size and surface properties of the materials (Batley and McLaughlin, 2010), properties water carrying them etc. However, due to their high surface-area-to-mass ratios, nanosized particles have the potential to sorb to soil and sediment particles (Oberdörster *et al.*, 2005). Abiotic degradation processes that may occur include hydrolysis and photocatalyzed reaction in surface water. Particles in the upper layers of aquatic environments, on soil surfaces, and in water droplets in the atmosphere are exposed to sunlight (EPA, 2007).

7.4. Biodegradation of Nanomaterials

The research on potential and possible mechanisms of biodegradation of nanosized particles have just initiated. However, a recent preliminary study found that C_{60} and C_{70} fullerenes were taken up by wood decay fungi after 12 weeks, suggesting that the fullerene carbon had been metabolized (Filley *et al.*, 2005).

Biodegradability in waste treatment and the environment may be influenced by a variety of factors. Recent laboratory studies on C_{60} fullerenes have indicated the development of stable colloid structures in water that demonstrate toxicity to bacteria under aerobic and anaerobic conditions (CBEN, 2005; Fortner *et al.*, 2005). Further studies are needed to determine whether fullerenes may be toxic to microorganisms under environmental conditions. In summary, not enough is known to enable meaningful predictions on the biodegradation of nanomaterials in the environment and much further testing and research are needed (EPA, 2007).

CONCLUSIONS

Research in crop protection has taken a giant leap from traditional ways through biotechnology to nanotechnology. Nanotechnological application in the area of crop protection is still in its nascent stage. Ultra-sensitive detection of plant pathogens, development of innovative pesticide formulations, smart delivery of compounds to the target site and many related works will benefit crop protection in the future. However, the fallout of imprudent use of nanoformulations may adversely affect the interest generated so far in this direction. Therefore, nanotechnological progress especially those for field application should be viewed with caution.

REFERENCES

Agrios, G.N. (2005). *Plant Pathology*. 5th Edition. Academic Press. pp 4.

Aitken, R.J., Creely, K.S. and Tran, C.L. (2004). Nanoparticles: An Occupational Hygiene Review Research Report 274. Prepared by the Institute of Occupational Medicine for the Health and Safety Executive, North Riccarton, Edinburgh, England.

Anderson, C B. (2009). "Regulating nanosilver as a pesticide", Environmental defense fund, February 12, 2009, available at http//blogs.edf.org/nanotechnology/2009/02/12/regulating–nano–silver–as–a–pesticide/

Bashir, R. (2004). BioMEMS: State–of–the–art in detection, opportunities and prospects. *Adv. Drug Delivery Rev.*, **56:** 1–22.

Batley, G.E. and. McLaughlin, M.J. 2010. Fate of Manufactured Nanomaterials in the Australian Environment. CSIRO Niche Manufacturing Flagship Report, pp. iii.

Bidleman, T.F. (1988). Atmospheric processes, wet and dry deposition of organic compounds are controlled by their Vapor-Particle Partitioning. *Environmental Science and Technology,* **22(4):** 361–67.

Brecht, M., Datnoff, L., Nagata, R. and Kucharek, T. (2003). The role of silicon in suppressing tray leaf spot development in St. Augustine grass. *Publication in University of Florida*, pp. 1–4.

CBEN. (2005). Center for biological and environmental nanotechnology, Rice University. Information about the center and current research summaries are available online: http://cohesion.rice.edu/centersandinst/cben/.

Colvin, V. 2003. The potential environmental impact of engineered nanoparticles. *Nature Biotechnology,* **21(10):** 1166–70.

Corredor, E., Testillano, P.S., Coronado, M.J., González–Melendi, P., Fernández Pacheco, R., Marquina, C., Ibarra, M.R., de la Fuente, J.M., Rubiales, D., Pérez–de–Luque,

A. and Risueño, M.C. (2009). Nanoparticle penetration and transport in living pumpkin plants: *in situ* subcellular identification. *BMC Plant Biology*, **9:** 45.

Environmental Protection Agency (US), 2007. EPA Nanotechnology White Paper, pp. 33–36.

Filley, T.R., Ahn, M., Held, B.W. and Blanchette, R.A. (2005). Investigations of Fungal Mediated (C_{60}–C_{70}) Fullerene Decomposition. Preprints of extended abstracts presented at the ACS national meeting, American Chemical Society. *Division of Environmental Chemistry,* **45(1):** 446–50.

Fortner, J.D., Lyon, D.Y., Sayes, C.M., Boyd, A.M., Falkner, J.C., Hotze, E.M., Alemany, L.B., Tao, Y.J., Guo, W., Ausman, K.D., Colvin, V.L. and Hughes, J.B. (2005). C60 in water: Nanocrystal formation and microbial response: *Environmental Science and Technology,* **39:** 4307–16.

Gajbhiye, M., Kesharwani, J., Ingle, A., Gade, A. and Rai, Mahendra. (2009). Fungus–mediated synthesis of silver nanoparticles and their activity against pathogenic fungi in combination with fluconazole. *Nanomedicine: Nanotechnology, Biology and Medicine*. **5(4):** 382–86.

Garg, J., Poudel, B. and Chiesa, M. (2008). Enhanced thermal conductivity and viscosity of copper nanoparticles in ethylene glycol nanofluid. *Journal of Applied Physics*, **103:** 074301.

Garver, T.L.W., Thomas, B.J., Robbins, M.P. and Zeyen, R.J. (1998). Phenyalanine ammonia–lyase inhibition, autofluorescence, and localized accumulation of silicon, calcium and manganese in oat epidermis attacked by the powdery mildew fungus *Blumeria graminis* (DC) speer. *Physiological and Molecular Plant Pathology*, **52:** 223–43.

Gogoi, R., Dureja, P. and Singh, P.K. (2009). Nanoformulations—a safer and effective option for agrochemicals. *Indian Farming*, **59(8):** 7–12.

Gonzalez–Melendi, P., Fernandez–Pacheco, R., Coronado, M.J., Corredor, E., Testillano, P.S., Risueño, M.C., Marquina, C., Ibarra, M.R., Rubiales, D. and Perez–de–luque, A. (2008). Nanoparticles as smart treatment delivery systems in plants: Assessment of different techniques of microscopy for their visualization in plant tissues. *Annals of Botany*, **101:** 187–195.

Guan, H., Chi, D., Yu, Jia. and Li, X. (2008). A novel photodegradable insecticide: Preparation, characterization and properties evaluation of nano-Imidacloprid. *Pesticide biochemistry and physiology,* **92(2):** 83–91.

Himmelweit, F. (*Ed*). (1960). The collected papers of Paul Ehrlich, Vol. 3. London: Pergamon Press.

http://www.agronano.com, site visited on 02-06-11.

http://www.unofortune.com.tw/index.htm, site visited on 09-04-2010

Jo, Y.K., Kim, B.H. and Jung, G. (2009). Antifungal activity of silver ions and nano-particles on phytopathogenic Fungi. *Plant Disease*, **93(10):** 1037–43.

Kalpana Sastry, R., Rashmi, H.B. and Rao, N.H. (2010). Nanotechnology patents as RandD indicators for disease management strategies in agriculture. *Journal of Intellectual Property Rights*. **15:** 197–205.

Kanto, T., Miyoshi, A., Ogawa, T., Maekawa, K. and Aino, M. (2004). Suppressive effect of potassium silicate on powdery mildew of strawberry in hydroponics. *Journal of General Plant Pathology*, **70:** 207–11.

Kennedy, R. and Wakeham, A. (2008). Development of detection systems for the sporangia of *Peronospora destructor*. *European Journal of Plant Pathology*, **122:** 147–55.

Khodakovsky, A., Schröder, P. and Sweldens, W. (2000). Progressive geometry compression, in Siggraph 2000. Computer Graphics Proceedings, pp. 271–78.

Kim, S.G., Kim, K.W., Park, E.U. and Choi, D. (2002). Silicon–induced cell wall fortification of rice leaves: a possible cellular mechanism of enhanced host resistance to blast. *Phytopathology*, **92:** 1095–1103.

Kim, T.N., Feng, Q.L., Kim, J.O., Wu, J., Wang, H., Chen, G.C. and Cui, F.Z. (1998). Antimicrobial effects of metal ions (Ag^{+}, Cu^{2+}, Zn^{2+}) in hydroxyapatite. *Journal of Materials Science: Materials in Medicine*, **9:** 129–34.

Kua, C.H., Lam, Y.C., Yang, C. and Toumi, K.Y. (2005). Review of bio-particle manipulation using dielectrophoresis. Report on innovation in manufacturing systems and technology. http://hdl.handle.net/1721.1/7464 (accessed January 15, 2005).

Kukowska–Latallo, J.F., Candido, K.A., Cao, Z., Nigavekar, S.S., Majoros, I.J., Thomas, T.P., Balogh, L.P., Khan, M.K. and Baker, J.R.Jr. (2005). Nanoparticle targeting of anticancer drug improves therapeutic response in animal model of human epithelial cancer. *Cancer Research,* **65:** 5317–24.

Lane, C.R., Beales, P.A., Hughes, K.J.D., Tomilson, J. and Boonham, N. (2006). Diagnosis of *Phytophthora ramorum*–evaluation of testing methods. *OEPP / EPPO, Bulletin,* **36:** 389–92.

Liu, W.T. (2006). Nanoparticles and their biological and environmental applications. *Journal of Bioscience and Bioengineering*, **102(1):** 1–7.

Ma, J.F., Goto, S., Tami, K. and Ichii, M. (2001). Role of root hairs and lateral roots in silicon uptake by rice. *Plant Physiology*, **127:** 1773–80.

McNeil, S.E. (2005). Nanotechnology for the biologist. *Journal of Leukocyte Biology,* **78:** 585–94.

Meng, Y., Li, Y., Galvani, C.D., Hao, G., Turner, J.N., Burr, T.J. and Hoch, H.C. (2005). Upstream migration of *Xylella fastidiosa via.,* pilus–driven twitching motility. *Journal of Bacteriology*, **187(16):** 5560–67.

Nugaeva, N., Gfeller. K.Y., Backmann, N., Lang, H.P., Duggelin, M. and Hegner, M. (2005). Micromechanical cantilever array sensors for selective fungal immobilization and fast growth detection. *Biosensors and Bioelectronics,* **21(6):** 849–56.

O'Neill, M., Vine, M.G., Beezer, G., Bishop, A.E., Hadgraft, A.H., Labetoulle, J., Walkeer, M. and Bowler, P.G. (2003). Antimicrobial properties of silver–containing wound dressings: a microcalorimetric study. *International Journal of Pharmaceuties,* **263:** 61–68.

Oberdörster, G., Oberdörster, E. and Oberdörster, J. (2005a). Nanotoxicology: An emerging discipline evolving from studies of ultrafine particles. *Environmental Health Perspectives,* **113(7)**: 823–39.

Park, H.P., Kim, S.H., Kim, H.J. and Choi, H.S. (2006). A new composition of nanosized silica-silver for control of various plant diseases. *Plant Pathology Journal*, **22(3):** 295–302.

Perez–de–Luque, A. and Rubiales, D. (2009). Nanotechnology for parasitic plant control. *Pest Management Science*, **65(5):** 540–45.

Rajan, M.S. (2004). Nano: The next revolution. National Book Trust, India.

Rosi, N.L. and Mirkin, C.A. (2005). Nanostructures in biodiagnostics. *Chemical Reviews,* **105:** 1547–62.

Sason Y, Ruso, G.V., Tolded, O. and Ishaaya, I. (2007). Nanosuspensions: Emerging novel agrochemicals formulations. *In*: Insecticides design using advanced techniques *Eds* by Ishaaya, Isaac. Ralf Nauen, A. and Rami, Horowitz. Springer Publications, Netherlands, pp. 1–40.

Sharon, M. and Sharon, M. (2008). Carbon Nanomaterials: Applications in physico-chemical and bio-systems. *Defense Science Journal*, **58(4):** 5491–5516.

Sharon, M., Choudhary, A.K. and Kumar, R. (2010). Nanotechnology in agricultural diseases and food safety. *Journal of Phytology*. **2(4):** 83–92.

Shrestha, S., Yeung, C.M.Y., Nunnerley, C. and Tsang, S.C. (2007). Comparison of morphology and electrical conductivity of various thin films containing nano-crystalline praseodymium oxide particles. *Sensors and Actuators A: Physical*. **136:** 191–98.

Thomas, S. and McCubin, P. (2003). A comparison of the antimicrobial effects of four silver–containing dressings on three organisms. *Journal of Wound Care*, **12:** 101–107.

Thornton, C.R. (2008). Tracking fungi in soil with monoclonal antibodies. *European Journal of Plant Pathology*, **121:** 347–53.

Thornton, C.R., Groenhof, A.C., Forrest, R. and Lamotte, R. (2004). A one-step, immunochromatographic lateral flow device specific to *Rhizoctonia solani* and certain related species, and its use to detect and quantify *R. solani* in soil. *Phytopathology,* **94(3):** 280–88.

Torney, F., Trewyn, B.G., Lin V.S.Y. and Wang, K. (2007). Mesoporous silica nanoparticles deliver DNA and chemicals into plants. *Nature Nanotechnology*. **2:** 295–300.

Wan, S.Q. (2005). Effect of action of mixture of two nano particles with two insecticides to pest mite (*Epitrimerus pyri*), *Chinese Journal of Pesticides,* **44(12):** 570–72.

Wang, Y.A., Li, J.J., Chen, H.Y. and Peng, X.G. (2002). Stabilization of inorganic nanocrystals by organic dendrons. *Journal of the American Chemical Society*, **124:** 2293–98.

Yao, K.S., Li, S.J., Tzeng, K.C., Cheng, T.C., Chang, C.Y., Chiu, C.Y., Liao, C.Y., Hsu, J.J. and Lin, Z.P. (2009). Fluorescence silica nanoprobe as a biomarker for rapid detection of plant pathogens. *In*: Advanced materials research, Volume–*Multi-Functional Materials and Structures*, II, **79–82:** 513–16.

3

Information Technology in Plant Protection

P.P. JAMBHULKAR[1] AND MURLIDHAR SADAWARTI[2]

ABSTRACT

Information of the required quality always has the potential of improving efficiency in all spheres of agriculture. The personnel, who work for the welfare of Indian farmers, such as plant protectionists, extension workers, do not have access to latest information which hinders their ability to serve the farming community effectively. Rapid developments in scientific techniques, ideas and information, such as those in the field of molecular biology, have made electronic systems of storage essential to the laboratory scientist. The Internet is becoming increasingly important as a source of plant protection information. The range of information available on the web is increasing, with more organizations moving their databases and publications to the web. Plant protectionists now have a vast array of tools to manage information and access to a wide range of sources of information. Decision Support Systems (DSS), Expert systems are now being widely used in various sectors of agriculture. Expert systems and DSS for pest control and crop protection comprises one of the most important and commonly used types of agricultural systems for plant protection. The scale and diversity of this information can seem quite daunting. This chapter attempts to provide a brief overview of the main uses of Information Technology (IT) in relation to plant protection and some of the future prospects for developments in this field.

***Key words*:** Plant protection, Information technology, Expert system, Decision Support System, Database, Web.

[1] Agricultural Research Station, Borwat Farm, MPUAT (Udaipur) Banswara, Rajasthan - 327 001.

[2] Central Potato Research Station, Gwalior (M.P.)

Corresponding author: E-mail: prashant_pj@rediffmail.com

1. INTRODUCTION

Recent developments in technologies and instrumentation, which allow large scale as well as nano-scale diagnosing of biological samples, are generating an unprecedented amount of digital data. This sea of data is too much for the human chain to process and thus there is an increasing need to use computational methods to process and contextualize these data. The application of information technology (IT) can now provide a variety of tools for plant protectionist, whether they are working in laboratory or field or involved in making decisions or guiding policy.

Databases can be used routinely for storing experimental data, recording field conditions, mapping the distribution of pests, searching the literature, and in making decisions for quarantine and pest–risk analysis. A database can be defined as a collection of structured data independent of any particular application and can comprise a simple 'electronic card index' or a sophisticated multimedia tool for managing knowledge and exchanging information (Bridge *et al.,* 1998). Wider accessibility to computers and related software, both in terms of cost and user-friendliness, combined with continuous improvements in speed, power and memory have increased the importance of electronic data as an information resource for plant protectionists. Such databases enable users to organize stored information in a manner that can be easily searched, retrieved, analysed and updated. These attributes have become invaluable in monitoring constantly changing variables, such as pathogen nomenclature, host range, distribution and molecular biology. Rapid developments in scientific techniques, ideas and information, such as those in the field of molecular biology, have made electronic systems of storage essential to the laboratory scientist. The shift towards the publication of databases and information on the World Wide Web will, by facilitating communication between colleagues in different disciplines and countries, inevitably further accelerate the pace of both research and the development of more sophisticated software tools to order information and enable its exchange. The practical implementation of IT helps not only those involved directly in research. The transformation of methods of recording and interpreting research data into knowledge-based systems has led to improvements in the quality and ease of access of information available to scientists working at the decision-making level, so that policies can be based on real events and the prediction of potential outcomes.

It is readily accepted that increased information flow has a positive effect on the agricultural sector and individual firms. However, collecting and disseminating information is often difficult and costly. Information Technology (IT) offers the ability to increase the amount of information provided to all participants in the agricultural sector and to decrease the cost of disseminating the information. An understanding of the factors associated with IT adoption and use in plant protection will enable the development of strategies to promote IT adoption and increase the effectiveness and efficiency of information used in agriculture.

Plant protectionists now have a vast array of tools to manage information and access to a wide range of sources of information. The scale and diversity of this information can seem quite daunting. This chapter attempts to provide a brief overview of the main uses of IT in relation to plant protection and some of the future prospects for developments in this field.

2. ROLE OF INFORMATION TECHNOLOGY IN AGRICULTURE

India is on the threshold of an information revolution. Agricultural technology is constantly subjected to metamorphosis over years, and farmers are swamped with many new cultivars, pesticides, farm machines and farming techniques. Coping with this ever-growing complexity is overwhelming. It is at this juncture that the development of telecommunications and computer based information technology in the era of globalization poses the best alternative and means for a sea change in extension. Information and Communication Technology based system (ICTS) can be exploited to design cost-effective systems to provide expert advice particularly to rural communities, helping to increase productivity and livelihoods (Swaminathan, 2003).

Access to information holds the key for successful development. Improved communications and information access is directly related to socio-economic development of any nation. Agriculture is one of the prospective areas in which IT can effectively be applied particularly for the social and economic development of the Indian agrarian community.

IT has found varied applications in agriculture; specifically in crop production and protection (Table 1). Agricultural information and communication systems have to be developed keeping in mind the large number of small farms, the ownership and localized distribution of the farms, the lack of education, infrastructure and access to markets and knowledge. Information has value when it is disseminated in such a way that the end users get the maximum benefit out of it (Weiss *et al.*, 2000). Broadband, wireless, internet connections allow users to be connected wherever they are. IT allows farmers to save time on order and delivery and getting feedback. According to Thysen (2000) the changes in agriculture have mainly been driven by technological developments in farm machinery and equipment, improved crop and animal genetics and improved feeding, fertilization and plant protection practices. Kuhlmann (2005) has proposed that information technology may support the production process in agriculture by generating information output from data input by means of models. IT may also support procurement of data as necessary model input and also the transformation of data and information over space and time. Development of knowledge based bio-economic models is the major application of IT in agriculture which will contain appropriate input output production functions to take into account space and time variability by incorporating the relevant, un-controllable yield factors and also contain biological and technological as well as economic components, in order to provide effective decision support for farmers (Kuhlmann, 2005). In India, the penetration of information and communication technologies is low as large chunk of farmers live in rural areas. In order to reduce this digital divide we need to create IT infrastructure in nook and corner of India so that every farmer could access to information resources. Thus, there is an urgent need to develop the following resources:

(i) Farmers' crop database must be managed. The database includes the kinds of crops, the size of cultivated area, time of harvest and yield. Farmers or the extension personnel transmit those data via the internet to database server. Further, information provides the farmer with an important instrument for decision-making and taking action.

Table 1: Some IT applications that have already been made in plant protection service at grass root level in India

Project	*Organization*	*Coverage*
Warna Wired village project	NIC, Warna Cooperative Complex, Warna, Maharashtra (Public Cooperative)	Cultivation practices, pests and diseases, marketing, processing
Village knowledge centres	MS Swaminathan Research Foundation, Chennai (NGO)	Agronomic practices and other things
e–choupals	ITC, Madhya Pradesh (Private)	Cultivation practices and other things
Rural Information Kiosks	EID Parry, Tamil Nadu (Private)	Farm practices
I–kisan portal	Nagarjuna group, Andhra Pradesh (Private)	Agricultural practices, plant protection etc.

(ii) Crops information service system should be created. This system analyzes the crop data to create some statistical tables. Farmers can access these statistical data by browsing the homepage and make their production plan. Changes within the structure of agriculture will probably have an impact on the selection and types of acquisition of software and other integrated systems made by the farmers.

(iii) Production techniques and information inquiry system should be created. This system integrates the production techniques and information, which are developed by experimental agricultural institutes and agricultural improvement stations. Farmers can find out relevant production information through this inquiry service system.

(iv) Field equipment's inquiry service system should be created. This system collects information from the companies of seeds and crop production equipment to build the production equipment's inquiry service system. At the same time, allow relevant companies to access this system and enter their own data. Therefore, farmers can order the needed items through this system. Good communication system and information system reinforce commitments to sustainable productivity. The Government of India is giving more thrust on agriculture, food and information technology sectors towards achievement of economic reforms to achieve high growth rate in production in the years to come.

3. ROLE OF INFORMATION TECHNOLOGY IN PLANT PROTECTION

Legislation and limitations are being imposed on the use of pesticides in many countries including India. Recently Supreme Court of India imposed a ban on Endosulphon by declaring it unsafe for use. In near future a lot chemical pesticides will be banned. Based on these national action plans, it has become important to encourage a low pesticide input strategy. Similar proposals to eliminate unnecessary use are also stated in the EU's Thematic Strategy on the Sustainable Use of Pesticides (Anonymous, 2006). In order to eliminate unnecessary use of fungicides, precise knowledge of the disease risk of an epidemic at field level is essential. In order to achieve this, it is very important to provide farmers with simple and robust methodologies to assess disease occurrence or risk in order to judge whether fungicides should be applied (Jørgensen, *et al.*, 2003; Verret, *et al.*, 2000). Plant protection tasks are part of crop management of any crop which need expertise that can be better represented in symbolic knowledge representation schemes rather than other tasks, like irrigation and fertilization which includes mathematical models more than symbolic knowledge. In addition to that annual yield losses due to pests are estimated to be in billions of dollars worldwide. Farmers expect specialised expertise from plant protection experts regarding decision about treatment, treatment selection and time of its application. Thus by combining field observations, current weather data, grain field conditions, pesticide application, a computer based decision support system can be developed. Such systems for pest control and crop protection constitute a very significant class of agricultural expert systems. The development of thresholds, models and decision support systems (DSS) to farmers and advisors covering this need started in the late 1980s. EPIPRE was one of the first systems to be developed (Zadoks, 1983). Several other systems have been developed since and new systems are still entering the field today (Röhrig, 2006). Among the systems that have been on the market for the longest time are Pro_Plant and Crop Protection Online (CPO); both systems are today Internet-based (Rydahl, 2003; Volk *et al.*, 2003). Pest management and crop protection includes a large number of techniques using varied knowledge in entomology, plant pathology, nematology, weeds and vertebrate pests. This knowledge, gathered over the years, is voluminous and diverse. This makes the field implementation of techniques required to deal with a particular disease of a given crop quite difficult. It has been discovered that the knowledge can be efficiently structured and organised only using a computerized expert system. Similarly a web based domain specific tool for building plant protection expert system has been developed for barley protection (Rafea, 2010). The main such tool system has few characteristics such as share knowledge acquisition through availing the tool on internet, enable human experts to cooperate with knowledge engineer through using the tool and represent knowledge in XML (Extensible Markup Language) which is standard data language to facilitate knowledge verification and future upgrading. There is still more work to be done to prepare more user friendly tool to support more crop protection expert system.

4. APPLICATION OF WORLD WIDE WEB AND DATABASE TECHNOLOGY IN PLANT PROTECTION

(i) *Role of Web Services in Data Sharing*: Web services are building blocks for creating open distributed systems. In other words web services are defined as a set of programmatic interface *e.g.*, Simple Object Access Protocol (SOAP), Web Service Description Language (WSDL) and Uniform Data Discovery Interface (UDDI) platform independent application-to-application communication via Internet. Gartner, Inc. defines web service as "software components that employ one or more of three technologies—SOAP, WSDL and UDDI—to perform distributed computing. Use of any of the basic technologies constitutes web services. Use of all of them is not required."

- SOAP (Simple Object Access Protocol) is a simple protocol for exchange of information. It is based on XML and consists of three parts: a SOAP envelope (describing what's in the message and how to process it); a set of encoding rules, and a convention for representing RPCs (Remote Procedure Calls) and responses.

- UDDI (Universal Description, Discovery, and Integration) is a specification designed to allow businesses of all sizes to benefit in the new digital economy. Searches can be performed by company name, specific service, or types of service. This allows companies providing or needing web services to discover each other, define how they interact over the Internet and share such information in a truly global and standardized fashion.
- WSDL (Web Services Description Language) defines the XML grammar for describing services as collections of communication endpoints capable of exchanging messages. Companies can publish WSDLs for services they provide and others can access those services using the information in the WSDL. Links to WSDLs are usually offered in a company's profile in the UDDI registry.
- Towards the ultimate goal of sustainability and increasing capacity for plant disease and insect pest diagnostics, with emphasis on important staple and export crops, the International Plant Diagnostic Network (IPDN) was established in 2006. The IDPN was modelled after the United States National Plant Diagnostic Network (NPDN), which operates as a 'hub and spoke' system in five regions (Stack *et al.*, 2006). A hub and spoke system is an operational structure, in which a central laboratory (hub) coordinates activities amongst other laboratories (spokes) within the network. In East Africa, the National Agricultural Research Laboratories of the Kenya Agricultural Research Institute (KARI–NARL) was designated the hub laboratory for the IPDN East Africa region. Laboratories in other KARI centres, as well as other institutions and laboratories in Uganda, Tanzania, and Rwanda joined the program as spoke laboratories. In West Africa, the International Institute of Tropical Agriculture at Benin (IITA–Benin), with backstopping from IITA–Nigeria, was designated the hub laboratory, with spoke laboratories in Ghana, Senegal and Mali. Diagnostic laboratories in each of these countries have access to the Distance Diagnostic and Identification System/Clinic Information System (DDIS/CIMS; see below), as well as training opportunities (Miller *et al.,* 2010).

(ii) *Role of Web Services in Plant Protection*: Web services are usually seen as Web-based enterprise—wide or inter-organizational applications that use open standards (mostly based on XML) and transport protocols to exchange data with clients, thus forming a loosely–coupled information systems architecture (Ferris and Farrell, 2003). Web services, as specified by the World Wide Web Consortium (W3C) [World Wide Web Consortium 2004], use the Hypertext Transfer Protocol (HTTP) in the transport layer. This protocol only allows the client to perform synchronous calls. This characteristic represents a problem when dealing with delayed–time transactions and call resume. Plant protectionists now have a vast array of tools to manage information and access to a wide range of sources of information. The scale and diversity of this information can seem quite daunting.

- Numerous web sites have attempted to collect and store data on pests and their management. The process of collecting and managing these data by manually querying the web using the information into a meaningful result set is time consuming and labour intensive task. Web services are an emerging technology thus they are slow to be adopted within much of the scientific community. There are, however few instances where information relating to IPM, invasive species and species diversity has been made available using web services. The following examples illustrate the application of web service technology towards achieving data interoperability and data sharing.
- One example of a consumer provided web service is the integration of plant database (USDA, NRCS, 2004) and the USDA ARS Systematic Botany and Mycology Laboratory (SBML) (*http://nt.ars–grin.gov/SB MLWeb/homehtml.cfm*). The plant database exposes their plant taxonomic data via Web services to SBML.
- Reactome is a human–curated knowledgebase of biological pathways and reactions. The information in this knowledgebase is authored by biological researchers with expertise in their fields, maintained by the Reactome editorial staff, and cross-referenced with PubMed, Gene Ontology, NCBI, Ensembl, UniProt, OMIM and other databases. Reactome is a joint effort among Cold Spring Harbor Laboratory, European Bioinformatics Institute and Gene Ontology Consortium (Stein and Wu, 2006).
- The Distance Diagnostic and Data Management System DDIS/CIMS, developed at the University of Florida, is a web-based client/server application that provides a collaboration environment for specialists, first detectors, and experts for identification of pest and disease problems (Xin *et al.,* 2005). The system uses textural data and microscopic digital media with symptoms of pest and diseases, and shares the media with specialists to rapidly detect and identify diseases and pathogens. First detectors submit digital samples of pests and plant diseases to a clinic or specialists in the network. Specialists or laboratory technicians make a diagnosis based on the digital samples. The diagnostic network allows specialists and laboratory technicians to invite external experts around the world to assist with the diagnosis. This is particularly important for those developing countries lacking diagnostic expertise. Throughout the interaction, a

pest or disease problem can be quickly addressed or screened. Laboratory results can be submitted along with the digital sample. Digital diagnostics may help to overcome problems in networking and utilization of limited expertise resources, due to restrictions on trans-border shipments of live insect pest and disease samples. The archived diagnostic database and a lab digital media library can be used for research and educational purposes.

- In India, AGROWEB—a Digital Dissemination System for Indian Agricultural Research (ADDSIAR) has been created by NBPGR, New Delhi in collaboration with few ICAR institutions. This system proved to be a 'one-stop window' for getting access to all the information about National Agricultural Research and Education Systems in India. From the plant protection point of view this system provides database on key pests and their identification, picture and video clippings, make available management options for major crop diseases and insect pests and also maintain database of plant protection/IPM specialists in India.

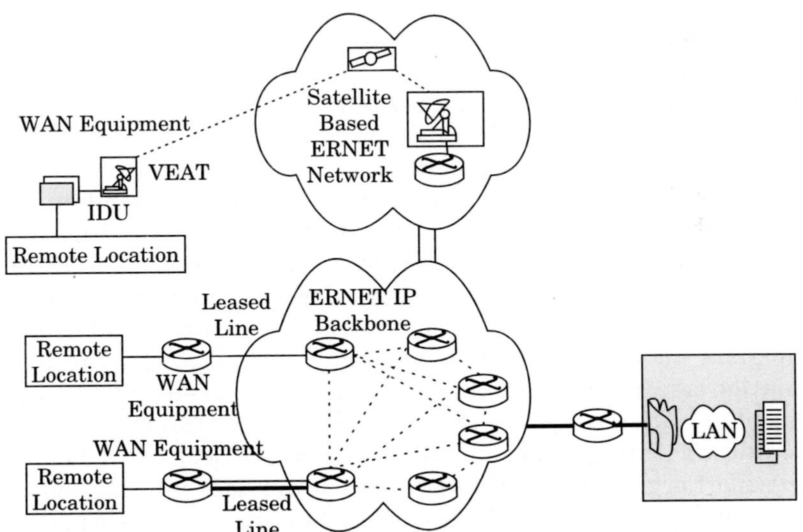

Fig. 1: Schematic Diagram of Data Centre of ICAR being developed by ERNET at New Delhi for hosting Agroweb

Source: *Brochure*: AGROWEB—Digital dissemination system for Indian Agricultural Research. NAIP–ICAR.

- An another project of "National Information System for Pest Management" (NISPM, Bt cotton), coordinated by National Center for Integrated Pest Management, New Delhi. The subcenters of the project collect data of the incidence and population of diseases and insect pests of Bt cotton crop. The data collected by scouts are fed in the computers by opening the website of the NISPM (http://www.ncipm.org.in/NISPM/ReportSystem/Login.aspx). The data are then evaluated and displayed on the website of NISPM (Bt cotton).

- Kisan Call Centers were launched by Government of India across the country to deliver extention services to the farming community. Under this service farmers can solve any kind of farming query just by dialing 1800–180–1551.

5. WORLD WIDE WEB RESOURCES FOR REPORTING, UNDERSTANDING AND EVALUATING PLANT DISEASES

The World Wide Web has a number of valuable resources that can be used to keep track of plant diseases. Many countries have government sites and information resources devoted to the problem of reporting plant diseases. Collectively these agencies and online connectivity and notification resources have the potential to provide an excellent means of communication. The Internet is becoming increasingly important as a source of plant protection information. The range of information available on the web is increasing, with more organizations moving their databases and publications to the web. Member societies such as the British Society for Plant Pathology (www.bspp.org.uk/) and the American Phytopathological Society (www.apsnet.org/) aim to create 'virtual communities' that complement their publishing and conference activities. The International Society for Plant Pathology (www.isppweb.org/) maintains a searchable database of plant pathologists around the world and links to national plant pathology societies. There are also sites available on the web advising plant pathologists on useful links to help navigate through the array of information available (Kraska, 2001). The University of Florida, the Smithsonian National Museum of Natural History, USA and the FAO are developing an Internet-based access portal to Ecological Knowledge (www.ecoport.org/). Among other features, it provides information on host plants (including uses, ethno-botany, distribution and recorded pests) and on pests, detailing each pest's symptoms, ecology, dispersal/vectors, control and distribution.

The U.S. Department of Agriculture's (USDA) Animal and Plant Health Inspection Service (APHIS) provides much useful information through its Plant Protection Quarantine (PPQ) web site. The APHIS/PPQ website provides links to a number of useful information sources including state plant protection resources and various biological control facilities.

The ProMed component known as ProMed–Plant is one of the most innovative and useful tool for tracking emerging plant diseases, especially on world-wide basis. It is internationally–based and receives reports of diseases worldwide. ProMed–plant has addressed a number of issues related to crop biosecurity. These topics can be reviewed by a search of the archives at the ProMed home page using the keyword, "plant." Through interactions with the National Agricultural Library and the Agricultural Research Service (ARS), the USDA maintains the Agriculture Network Information Center (AgNIC). In turn AgNIC collaborates with ProMed (see above) to maintain a sophisticated searchable archive of emerging plant diseases (Yu, 2004).

In Canada the Plant Disease Survey reports occurrences and other information about plant diseases in Canada. The survey can be downloaded from a number of servers connected with Agriculture and Agri-Food Canada, including the Southern Crop Protection and Food Research. The Canadian Phytopathological Society and Canadian News Service provide links to a number of information resources. The European and Mediterranean Plant Protection Organization (EPPO) also provides plant disease reporting information, with the intent to keep Europe and Russia informed on new outbreaks and emerging disease problems. This information can easily be obtained through the Plant Quarantine information and the EPPO Early Alert List.

6. DATABASE TECHNOLOGY IN PLANT PROTECTION

Traditionally biologists relied on textbooks and research articles published in scientific journals as the main source of information. This has changed dramatically in the past decade as the Internet and Web browsers became commonplace. Today, the internet is the first place researchers go to find information. Databases are available via the web also became an indispensible tool for biological research.

6.1. Types of Database

Three types of databases have been established and are developed: large scale public repositories, community specific databases and project specific databases. Nucleic Acid Research (http://nar.oxfordjournals.org) publishes a database issue in January of every year. Recently Journal of Plant Physiology started publishing articles describing databases (Rhee and Crosby, 2005). Large scale public repositories are usually developed and maintained by government agencies or international consortia and are placed for long term data storage. Examples include GenBank for sequences, Uniprot for protein information, Protein Data Bank for Protein structure information and Array express and Gene expression omnibus (GEO) for microarray data. There are number of community specific databases, which typically contain information curated with high standards and address the needs of a particular community of researchers. The concept of community databases is subject to change as researchers are widening their scope of research. For example, databases focused on comparing genome sequence recently emerged.

The third category of databases includes smaller scale, and often short lived, databases that are developed for project data management during the funding period. Often these databases and web resources are not maintained beyond the funding period of the project and currently there is no standard way of depositing or archiving these databases after the funding period.

The following are some examples of large databases that may be of use to those working in plant pathology research:

(i) ***Nomenclature Systems*:** Databases of pest names and synonymy are very important to plant protectionists searching the literature and embarking on research projects. Pest nomenclature is a constantly changing field. "Species 2000" (www.sp2000.org/) has the objective of enumerating all known species of plants, animals, fungi, bacteria and microbes on earth as the baseline dataset for studies of global biodiversity. It will also provide a simple access point enabling users to link to other data systems for all groups of organisms, using direct species links.

(ii) ***Culture Collections*:** Sources of culture collections are available electronically and are valuable to pathologists looking for sources of cultures for their research. For example, databases developed at the US National Fungus Collections (http://nt.ars–grin.gov/) provide access to information about fungi, primarily those associated with plants or otherwise of agricultural importance. The United Kingdom National Culture Collection (www.ukncc.co.uk/) coordinates the activities of the UK national service collections. Index Herbariorum (www.nybg.org/bsci/ih/), a joint project of the International Association for Plant Taxonomy and The New York Botanical Garden, is a detailed directory of over 3000 public herbaria of the world and the 8000 staff members associated with them.

(iii) ***Molecular Data*:** Access to molecular biological data, *e.g.*, protein and DNA sequences and structural models, has proved extremely valuable in allowing plant pathologists to keep abreast of the rapid developments in molecular biology. See Kraska (2001) for an up-to-date guide to further sources of information.

(iv) ***Quarantine Databases*:** Several computerized relational databases, developed in the 1980s and 1990s to compile data on pest distributions and make these searchable by pest, host or country basis, are now available to plant pathologists. The European and Mediterranean Plant Protection Organization (EPPO) database PQR is one example of this type of database. It gives access to data on all the quarantine pests of the EPPO A1 and A2 lists, contains data on many other quarantine pests of interest to other regional plant protection organizations, and is updated regularly. For further information, see EPPO (2000). Information relating to international aspects of plant quarantine can be found at the International Plant Protection Convention (IPPC) area of the Food and Agriculture Organization (FAO) web site (www.fao.org).

(v) ***Bibliographic Databases*:** Bibliographic databases usually include references and abstracts, indexed by subject–specific keywords so users can retrieve searches for a particular topic or species of interest. For example, the CAB ABSTRACTS database, published by CAB *International*, provides abstracts of internationally published scientific research literature in agriculture and the biosciences. The Review of Plant Pathology is compiled from this database and is available as a CD–ROM, online and as a printed journal (see www.cabi.org or http://pest.cabweb.org/ for further information). AGRICOLA (Agricultural Online Access) is a machine-readable database of bibliographic records created by the National Agricultural Library of the US Department of Agriculture and its co-operators (See www.nalusda.gov/ for further information). Some bibliographic databases include the full text of the original paper, allowing users greater access to the world's literature. Software packages are also available to help scientists manage databases of references they need for writing reports.

7. INTERPRETING INFORMATION TO PRODUCE KNOWLEDGE

The interpretation of stored scientific data leads to knowledge-based systems. There is wide scope for the development of these systems in the future and for their ultimate transfer to the web. Examples of such systems include the following:

(i) ***Taxonomic Information Systems*:** One example of a taxonomic information system is the Universal Virus Database which uses DELTA (the Description Language for Taxonomy), a method of coding the descriptions of taxa, their characters and states in a data matrix using ASCII text (Dallwitz *et al.*, 1993). The resulting database is authorized by the International Committee on Taxonomy of Viruses and is available on the Internet (www.ncbi.nlm.nih.gov/ICTVdb/).

(ii) ***Molecular Information Systems*:** These systems record, interpret and model molecular information. Molecular structures, produced from primary data, are available. Examples of this type of system are the Protein Data Bank (maintained by the Research Collaboratory for Structural Bioinformatics; see www.rcsb.org/pdb/) and the European Molecular Biology Laboratory's EMBL Nucleotide Sequence Database (Stoesser *et al.*, 1999; see also www.ebi.ac.uk/embl/).

(iii) ***Geographical Information Systems*:** Geographical Information Systems (GIS) present digital information on a geographical map of the area to which the information pertains. CLIMEX, a program developed by CSIRO Division of Entomology (Australia) in 1985 and updated since, is an example of a computer-based system for estimating or predicting the potential distribution and relative abundance of species in relation to climate (See www.ento.csiro.au/research/climex/climex.htm for further information).

(iv) ***Multimedia Systems*:** Multimedia systems incorporate various forms of media in a single program; such systems may resemble 'encyclopaedias', 'picture galleries' or 'film archives'. Within the discipline of crop protection there is considerable scope for the further development of multimedia systems to help diagnose and manage pest and disease problems. Many such systems are already available; some focus on specific crop hosts, whereas others aim to present a broader coverage. The *Crop Protection Compendium* is an example of one such multimedia resource that will also be regularly updated. Currently available on CD-ROM (with options to link to Internet information) and *via.*, the Internet, it incorporates a hierarchical taxonomic structure, text, images, GIS, identification keys, diagnostic aids, glossaries, details on pesticides and biopesticides and bibliographic information. The compendium was developed by *CAB International*, in partnership with many organizations from both the public and private sectors and hundreds of international pest specialists (Sweetmore *et al.*, 1998). Users can search the relational database for data sheets on specific pests using a list of names, including scientific names, synonyms and common names in a number of languages. Data sheets on pests, diseases, weeds, nematodes and natural enemies each consist of comprehensive texts, taxonomic information, pictures and a distribution map. Crop and country data sheets link to information on crop production from the FAO. Country data sheets contain information on climate, land area, population, etc. and selected statistics from the World Bank including gross national product (GNP) per capita and gross domestic product (GDP) from agriculture. As an encyclopaedic source of worldwide information on pests and crops, with links to many other information sources, the compendium is an extensive, authoritative knowledge base with potential as an aid for research, training and decision-making. See www.pest.cabicompendium.org/cpc/ for further details.

8. ROLE OF EXPERT SYSTEM IN PLANT PROTECTION

Developing an expert system for crop protection is not an easy task. Several issues are required to be considered during the design and implementation stages. Some of these issues are quite different from the issues in developing expert systems in other domains (Chakraborty and Chakrabarti, 2006). There are three types of expert systems for crop protection *viz.* crop non-specific, crop specific and disease non-specific, and crop specific and disease specific.

"Pesticide advisor", a non specific expert system, developed to choose the most effective and safest pesticide and apply them correctly in optimum dose. It has detailed information about pesticides including their target pests, shelf life, manufacturers, chemical formulae and safety measures. Expert system also proposes to use safe biopesticides instead of hazardous chemicals. The Expert System for Pest and Disease on Different Field Crops in India (ESPDDFCI) is a crop non-specific expert system for crop protection. It has been developed for identifying pests and diseases on the basis of various distinguishable symptoms. After the diagnosis, the ESPDDFCI proposes well practiced scientific control measures for the diagnosed pest and disease. The ESPDDFCI has the unique feature of supporting several local Indian languages (Chakraborty and Chakrabarti, 2008).

The Indian Cotton Insect Pest Management, ICOTIPM, a crop specific and disease non-specific expert system for crop protection diagnoses the pests either by description of the damages done to the plants or by the photographs of the insects. After the diagnosis, the expert system proposes a concrete pest management strategy specifying the pesticides to be used, their doses and application techniques (Chakraborty and Chakrabarti, 2008).

The Expert System for Management of Malformation Disease of Mango (ESMMDM), a crop and disease specific expert system developed by Chakrabarti and Chakraborty (2006) was tested in North and North-central India to predict the disease occurrence and suggest an appropriate crop protection and pest management strategy. The ESMMDM expert system is an interactive software tool with a graphic user interface. It asks some simple multiple choice questions about the test case. A user is required to select one of the options from an interactive control like a radio button or a drop down list box. Photographs have been used to help the user to accurately identify the symptoms. At the end of the diagnosis, the expert system generates a descriptive report of the present case. The report includes the particulars of the symptoms detected and the crop protection and pest management strategy prescribed. If required, the report can be printed. The ESMMDM expert system can be used by extension workers as well as farmers with or without any experience of using computers.

Apart from native expert system, some exotic expert systems are also in use in India. Jones *et al.* (1980) have earlier developed a microcomputer based system to predict primary apple scab infection periods. This system has now been modified as per Indian climatic conditions and adopted successfully for forecasting primary scab infections in apple orchards in India (Sharma and Gupta, 1995). Krause *et al.* (1975) developed a computerized system, called Blitecast, for forecasting the occurrence of the late blight disease of potato and for scheduling timely application of appropriate fungicides. Using the principle of Blitecast, a decision support system has been developed in India for forecasting the outbreak of the late blight disease in potato and prescribe suitable fungicides.

9. USE OF INFORMATION TECHNOLOGY IN DECISION SUPPORT SYSTEM (DSS) FOR PLANT PROTECTION

(i) ***Decision Support System (DSS)*:** Agriculture in India is a gamble with weather. Pest associated crop losses are higher in the tropics which are currently estimated between 10–30% of the total agricultural production valued at $1400 billion in 2010. There are additional costs in the form of pesticides applied for pest control, currently valued at $68 billion. Knowledge and information is the key to correct pest management decisions. Integrated Pest Management (IPM) is a system that emphasizes appropriate decision-making, depends heavily on intensive, accurate and timely information for field implementation by practitioners. Pest forecast is an important component of the broad IPM philosophy. Forewarning models provide lead time for managing impending pest attacks and thus minimise crop loss and optimize pest control leading to reduction in cost of cultivation. A Decision Support System (DSS) integrates a user-friendly front end to often–complex models, knowledge bases, expert systems, and database technologies. DSSs have emerged as essential tools to bridge the gap between science–based technology and end-users who make day-to-day management decisions. An IPM–DSS provides user all necessary information on pest developmental models including sampling, decision-making criteria and management options. Such systems are essential to enhance the adoption and spread of IPM in the country. Decision support systems are a class of computer-based information systems including knowledge based systems that support decision-making activities. DSS is a computerized system for helping make decisions. A decision is a choice between alternatives based on estimates of the values of those alternatives. Supporting a decision means helping people working alone or in a group gather intelligence, generate alternatives and make choices. Supporting the choice making process involves supporting the estimation, the evaluation and/or the comparison of alternatives. In practice, references to DSS are usually references to computer applications that perform such a supporting role. DSS are "interactive computer-based systems that help decision makers utilize data and models to solve unstructured problems" (Janakiraman and Sarukeshi, 2001).

(ii) ***Operation of DSS*:** In plant protection selection of an appropriate DSS for a given cropping situation often depends upon pathogen/pest complex as it interacts with crop and grower preference factors. Decision Support Systems (DSS) integrate all relevant information to generate spray recommendations. While there is room for improvement in DSS, they operate at such a technically high level that we estimate any effect will be small. More can be gained by increasing the use of DSS (or parts of DSS) by farmers and advisers. It is important to understand that information from DSS will increase the efficacy of farmers' control strategies without increasing risk. In other words, DSS should primarily not aim at a large reduction in the number of sprays but should aim at effective control of the disease and insect pests (including a large enough safety margin). DSS can also be used to justify the input of fungicides/insecticides and as a source of advice in situations when the number of sprays is limited by legislation.

- The focus on pesticide risk reduction has created a demand for scientific information and advice on pest management with emphasis on how the practical farmers can control pest, diseases or weeds by using combinations of alternative methods combined with minimum input of pesticides. This has in some cases led to creation of Decision Support Systems including all relevant information for growers or advisors on a specific crop. The Danish system PC-Plant Protection is an example of such a system (Secher *et al.*, 1995).

(iii) ***Application of DSS*:** A NAIP funded project, "Development of Decision Support Systems for Insect Pests of Major Rice and Cotton based Cropping Systems" coordinated by CRIDA, Hyderabad, has developed decision support system in rice and cotton for use at micro and macro level. They also developed a database of historical data on crop–pest–weather related to rice (nine insect pests, two diseases) and cotton (10 insect pests and two diseases). The retrievable historical database is available as a web enabled interface and a compact disc (CD) product is in the pipeline.

- National Centre for Integrated Pest management, New Delhi has developed "e-National Pest reporting and alert system" which can be accessed at http://www.ncipm.org.in/A3P/UI/ HOME/Login.aspx. The structure of this online reporting system is based on review and capabilities of available Decision Support System (DSS) in agricultural system and mobile communication technologies. The information used has been collected directly from the farmer's fields and real time data have been selected carefully in the pilot case so that the system can be of immediate use not only by Indian farmers but also by members of the farming community faced with similar problems. The above system was developed using, three tier architecture with facilities for "online" data entry, reporting and advisory to farmers through short messaging service (SMS) in their own language.
- ENDURE's Potato Case Study has considered all DSS in Europe, where all potato growing regions have one or more regional DSS available. ENDURE is the European Network for the Durable Exploitation of Crop Protection Strategies. ENDURE is a Network of Excellence (NoE) with two key objectives: restructuring European research and development on the use of plant protection products, and establishing ENDURE as a world leader in the development and implementation of sustainable pest control strategies through:
  - (a) Building a lasting crop protection research community
  - (b) Providing end-users with a broader range of short-term solutions
  - (c) Developing a holistic approach to sustainable pest management
  - (d) Taking stock of and informing plant protection policy changes.
- A network of European scientists has developed an innovative technology for implementing crop protection strategies in maize crop. The model developed for an innovative crop protection system (ICPS) named ENDURE–DR 2–21, is a state-of-the-art system for management of diseases, pests and weeds in maize so that high quality product can be grown with minimal use of agrochemicals pesticides (Meissle, 2010).

(iv) ***Components of DSS***: A DSS is often considered to be simply a piece of computer software or a tool but it is important to remember that it is also a methodology. The four components of a DSS are the crop environment, data, analysis and decision. Each component can be thought of as a method with a set of associated tool. For example, the data component is associated with the collection method, which has several tools including automated weather station and site-specific weather products. The data need to be analyzed with respect to a pathogen's life cycle to determine likely disease incidence and severity, if no action is taken. Analysis, interpretation is the formulation of appropriate management option for a disease or the spectrum of diseases and pests. Finally, the decision step is choosing between the management options. In the following discussion, we will briefly describe each step of the DSS process and then describe methods for the delivery of DSS (Fig. 2).

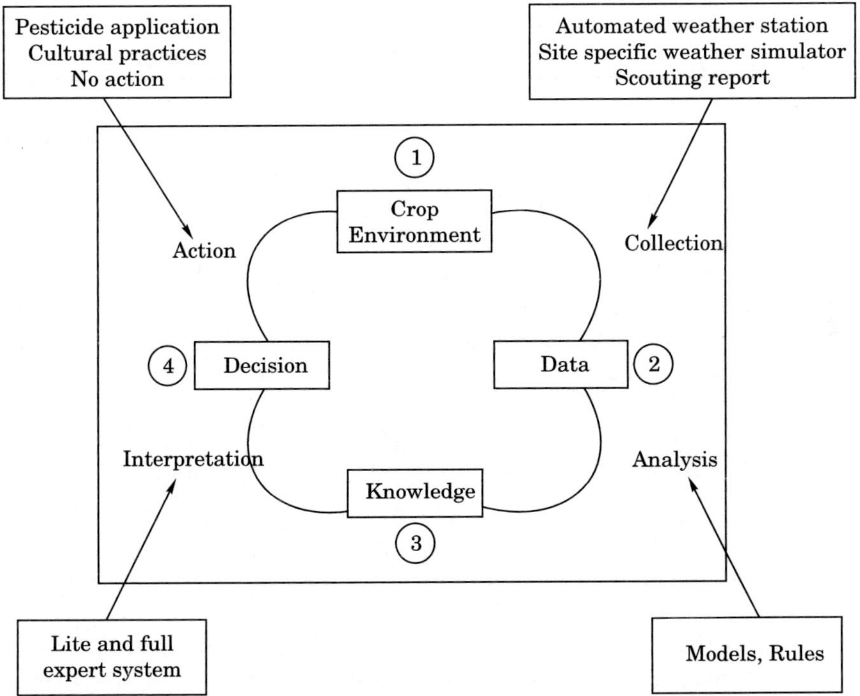

Fig. 2: An idealized Decision Support System for plant disease management showing components (red font), methodology (black font), and tools (blue font) (Magarey *et al.*, 2002).

(a) ***Data collection*:** The DSS process begins with the collection of data in the crop disease incidence and insect pest population. Field observations collected by a scout represent one of the most effective methods of data collection. The use of well trained personnel data assistants equipped with global positioning provides an efficient mechanism for survey data collection. However, for plant diseases, some significant epidemiological events cannot be observed without a trained technician and a microscope (Seem and Russo, 1984). Weather data can be used as a surrogate to observation for the predication of many pathogenic life stages or processes. Commonly collected variables include air temperature, leaf wetness, relative humidity and precipitation. Since the mid-1980s, automated weather stations (AWS) have become more frequently used as an alternative to manual weather data collection. Although a great improvement upon the manual process, many AWSs are expensive, some require considerable maintenance, and there are problems associated with the calibration and standardization of sensors (Magarey *et al.*, 2002).

(b) ***Analysis:*** After data have been collected it needs to be analyzed. It is important to understand the epidemiological basis by which a decision support system leads to the correct biological interpretation. A simplified representation of a pathogenic life cycle begins with survival of spores in the soil, in leaf litter, in a vector or on a plant part. For pathogens, preconditioning may be related to degree accumulation for fruiting structures to mature. Once the spores get discharged they can be dispersed to a growing plant surface where infection takes place. Infection is the process by which the pathogen enters the plant. For most pathogens, infection is limited first by moisture and second by temperature. Often rain is required for dispersal to susceptible host surfaces and a film of wetness is required for infection itself. The duration of moisture required for infection is dependent upon temperature and can vary from a few hours to several days (Huber and Gillespie, 1992). Another important component of the life-cycle is the overwintering (or over summering) survival spores. Usually these spores are in a dormant or inactive state but this is not always the case. The density of survival spores is mostly determined by the severity of disease in the previous season. For this reason, maintaining a clean field, orchard or vineyard is one of the most effective methods for disease control.

(c) ***Interpretation:*** After analysis of the biological data, interpretation will lead to the best management options. Interpretation depends upon host factors including: (i) phenological stage and canopy size, (ii) remaining time to harvest or ontogenic resistance, and (iii) cultivar susceptibility; pathogenic factors like (iv) disease history, (v) scouting observations, (vi) destructiveness of disease; environmental factors like (vii) predicted infection periods and (viii) soil type and accessibility after rain; management factors like (ix) growers perception and willingness to take risk or manipulate spray timing, (x) effectiveness of cultural practices, (xi) ease of arresting disease development, (xii) pesticide label restrictions, (xiii) safety considerations (*e.g.*, re-entry time), (xiv) fungicide price and (xv) crop value and end use (Bailey, 1999; Travis and Latin, 1991). The automated interpretation of some host, environmental, and disease factors by a DSS may save a user considerable time. However many management factors with the exception of fungicide timing may be too complex to incorporate into a DSS with existing technologies and they may be more easily interpreted manually. It is important for the DSS to display the phenological stages along with which incidences as observed by a field scout and predicted incidence as measured by the cumulative number of infection periods. The number 'missed' infection periods may also be useful indicator of likely disease incidence.

(d) ***Delivery*:** To facilitate decision-making, a user needs to be able to see the analysis in visual form. Complexity should be avoided since many users have limited time allotment for the use of a DSS (Magarey *et al.*, 2002). However, the flip side is that some DSSs are so simple that they do not answer questions that real users want to know. The time it takes to use a DSS is also important. Many users want to be able to interrogate a DSS in a very short time. Information in a DSS is often summarized in text messages, tables, graphs, maps or calendars. For many growers simple text messages including the last effective spray date may be appropriate. However, text messages are often limited in the amount of information that can be quickly absorbed by a user.

- Tables may often present too much information and confuse users. Often it is best to use simple symbols rather than complex arrays of numbers. Graphs are a nice method to present information but may have limited application for certain types of information. Maps need to be interpreted with care since there may be greater spatial uncertainty in disease prediction than those associated with point forecasts. Nevertheless maps often provided a good method to present disease forecasts at a regional scale or to provide high resolution site-specific information. Where maps are used they can be linked to a calendar providing a quick, easy and familiar method for a user to move through space and time. With the increasing importance of precision and site-specific agriculture it is possible that the map and calendar approach will become one of the most valuable tools for presenting disease forecasts.

- The Web represents one of the most popular methods of delivering information but there are also many other methods. The choice of delivery system depends upon the choice of local or regional scale; information content site-specific or generic; and accessibility push, pull or plug (Russo, 2000). Push and pull systems provide DSS information from a remote source. Push systems deliver information to the user, while pull systems require the user to request the information. Push systems work best for supplying daily information, updates and other types of dynamic information. Push systems provide DSS information built into an on-site device, for example a disease predictor (Magarey and Western, 1999). They allow users to maintain hands-on contact with the data source. Push is analogous to turning on a faucet (tap), while pull is drawing water from the reservoir with a pump and plug is like installing bottled water.

(v) ***Limitations and Future Development in DSS*:** Historically, the application and use of decision support systems has been limited by several factors including:

(a) The lack of sufficient computer power for data analysis and interpretation,

(b) The cost of collecting site-specific weather data and disease observations,

(c) Cost and speed of the communications, and

(d) The ease of DSS use.

Technological advances have greatly reduced many of these problems but the cost of collecting disease observations may continue to be a significant limitation for some DSS applications. Remote sensing in the future will become a more useful method but its application at present has been limited by the spatial and temporal resolution of the images. For many diseases, the action threshold is so low that the use of image technology is especially problematic. There is great need for the development of a new range of sensors to enable the precision treatment of crops. These sensors could be mounted on a tractor or on robotic droid.

As a final word, the cost of developing and maintaining DSS remains an important problem (Rajotte *et al.*, 1992). Funding for system development will be provided by small or large agricultural supply companies or cooperatives. In exchange, the DSS provides a mechanism for the company to market their products or coordinate their management protocols.

10. CONSTRAINTS AND REMEDIES FOR EFFECTIVE DISSEMINATION

Some of the major constraints delaying the spread of information technology to rural India are listed below:

(i) ***Haphazard Development***: It is observed that some initiatives have already been made to provide IT based services to rural community. However, duplication of efforts is witnessed as most of the services revolve around limited subjects. Keeping in view the giant task involved, it is necessary to form a coordination mechanism to strive for a concerted effort to support farming community in the country. Such a coordination agency may only have advisory powers such as user interface, broad design, delivery mechanism of the content, standards for setting up kiosks.

(ii) ***User Friendliness***: The success of this strategy depends on the ease with which rural population can use the content. This will require intuitive graphics based presentation. Touch Screen kiosks are required to be set up to encourage greater participation.

(iii) ***Local Languages***: Regional language fonts and mechanisms for synchronisation of the content provide a challenge that needs to be met with careful planning.

(iv) ***Restrictions***: Information content based on remote sensing and geographical information systems can provide timely alerts of possible incidence of disease and insect pest to the farmers so that prior precautionary measures can be taken. These applications can have a major impact on the plant protection practices and help farmers to appreciate the potential of information technology. However, government's map restriction policies often threaten to stifle the optimal utilisation of these tools.

(v) ***Power Supply***: In most of the rural India, power supply is not available for long hours. This will reduce the usefulness of the intended services. Since almost entire country receives sunshine for most part of the year, it is useful to explore solar power packs for UPS as well as for supply of power. The Ministry of Non-conventional Energy Sources may pay special attention in this area which can be a major contributor to the growth of IT in villages.

(vi) ***Connectivity***: Despite the phenomenal progress made in the recent years, the connectivity to rural areas still requires to be improved. Reliable connectivity is a prerequisite for a successful penetration of IT into rural areas. Many private ISPs are setting up large networks connecting many major towns and cities. Since some of these networks pass through rural areas, it is possible to provide connectivity to a large number of villages. Several technologies exist that can be utilised for connecting rural areas. Cable network is a possible medium for providing the last mile connectivity to villages.

(vii) ***Bandwidth*:** Even in areas where telephone and other communication services exist, the available bandwidth is a major constraint. Since internet based rural services require substantial use of graphics, low bandwidth is one of the major limitations in providing effective e-services to farmers. As already stated, networks with high bandwidth are being set up by several companies passing through rural segments which can be utilised. Until this materialises, a two pronged strategy of storing static information at the kiosks and providing dynamic information from remote locations can be examined. The graphic oriented content which does not change frequently, such as, demonstration clips for farmers, can be stored on the local drives at the kiosks and arrange for periodic upgradation of this information over the network during non-peak hours. The dynamic information which changes more frequently can be accessed from remote locations to obtain the latest status.

(viii) ***Dissemination Points*:** Mass deployment of information kiosks is critical for effective use of the Internet based content and services. In order to ensure that the information kiosks are economically feasible, it is necessary to make the proposition sustainable and viable. This requires a major focus on a viable revenue model for such kiosks. In the new information era, the kiosks should be designed to become electronic super markets that can, in addition to being information sources, handle other services of use to the people living in rural areas. The revenue available through such sources can make a kiosk attractive for prospective investors. The government can provide finance facilities to unemployed rural agricultural graduates who can be expected to have greater commitment and at the same time act as an efficient interface for less educated rural visitors. The objective should be to transform rural information kiosks into 'clicks and mortar' gateway to rural India for 'Bricks and mortar' industry. Some of the sources that can generate revenue for rural kiosks are:

(a) ***Distance education*:** A large number of people travel substantial distances to attend educational courses at Krishi Vigyan Kendras (KVKs). It is possible to set up virtual class room right in their villages.

(b) ***Training*:** People living in rural areas require training and a means for upgrading their skills in area of plant protection. It is possible to provide quality education right at their door steps with facilities for online interaction with experts. For example, a village worker or agriculture supervisor or Assistant Agriculture Officers staff can keep abreast of latest developments without disturbing his/her routine. Similarly, training can be imparted on various aspects of agriculture such as correct practices, irrigation practices, efficient utilisation of tools used in farming such as tractors.

(c) ***Insurance*:** The advent of private players into insurance has brought about advanced IT systems that can render services over networks. The kiosks can be insurance agents for insurance firms which, in turn, can compensate the kiosk operators for online transactions for new business as well as maintaining the old.

(d) ***Local agent*:** Many companies have difficulty in working out logistics for their supplies to rural outlets. A rural kiosk can act as conduit for such 'bricks and mortar' companies. This has the potential of transforming a rural kiosk into a profitable venture.

(e) ***Rural post office*:** The kiosks can facilitate sending and receiving e–mails, facilitate 'chats' with experts. Several successful rural kiosks are already available in many states which run essentially on this model.

(f) ***e–Governance*:** Rural kiosks are the stepping stones for effective implementation of e-governance. Details related to central/state/ local governments, formats and procedures, status verification such as case listings in courts, filing of applications in electronic format where admissible, etc. are some of the areas where kiosks can be of major use.

(ix) ***Who should take up the task?*** At present, several initiatives have been taken in the form of websites/portals targeting rural India. These are at best sketchy information sources catering to pockets of rural India. It is to be noted that strong inter-linkages exist within entire rural India and concerted and coordinated effort is required for carrying the benefits of IT to rural India. The magnitude of the task is such that no single institution or organisation can accomplish it. It is necessary for stake holders in rural India, such as fertiliser industry, to come together to provide adequate thrust to the effort initially. The fertiliser industry distributes more than 15 million tonnes of nutrients per annum in the country involving complex production, logistics and storage operations. A small savings made possible through better management of information up to the point of delivery to farmers can mean significant savings. The success of e-powering Indian agriculture is high if fertiliser industry makes a concerted and coordinated effort to set up Business to Business (B–B) market place with dealer/cooperative networks. The consumer industry also benefits from efficient operations in rural India. The corporate India may be willing to participate in a joint effort that proves beneficial to them as well as the rural India. The Government of India may, as outlined above, initiate a coordinating agency where various stake holders can join hands to spread e-culture to rural India and at the same time benefit from efficient operations.

CONCLUSIONS

The Indian farmer and those who are working for their welfare need to be e-powered to face the changing scenario of variability in insect pests and disease incidence. Wide spread unawareness about quality seed, good farming practices, potential plant protection measures are the major challenges to get higher yield. The quality of rural life can also be improved by quality information inputs which provide better decision-making abilities. IT can play a major role in facilitating the process of transformation of rural India to meet these challenges and to remove the fast growing digital divide.

The rapid changes in the field of information technology make it possible to develop and disseminate required electronic services to rural India. The existing bottlenecks in undertaking the tasks need to be addressed immediately. A national strategy needs to be drawn for spearheading IT penetration to rural India. A national coordinating agency with an advisory role can act as a catalyst in the process. No single institution or organisation alone can succeed in the task of e-powering farmers and rural India. At the same time, scattered and half-hearted attempts cannot be successful in meeting the objective. Industries with major stake in villages, such as fertiliser sector, should come together to provide the initial impetus.

The success of any IT based service to rural India hinges on evolving a proper plant protection model for the dissemination points. The 'clicks and mortar' rural kiosks should be integrated with the 'bricks and mortar' industry to make them sustainable ventures by making them a business gateway to rural India. The information kiosks can draw information from the subject matter specialists by providing and disseminating required services. Once these dissemination points prove to be economically viable, the IT revolution in rural India will require no crusaders.

ROAD AHEAD IN FUTURE

There are number of IT based DSS and Expert systems in user domain but they are not easily accessible to the farmers. Technologically it is possible to develop suitable systems, to cater to the information needs of Indian farmers. User friendly systems, particularly with content in local languages, can generate interest in the farmers and others working at the grassroots. It is possible to create dedicated networks or harness the power of Internet to make these services available to all parts of the country. There is gradual change in disease and insect pest scenario on major crops which can be easily observed due to change in global climatic conditions. This makes the minor pests and diseases of earlier years major and more severe in today's crops. The task of creating application packages and database to support newer precision plant protection equipments is a giant task. This needs long term plant protection policy which will provide an exhaustive list of all major and minor pests and localised infested areas. This can be taken as guiding list to evolve design and develop suitable technological system to cater each of the specified area. Our country has wide network of specialised institutions of ICAR and CSIR in place to cater various aspect of plant protection. These institutions can play a crucial role in designing the necessary applications and databases and services. This will facilitate modularisation of the task, better control and help in achieving quick results. As it is, several institutions have already developed systems related to their area of specialisation. For quick results, it may be useful to get the applications outsourced to software companies in India. This will facilitate quick deployment of applications and provide boost to the software industry in India. In order to avoid duplication of efforts, it may be useful to consider promoting a coordinating agency which will have an advisory role to play in evolving standard interface for users, broad design and monitoring of the progress.

In the post–WTO regime, it is suggested that it is useful to focus more on some agricultural products to maintain an unquestionable competitive advantage for exports. This will call for urgent measures to introduce state of the art technologies such as remote sensing, geographical information systems (GIS), bio-engineering, etc. in support with quarantine network. India has made rapid strides in satellite technologies. It is possible to effectively monitor agricultural performance using remote sensing and GIS applications. This will not only help in planning, advising and monitoring the status of the insect pest population and disease incidence on crops but also will help in responding quickly to any crop stress conditions and natural calamities. Challenges of insect pests, diseases, soil problems, natural disasters can be tackled effectively through these technologies.

While developing these systems it is necessary to appreciate that major audience that is targeted is not comfortable with computers. This places premium on user friendliness and it may be useful to consider touch screen technologies to improve user comfort levels. It is often observed that touch screen kiosks, with their intuitive approach, provide a means for quick learning and higher participation. It is also necessary to provide as much content as possible in local languages.

Under Indian conditions the required technological application systems and databases will be ready in no time but a major challenge is with respect to dissemination of the information. The Krishi Vigyan Kendras, NGOs and cooperative societies may be used to set up information kiosks. Private enterprise like e-choupal, Rural Information Kiosks have to be strengthened up to disseminate the knowledge. These kiosks should provide information on other areas of interest such as proper dose of agrochemicals, method of application, residual toxicity of chemicals, time of application, modern farm equipments, etc. Facilities for e-mail, raising queries to experts, uploading digital clips to draw the attention of experts to location specific problems can be envisaged.

REFERENCES

Anonymous. (2006). A thematical strategy on the sustainable use of pesticides COM 2006, p. 372.

Bailey, J.E. (1999). Integrated method of organizing, computing and deploying weather-based decision advisories for selected peanut diseases. *Peanut Science,* **26:** 74–80.

Bridge, P., Jeffries, P., Morse, D.R. and Scott, P.R. (1998). *Information Technology, Plant Pathology and Diversity*. CAB International, Wallingford, UK.

Chakrabarti, D.K. and Chakraborty, P. (2006). Expert system for management of malformation disease of mango. *ICAR News,* **12(1):** 18.

Chakraborty, P. and Chakrabarti, D.K. (2008). A brief survey of computerized expert systems for crop protection being used in India. *Progress in Natural Science,* **18:** 469–73.

Dallwitz, M.J., Paine, T.A. and Zurcher, E.J. (1993). *DELTA User's Guide: A General System for Processing Taxonomic Descriptions*, 4th edition. CSIRO, Canberra, Australia, p. 136.

Ferris, C. and Farrell, J. (2003). What are web services? *Communications of the ACM,* **46(6):** 31.

Huber, l. and Gillespie. T.J. (1992). Modeling leaf wetness in relation to plant disease epidemiology. *Annual Review of Phytopathology,* **30:** 553–77.

Janakiraman, V.S. and Sarukeshi, K. (2001). Decision Support Systems–*Prentice–Hall of India Private Limited, New Delhi*, 2001, pp. 26–77.

Jones, A.L., Lillevik, S.L. and Fisher, P.D. *et al.* (1980). A microcomputer based instrument to predict primary apple scab infection periods. *Plant Dis.,* **64(1):** 69–72.

Jørgensen, L.N., Hagelskjer, L. and Nielsen, G.C. (2003). Adjusting the fungicide input in winter wheat depending on variety resistance. *BCPC conference on Crop Science and Technology, Glasgow*, pp. 1115–20.

Kraska, T. (2001). *The plant pathology internet guide book*. Institute for plant diseases, University of Bonn, Bonn, Germany. www.ifgb.uni hannover.de/extern/ppigb/ppigb.htm

Krause, R.A., Massie, L.B. and Hyre, R.A. (1975). Blitecast: A computerized forecast of potato blight. *Plant Dis. Rep.*, **63(1):** 21–25.

Kuhlmann, F. (2005). IT Applications in Agriculture: Some developments and perspectives. http://departments.agri.huji.ac.il/economics/gelb–appli–2.pdf

Magarey, P.A. and Western, M.D. (1999). The Model T Met Station: A weather station disease predictor. *In*: *Proceedings of the 12th Biennial Australsian Plant Pathology Conference,* 27–30 September, 1999, Canberra, ACT, Australia, p. 312.

Magarey, R.D., Travis, J.W., Russo, J.M., Seem, R.C. and Magarey, P.A. (2002). Decision support systems: Quenching the thirst. *Plant Disease* **86(1):** 4–14.

Meissle, M., Mouron, P., Musa, T., Biger, F., Pons, X., Vasileiadis, P., Otto, S., Antichi, D., Kiss, J., Palinkas, Z., Dorner, Z., Van der Weide, R., Groten, J., Czembor, E., Adamczyk Thobord, J.B., Melander, B., Cordsen Nielsen, G., Poulsen, R.T., Zimmerman, O., Verschwele, A. and Oldenburg, E. (2010). Pests, pesticide use and alternative options in European maize production: current status and future prospects. *Journal of Applied Entomology* doi: 10.1111/j.1439–0418.2009.01491.x.

Miller, S.A., Kinyua, Z.M., Beed, F., Harmon C.L, Xin, J., Vergot, P., Momol, T., Gilbertson, R. and Garcia, L. (2010). The International Plant Diagnostic Network (IPDN) in Africa: Improving capacity for diagnosing diseases of banana (*Musa* spp.) and Other African Crops. *Proc. International committee on Banana and Plantain in Africa*. *Ed*.: Dubois, T. *et al. Acta Hort.* **879:** 341–48.

Rafea, A. (2010). Web based domain specific tool for building plant protection expert systems. *Expert Systems*: *Ed*. Petrica Vizureanu, pp. 193–202.

Rajotte, E.G., Bowser, T., Travis, J.W., Crassweller, R.M., Muster, W., Laughland, D. and Sachs, C. (1992). Implementation and adoption of an agricultural expert system: the Penn State Apple Orchard Consultant (PSAOC). *Acta Horticulturae,* **313:** 227–31.

Rhee, S.Y. and Crosby, B. (2005). Biological databases for plant research. *Plant Physiol.* **138:** 1–3.

Röhrig, M. (2006). http://www.isip.de, online plant protection information in Germany. Abstract from EPPO conference on computer aids for plant protection. Wageningen, 17–19 Oct. 2006.

Russo, J.M. (2000). Weather forecasting for IPM. *In*: Emerging technologies for integrated pest management–concepts, research, and implementation. Kennedy, G.C. and Sutton, T.B. (*Eds*.). APS Press, St. Paul, MN, USA. pp. 453–73.

Rydahl, P. (2003). A web–based decision support system for integrated management of weeds in cereals and sugarbeet. *EPPO Bulletin*., **33:** 455–60.

Secher, B.J., Jørgensen, L.N., Murali, N.S. and Boll, P.S. (1995). Field validation of a decision support system for control of pests and diseases in cereals in Denmark. *Pesticide Science*. **45:** 195–99.

Seem, R.C. and Russo, J.M. (1984). Simple decision aids for practical control of pests. *Plant Disease,* **6:** 656–60.

Sharma, J.N. and Gupta, V.K. (1995). Studies on apple scab forecasting in Himachal Pradesh. *Indian Phytopathology*, **48(3):** 325–30.

Stack, J.P., Cardwell, K., Hammerschmidt, R., Byrne, J., Loria, R., Snover–Clift, K., Baldwin, W., Wisler, G., Beck, H., Bostock, R., Thomas, C. and Luke, E. (2006). The national plant diagnostic network. *Plant Dis*., **90:** 128–36.

Stein, L. and Wu, G. (2006). Reactome data sharing web services API. User's Guide. http://www.reactome.org p. 5.

Stoesser, G., Tuli, M.A., Lopez, R. and Sterk, P. (1999). The EMBL nucleotide sequence database. *Nucleic Acids Research*. **27(1):** 18–24.

Swaminathan, M.S. (2003). A knowledge revolution in rural India, The Hindu, 20th July.

Sweetmore, A., Schotman, C.Y.L., Zhang, B.C., Rudgard, S.A. and Scott, P.R. (1998). Integrated information management: a multimedia system for crop protection. *In*: Bridge, P., Jeffries. P., Morse, D.R. and Scott, P.R. (*Eds*) *Information Technology, Plant Pathology and Biodiversity*. CAB International, Wallingford, UK, pp. 117–28.

Thysen, I. (2000). Agriculture in the Information society. *J. Agri. Engg. Res.*, **76:** 297–303.

Travis, J.W. and Latin, R.X. (1991). Development, implementation and adoption of expert systems. *Annual Review of Phytopathology,* **29:** 343–60.

Verret, J.A., Klink, H., and Hoffmann, G.M. (2000). Regional monitoring for disease prediction and optimisation of plant protection measures: The IPM Wheat Model. *Plant Disease,* **84:** 816–26.

Volk, T., Johnen, A., Newe, M. and Meier, H. (2003). ProPlantexpert.com—The online consultation system on crop protection in cereals, rapeseed, potatoes and sugar beet: Experiences with cereal disease control in the region and possibilities for regional adaptations. *Crop Protection Conference for the Baltic Sea Region*, April 2003, Poznan, Poland, pp. 103–13.

Weiss, A., Van Crowder, L. and Bernadi, M. (2000). Communicating agrometeorological information to farming communities. *Agril. and For. Met.* **103:** 185–96.

Xin, J., Christy, G.S. and Barfield, B. (2005). Development of a Web-based state agricultural emergency response system. *Proc. of WCCA/EFITA*, Portugal.

Yu, V.L. (2004). ProMED–mail: An early warning system for emerging diseases: Surfing the Web. *Clinical Infectious Diseases,* **39:** 227–32.

Zadoks, J.C. (1983). An integrated disease and pest-management scheme, EPIPRE, for wheat. *Ciba Foundation Symposia,* **97:** 116–29.

4

Recent Biotechnological Achievements in Plant Disease Management

A.K. Mandal[1], Prem L. Kashyap[2], Malkhan Singh Gurjar[3] Gulzar S. Sanghera[4], Sudheer Kumar[5] and Sunil C. Dubey[6]

ABSTRACT

Despite substantial advances in crop protection strategies, global food supply is still in jeopardy by a multitude of pathogens and insect pests. This changed scenario warrants us to act more efficiently and effectively to tackle these problems. The situation demands judicious amalgamation of conventional, unconventional and frontier technologies. In this sense, plant biotechnology has emerged as one of the most promising tool for enhancing the innate immunity in plants to combat various fungal, bacterial and viral diseases causing huge losses in agriculture. The information and critical analyses provided in this chapter can give a glimpse of the role of plant biotechnology in relation to sustainable crop protection.

Key words: Elicitor, Pathogens, Quorum quenching, Resistance gene, Transgenics.

Plant pathogens continue to impact on the productivity of crops and quality of crop products worldwide despite decades of research and development

1 Directorate of Seed Research (DSR), Mau, Uttar Pradesh - 275 101, India.
2 National Bureau of Agriculturally Important Microorganisms, Mau, Uttar Pradesh - 275 101, India.
3 Central Potato Research Station (CPRI), Shillong - 793 009, India.
4 SKUAST (K) Rice Research and Regional Station, Khudwani, Anantnag - 192 102, J&K, India.
5 National Bureau of Agriculturally Important Microorganisms, Mau, Uttar Pradesh - 275 101, India.
6 Division of Plant Pathology, Indian Agricultural Research Institute, New Delhi - 110 012, India.
Corresponding author: E-mail: asitmandal.iari@gmail.com

on improved methods for their control. The losses may be significantly greater in subsistence agriculture, where crop protection measures are often not applied. In the former scenario, the problem is that biotic agents of disease are moving targets that evolve in response to agricultural practices and environmental change. The emergence and spread of new pests and diseases, or more aggressive or pesticide–resistant biotypes are examples of such evolution. The key issues facing crop protection scientists in the 21st century are therefore twofold. First to devise disease control systems that are sustainable and not compromised by the evolution of insect pest and pathogen strains able to overcome crop resistance or chemicals, and second to devise appropriate crop protection technologies, as well as mechanisms for their use. Given the projected global need to produce 40% more food grains, there is increasing pressure to optimize inputs, reduce environmental impact but at the same time minimize the risk of widespread crop failure due to biotic stress (Beddington, 2010). Increasing knowledge of plant defense has led to more sophisticated biotechnological approaches to enhance resistance. The number of candidate genes put forward by genomics, transcriptomics, metabolomics, proteomics and protein interaction studies gives us a large choice of genes to be used. Potentially these genes can be manipulated by overexpression, induced expression, tissue–specific expression, and stable gene knockouts or silencing by RNA interference (RNAi) (Sanghera *et al.*, 2011). They can be dervied from plant itself, from other plants and from a microbe or even they could be completely synthetic. The addition of new antimicrobial secondary metabolites to a species, which can be achieved by adding gene(s) encoding the appropriate biosynthetic enzymes, detoxification, quenching pathogen signals and use of pathogen–inducible promoters to regulate antimicrobial factors etc. is the simplest approach to enhance resistance in plants. Second approach concerns with the manipulation of the regulatory network related to plant's own antimicrobial defenses (induced, basal and race–specific resistance) and exploit the recognition processes as well as the regulatory signal transduction pathways. Pathogen mimicry is the third approach by which one can manipulate or prime the plant to recognize a specific pathogen. Mechanistically, this is also termed as pathogen derived resistance (genetic vaccination). With these novel technologies, new strategies aimed at improving disease resistance can be devised. The chapter highlights feasibility of using these biotechnological approaches for enhancing diseases resistance in most effective, efficient and durable manner to address key challenges imposed by plant pathogens for sustainable crop protection.

1. STRATEGIES FOR CROP PROTECTION

There are several biotechnological strategies for crop protection (Fig. 1) which are summarised in following sections.

Strategy 1—Intervention with pathogenicity: Constitutive expression of antimicrobial factors and pathogen–induced gene expression.

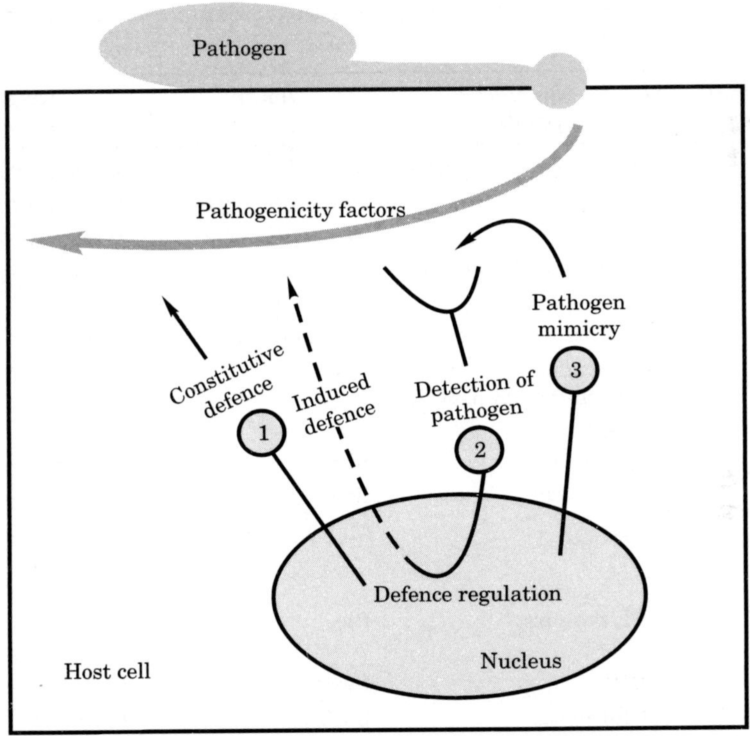

Fig. 1: Biotechnological strategies for plant protection from pathogens. (1) Strategy–I concerns direct interference with pathogenicity or inhibition of pathogen physiology. Thus it involves constitutive expression of antimicrobial factors and pathogen–induced expression of one or more genes in the transgenic plant; (2) Strategy-II concerns the regulation of the natural induced host defences and altering recognition of the pathogen (*R-genes*) and concerns downstream regulatory pathways (SAR) and transcription factors; and (3) Strategy-III is pathogen mimicry (genetic vaccination) concerns manipulation of plant to prime recognition of a specific pathogen through pathogen derived gene sequences (Modified after Collinge *et al.*, 2008).

Plants protect themselves against microbial pathogens by various constitutive and inducible defense responses, which include the production of antimicrobial peptides, secondary metabolites, lytic enzymes, membrane-interacting proteins and reinforcement of cell walls (Scheel, 1998). During the past few years a wide spectrum of antimicrobial factors has been detailed, and enhanced resistance has been obtained by introducing the corresponding genes into crop species to produce transgenic crops (Table 1).

Table 1: Engineering crops using constitutive expression of antimicrobial factors

Transgenic crop	***Gene/gene product***	***Donor***	***Target pathogen***
Rice	Allene oxide synthase	Rice	*Magnaporthe grisea*
	Cercosporin A	Giant silk moth	*Magnaporthe grisea*
	AFP	*Aspergillus giganteus*	*Magnaporthe grisea*
	*ech*42, *nag* 70, *gluc* 78	*Trichoderma atroviride*	*Magnaporthe grisea*
	Glucose oxidase gene	Fungal gene	Multiple pathogen
	Puroindolines (antimicrobial peptide gene)	Wheat	*Magnaporthe grisea, Rhizoctonia solani*
	Chitinase	Rice	*Rhizoctonia solani*
	TL protein	Rice	*Rhizoctonia solani*
	RIP	Maize	No effect on *M. grisea* or *R. solani*
Wheat	TL protein	Rice	*Fusarium graminearum*
	RIP	Barley	*Blumeria graminis*
	Chitinase	Barley	*Blumeria graminis, Puccinia recondita*
	Ace–AMP1	Onion	Enhanced antifungal activity
Tobacco	*GAFP* (gastrodia antifungal gene)	*Gastrodia* (orchid)	*Rhizoctonia* spp., *Phytophthora* spp.
	Cercopin–A–melittin peptide gene	Hybrid peptide	*Fusarium solani*
	Mannitol dehydrogenase	Celery	*Alternaria alternata*
	Spi–2 (peroxidase gene)	Norway spruce	*Phytophthora* spp.
	bO/Bacterio–opsin (BO)	*Halobacterium halobium*	*Pseudomonas syringae* pv. *tabaci*
	expI/N–oxoacyl–homoserine lactone (OHL)	*Erwinia carotovora*	*Erwinia carotovora*

Table 1: (*Contd...*)

Table 1: (*Contd...*)

Transgenic crop	***Gene/gene product***	***Donor***	***Target pathogen***
	aiiA / Acyl–homoserine lactonase	*Bacillus* sp. *240B1*	*Erwinia carotovora*
	argK / ROCT ornithine carbamoyltransferase	*Pseudomonas syringae*	*Pseudomonas syringae* pv. *phaseolicola*
	Sarcotoxin peptide gene	*Sarcophaga peregrine*	*Rhizoctonia solani*, *Pythium aphanidermatum*, *Phytophthora nicotianae*
	Chloroperoxidase	*Pseudomonas*	*Colletotrichum destructivum*
	Magainin analog	Frog	*Peronospora tabacina*
	Chitinase	*Baculovirus*	*Alternaria alternata*
	Salicylic acid synthase	Bacterial origin	*Oidium lycopersicon*
	Antimicrobial peptide	Synthetic	*Colletotrichum destructivum*
	scFv antibodies	*Tomato bushy stunt virus* (TBSV)	Tomato bushy stunt
	hrp N	*Erwinia amylovora*	*Botrytis cinerea*
	PAPII	*Phytolacca americana*	Broad spectrum resistance to viral and fungal pathogens
	scFv antibodies	*Cucumber mosaic virus*	Cucumber mosaic
Arabidopsis	*Fusarium* specific antibody linked to antifungal peptides	Fusion proteins	Multiple fungal pathogens
Pearl millet	*Afp*	*Aspergillus giganteus*	Rust and Downy mildew
Sunflower	Oxalate oxidase	Wheat	*Sclerotinia sclerotiorum*
Tobacco and banana	MSI–99 peptide	African clawed frog	*Alternaria*, *Botrytis*, *Mycosphaerella musicola*
Poplar	Oxalate oxidase	Wheat	*Septoria musiva*
Grape	Chitinase	Rice	*Uncinula necator*, *Elsinoe ampelina*

Table 1: (*Contd...*)

Table 1: (*Contd...*)

Transgenic crop	***Gene/gene product***	***Donor***	***Target pathogen***
	Endochitinase	*Trichoderma harzianum*	*Botrytis cinerea*
	Polygalacturoase inhibiting protein	Pear	*Botrytis cinerea*
Tomato	*Gene1–2*	Tomato	*Fusarium* spp.
	Defensin	Radish	*Alternaria solani*
	Oxalate decarboxylase	*Collybia velutipes*	*Sclerotinia sclerotiorum*
	scFv antibodies	*Cucumber mosaic virus*	Cucumber mosaic
Alfalfa	Resveratrol synthase	Peanut	*Phoma medicaginis*
Carrot	Chitinase	Tobacco	*Alternaria dauci, A. radicina, Colletotrichum corotae*
	Human lysozyme	Human	*Erysiphe heraclei, Alternaria dauci*
Chrysanthemum	Chitinase	Rice	*Botrytis cinerea*
Geranium	Antimicrobial protein	Onion	*Botrytis cinerea*
Potato	scFv antibodies	*PVY*	Potato mosic
	Endochitinase	*Trichoderma harzianum*	Foliar and soil borne fungal pathogen
	34–aa chimaeric peptide MsrA1+ melittin	Bee venom	*E. carotovora* ssp. *atroseptica*
	sarco gene coding for sarcotoxin IA	*Sarcophaga peregrina*	*Erwinia carotovora, P. syringae* pv. *lachrymans* and *R. solanacearum*
	Lactoferrin	Human	Not tested
	Defensins (alfAFP)	Alfalfa	*Verticillium dalhiae*
	Cercosporin–melittin cationic peptide	Synthetic gene	Multiple pathogens
	Osmotin gene	Tobacco	*Phytophthora infestans*
Apple	Endochitinase, Exochitinase	*Trichoderma harzianum*	*Venturia inaequalis*
	attacin E gene (ate)	*Sarcophaga peregrina*	*Erwinia amylovora*

1.1. Plant Derived Antimicrobial Proteins

Plants represent a storehouse of various proteins with unique lysozyme activity like a chitinase from cucumber (Métraux *et al.*, 1989) or hevamine from *Hevea brasiliensis* (Jekel *et al.*, 1991). However, these proteins have not been expressed so far in transgenic plants in order to control antimicrobial infections. The best characterized plant antimicrobial proteins are thionins (Florack and Stiekema, 1994), which are able to inhibit a broad range of pathogenic bacteria *in vitro* (Molina *et al.*, 1993). Of these, the expression of a barley μ–thionin gene in transgenic tobacco enhanced resistance to *P. syringae* in laboratory assays (Carmona *et al.*, 1993), while synthetic hordothionin genes were not secreted into the intercellular space in transgenic tobacco plants (Florack *et al.*, 1994). Unfortunately, most thionins can be toxic to plant cells (Reimann–Philipp *et al.*, 1989), and thus, may not be ideal for expression in transgenic plants. The plants expressing *Dm–AMP1*, a defensin from *Dahlia merckii* showed enhanced resistance to *M. oryzae* and *R. solani*, by 84 and 72%, respectively. Interestingly, transgenic expression of *Dm–AMP1* was not accompanied by an induction of *PR* gene expression (Jha *et al.*, 2009). Similarly, transgenic rice over-expressing the wasabi defensin or *Mirabilis jalapa* antimicrobial protein *Mj–AMP2* gene exhibited resistance to rice blast (Prasad *et al.*, 2008). The average size of disease lesions of the transgenic plants was reduced to about half of that of non–transgenic plants (Kanzaki *et al.*, 2002). Another class of antimicrobial peptides called fabatins, which are active against both Gram–negative and Gram–positive bacteria (Zhang and Lewis, 1997), a defensin–type pseudothionin from potato (Moreno *et al.*, 1994) or plant defensins from various seed plants (Broekaert *et al.*, 1995) are very suitable for this purpose. Based on results obtained in transgenic tobacco against *P. syringae* pv. *tabaci* (Molina and García–Olmedo, 1997), lipid transfer proteins and snakins are emerging as a good candidates for use against plant pathogens.

Puroindolines are small proteins reported to have *in vitro* antimicrobial properties. Transgenic rice plants that constitutively express the wheat puroindoline genes (*PinA* and/or *PinB*) throughout the plants were produced. Puroindolins in leaf extracts of the transgenic plants reduced *in vitro* growth of *M. grisea* and *R. solani* by 35–50%. Transgenic rice expressing *PinA* and/or *PinB* showed elevated level of resistance to *M. grisea* and *R. solani* (Krishnamurthy *et al.*, 2001). Thus, expression of defensin in transgenic rice has the potential to provide broad–spectrum disease resistance against fungal pathogens.

Snakin–1 (SN1), a cysteine–rich peptide is an interesting candidate gene for engineering plants to confer broad–spectrum crop protection. This strategy was adopted by Almasia *et al.* (2008) to transform potato plants (*Solanum tuberosum* subsp. *tuberosum* cv. Kennebec) via *Agrobacterium tumefaciens* with a construct encoding the *S. chacoense SN1* gene under the regulation of the ubiquitous *CaMV 35S* promoter. As a result, transgenic plants showed significant protection against *Erwinia carotovora* and *R. solani*.

1.2. Ribosome–Inactivating Proteins (RIP)

Ribosome–inactivating proteins (RIPs) are plant enzymes that have 28S rRNA N–glycosidase activity, which depending on their specificity can inactivate foreign ribosomes, thereby shutting down protein synthesis. Plant RIPs inactivate foreign ribosomes of distantly related species and of other eukaryotes including fungi. A purified RIP from barley inhibits growth of several fungi *in vitro*. Tobacco plants constitutively expressing a RIP encoding DNA sequence of barley showed better resistance to *R. solani* (Logemann *et al.*, 1993). Resistance levels improved when RIP was used in combination with either *PR2* or *PR3*.

1.3. Phytoalexins

Phytoalexins are antimicrobial low molecular weight secondary metabolites produced in plants following pathogen attack and are believed to have a role in plant defense (Hammerschmidt, 1999). Biosynthesis of phytoalexins is often complex involving many pathways and hence, several substrates and enzymes. Nevertheless, there has been success in developing transgenics which synthesize new phytoalexins by simply introducing the gene for the last enzyme of the pathway. As an example, Hain and co-workers introduced the gene encoding stilbene synthase from grape vine (*Vitis vinifera*) into tobacco plants (Hain *et al.*, 1993). In tobacco the substrate for stilbene synthase is available, but the enzyme *per se* is absent. The expression of stilbene synthase (or resveratrol synthase) gene resulted in the production of resveratrol, a stilbene–type phytoalexin. Such transgenics showed enhanced resistance to *B. cinerea*. Similar transgenic plants were developed in rice, tomato, barley and wheat and were shown to have increased resistance to *M. grisea*, *P. infestans* and *B. cinerea* respectively (Thomzik *et al.*, 1997; Stark–Lorenzen, 1997; Leckband and Lorz, 1998). *Arachis hypogea* resveratrol synthase cDNA when expressed under 35S CaMV promoter in transgenic alfalfa confers resistance to *Phoma medicaginis* (Hipskind and Paiva, 2000).

1.4. Non-Plant Derived Antimicrobial Proteins

Antibmicrobial proteins are important components of the overall antimicrobial defense mechanisms of many groups of animals, including arthropods (particularly insects), amphibians and mammals. Probably acting in synergy, they have an antimicrobial action against several plant pathogens. Genes encoding antibacterial proteins that have been cloned and expressed in plants in an attempt to confer resistance to plant pathogens are described in following sections.

1.5. Lytic Peptides

Lytic peptides are small proteins with an amphipathic α–helical structure which makes pores in membranes resulting in the lysis, for example, of the bacterial cell membrane (Boman, 1991). Cecropins are antibacterial lytic peptides native to the hemolymph of *Hyalophora cecropia,* the giant silk moth. Transgenic tobacco plants expressing cecropins have increased resistance to *Pseudomonas syringae* pv. *tabaci* (Huang and McBeath, 1997). The Cecropin A–expressing transgenic rice plants accumulated biologically active cecropin A protein and exhibited resistance to rice blast without an induction of PR gene expression (Coca *et al.*, 2006). Similarly, transgenic rice plants over-expressing cecropin B gene showed a significant reduction in development of lesions caused by *Xanthomonas oryzae* pv. *oryzae* (Coca *et al.*, 2004). Synthetic lytic peptide analogs, Shiva–1 and SB–37, produced from transgenes in potato plants reduce bacterial infection caused by *E. carotovora* subsp. *atroseptica* in transgenic potato plants (Arce *et al.*, 1999). Transgenic apple expressing the SB–37 lytic peptide analog showed increased resistance to fire blight, a bacterial disease of apple caused by *E. amylovora* in field tests (Norelli *et al.*, 1998).

Attacins are another group of lytic peptides produced by *H. cecropia* pupae (Hultmark *et al.*, 1983). The mechanisms of antibacterial activity of this protein are to inhibit the synthesis of the outer membrane protein in Gram–negative bacteria (Carlsson *et al.*, 1998). Attacin expressed in transgenic potato enhanced its resistance to bacterial infection by *E. carotovora* subsp. *atroseptica* (Arce *et al.*, 1999). Transgenic pear and apple expressing attacin genes have significantly enhanced resistance to *E. amylovora* in *in vitro* and growth chamber tests (Ko *et al.*, 2000). In field tests, reduction of fire blight disease has been observed in transgenic apples expressing attacin genes (Norelli *et al.*, 1999). Transgenic apple expressing attacin targeted to the intercellular space, where *E. amylovora* multiplies before infection, has significantly reduced fire blight, even in apple plants with low attacin production levels (Ko *et al.*, 2000).

1.6. Lysozymes

Lysozymes are enzymes that hydrolyze the peptidoglycan layer of the bacterial cell wall. Hen egg–white lysozyme (*HEWL*), T4 lysozyme (*T4L*), and human lysozyme genes have been cloned and transferred to enhance plant bacterial or fungal resistance. These lysozyme genes have been used to confer disease resistance against plant pathogenic bacteria in transgenic tobacco plants (Kato *et al.*, 1998). *T4L*, from T4–bacteriophage, also has been reported to enhance resistance in transgenic potato against *E. carotovora*, which causes bacterial soft rot (Düring *et al.*, 1993). Transgenic apple plants with the T4L gene showed significant resistance to fire blight infection (Ko, 1999). Human lysozyme transgenes have conferred disease resistance in tobacco through inhibition of fungal and bacterial growth, suggesting the possible use of the human lysozyme gene for controlling plant disease (Nakajima *et al.*, 1997).

1.7. Lactoferrin

Lactoferrin is an iron–binding glycoprotein known to have antimicrobial properties (Nguyen *et al.*, 2011). Mitra and Zhang (1994) showed that human lactoferrin expressed in tobacco calli inhibited several phytopathogenic bacteria *in vitro*. The tachyplesin gene, which codes for a lytic peptide isolated from the horseshoe crab, has also been expressed in transgenic potatoes, resulting in a reduced amount of the tuber rot caused by *E. carotovora* (Allefs *et al.*, 1996). In addition, transgenic tobacco and tomato plants expressing lactoferrin significantly delayed wilt symptom development caused by *Ralstonia solanaceraum*, in a dose–dependent manner (Zhang *et al.*, 1998; Lee *et al.*, 2002). Similarly, expressed lactoferrin in transgenic pear showed an increased resistance against *Erwinia amylovora* (Malnoy *et al.*, 2003). Takase *et al.* (2005) evaluated transgenically expressed human lactoferrin and lactoferricin in rice against disease–causing *Burkholderia plantarii*, rice dwarf virus and *Pyricularia oryzae*. However, they found significant resistance only against *B. plantarii*. Recently, Nguyen *et al.* (2011) showed transgenic expression of bovine lactoferrin (*BLF*) gene in *Nicotiana tabacum* and *Arabidopsis thaliana* imparts considerable resistance and also delays the onset and progress of disease development by *R. solani*. Furthermore, the recent advances in our knowledge of the structures and properties of these proteins could result in the construction of synthetic molecules with enhanced expression and stability in plant tissues.

1.8. Hydrolytic Enzymes

The most widely used approach has been to overexpress chitinases and glucanases, and have been shown to exhibit antifungal activity *in vitro* (Giannakis *et al.*, 2005). Since chitins and glucans comprise major components of the cell wall in many groups of fungi, the over-expression of these enzymes in plant cells is postulated to cause the hyphae to lyse and thereby reduce fungal growth (Mauch and Staehelin, 1989). The effectiveness of the chitinase gene of fungal origin in transgenic plants has been demonstrated by the reduced rate of lesion development and reduction of overall size and number of lesions upon challenge with pathogenic fungi

(Islam, 2006). For instance, Terakawa *et al.* (1997) transferred chitinase encoding gene (*chi1*) of *Rhizopus oligosporus* responsible for causing cell autolysis of the filamentous fungus to tobacco and obtained a significant level in the reduction of the foliar symptoms imposed by *Sclerotinia sclerotiorum* and *Botrytis cinerea*. Another approach was followed by Lorito *et al.* (1998) who introduced endochitinase–encoding gene of *Trichoderma harzianum* in tobacco and potato to achieve tolerance against variety of phytopathogenic fungi. The transgenic expression of the *T. harzianum chit42* gene in tobacco and potato conferred almost complete resistance to *A. alternata, A. solani, B. cinerea, R. solani* and *S. sclerotiorum* (Lorito and Scala, 1999). Interestingly, biochemical and molecular characterization of the progeny indicated that in addition to the direct antifungal effects of the transgenic chitinase, other mechanisms of the plant defence system also activated which help transgenics to attain sustained and high level of disease resistance. Following this successful approach, *T. harzianum* genes have been transferred into a variety of crops, including potato (Lorito *et al.*, 1998), apple (Bolar *et al.*, 2001), petunia (Esposito *et al.*, 2000), grape (Kikkert *et al.*, 2000), broccoli (Mora and Earle, 2001), cotton (Emani *et al.*, 2003), citrus (Distefano *et al.*, 2008) and rice (Zhang *et al.*, 2009). The antifungal effect against *Venturia inaequalis* was further enhanced by the synergistic action of a co-expressed exochitinase (N-acetylhexosaminidase) from *T. harzianum* in transgenic apple (Bolar *et al.*, 2001). No effect of the fungal endochitinase was observed on the life cycle of the agriculturally beneficial micro-organisms including symbiotic mycorrhiza *Gigantus margaritus* (Lorito *et al.*, 1997). Recently, enhanced resistance to *Phoma tracheiphila* and *Botrytis cinerea* in transgenic lemon plants expressing a *Trichoderma harzianum* chitinase gene (*Chit 42*) was obatined (Gentile *et al.*, 2007). Similarily, transgenic carrot plants were generated by expressing *CHIT36* endochitinase gene of *T. harzianum* in carrot plants, showing tolerance to *Alternaria dauci*, *Alternaria radicina* and *B. cinerea* (Baranski *et al.*, 2007). Later, *chit1* gene encoding endochitinase *CHIT42* from the entomopathogenic fungus *Metarhizium anisopliae* was engineered in tobacco plants to achieve resistance against *R. solani* (Kern *et al.*, 2010).

Although, chitinase genes of bacterial origin have not been used extensively to engineer plants; this may be because they are exochitanases and presumed to be less active against fungi (Roberts and Selitrennikoff, 1988). The first attempt to use a bacterial origin chitinase gene to enhance plant defense were made by using *chiA* from *Serratia marcescens*. Jones (1988) and Suslow *et al.* (1988) transferred the chitinase gene of *S. marcescens* into tobacco; the transgenic plants exhibited elevated chitinase activity and increased resistance to *Aspergillus longipes*. Later, Howie *et al.* (1994) found that the same gene provided tolerance to *R. solani* in the field. The protective role of bacterial chitinases was also confirmed by Toyoda *et al.* (1991) who injected chitinase from *Streptomyces griseus* into barley epidermis cells, and found that the enzyme digested the haustoria of *Erysiphe graminis*, the powdery mildew pathogen. Moreover, a bacterial family 19 chitinase *ChiC* from *Streptomyces griseus* showed clear inhibition on fungal hyphal extension. Ninety percent of transgenic rice plants expressing *ChiC* had higher resistance against *Magnaporthe grisea* than non-transgenic plants (Itoh *et al.*, 2003).

1.9. Other Antimicrobial Factors

Antimicrobial proteins found in many organisms other than plants have been exploited in transgenic strategies. Transgenic wheat has been prepared with a gene encoding the protein KP4 from a virus, which infects the smut fungus, *Ustilago maydis*. These plants exhibited 10–30% protection against the smut (*Tilletia caries*) in field and greenhouse tests (Clausen *et al.*, 2000; Schlaich *et al.*, 2007). Antimicrobial peptides do not need to be natural. For example, transgenic cotton, prepared using a synthetic peptide, D4E1 (derived from an insect antimicrobial peptide), exhibited enhanced resistance against the fungus *Thielaviopsis basicola* (Rajasekaran *et al.*, 2007). Interestingly, the same peptide provided resistance to bacterial pathogens in transgenic poplar (Montesinos, 2007).

A synthetic cationic peptide chimera (ce-cropin–melittin) with a broad-spectrum antifungal activity has been produced (Osusky *et al.*, 2000). When expressed in transgenic potato and tobacco, these synthetic peptides have provided enhanced resistance against a number of fungal pathogens, including *Colletotrichum, Fusarium* and *Phytophthora*. These peptides may demonstrate lytic activity against fungal hyphae, inhibit cell wall formation, and enhance membrane leakage. The ability to create synthetic recombinant and combinatorial variants of peptides that can be rapidly screened in the laboratory could provide additional opportunities to engineer resistance to a range of pathogens simultaneously (Dhekney *et al.*, 2007). *MsrA2* peptide was shown to control Fusarium head blight (FHB) of wheat and barley grains caused by *Fusarium graminearum*. Trichothecenes genes–the virulence factors produced by the fungus were introducing into wheat in order to increase the FHB defense mechanism in wheat spikes and reduced the initial infection (Osusky, 2004).

Another anti-viral approach that may have considerable applicability to ornamentals is the expression of dominant interfering peptides ('aptamers') that interact with essential viral proteins. The first example of this approach is the expression of a 29 amino acid peptide selected for interference with multiple tospovirus N-proteins in *N. benthamiana* (Rudolph *et al.*, 2003). The transgenic plants were highly resistant to *Tomato spotted wilt virus* (TSWV), *Groundnut ringspot virus* (GRSV), and

Chrysanthemum stem necrosis virus, while a lower level of resistance was observed with *Tomato chlorotic spot virus*, and delayed disease development with *Impatiens necrotsis spot virus* (INSV) (Rudolph *et al.*, 2003).

1.10. Cell Wall Reinforcement

One of the first barriers among plant pathogens, in particular necrotrophs, encounter during penetration and subsequent colonization is the plant cell wall. Fungal pathogens secrete a number of enzymes to degrade the major plant cell-wall polymers. The major enzymes used for this purpose are cutinases, endopolygalacturonases and pectate–lyases. In response, higher plants apply a number of strategies to inhibit penetration of the pathogen through the cell wall. One strategy is cell wall reinforcement. This is a complex process that involves the rapid synthesis of phenolic compounds leading to the accumulation of lignins in the cell wall and the synthesis of hydroxyproline–rich and other glycoproteins to strengthen the extracellular matrix. Another protective measure is initiated by the degradation of cell walls. Cell wall components, such as sugar oligomers, which are released by fungal pectinolytic enzymes, can serve as elicitors to activate plant defense reactions. It is possible, therefore, that the engineered expression of a pectinolytic enzyme may result in the transgenic plant having an activated defense status. Wegener (2001) found that when the pectate–lyase gene from the bacterium *Erwinia carotovora* was expressed in potato tubers, the transgenic plants were resistant to infection by this bacterium.

Another strategy is based on iron deprivation in plant tissues. *E. chrysanthemi*, which produces a number of pectinolytic enzymes that cause soft rot, synthesizes a siderophore called chrysobactin, which provides the bacteria with iron under the iron–limited conditions that prevail *in planta* and is required for the bacteria to provoke systemic symptoms in the plant. Engineering plants with genes encoding transferrins in order to deprive the bacteria of nutritional iron during infection has consequently been suggested as a method for promoting disease resistance (Expert, 2005).

Lignification of plant cells around sites of infection or lesions has been reported to be a defense response of plants that can potentially slow down pathogens spread (Nicholson and Hammerschmidt, 1992). The enzyme peroxidase is required for the final polymerization of phenolic derivatives into lignin and may also be involved in suberization or wound healing. A decrease in polyphenolic compounds, such as lignin, in potato tubers by redirection of tryptophan in transgenic plants through expression of tryptophan decarboxylase rendered tissues more susceptible to *Phytophthora infestans* (Yao *et al.*, 1995), illustrating the role of phenolic compounds in defense. Lignin levels were significantly higher following expression of the H_2O_2–generating enzyme glucose oxidase in transgenic potato and by expression of the hormone indoleacetic acid (IAA) in transgenic tobacco (Sitbon *et al.*, 1999). In the former case, tolerance to several fungal pathogens was enhanced. A reduction in large callose deposits surrounding haustoria of *Peronospora parasitica* infecting *Arabidopsis thaliana* was indirectly achieved in transgenic plants not accumulating salicylic acid (SA) by expression of the enzyme salicylate hydroxylase (Donofrio and Delaney, 2001). These plants also had reduced expression of the *PR1* gene and exhibited significantly enhanced susceptibility to the pathogen, suggesting that callose deposition during normal defense responses of the plant was influenced by the reduced levels of SA.

1.11. Inhibition of Virulence Factors

Expression by the plant of any mechanism inhibiting virulence/pathogenecity factors of plant pathogens should, in theory, lead to resistance or reduced susceptibility. However, even though several of these factors (*e.g.*, toxins, pectic enzymes, exopolysaccharides and hormones etc.) are well known for many microbe species and hence this knowledge has been exploited to develop strategies for engineering disease resistance.

1.11.1. *Detoxification*

Necrotrophic pathogens, in contrast to biotrophs, use pathogenicity factors such as toxins and hydrolytic enzymes to effect successful infection. Indeed, without effective production of the toxin, the pathogen is often incapable of causing infection. Some toxins have the unfortunate side effect of being toxic to mammals and not just the target plant tissue, in which case they fall into the category of mycotoxins. Toxins often accumulate in biologically active concentrations in tissues remote from the site of infection. In some cases, the toxins are therefore a significant factor in crop spoilage disproportionate to actual loss of yield, especially where they are distasteful or poisonous to the consumer of the crop. One strategy for enhancing disease resistance in plants is to target the toxin. Transgenic maize, where the levels of the *Fusarium*–toxin zearalerone were reduced to 10% of the wild-type levels, proves that the approach is feasible (Igawa *et al.*, 2007). A concern for this approach is that it may lead to the accumulation of breakdown products of which little is currently known. This is not likely to be a major problem where removal of the toxin from the system arrests pathogen development.

In case of *P. syringae* pv. *tabaci* bacterium, *ttr* (tabtoxin resistance) gene encodes an enzyme that inactivates the tabtoxin, a glutamine synthetase inhibitor. The cloned gene has been expressed in tobacco and the resulting transgenic plants developed complete resistance to the bacteria (Anzai *et al.*, 1989). A second mechanism is based on the synthesis of a target that is insensitive to the toxin *P. syringae* pv. *phaseolicola* synthesizes the only ornithine carbamoyltransferase (involved in arginine biosynthesis) that is known to be resistant to its own phaseolotoxin. The corresponding gene (*argK*) has been cloned and used to transform bean plants, which appeared to be totally resistant to infection by *P. syringae* pv. *phaseolicola* (De la Fuente–Martinez *et al.*, 1992)

Oxalic acid has an important role as a toxic pathogenicity factor in several species of necrotroph, of which *Sclerotinia sclerotiorum* is a particular problem in many dicotyledonous species, for example, oil seed rape (*Brassica napus*) and sunflower (*Helianthus annuus*). Several studies have therefore taken the approach of constitutively expressing a heterologous oxalate oxidase gene in a target crop in order to neutralise the oxalic acid produced by the pathogen. The products of the enzyme include the reactive oxygen species hydrogen peroxide, which itself has an important role in disease resistance (Shetty *et al.*, 2008). Examples where partial resistance has been obtained in the laboratory include sunflower and soybean (*Glycine max*) against *Sclerotinia sclerotiorum* (Hu *et al.*, 2003) as well as poplar (*Populus euramericana*) against *Septoria musiva* (Liang *et al.*, 2004). Transgenic maize with *Bt* toxin genes (Cry1Ab protein) from *Bacillus thuringiensis* are not just insect–resistant but also consistently less attacked by *Fusarium* spp. and contain consistently reduced levels of toxin (Hammond *et al.*, 2004).

1.11.2. *Polygalacturonase Inhibiting Proteins (PGIP)*

Polygalacturonase inhibiting proteins (PGIP) are glycoproteins present in the cell wall of many plants and that can inhibit the activity of fungal endopolygalacturonases (Desiderio *et al.*, 1997). It is believed that PGIPs only slightly inhibit fungal polygalacturonases and this leads to the production of longer, oligomeric degradation products, which are large enough to act as elicitors of plant defense responses (Cervone *et al.*, 1989). A gene encoding PGIP has been isolated from bean (Toubart *et al.*, 1992) and it became possible to test if PGIP expression in transgenic plants does confer resistance. However, the transgenic tomato expressing the bean PGIP was not resistant to *F. oxysporum*, *Botrytis cinerea* or *Alternaria solani* and the purified recombinant PGIP did not inhibit fungal polygalacturonases (Desiderio *et al.*, 1997). The authors concluded that the expression of more than one PGIP might be required to confer resistance to fungi. In contrast, transgenic expression of a pear PGIP in tomato resulted in inhibition of polygalacturonase activity from *B. cinerea* and delayed symptom developments (Powell *et al.*, 2000). Recently, Ferrari *et al.* (2008) have characterized two differentially regulated *PGIP* genes in *Arabidopsis* and demonstrated that their overexpression in *Arabidopsis* significantly reduced the disease symptoms caused by *B. cinerea*.

1.11.3. *Plantibodies*

The expression of monoclonal antibodies by transgenic plants (plantibody) is an alternative approach that has already been successfully applied to protect plants from viral infection (Schillberg *et al.*, 2001). The antibodies can be expressed inter- or extra-cellularly and can bind to and inactivate enzymes, toxins, or other pathogen factors involved in disease development. Single-chain antibodies are particularly suitable, as they do not have to be assembled as complete antibodies do, and can be targeted to different compartments of the cell. The first successful use of plantibodies described reduced susceptibility to *Artichoke mottle crinkle virus* by means of a scFv directed against the CP of the virus (Tavladoraki *et al.*, 1993). Subsequently, only little progress was made for many years. It became clear that the main problem for a broad application of this technique was the instability of many scFvs in plant cells. As most viruses replicate in the cytoplasm, it is necessary to direct scFvs into this cell compartment, which presents problems for the correct folding of scFvs. Fecker *et al.* (1997) demonstrated that scFvs directed against the coat protein and the non-structural protein p25 of *Beet necrotic yellow vein virus* could be expressed in *Nicotiana benthamiana* plants. However, expression was only possible if the scFvs were targeted to the ER, which seems in contrast to their place of function in the cytosol. TMV infection of transgenic tobacco plants expressing scFvs against the intact CP of TMV can be reduced, even with low expression levels of the scFv (Zimmermann *et al.*, 1998), and lead to immunity upon stabilized expression (Bajrovic *et al.*, 2001). Xiao *et al.* (2000) generated an scFv from an existing broad–range monoclonal antibody (MAb) obtained against *Johnson grass mosaic virus*, which reacts with several potyviruses. Transformed tobacco plants had lower susceptibility to two potyviruses, PVY and *Clover yellow vein virus*.

A different approach is based on scFvs directed against nonstructural proteins that play an important role in virus replication, such as the viral replicase. Using scFvs generated against the RdRp of *Tomato bushy stunt virus* (TBSV), Boonrod *et al.* (2004) obtained *N. benthamiana* plants with high levels of resistance not only to TBSV, but also to other members of the family Tombusviridae: *Red clover necrotic mosaic virus*, *Cucumber necrotic virus* and *Turnip crinkle virus*. Gargouri–Bouzid *et al.* (2006) demonstrated that high levels of expression of an scFv directed against the NIa of PVY led to complete protection against PVY. Apart from a structural protein associated with the viral RNA, the nucleocapsid proteins of tospoviruses are known to regulate viral transcription/replication and cell-to-cell movement, and thus might be particularly sensitive to inactivation by an scFv. Indeed, it was demonstrated that specific scFvs can interrupt virus replication at an early stage of infection (Prins *et al.*, 2005). For the future it is expected that increasing knowledge of the structure of antibodies will provide the opportunity to improve their stability (Ewert *et al.*, 2004). In addition, new protein scaffolds will be developed that can be used as protein-binding alternatives to antibodies, which may circumvent the shortcomings of scFvs (Binz and Plückthun, 2005).

Currently, there are no published reports on the expression of antifungal antibodies in transgenic plants that have led to a reduction in disease. However, it has been demonstrated that anti-lipase antibodies inhibited infection of tomato by *Botrytis cinerea*, when mixed with spore inoculum, by preventing fungal penetration through the cuticle (Comménil *et al.*, 1998). Similarly, infection by *Colletotrichum gloeosporioides* on various fruits was inhibited using polyclonal antibodies that bound to fungal pectate lyase (Wattad *et al.*, 1997). For the successful protection of plants against other pathogens, high-affinity antibodies specific for one of the bacterial or fungal pathogenecity or virulence factors should be used. The production of antibodies directed against the pectinolytic enzymes of *Erwinia chrysanthemi* and *E. carotovora* has already been suggested (Düring, 1993).

1.11.4. *Type III Secretion System and Harpin Proteins*

Plant pathogenic bacteria must overcome plant defenses and gain access to nutrients trapped inside the plant cell in order to colonize the host, multiply and cause disease. Many phytobacteria require a type III secretion system (TTSS) to mount a successful attack on a host (Fig. 2). The TTSS functions like a syringe and is used by phytobacteria to inject virulence proteins like harpins into the host cell (Tampakaki *et al.*, 2010). Harpins are glycine–rich, protease–sensitive, heat–stable, acidic proteins and are capable of eliciting HR and/or stimulating defense gene expression. To date, a few harpins or harpin–like proteins, such as $hrpZ_{Pss}$ from *Pseudomonas syringae* pv. *syringae* (He *et al.*, 1993), $hrpN_{Ea}$ from *Erwinia amylovora* (Kim and Beer, 1998), $hpa1_{Xoo}$ from *Xanthomonas oryzae* pv. *oryzae* (Peng *et al.*, 2003) and *hrpZ* from *Pseudomonas syringae* pv. *phaseolicola* (Pavli *et al.*, 2011) have been characterized.

The feasibility of using these harpin protein genes from various infection–induced HR, as a new strategy of engineering fungal disease resistance was explored by Li and Fan (1999). By utilizing harpin protein gene from *Erwinia amylovora* and potato *prp1–1* promoter as main DNA elements, they for the first time able to develop transgenic potato showing enhanced level of resistance towards *Phytophthora infestans*. Belbahri *et al.* (2001) used the *popA* gene from the phytopathogenic bacterium *Ralstonia solanacearum* showing sequence similarity with $Hpa1_{Xoo}$ to develop transgenic tobacco plants resistant to *Phytophthora parasitica* var. *nicotianae* under the control of the hypersensitivity–related plant promoter *hsr203J*. The developed plants accumulating the PopA protein showed localized HR and increased resistance to *P. parasitica*. Recently, Miao *et al.* (2010) showed that the transformation of *hpa1*$_{Xoo}$ into cotton conferred the improved resistance to *Verticillium dahliae* in the transgenic cotton. Similarly, a high level of resistance to all predominant races of *Magnaporthe grisea* in China was obtained in the rice line transformed with *hpa1*$_{Xoo}$ from *Xanthomonas oryzae* pv. *oryzae* (Shao *et al.*, 2008). Several other *hrp* genes have also been successfully transformed into different plant species including pear (Malnoy *et al.*, 2005) and tobacco (Sohn *et al.*, 2007). These results highlighted that engineering harpins of phytopathogenic bacteria in plants was a promising approach to produce disease resistant genotype. Unfortunately, the defense responses elicited by harpins and their active sites in hosts have not been fully understood and need to be dissected in future to harness their real potential in plant engineering.

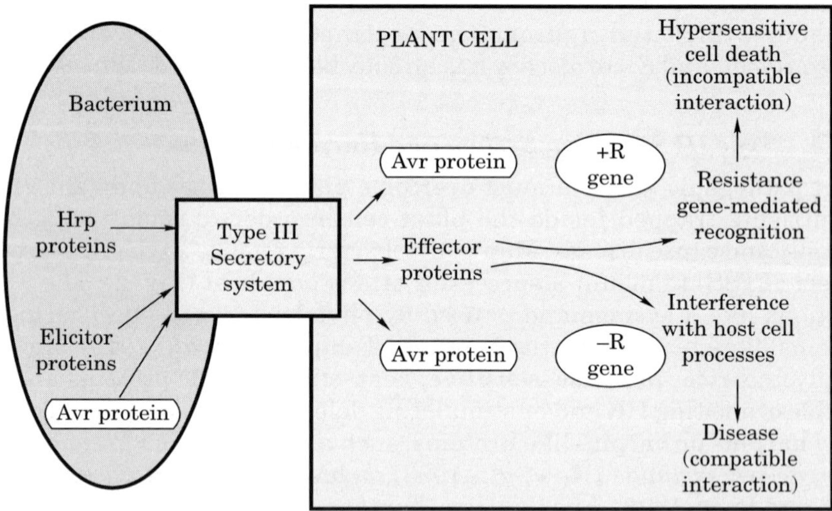

Fig. 2: Engineering disease resistance via over-expression of harpin proteins or manipulation of TTSS. Bacterial effector proteins (*e.g.*, *popA*, the Avr protein of *Ralstonia solanacearum*) translocate into plant cell via a Hrp type III secretion system by way of Hrp proteins. The *Avr genes* of the bacterium encode for the bacterial effector proteins and render the pathogen non-virulent to plants carrying the corresponding *R* gene. Once inside the cell, the effector proteins interfere with the normal cellular processes, such as signal transduction or metabolic pathways. Providing the host plant holds the complementary *R* gene (+*R*), it will launch a defence response resulting in HR cell death, ultimate death of the pathogen and survival of the host.

1.11.5. *Quroum Quenching: Interference with Quorum–sensing*

A new approach to protect plants against bacterial diseases is based on interference with the communication system (quorum–quenching) (Fig. 3), used by several phytopathogenic bacteria to regulate expression of virulence genes according to population density (Cui and Harling, 2005). The enzyme, *AiiA*, isolated from bacterial strain, *Bacillus* sp.240B1, was found to degrade the quorum–sensing signalling molecule of the soft rot pathogen, *Erwinia carotovora*, and thereby rendering the bacteria incapable of infecting the host (Dong *et al.*, 2000). Transgenic expression of *AiiA in planta* was subsequently demonstrated to provide significant enhancement of resistance against soft rot in potato (Dong *et al.*, 2001). The strategy looks technically very promising since the microbial target is likely to be strongly conserved. However, since similar quorum–sensing is also known for bacterial pathogens of humans, this strategy also raises the concern that there is a risk that control of bacterial infection in humans will be impaired.

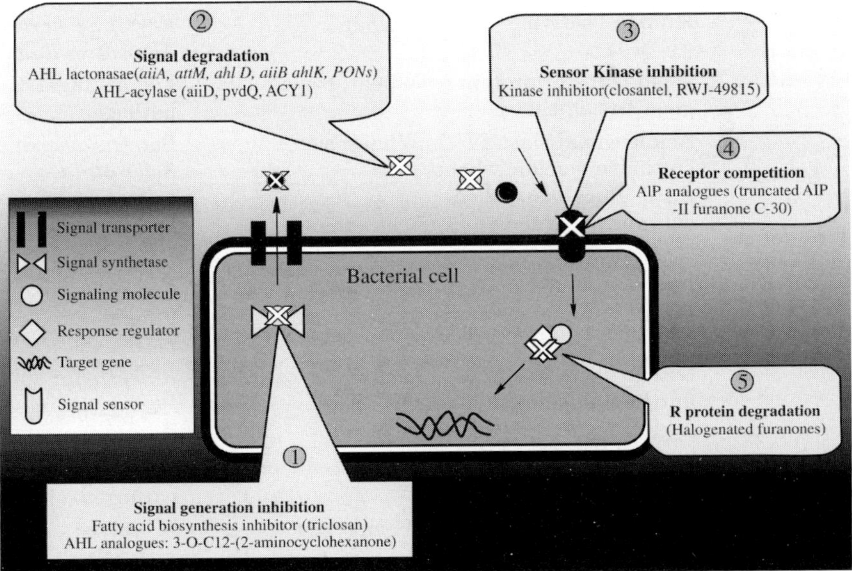

Fig. 3: Overview of different quorum quenching strategies (interruption with QS signal generation, signal dissemination and signal detection) to manage phytopathogenic bacterium population (Source: Kashyap *et al.*, 2010)

Strategy 2: Plant defense regulation: Alteration of induced plant defence, pathogen recognition and downstream regulatory network.

The molecular steps leading to the activation of plant defense responses are under intense investigation, and recent advances provide a variety of possibilities for improving crop resistance by altering natural induced plant defence, pathogen recognition and downstream regulatory network including transcription factors. Several examples, where these strategies are used to engineer plants against plant pathoens have been described in Table 2.

Table 2: Engineering crops by altering induced plant defence, pathogen recognition and downstream regulatory network during plant pathogen interactions

Transgenic crop	***Gene/gene product***	***Donor***	***Target pathogen***
Cotton	*NPR1*	*Arabidopsis*	*Verticillium dahliae*, *Fusarium oxysporum* f. sp. *vasinfectum*, *Rhizoctonia solani*, and *Alternaria alternata*
Carrot	Human lysozyme	Human	*Erysiphe heraclei*, *Alternaria dauci*
	Lipid transfer protein gene and chitinase	Wheat, barley	Foliar fungal pathogen
	Acidic wheat class IV chitinase + acidic wheat β 1,3–glucanase + rice cationic peroxidase (*POC1*)	Wheat, rice	*Botrytis cinerea* and *Sclerotinia sclerotiorum*
Rice	*PR3*	*T. viride*	*R. solani*
	PRm5	Maize	Fungal pathogen
Tobacco	*GAFP* (gastrodia antifungal gene)	*Gastrodia* (orchid)	*Rhizoctonia* spp., *Phytophthora* spp.
	PR3	Bean	*R. solani*
	WRKY 1	Grape	Multiple fungal pathogen
	PR1	*W. japonica*	*B. cinerea*
	Salicylic acid synthase	Bacterial origin	*Oidium lycopersicon*
	β–cryptogein elicitor	*Phytophthora cryptogea*	*Phytophthora parasitica*
Arabidopsis	*SAR 8.2* gene (*CASAR82A*)	Pepper	*Fusarium* and *Botrytis*
Wheat	*PR5*	Barley	*Erysiphe graminis*
	NPR 1	*Arabidopsis*	*Fusarium graminearum*
	Stilbene synthase	Grape	*Puccinia recondita*

Table 2: (*Contd...*)

Table 2: (*Contd...*)

Transgenic crop	***Gene/gene product***	***Donor***	***Target pathogen***
Wheat (spring)	Thaumatin like protein, chitinase, glucanase	Wheat (Sumai–3 cultivar)	*Fusarium graminearum*
Tomato	*NPR1*	*Arabidopsis*	Fungal and bacterial disease
	PR2	Tobacco	*F. oxysporum*
Peanut	*PR2*	Alfalfa	*Sclerotinia minor*
	PR3	Rice	*Sclerotinia minor*
Grape	Chitinase	Rice	*Uncinula necator*, *Elsinoe ampelina*

2. ELICITORS

The signals recognized by plant cells that trigger defense mechanisms are called elicitors. They are either produced by the pathogen itself or derived from plant components such as the cell wall after the hydrolyzing action of the pathogen (Montesano *et al.*, 2003). The production of elicitors *in planta* as a consequence of recombinant DNA technology is an attractive approach to increase disease resistance and this has already been attempted in potato (Wegener *et al.*, 1996). Transgenic tubers producing a pectinolytic enzyme of *E. carotovora* were significantly less susceptible to the bacteria than untransformed tubers. In this case, the enzyme was produced intracellularly and elicitors (cell-wall sugar oligomers) were released only upon infection, as the result of enzyme liberation by tissue wounding during the infection process. A gene encoding the elicitor cryptogein from *Phytophthora cryptogea* was cloned and expressed in transgenic tobacco under control of a pathogen–inducible promoter (Keller *et al.*, 1999). Challenge inoculation with a range of fungi induced the HR as well as several defense genes, and growth of the pathogens was concomitantly restricted. Resistance to the pathogens was not complete, possibly because of the time needed for production of the transgenic elicitor following initial infection (Keller *et al.*, 1999).

Another elicitor, *INF1*, was shown to act as an Avr factor in the tobacco–*Phytophthora infestans* interaction and triggered the onset of the HR (Kamoun *et al.*, 1998). Expression of the gene encoding the AVR9 peptide elicitor from *Cladosporium fulvum* in transgenic tomatoes containing the *Cf–9* gene resulted in a necrotic defense response (Honée *et al.*, 1995). Recently, Qiu *et al.* (2009) isolated an elicitor–encoding gene (*pemG1*) from *Magnaporthe grisea* and introduced in rice (*Oryza sativa* L. cv. Nipponbare) under the constitutive control of maize ubiquitin promoter. The *pemG1*–

expressing plants showed enhanced resistance against rice suggesting that the elicitor–encoding gene could be a novel approach to enhance disease resistance in various crops. These reports not only demonstrates that single gene can produce broad spectrum resistance, but also implies that when transgenes are introduced into heterologous system the effectiveness of the gene needs to be assessed very cautiously as introduced gene may produce resistance towards many more untargeted, non-related species.

2.1. Pathogen–Associated Molecular Patterns (PAMPs)

Plants have broader perception systems that include receptors that recognize pathogen–associated molecular patterns (PAMPs) (Nürnberger *et al.*, 2004). A good example is the flagellin receptor *FLS2* that recognizes a conserved 22-amino acid portion of bacterial flagellin. This mode of pathogen recognition resembles pathways of innate immunity response (Navarro *et al.*, 2004). Treatment of plants with flagellin induces the expression of numerous defense related genes and triggers resistance to pathogenic bacteria in wild-type plants. Two strategies using receptor like kinases (RLKs) show promise. First, flagellin treatment up-regulates the expression of *FLS2* and numerous other RLKs (Zipfel *et al.*, 2004). This up-regulation suggests that transgenic approaches that up-regulate RLK expression might lead to increased resistance. Second, RLK genes can be introduced into species or ecotypes that are deficient in them to confer the ability to recognize a pathogen and impart resistance. For example, *Arabidopsis* plants transformed with a functional *FLS2* gene, under the control of its native promoter, showed higher responsiveness to flagellin and became less susceptible to *Pseudomonas syringae* (Zipfel *et al.*, 2004). Recently, flagellin gene from *Acidovorax avenae* strain N1141, was introduced into rice and the resultant transgenic rice plants accumulated flagellin at various levels. The transgenic plants showed increased expression of defense genes, H_2O_2 production and cell death, suggesting that the flagellin triggers innate immune responses in the transgenic rice. When inoculated with *M. grisea*, the transgenic plants exhibited enhanced resistance, suggesting that the flagellin approach might provide a new strategy for developing genetically engineered disease-resistant rice varieties (Takakura *et al.*, 2008). This demonstrates that pathogen perception and increased resistance could be engineered by the introduction of a RLK transgene.

2.2. Resistance (R) Gene

Most important component of plant surveillance system is resistance (R) genes (McDowell and Woffenden, 2003). They directly or indirectly recognize the pathogen and as a result trigger diverse array of defense mechanisms. The degree to which a plant recognizes a pathogen determines its level of resistance, although the pathogen can also influence the outcome by avoiding or actively suppressing the host defenses. There are five major classes of *R* genes, the NB–LRR (nucleotide binding leucine rich repeat) genes, Ser/Thr kinases such as *Pto*, receptor–like kinases (RLKs) and receptor–like proteins (RLPs) (Tor *et al.*, 2004). These *R genes* recognize pathogen avirulence (*Avr*) determinants and bring about resistance based on *gene–for–gene* hypothesis. Briefly, when corresponding *R* and Avr genes are present, the result is disease resistance. A classic strategy for producing plants with enhanced disease resistance involves the manipulation of *R* genes. The idea is to introduce an *R* gene and thereby confer on the plant the ability to recognize the pathogen and mount an effective defense. There have been some notable reports of success. Resistance (*R*) genes have been widely employed for many years, through conventional breeding programmes, with great success. These genes control many plant diseases caused by biotrophic pathogens such as rusts and powdery mildew fungi. These genes have the advantage of conferring complete resistance against specific races of the pathogen. Many specific resistance genes are available in the major crops (Hovmøller, 2007). Typically, 40–70 specific *R*-genes and alleles have been described for the rust and powdery mildew diseases of major crops. Most *R*-genes encode nucleotide binding–site, leucine–rich repeat (NBLRR) proteins, which activate down-stream defence to combat the disease, when the pathogen has a specific avirulence-gene (*Avr*) that corresponds to the specific *R*–gene (McHale *et al.*, 2006). However, resistance obtained by introgression of these types of gene generally has the drawback that pathogen populations eventually adapt to their presence and overcome them (McDonald and Linde, 2003). In other words, when the *Avr*-gene in the pathogen is inactivated by mutation, the resistance is no longer functional. As the *Avr*-genes often encode effector proteins which have evolved to function in pathogenicity, there is strong balancing selection in natural plant and pathogen populations for polymorphism at the genetic loci in host and pathogen. This means that many *Avr* alleles are present in natural pathogen populations. However, genotypes carrying a virulent allele of any *AvR*–gene locus will eventually migrate to and invade the resistant plant population, leading to reduced efficiency of the specific resistance gene. These types of resistance genes operate at the recognition stage of an interaction and generally against biotrophic pathogens, where the expression of resistance is often associated with hypersensitive response (HR). In some cases, *R*-genes can provide effective protection against pathogens when transformed into new species and even into new genera, and this protection can be broad spectrum (Tai *et al.*, 1999). Recently, *Rxo1* gene from maize (*Zea mays*) was successfully transformed into rice (*Oryza sativa*) and shown to confer resistance against *Xanthomonas oryzae* pv. *oryzicola* (Zhao *et al.*, 2005). Thus, the potential of using *R*-genes as transgenes across natural breeding barriers exists. However, inter-species differences may radically influence *R*-gene function

(Ayliffe *et al.*, 2004) and therefore it is preferable to use *R*-genes from closely related species. The transgenic approach circumvents tedious backcrossing and has successfully been accomplished in rice for the *R*-gene, *Xa21*, conferring broad, but nevertheless race-specific resistance to the bacterial leaf blight disease (Wang *et al.*, 2007). *Xa21* has subsequently been transformed into a restorer line for hybrid rice and shown to provide resistance without compromising elite traits (Zhai *et al.*, 2002). Field tests of *Xa21* transgenic rice in Philippines, China and India have shown satisfactory results (Datta, 2004). However, deregulation of transgenic *Xa21* rice for large scale cultivation is still pending. It should be noted that conventional breeding assisted by the use of molecular marker techniques has already provided hybrids containing *Xa21*, pyramided with other resistance genes (Zhang *et al.*, 2006), thereby creating a competitive alternative to the transgenic approach.

An NB–LRR *R*–gene, *Rpi–blb2*, derived from the wild potato relative (*Solanum bulbocastanum*), confers broad-spectrum race-non-specific resistance in potato (*Solanum tuberosum*) against *Phytophthora infestans* (van der Vossen *et al.*, 2003). A representative of the Cf family of *R*-genes, *Vf* was cloned from the wild apple species *Malus floribunda*, and transferred to cultivated apples (*Malus domestica*) where resistance to a presumably mixed population of *Venturia inaequalis* isolates was demonstrated (Belfanti *et al.*, 2004). Compared to conventional breeding, the transgene approach will facilitate introduction of more *R*-genes into a crop at the same time (pyramiding). This will extend the durability of the resistance concerned. The transgenic strategy using *R*-genes can, however, have negative side effects. For example, over-expression of the *Pto* gene from tomato (*Lycopersicon esculentum*) resulted in a lesion mimic phenotype in the mesophyll tissue (Tang *et al.*, 1999). Some necrotrophic pathogens are adapted to deliberately use the *R-gene*-type of recognition in order to activate PCD (Programmed Cell Death) through the use of specific toxins that provoke the *R*-gene signalling (Glazebrook, 2005). *R*-genes effective against these types of necrotrophic pathogens are simply unknown. This statement needs to read in the light of the fact that many apparently necrotrophic pathogens are really hemibiotrophic and exhibit albeit brief biotrophic or endophytic phase during early stages of infection, where *R*-genes can be effective, *e.g.*, *Bipolaris sorokiniana*, *Magnaporthe oryzae*, *Rhynchosporium secalis* and *Phytophthora infestans* etc. (Parlevliet, 2003).

2.3. Gene Pyramiding

The resistance in newly released cultivars can be lost quickly due to the high level of instability in the pathogen population. One way to circumvent this problem is to develop genetically engineered cultivars with (i) a combination of genes encoding/controlling interdependent or synergistic subcomponents of the disease–resistant trait to realize effective resistance against a particular disease or (ii) a combination of genes associated with different diseases to realize a broad–spectrum resistance. Pyramiding *Xa21* gene, a chitinase gene, and a *Bt*–fusion gene into IR72 through conventional crossing of two independent transgenic homozygous rice lines confers multiple resistance against *X. oryzae* pv. *oryzae* and *R. solani* (Datta *et al.*, 2002). Combination of marker–assisted breeding and genetic transformation yielded rice lines resistant to blast and leaf blight diseases by pyramiding *Pi1*, *Piz5* and *Xa21* (Narayanan *et al.*, 2004). Genetic engineering of rice to pyramid a maize ribosome inactivating protein gene and a rice basic chitinase gene confers enhanced resistance of the transgenic rice lines against three fungal pathogens, *R. solani*, *Bipolaris oryzae*, and *M. grisea* (Kim *et al.*, 2003). Co–expression of rice chitinase and thaumatin–like protein in an elite indica rice line resulted in significant higher level of resistance against *R. solani* (Kalpana *et al.*, 2006). Transgenic plants pyramided with *chi11*, *tlp* and *Xa21* showed an enhanced resistance to both sheath blight and bacterial blight diseases (Maruthasalam *et al.*, 2007). Combined expression of chitinase and β–1, 3-glucanase genes in indica rice enhanced resistance against *R. solani* (Sridevi *et al.*, 2008), while transgenic lines expressing four antifungal genes including *RCH10*, *RAC22*, *β–Glu* and *B–RIP* showed not only high resistance to *M. grisea* but also enhanced resistance to rice false smut *(Ustilaginoidea virens)* and rice kernel smut (*Tilletia barclayana*) (Zhu *et al.*, 2007). Recently, Esfahani *et al.* (2010) used chitinase gene (*chit42*) from *Trichoderma atroviride* and the β–1, 3-glucanase gene (*bgn13.1*) from *Trichoderma virens* to prepare a double gene construct containing these two genes. The transformed potato (*Solanum tuberosum* cv. Savalan) showed enhanced resistance to *Rhizoctonia solani* AG–3, responsible for causing stem and root rot diseases. Therefore, an ingeniously planned genetic engineering involving a well-balanced expression of transgenes with different modes of action would ensure broad-spectrum and durable resistance against pathogens.

3. INDUCTION OF DOWNSTREAM SIGNALING NETWORK

Plant resistance to pathogen infection can be achieved by systemic regulation of the defense–related genes that respond to specific systemic signals. One activator of host defense responses is elicitor molecules from an invading pathogen. These can trigger a network of signalling pathways that coordinate the defense responses of the plant, including PR protein, and phytoalexin production (McDowell and Dangl, 2000). Hypersensitive response (HR)–mediated cell death is triggered sequentially through an increase in the intracellular cytosolic Ca^{2+} concentration by an influx of external Ca^{2+} and the secretion of Ca^{2+} from the calcium stores into the cytoplasm (Chung *et al.*, 2004), a burst of reactive oxygen species, changes in the extracellular pH and membrane potentials, and variations in protein phosphorylation patterns (Tsukamoto *et al.*, 2005). Finally, key mediators such as H_2O_2, salicylic acid (SA), jasmonic acid (JA) and ethylene accumulate and thereby triggering defense responses in plants (Pieterse and Van Loon, 2004).

3.1. Ion Fluxes

Ion fluxes are one of the early events in incompatible plant pathogen interactions. Therefore, changes in proton translocation by altered expression of proton pumps can lead to plant defense responses even without pathogen infection. The successful example to exploit such type of approach to engineer plants is based on the expression of a bacterial gene (*bO*) encoding a proton pump, the bacterio–opsin, derived from *Halobacterium halobium*. The constitutive expression of *bO* gene in potato plants produces a lesion mimic phenotype in which necrotic lesions are displayed and local and systemic defence responses are activated in the absence of pathogens. Tobacco plants expressing *bO* had increased levels of salicylic acid and were more resistant to *P. infestans* isolates showing A1 mating type (Mourgues *et al.*, 1998).

3.2. Hydrogen Peroxide

Hydrogen peroxide (H_2O_2) is one of the activator molecules interacting with signal cascade and exerting direct antimicrobial activity, diffusing the signal for activation of cellular defence genes and reinforcing the plant cell wall (Shetty *et al.*, 2008). The direct evidence that ROS (Reactive Oxygen Species) are involved in conferring disease resistance was provided by the constitutive expression of glucose–oxidase gene (*GOX*) from *Aspergillus niger* in transgenic potato. Wu *et al.* (1997) reported that by introducing *GOX* gene in potato, delayed lesion development by *Phytophthora infestans*, reduced disease development by *Alternaria solani* and *Verticillium dahliae* can be achieved. Enhanced resistance of transgenic rice plants to *M. grisea* was also demonstrated and it is correlated with constitutive and pathogen–induced expression of an *Aspergillus niger GOX* gene, responsible for elevating the endogenous levels of H_2O_2, which in turn caused typical cell death and activated the expression of several defense genes (Kachroo *et al.*, 2003). However, it should be done with caution since high levels of H_2O_2 in transgenic plants may cause metabolic disturbances interfering with normal growth and development. Using pathogen–inducible expression of H_2O_2 generating genes may be an effective way to confer broad-spectrum resistance in rice without the penalty of causing any developmental abnormalities or metabolic errors. Thus, the expression of H_2O_2 generating enzymes in transgenic plants represents a novel strategy for engineering broad-spectrum resistance to pathogens.

Expression of elevated levels of H_2O_2 in transgenic cotton, tobacco, and potato reduced disease development due to a number of different fungi, including *Rhizoctonia*, *Verticillium*, *Phytophthora*, and *Alternaria*; high levels can, however, be phytotoxic (Murray *et al.*, 1999). Certain bacteria, such as *Escherichia coli* and *Pseudomonas fluorescence*, have well-characterized pathways for the production of salicylic acid. Verberne *et al.* (2000) developed transgenic tobacco plants that constitutively expressed both *entC* from *E. coli* and *pmsB* from *P. fluorescence* in the chloroplast. The transgenic tobacco plants expressing these genes showed accumulation of salicylic acid that were up to 1000 times higher than in wild-type tobacco. When challenged with the fungus *Oidium lycopersicon*, the transgenic tobacco plants showed increased levels of resistance compared to the wild-type plants. Rajasekaran *et al.* (2000) reported that the expression of a chloroperoxidase gene from *Pseudomonas pyrrocinia* (*cpo–p*) in transgenic tobacco resulted in significant inhibition of *Aspergillus flavus* hyphal growth and reduced leaf anthracnose lesions caused by *Colletotrichum destructivum*. Moreover, expression of canola plants catalase *katE*–transformed *via Agrobacterium*–mediated in the chloroplasts also enhanced defense gene expression leading to SAR and enhanced resistant to *Peronospora parasitica* and *Erysiphe polygoni* (El–Awady *et al.*, 2007). Recently, Niu *et al.* (2009) also demonstrated the successful insertion and expression of the *cpo–p* gene derived from *Pseudomonas pyrrocinia* into *A. hypogaea*. The developed transgenic groundnut plants showed inhibition of *Aspergillus flavus* hyphal growth, which resulted in the reduction of aflatoxin contamination in peanut seed.

3.3. Pathogenesis–related Protein

Pathogenesis–related (PR)–protein encoding genes when over-expressed in crop plants have been shown to enhance resistance to many fungal diseases (Punja, 2006). These proteins play a direct role in defense by attacking and degrading pathogen cell wall components. Currently, 17 PR–protein families that mark active defense are recognized and include antifungal proteins, protease inhibitors, defensins, and other small peptides (van Loon *et al.*, 2006). Osmotin has been shown to have antifungal activity *in vitro* and, when tested in combination with chitinase and β–1, 3-glucanase, showed enhanced lytic activity (Lorito *et al.*, 1996). When expressed in transgenic potato, osmotin was shown to delay expression of disease symptoms caused by *Phytophthora infestans*. Thaumatin–like proteins are also expressed in plants in response to a range of stress conditions and were demonstrated to have antifungal activity *in vitro* against several pathogens, including *Botrytis*, *Fusarium*, *Rhizoctonia*, and *Sclerotinia* (Koiwa *et al.*, 1997). Recently, Altpeter *et al.* (2005) attempted epidermis–specific expression of a PR protein in transgenic wheat. They developed a novel epidermis–specific promoter by fusion of 2.3 kb fragment of wheat GstA1 promoter with an intron–containing part of wheat *WIR1a* gene. When this promoter–intron combination was used to over-express *TaPERO* peroxidase, it showed enhanced resistance against powdery mildew. This suggests that wheat *GstA1* promoter in combination with *WIR1a* intron is a useful approach to confer fungal disease resistance in cereals.

3.4. Over–expression of Signal Regulators

The discovery of systemic acquired resistance pathways in plants opened the possibility of activating plant defense pathways by over-expression of a regulatory protein such as NPR1. Over-expression of the *NPR1* gene in *Arabidopsis* induced the SAR response and potentiated resistance to diseases caused by biotrophs (Friedrich *et al.*, 2001). Subsequently, expression of the Arabidopsis *NPR1* gene (*AtNPR1*), or native homologues of *NPR1*, in crops, has been shown to boost defence against diverse pathogens. Examples include transgenic wheat expressing the *AtNPR1* gene that exhibits enhanced resistance to *Fusarium* head blight (Makandar *et al.*, 2006), and constitutive over-expression of an apple *NPR1* homologue in apple cultivars (Malnoy *et al.*, 2007). Transformed lines had significantly enhanced resistance to the bacterial disease fire blight, as well as to apple scab and rust diseases. Transgenic rice plants over-expressing the rice *NPR1* orthologue (*NH1*) acquire high levels of resistance to *Xanthomonas oryzae* pv. *oryzae* (Chern *et al.*, 2005). Moreover, constitutive expression of *AtNPR1* in transgenic rice was shown to improve resistance to fungal and bacterial pathogens, but increased susceptibility to *Rice yellow mottle virus*, as well as sensitivity to salt and drought stress (Quilis *et al.*, 2008).

3.5. Mitogen Activated Protein Kinase (MAPK)

A mitogen–activated protein kinase (MAPK) cascade plays a pivotal role in plant innate immunity (Nakagami *et al.*, 2005). They utilize evolutionally conserved signaling pathways to transduce extracellular stimuli into intracellular responses. In these protein kinase cascades, MAPK kinase (MAPKK) is activated by upstream MAPKK kinase (MAPKKK) and finally activates MAPK (Asai *et al.*, 2002). Yamamizo *et al.* (2006) showed that the transgenic potato plants that carry a constitutively active form of MAPK kinase driven by a pathogen–inducible promoter of potato showed high resistance to early blight pathogen *Alternaria solani* as well as *P. infestans*.

3.6. Transcription Factors (TFs)

Transcription factors (TFs) naturally act as master regulators of cellular processes and are emerging candidates for modifying complex traits in crop plants. TFs are proteins that act together with other transcriptional regulators, including chromatin remodeling/modifying proteins, to employ or obstruct RNA polymerases to the DNA template (Udvardi *et al.*, 2007). Plant genomes assign approximately 7% of their coding sequence to TFs, which proves the complexity of transcriptional regulation (Udvardi *et al.*, 2007). The TFs interact with *cis*–elements in the promoter regions of several stress–related genes and thus regulate banks of target genes that could boost signaling through large portions of the pathogen induced signaling network and thereby lead to an increase in disease resistance. For instance, *Arabidopsis* plants expressing the tomato (*Lycopersicon esculentum*) ethylene–response factor (ERF) *Pti4* displayed increased resistance to the fungal pathogen *Erysiphe orontii*, and increased tolerance to the bacterial pathogen *Pseudomonas syringae* pv. *tomato* (Wu *et al.*, 2002). In addition, *wrky70* mutants are compromised in both basal defense and full *RPP4*–mediated disease resistance to *Hyaloperonospora parasitica* (Knoth *et al.*, 2007). Disruption of *WRKY33* enhanced susceptibility to the necrotrophic fungal pathogens *B. cinerea* and *Alternaria brassicicola* (Zheng *et al.*, 2006). Virus–induced silencing of three *WRKY* genes in tobacco compromised *N*–gene–mediated resistance to *Tobacco mosaic virus* (Liu *et al.*, 2004). Overexpression of *OsWRKY23* or *OsWRKY45* in *Arabidopsis* conferred enhanced disease resistance against *Pseudomonas syringae* (Qiu and Yu, 2009). Thus, certain pathogen–induced WRKY proteins can positively regulate plant disease resistance. In plants, five major families of transcription factors (*bZIP*, *WRKY*, *MYB*, *EREBF*, and homeodomain proteins) are involved in the defense against attack by phytopathogens such as bacteria (Dong *et al.*, 2003), fungi (Kalde *et al.*, 2003), and viruses (Liu *et al.*, 2004). A good example is *WRKY* transcription factors. It is generally assumed that *WRKY* transcription factors act as major regulatory proteins by binding to the W-box, a common promoter element contained in several SAR gene promoters (Maleck *et al.*, 2000). This class of transcription factors belongs to a large family of proteins mainly present in plants, and is characterized by their highly conserved DNA–binding region termed the *WRKY* domain. Since the first demonstration that *WRKYs* are involved in plant defence (Rushton *et al.*, 1996) much evidence has emerged to show that they play many crucial roles (Navarro *et al.*, 2004) and encouragingly, *WRKY* transcription factors have also been shown to be important in quantitative resistance to pathogens such as *Phytophthora infestans* (Trognitz *et al.*, 2002). One problem with using transcription factors to improve crops is the identification of the best candidate gene(s) for manipulation because many consist of large multigene families (Eulgem *et al.*, 2000). Attempts to assign a function to each gene are hindered by functional redundancy with knockouts of single genes often having no observable phenotypes. Despite this, good candidate *WRKY* genes have been identified, including the *Arabidopsis* genes *WRKY18*, *WRKY29* and *WRKY70*. When these genes were overexpressed the resultant plants showed enhanced resistance to *P. syringae* and, in the case of *WRKY70*, also to *Erwinia carotovora* subsp. *carotovora* (Asai *et al.*, 2002). Several transcription factor families that have roles in plant defence could yield useful master switch genes. Overexpression of genes such as *ERF1*, *Pti4* and *MYB30* shows promise for increasing resistance to pathogens (Hammond–Kosack and Parker, 2003) and exciting new candidates including the Whirly factor Why1 (Desveaux *et al.*, 2004), the CGCG box-binding proteins SR1–6 (Yang and Poovaiah, 2002) and the tobacco DNA-binding protein *DBP1* (Carrasco *et al.*, 2003) can also be tested.

Another source of potential master–switch genes are protein kinases. MAP kinase (MAPK) signalling cascades are integral parts of many defence–signalling pathways, such as the response to flagellin (Asai *et al.*, 2002), *Pep–13* (Lee *et al.*, 2004) and *N*-mediated resistance to *Tobacco mosaic virus* (TMV) (Liu *et al.*, 2004). Among their targets are *WRKY*, *MYB* and *TGA* transcription factors (Liu *et al.*, 2004) and *NPR1* (Ekengren *et al.*, 2003). Overexpression of the tobacco *MAPK* and *SIPK* illustrates their potential as it led to activation of defence responses and hypersensitive response (HR)–like cell death (Zhang and Liu, 2001). Additionally, transient overexpression of *MKK4a*, *MKK5a* or constitutively active *MEKK1* resulted in enhanced resistance to virulent *P. syringae* and *Botrytis cinerea* (Asai *et al.*, 2002). Other protein kinases could also be employed and some of the best candidates could be calcium dependent protein kinases (CDPKs) because they act as calcium sensor proteins that link changes in cytosolic Ca^{2+} to defence responses (Romeis, 2001). In addition to kinases and transcription factors, other signalling molecules such as NPR1, *NDR1*, *EDS1*, *PAD4*, *SGT1*, *COI1* and *JAR1* that might represent important nodes in the signaling networks are candidates for this approach. For example, *NPR1* is an important master switch gene because it constitutes a node that links SAR, ISR, *R*–gene, SA, JA and ethylene–mediated resistance (Pieterse and Van Loon, 2004). It can activate defence gene expression through interaction with members of the TGA family of *bZIP* transcription factors in the nucleus (Zhang *et al.*, 1999). In the case of SAR, induction of SAR leads to more reducing conditions in the cell and as a result, *NPR1* molecules present as an inactive oligomeric complex in the cytoplasm are converted into active monomers that become nuclear localized and trigger gene expression via interaction with TGA factors (Mou *et al.*, 2003). In *Arabidopsis*, overexpression of *NPR1* led to enhanced resistance to diverse pathogens (Chern *et al.*, 2001) and, crucially, this was achieved without a substantial yield penalty. The reason for this appeared to be that the *NPR1*–over-expressing plants did not constitutively turn on their defences but rather appeared to be primed to respond to pathogen attack. Recently, however, similar rice plants have shown a lesion mimic or cell death phenotype (Fitzgerald *et al.*, 2004). This would reduce yields. It seems that care must be taken with the level and location of expression of *NPR1*. As an increasing number of important signaling components are discovered, so the list of candidate genes for manipulation grows. One exciting new discovery is SA binding protein 2 (SABP2) that specifically binds SA and displays lipase activity. SABP2 might be a receptor for SA because lipase activity is stimulated by SA binding and this could generate a lipid derived signal that is important in defence signalling (Kumar and Klessig, 2003). Several other new discoveries such as *DIR1* (Maldonado *et al.*, 2002) and *SFD1* (Nandi *et al.*, 2004) implicate lipid–derived signals in SAR. Negative regulators of defence also represent good candidates for manipulation. Mutations or knockouts of *mlo* and *edr1* might impart resistance even though the loss of activity is felt in all cells of the plant (Peterhänsel *et al.*, 1997). For example the *mlo* mutation of barley has conferred durable resistance to all *B. graminis* isolates for decades (Wolter *et al.*, 1993). The list of candidates for manipulation also includes proteins such as *EDS1*, *PAD4*, *SGT1*, *NDR1*, *ETR1*, *RIN4* and *SNI1* (Hammond–Kosack and Parker, 2003). The disadvantage with this approach is that manipulation of some master switch genes could be detrimental to plant development.

4. PATHOGEN MIMICRY: PATHOGEN DERIVED RESISTANCE

Pathogen mimicry or Pathogen–derived resistance (PDR) is mediated either by the protein encoded by the transgene (protein–mediated) or by the transcript produced from the transgene (RNA–mediated) (Prins *et al.*, 2008). Extensive research with genes from viruses and other sources has documented the efficacy of viral sense or antisense genes (coat protein, replicase, satellite RNAs, defective interfering RNAs) in protecting plants against virus infection following transfer and expression of these genes in plants (Goldbach *et al.*, 2003). Using various coat protein (CP) sense, CP antisense, or replicase sense viral genes, several groups worked to introduce virus resistance into various crops (Table 3).

4.1. Coat Protein Mediated Resistance

Coat proteins (CP) of viruses are involved in many aspects of virus–related biology, including encapsidation, virus replication, dissemination, and cell-to-cell and/or systemic movement (Callaway *et al.*, 2001). The RSV (*Rice stripe virus*) CP gene was introduced into two japonica varieties and the resultant transgenic plants expressed the CP at high levels and exhibited a significant level of resistance to virus infection (Hayakawa *et al.*, 1992). *Rice tungro spherical virus* (RTSV) CP genes *CP1*, *CP2* and *CP3* were introduced individually or together into rice and transgenic plants accumulated transcripts of the chimeric CP genes. Moderate levels of protection to RTSV infection, ranging from 17 to 73% of reduction from virus infection, and a significant delay of virus replication under greenhouse conditions in transgenic plants expressing the RTSV–*CP1*, –*CP2* and

–*CP3* genes singly or together were observed (Sivamani *et al.*, 1999). Transgenic plants expressing genes that encode wild–type CP, deleted CP, mRNA of the CP, or antisense CP sequences showed two types of reactions when challenged with RYMV (*Rice yellow mottle virus*). Most of the transgenic rice plants expressing antisense sequences of the CP and untranslatable CP mRNA of RYMV exhibited a delay in virus accumulation of up to a week. However, transgenic plants expressing wild-type CP gene accumulated high level of virus particles. These suggest that antisense CP and untranslatable CP mRNA induced moderate resistance, whereas transgenic CP enhanced virus infection (Kouassi *et al.*, 2006). The RHBV (*Rice hoja blanca virus*) nucleocapsid protein *N* gene was introduced into rice and the resultant transgenic plants had a significant reduction in disease development and a significant increase in performance for important agronomic traits when challenged with RHBV. The *N* gene and RHBV resistance in the transgenic plants were inherited in a stable manner. These transgenic lines could become new genetic resources in developing RHBV–resistant cultivars (Lentini *et al.*, 2003).

Table 3: Engineered transgenic plants using pathogen derived resistance (PDR)

Strategy	***Transgenic plant product***	***Source/gene***	***Virus***
RNA interference (RNAi)	Common bean	Replication initiator protein (*rep*; *AC1*), transactivator protein (*TrAP*; *AC2*), replication enhancer protein (*REn*; *AC3*) and movement protein (*BC1*)	*Bean golden mosaic virus* (BGMV)
Coat protein-mediated resistance	Tomato	*N* gene	*Tomato spotted wilt virus* (TSWV)
	Tobacco	*N* gene	*TSWV*, *Impatiens necrotic spot virus* (INSV), *Groundnut ringspot virus* (GRSV)
	Tobacco	Coat protein (CP)	*Cucumber mosaic virus* (CMV) *sub group IB*, *Tobacco mosaic virus*, *Cowpea aphid–borne mosaic virus* (CABMV), *Alfalfa mosaic virus*

Table 3: (*Contd...*)

Table 3: (*Contd...*)

Strategy	***Transgenic plant product***	***Source/gene***	***Virus***
	Papaya	CP *virus*	*Papaya ringpost*
	Cucumber	CP	CMV
	Melon	CP	CMV
	Tomato	CP	CMV
	Squash	CP	CMV
	Pepper	CP	CMV
RNA dependent RNA polymerase–mediated resistance	Tobacco	Gene encoding viral RNA dependent RNA–polymerases (RdRps)	*Tobacco mosaic virus*
Replicase–mediated resistance	Tobacco	Modified tobacco mosaic virus replicase transgene	Broad spectrum resistance to Tobamoviruses
	Tobacco	replicase gene	*Pea early browning virus RNA1, Potato virus X, Alfalfa mosaic virus,* CMV
	Tomato	replicase gene	CMV
RNA satellites	Tobacco	*Sat–I17N*	CMV
	Tobacco	Sat–RNA1, Sat–RNA1 + CP (CMV–O)	CMV
	Tobacco	Sat–S	CMV
	Pepper	*Sat–I17N*	CMV
	Tomato	Sat–S	CMV
Antisense RNAs	Tobacco	(CMV–D)	CMV
	Tobacco	(CMV–Q)	CMV
Hammerhead ribozyme	Tobacco	Conserved sequences of RNA1 and 2 of *CMV–Y*	CMV

4.2. Movement Proteins (MPs) Mediated Resistance

Movement proteins (MPs) allow viral infection to spread between the adjacent cells (cell to cell) as well as systemically (long distance). Transgenic plants that contain mutant MPs from TMV show resistance to several TMVs as well as to AlMV (*Alfalfa mosaic virus*), *Cauliflower mosaic virus* (*CaMV*) and other viruses (Cooper *et al.*, 1995). Similar results were found for *Nicotiana occidentalis* plants expressing a movement protein (*P50*) and partially functional deletion mutants (*DeltaA* and *DeltaC*) of the *Apple chlorotic leaf spot virus* (ACLSV) showed resistance to *Grapevine berry inner necrosis virus* (GINV) (Yoshikawa *et al.*, 2006). The use of mutated MPs could, therefore, lead to transgenic plants that efficiently inhibit the local and systemic spread of many different viruses. Compared with CP– or replicase mediated resistance strategies, the expression of dysfunctional or mutant movement proteins (MP) has been reported to confer broader resistance.

4.3. Defective Interfering RNA and DNA

Defective interfering RNAs (DI–RNAs) or DNAs (DI–DNAs) are deletion mutations of the viral genome that are able to replicate in a parasitic fashion, utilizing the replicase complex of an active infection. DI–RNAs and DI–DNAs are not able to replicate on their own. In most cases DI–RNA reduces the replication level of the parental virus, resulting in reduced symptom expression, although at least one DI–RNA intensifies symptom expression (Li *et al.*, 1989). DI–DNAs have similar competitive effects on DNA viruses (Stanley *et al.*, 1990). Kollar *et al.* (1993) showed that a *Cymbidium ringspot virus* (CymRSV) DI–RNA protected transgenic *N. benthamiana* againt CymRSV infection, while Stanley *et al.* (1990) demonstrated protection against geminivirus infection from a DI–DNA. Stanley *et al.* (1997) have also shown the presence of a naturally occurring *DI–DNA* in *Ageratum yellow vein virus* infections of *Ageratum conyzoides*. Rubio *et al.* (1999) developed a DI–RNA from *Tomato bushy stunt virus* (TBSV) that conferred broad-spectrum protection against related tombusviruses in *N. benthamiana*; the DI–RNA was expressed at low levels in healthy transgenic plants, but was amplified to very high levels following TBSV infection, and resulting in plant recovery.

4.4. Ribozymes

Ribozymes are RNA molecules that autocatalytically cleave sequences complementary to their binding site. Ribozymes have potential to confer resistance against viruses if expression levels and activity are sufficient. Some viroids and viral satellite RNAs self-process from multimeric replicative forms by ribozyme activity, and some success has been achieved in protecting transgenic plants against specific pathogens. One line of transgenic melon expressing a ribozyme against *Watermelon mosaic virus 2* (WMV2) was found to have immunity to WMV2 (Huttner *et al.*, 2001).

4.5. Ribonucleases

It may be possible to introduce broad spectrum resistance against RNA viruses and viroids, based on the expression of ribonucleases specific for double–stranded RNA (dsRNA). dsRNA is a feature of the replication of

RNA viruses and viroids, but is not normally found in healthy plant cells. Two approaches that have been tested are the expression of the yeast dsRNA–specific RNase *Pac1*, and the mammalian interferon–induced 2′, 5′–oligoadenylate synthetase (2–5 A)/RNase L system.

4.6. Nucleic Acid Mediated Resistance

Nucleic acid mediated resistance has been achieved through the expression of virus sequences, the acquisition of resistance being dependent on the transcribed RNA. This RNA mediated virus resistance can be considered to be an example of post-transcriptional gene silencing (PTGS) in plants (Prins *et al.*, 2008). Napoli *et al.* (1990) firstly reported PTGS in Petunia hybrid transgenically expressing the chalcone synthase gene. They observed a co-ordinated and reciprocal inactivation of the host gene and the transgene encoding the same RNA. This process has been called RNA silencing or RNA interference (RNAi) and occurs in a variety of eukaryotic organisms (Mann *et al.*, 2008). The silencing process involves the cleavage of a dsRNA precursor into short (21–26 nucleotides) (nt) RNAs by an enzyme, Dicer, that has RNase III domains. These RNAs are known as short interfering RNAs (siRNA) and microRNAs (miRNAs). Both siRNA and miRNA are able to guide an RNA–induced silencing complex (RISC) to destroy single–strand cognate RNA (Naqvi *et al.*, 2009). In addition, longer siRNAs (24–26 nt) have been shown to result in methylation of homologous DNA causing chromatin remodeling and transcriptional gene silencing (TGS). Strategies using exogenously supplied dsRNA have already been used to combat virus infestation in plants. *E. coli* was used to produce large amounts of dsRNA coding for partial sequences of two different viruses, *Pepper mild mottle virus* (PMMoV) and *Plum pox virus* (PPV) (Tenllado *et al.*, 2003). Simultaneous injection of dsRNA together with purified virus particles resulted in the inhibition of both viruses. Interestingly, resistance to infection was also observed when the crude bacterial preparations were sprayed onto the *N. benthamiana* leaves. These data suggest a simple, economic and effective application of RNA silencing technology.

Indeed, many virus resistant plants (including melon, squash, tomato and tobacco) have been produced using these methods (Moffat, 2001). The most notable success has been in the development and commercialization of transgenic coat protein–protected papaya in Hawaii. The 1994 *Papaya ringspot virus* (PRSV) crisis in Hawaii led to the production and subsequent commercialization of transgenic papaya that express the PRSV coat protein and thereby eliminate expression of this essential protein upon infection (Fermin *et al.*, 2005). Based on sequence profile of silencing suppressor protein, HcPro, it was PRSV–HcPro that acts as a suppressor of RNA silencing through micro RNA binding in a dose–dependent manner. *In planta* expression of PRSV–HcPro affects developmental biology of plants, suggesting the interference of suppressor protein in micro RNA–directed regulatory pathways of plants. Besides facilitating the establishment of PRSV, it showed strong positive synergism with other heterologous viruses as well (Mangrauthia *et al.*, 2010). Therefore, resistance in transgenic papaya can be overcome by PRSV with distant homology to the transgene, or by PRSV strains with HcPro that can sufficiently suppress the silencing mechanism of transgenic papaya. It would therefore be important to develop transgenic papaya that could avoid the impact of these PRSV strains (Tripathi *et al.*, 2008).

RNA-mediated gene silencing is also tried as a reverse tool for gene targeting in plant diseases caused by fungal and bacterial pathogens. Homology–based gene silencing induced by transgenes (co-suppression), antisense RNA, or dsRNA has been demonstrated in many plant pathogenic fungi, including *Cladosporium fulvum* (Hamada and Spanu, 1998), *Magnaporthe oryzae* (Kadotani *et al.*, 2003), *Venturia inaequalis* (Fitzgerald *et al.*, 2004), *Neurospora crassa* (Goldoni *et al.*, 2004), *Aspergillus nidulans* (Hammond and Keller, 2005), and *Fusarium graminearum* (Nakayashiki *et al.*, 2005). Fitzgerald and colleagues (2004) using hairpin–vector technology, have been able to trigger simultaneous high frequency silencing of a green fluorescent protein (GFP) transgene and an endogenous trihydroxynaphthalene reductase gene (THN) in *V. inaequalis*. The GFP transgene acted as an easily detectable visible marker, while the trihydroxynaphthalene reductase gene (THN) played a role in melanin biosynthesis. Nakayashiki *et al.* (2005) developed a protocol for silencing the *mpg1* and polyketide synthase like genes. The *mpg1* gene is a hydrophobin gene that is essential for pathogenicity, as it acts as a cellular relay for adhesion and trigger for the development of appressorium (Talbot *et al.*, 1996). Nakayashiki *et al.* (2005) were successful in silencing the above–mentioned genes to varying degrees by pSilent–1–based vectors in 70–90% of the transformants. Ten to fifteen percent of the silenced transformants exhibited almost “null phenotype”. This vector was also efficiently able to silence a GFP reporter in another ascomycete fungus, *Colletotrichum lagenarium*.

4.7. Barnase–Barstar System

Several systems have been tested to mimic HR in transgenic plants as a mechanism of localized cell death and resistance to pathogen. Although this approach has the potential to provide a broad spectrum resistance to bacteria, fungi and viruses, it requires the use of specific promoters to restrict the effect of the transgene at the site of infection in order to avoid a deleterious generalized HR response and plant death. An interesting strategy has been applied by Strittmatter *et al.* (1995) who expressed in plant a bacterial ribonuclease gene (*barnase*) driven by a pathogen inducible promoter from potato (*prp1–1*). They also transferred to the transgenic potato a gene coding for an inhibitor of *Barnase* activity (*barstar*) to reduce detrimental effects of background activity of ribonuclease (RNase) on the non-infected tissues. When the progenies were challenged with *P. infestans* a strong *Barnase* activity was induced only at the infection site, which significantly reduced. This strategy, however, was not tested at the field level. The strategy of the two-component system was further adopted by Shengji *et al.* (2003) to obtain transgenic rice resistant to *Magnaporthe grisea*. In this study, two chimeric promoters, induced by rice blast fungus pathogen (*Magnaporthe grisea*), are fused with *Barnase* respectively to construct two plant expression vectors, pWBNBS and pPBNBS together with the *Barstar* driven by CaMV 35S promoter. The expression of *Barnase* is induced in rice leaves when inoculated with the spores of *Magnaporthe grisea* and the transgenic plant shows high resistance to the rice fungal blast disease. These results suggest that transgenic plants harboring this two component system may be exploited to acquire relatively broad spectrum and elevated resistance against fungal pathogen in various other agriculturally important crops.

CONCLUSIONS

The enormous potential of biotechnology as an applied science that uses biological knowledge to meet practical needs and the great advantages that will come to humankind from its applications, makes it likely that plant biotechnology will continue to flourish. It could be considered as the 'silver bullet' that will solve the problems of 21st century. To realize the full potential of this technology, concerted efforts in research and development are further needed to enhance the efficiency of various techniques used in different crops against multiple stresses. Recent advances in the field of molecular biology and transgenic technologies have enabled the plant tailors to know the pitfalls regarding the expression of transgenes, emerging trends involving fusion proteins, deployment of antimicrobial peptides, stacked genes and various genes regulating metabolic pathways require special attention. Further, large-scale field trials are needed to test whether expression of the introduced genes will affect yields, quality, or agronomic traits. Although the introduced genes are well-defined, the field trials also provide the opportunity to ascertain whether any unexpected or undesirable consequences have resulted from the transformation procedure. Genetically engineered crops are just beginning to make their way into the hands of breeders. Their potential for changing the characteristics of plants has already been demonstrated to a great extent. It remains for the market place to put a value on those traits and, ultimately, on the technology that makes these changes possible.

REFERENCES

Allefs, S.J.H.M., De Jong, E.R., Florack, D.E.A., Hoogendoorn, C. and Stiekema, W.J. (1996). Erwinia soft rot resistance of potato cultivars expressing antimicrobial peptide tachyplesin I. *Molecular Breeding,* **2:** 97–105.

Almasia, N.I., Bazzini, A.A., Hopp, H.E. and Vazquez–rovere, C. (2008). Overexpression of snakin–1 gene enhances resistance to *Rhizoctonia solani* and *Erwinia carotovora* in transgenic potato plants. *Molecular Plant Pathology,* **9:** 329–38.

Altpeter, F., Varshney, A., Abderhalden, O., Douchkov, D., Sautter, C., Kumlehn, J., Dudler, R. and Schweizer, P. (2005). Stable expression of a defense–related gene in wheat epidermis under transcriptional control of a novel promoter confers pathogen resistance. *Plant Molecular Biology,* **57:** 271–83.

Anzai, H., Yoneyama, K. and Yamaguchi, I. (1989). Transgenic tobacco resistant to a bacterial disease by the detoxification of a pathogenic toxin. *Molecular and General Genetics,* **219:** 492–94.

Arce, P., Moreno, M., Gutierrez, M., Gebauer, M., Dell'Orto, P., Torres, H., Acuna, I., Oliger, P., Venegas, A., Jordana, X., Kalazich, J. and Holuigue, L. (1999). Enhanced resistance to bacterial infection by *Erwinia carotovora* subsp. *atroseptica* in transgenic potato plants expressing the attacin or the cecropin *SB–37* genes. *American Journal of Potato Research,* **76:** 169–77.

Asai, T., Tena, G., Plotnikova, J., Willmann, M.R., Chiu, W.L., Gomez–Gomez, L., Boller, T., Ausubel, F.M. and Sheen, J. (2002). MAP kinase signaling cascade in *Arabidopsis* innate immunity. *Nature,* **415:** 977–83.

Ayliffe, M.A., Steinau, M., Park, R.F., Rooke, L., Pacheco, M.G. and Hulbert, S.H. (2004). Aberrant mRNA processing of the maize *Rp1–D* rust resistance gene in wheat and barley. *Molecular Plant–Microbe Interactions,* **17:** 853–64.

Bajrovic, K., Erdag, B., Atalay, E.O. and Cirakoclu, B. (2001). Full resistance to tobacco mosaic virus infection conferred by the transgenic expression of a recombinant antibody in tobacco. *Biotechnology and Biotechnology Equipment,* **15:** 21–27.

Baranski, R., Klocke, E. and Nothnagel, T. (2007). Enhancing resistance of transgenic carrot to fungal pathogens by the expression of *Pseudomonas fluorescence* microbial factor 3 (MF3) gene. *Physiological and Molecular Plant Pathology,* **71:** 88–95.

Beddington, J. (2010). Food security: Contributions from science to a new and greener revolution. *Philosophical Transactions of the Royal Society,* **365:** 61–71.

Belbahri, L., Boucher, C., Candresse, T., Nicole, Michel., Ricci, P. and Keller, H. (2001). A local accumulation of the *Ralstonia solanacearum* PopA protein in transgenic tobacco renders a compatible plant-pathogen interaction incompatible. *The Plant Journal,* **28:** 419–30.

Belfanti, E., Silfverberg–Dilworth, E., Tartarini, S., Patocchi, A., Barbieri, M. and Zhu, J. (2004). The *HcrVf2* gene from a wild apple confers scab resistance to a transgenic cultivated variety. *Proceedings of the National Academy of Sciences,* **101:** 886–90.

Binz, H.K. and Plückthun, A. (2005). Engineered proteins as specific binding reagents. *Current Opinion in Biotechnology,* **16:** 459–69.

Bolar, J.P., Norelli, J.L., Harman, G.E., Brown, S.K. and Aldwinkle, H.S. (2001). Synergistic activity of endochitinase and exochitinase from *Trichoderma atroviride* (*T. harzianum*) against the pathogenic fungus (*Venturia inaequalis*) in transgenic apple plants. *Transgenic Research,* **10:** 533–43.

Boman, H.G. (1991). Antibacterial peptides: Key components needed in immunity. *Cell,* **65:** 205–207.

Boonrod, K.J., Galetzka, D., Nagy, P.D., Conrad, U. and Krczal, G. (2004). Single-chain antibodies against a plant viral RNA–dependent RNA polymerase confers virus resistance. *Nature Biotech,* **22:** 856–62.

Broekaert, W.F., Terras, F.R.G., Cammue, B.P.A. and Osborn, R.W. (1995). Plant defensins: novel antimicrobial peptides as components of the host defense system. *Plant Physiology,* **108:** 1353–58.

Callaway, A. Giesman–Cookmeyer, D., Gillock, E.T., Sit, T.L. and Lommel, S.A. (2001). The multifunctional capsid proteins of plant RNA viruses. *Annual Review of Phytopathology,* **39:** 419–60.

Carlsson, A., Nyström, T., de Cock, H. and Bennich, H. (1998). Attacin–an insect immune protein–binds LPS and triggers the specific inhibition of bacterial outer-membrane protein synthesis. *Microbiology,* **144:** 2179–89.

Carmona, M.J., Molina, A., Fernandez, J.A., López–Fando, J.J. and García–Olmedo, F. (1993). Expression of the alpha–thionin gene from barley in tobacco confers enhanced resistance to bacterial pathogens. *Plant Journal,* **3:** 457–62.

Carrasco, J.L., Ancillo, G., Mayda, E. and Vera, P. (2003). A novel transcription factor involved in plant defense endowed with protein phosphatase activity. *EMBO Journal* **22:** 3376–84.

Cervone, F., Hahn, M.G., De Lorenzo, G. and Darvill, A. (1989). A plant protein converts a fungal pathogenesis factor into an elicitor of plant defence responses. *Plant Physiol*ogy, **90:** 542–48.

Chern, M., Fitzgerald, H.A., Canlas, P.E., Navarre, D.A. and Ronald, P.C. (2005). Overexpression of a rice NPR1 homolog leads to constitutive activation of defense response and hypersensitivity to light. *Molecular Plant–Microbe Interaction* **18:** 511–20.

Chern, M.S., Fitzgerald, H.A., Yadav, R.C., Canlas, P.E., Dong, X. and Ronald P.C. (2001). Evidence for a disease–resistance pathway in rice similar to the NPR1–mediated signalling pathway in *Arabidopsis*. *Plant J.*, **27:** 101–13.

Chung, E., Park, J.M., Oh, S.K., Joung, Y.H., Lee, S. and Choi., D. (2004). Molecular and biochemical characterization of the *Capsicum annuum* calcium–dependent protein kinase 3 (*CaCDPK3*) gene induced by abiotic and biotic stresses. *Planta,* **220:** 286–95.

Clausen, M., Krauter, R., Schachermayr, G., Potrykus, I. and Sautter, C. (2000). Antifungal activity of a virally encoded gene in transgenic wheat. *Nature Biotechnology,* **18:** 446–49.

Coca, M., Bortolotti, C., Rufat, M., Peñas, G., Eritja, R., Tharreau, D., del Pozo, A.M., Messeguer, J. and San Segundo, B. (2004). Transgenic rice plants expressing the antifungal AFP protein from *Aspergillus giganteus* show enhanced resistance to the rice blast fungus *Magnaporthe grisea*. *Plant Molecular Biology,* **54:** 245–59.

Coca, M., Penas, G., Gomez, J., Campo, S., Bortolotti, C., Messeguer, J. and Segundo, B.S. (2006). Enhanced resistance to the rice blast fungus *Magnaporthe grisea* conferred by expression of a cecropin A gene in transgenic rice. *Planta,* **223:** 392–406.

Collinge, D.B., Lund, O.S. and Thordal–Christensen, H. (2008). What are the prospects for genetically engineered, disease resistant plants? *European Journal of Plant Pathology,* **121:** 217–31.

Comménil, P., Belingheri, L. and Dehorter, B. (1998). Antilipase antibodies prevent infection of tomato leaves by *Botrytis cinerea*. *Physiological Molecular Plant Pathology,* **52:** 1–14.

Cooper, B., Lapidot, M., Heick, J.A., Dodds, J.A. and Beachy, R.N. (1995). A defective movement protein of TMV in transgenic plants confers resistance to multiple viruses wheras the functional analog increases susceptibility. *Virology,* **206:** 307–13.

Cui, X. and Harling, R. (2005). N–acyl–homoserine lactone–mediated quorum sensing blockage, a novel strategy for attenuating pathogenicity of Gram–negative bacterial plant pathogens. *European Journal of Plant Pathology,* **111:** 327–39.

Datta, K., Baisakh, N., Thet, K.M., Tu, J. and Datta, S.K. (2002). Pyramiding transgenes for multiple resistance in rice against bacterial blight, yellow stem borer and sheath blight. *Theoretical Applied Genetics,* **106:** 1–8.

Datta, S.K. (2004). Rice biotechnology: A need for developing countries. *AgBioForum,* **7:** 31–35.

De la Fuente–Martínez, J.M., Mosqueda–Cano, G., Alvarez–Morales, A. and Herrera–Estrella, L. (1992). Expression of a bacterial phaseolotoxin–resistant ornithyl transcarbamylase in transgenic tobacco confers resistance to *Pseudomonas syringae* pv. *phaseolicola*. *BioTechnology*, **10:** 905–09.

Desiderio, A., Aracri, B., Leckie, F., Mattei, B., Salvi, G., Tigelaar, H., van Roekel, J., Baulcombe, D., Melchers, L., De Lorenzo, G. and Cervone, F. (1997). Polygalacturonase–inhibiting proteins (PGIPs) with different specificities are expressed in *Phaseolus vulgaris*. *Molecular Plant Microbe Interaction,* **10:** 852–60.

Desveaux, D., Subramanian, R., Després, C., Mess, J., Lévesque, C., Fobert, P. and Dangl, J. (2004). A "Whirly" transcription factor is required for salicylic acid–dependent disease resistance in Arabidopsis. *Developmental Cell*, **6:** 229–40.

Dhekney, S.A., Z.T. Li., Van Aman, M., Dutt, M., Tattersall, J. and Gray, D.J. (2007). Genetic transformation of embryogenic cultures and recovery of transgenic plants in *Vitis vinifera*, *Vitis rotundifolia* and *Vitis* hybrids. *ACTA Horticulture,* **738:** 743–48.

Distefano, G., Malfa, S.L., Vitale, A., Lorito, M., Deng, Z. and Gentile, A. (2008). Defence–related gene expression in transgenic lemon plants producing an antimicrobial *Trichoderma harzianum* endochitinase during fungal infection. *Transgenic Research*, **17:** 873–79.

Dong, J., Chen, C.H. and Chen, Z.X. (2003). Expression profiles of the *Arabidopsis WRKY* gene superfamily during plant defense response. *Plant Molecular Biology,* **51:** 21–37.

Dong, Y.H., Xu, J.L., Li, X.Z. and Zhang, L.H. (2000). *AiiA*, an enzyme that inactivates the acylhomoserine lactone quorum–sensing signal and attenuates the virulence of *Erwinia carotovora*. *Proceedings of the National Academy of Sciences*, **97:** 3526–31.

Dong, Y.H., Wang, L.H., Xu, J.L., Zhang, H.B., Zhang, X.F. and Zhang, L.H. (2001) Quenching quorum–sensing–dependent bacterial infection by an N-acyl homoserine lactonase. *Nature,* **411:** 813–17.

Donofrio, N.M. and Delaney, T.P. (2001). Abnormal callose response phenotype and hypersusceptibility to *Peronospora parasitica* in defense–compromised *Arabidopsis nim 1–1* and salicylate hydroxylase–expressing plants. *Molecular Plant–Microbe Interaction,* **14:** 439–50.

Düring, K., Porsch, P., Fladung, M. and Lörz, H. (1993). Transgenic potato plants resistant to the phytopathogenic bacterium *Erwinia carotovora*. *Plant Journal,* **3:** 587–98.

Ekengren, S.K., Liu, Y., Schiff, M., Dinesh–Kumar, S.P. and Martin, G.B. (2003). Two MAPK cascades, NPR1, and TGA transcription factors play a role in pto–mediated disease resistance in tomato. *Plant J.,* **36:** 905–17.

El–Awady, M., Reda, E.A., Moghaieb, E.A., Haggag, W., Youssef, S.S. and El–Sharkawy, A.M. (2007). Transgenic canola plants over-expressing bacterial catalase exhibit enhanced resistance to *Peronospora parasitica* and *Erysiphe polygoni*. *Arabian Journal of Biotechnology*. **11:** 71–84.

Emani, C., Garcia, J.M., Lopata–Finch, E., Pozo, M.J., Uribe, P., Kim, D.J., Sunilkumar, G., Cook, D.R., Kenerley, C. M. and Rathore, K.S. (2003). Enhanced fungal resistance in transgenic cotton expressing an endochitinase gene from *Trichoderma virens*. *Plant Biotechnology Journal*. **1:** 321–36.

Esfahani, K., Motallebi, M., Zamani, M.R., Sohi, H.H. and Jourabchi, E. (2010). Transformation of potato (*Solanum tuberosum* cv. Savalan) by chitinase and β–1,3–glucanase genes of mycoparasitic fungi towards improving resistance to *Rhizoctonia solani* AG–3. *Iranian Journal of Biotechnology,* **8:** 73–81.

Esposito, S., Colucci, M.G., Frusciante, L., Fillippone, E., Lorito, M. and Bressan, R.A. 2000. Antifungal transgenes expression in *Petunia hybrida*. *Acta Horticulture,* **508:** 157–61.

Eulgem, T., Rushton, P.J., Robatzek, S. and Somssich, *I.E.* (2000). The WRKY superfamily of plant transcription factors. *Trends Plant Science,* **5:** 199–206.

Ewert, S., Honegger, A. and Plückthun, A. (2004). Stability improvement of antibodies for extracellular and intracellular applications: CDR grafting to stable frameworks and structure–based framework engineering. *Methods,* **34:** 184–99.

Expert, D. (2005). Genetic regulation of iron in *Erwinia chrysanthemi* as pertains to bacterial virulence. *In*: Barton L.L. and Abadia J. (*Eds*) Iron nutrition in plants and rhizospheric microorganisms. Springer, pp. 215–27.

Fecker, L.F., Koening, R. and Obermeier, C. (1997). *Nicotiana benthamiana* plants expressing beet necrotic yellow vein coat protein–specific scFv are partially protected against the establishment of the virus in the early stages of infection and its pathogenic effects in the late stages of infection. *Archives of Virology,* **142:** 1857–63.

Fermin, G., Tennant, P., Gonsalves, C., Lee, D. and Gonsalves, D. (2005). Comparative development and impact of transgenic papayas in Hawaii, Jamaica and Venezuela. *Methods Molecular Biology,* **286:** 399–430.

Ferrari, S., Galletti, R., Pontiggia, D., Manfredini, C., Lionetti, V., Bellincampi, D., Cervone, F. and Lorenzo, G.D. (2008). Transgenic expression of a fungal *endo*–polygalacturonase increases plant resistance to pathogens and reduces auxin sensitivity. *Plant Physiology,* **146:** 669–81.

Fitzgerald, A., Kha, J.A.V. and Plummer, K.M. (2004). Simultaneous silencing of multiple genes in the apple scab fungus *Venturia inaequalis*, by expression of RNA with chimeric inverted repeats. *Fungal Genet. Biol,* **41:** 963–71.

Fitzgerald, H.A., Chern, M.S., Navarre, R. and Ronald, P.C. (2004). Overexpression of (At) NPR1 in rice leads to a BTH– and environment–induced lesion–mimic/cell death phenotype. *Molecular Plant Microbe Interaction,* **17:** 140–51

Florack, D.E. and Stiekema, W.J. (1994). Thionins: properties, possible biological roles and mechanisms of action. *Plant Molecular Biology*, **26:** 25–37.

Florack, D.E.A., Dirkse, W.G., Visser, B., Heidekamp, F. and Stiekema, W.J. (1994). Expression of biologically active hordothionins in tobacco. Effects of pre- and pro-sequences at the amino and carboxyl termini of the hordothionin precursor on mature protein expression and sorting. *Plant Molecular Biology,* **24:** 83–96.

Friedrich, L., Lawton, K., Dietrich, R.A., Willits, M., Cade, R. and Ryals, J. (2001). *NIM1* overexpression in *Arabidopsis* potentiates plant disease resistance and results in enhanced effectiveness of fungicides. *Molecular Plant-Microbe Interaction,* **14:** 1114–24.

Gargouri–Bouzid, R., Jaoua, L., Rouis, S., Saidi, M.N., Bouaziz, D. and Ellouz, R. (2006). PVY–resistant transgenic potato plants expressing an anti–NIa protein scFv antibody. *Molecular Biotechnology,* **33:** 133–40.

Gentile, A., Deng, Z. and La Malfa, S. (2007). Enhanced resistance to *Phoma tracheiphila* and *Botrytis cinerea* in transgenic lemon plants expressing a *Trichoderma harzianum* chitinase gene. *Plant Breeding,* **126:** 146–51.

Giannakis, C., Bucheli, C., Skene, K., Robinson S. and Scott, N.S. (2005). Chitinase and β–1, 3-glucanase in grapevine leaves. *Australasian Journal Grape Wine Research,* **4:** 14–22.

Glazebrook, J. (2005). Contrasting mechanisms of defense against biotrophic and necrotrophic pathogens. *Annual Review of Phytopathology.* **43:** 205–27.

Goldbach, R., Bucher, E. and Prins, M. (2003). Resistance mechanisms to plant viruses: an overview. *Virus Research,* **92:** 207–12.

Goldoni, M., Azzalin, G., Macino, G. and Cogoni, C. (2004). Efficient gene silencing by expression of double stranded RNA in *Neurospora crassa. Fungal Genet. Biol.,* **41:** 1016–24.

Hain, R., Reif, H.J., Krause, E., Langebartels, R., Kindl, H., Vornam, B., Wiese, W., Schmelzer, E., Schreier, P.H. and Stocker, R.H. (1993). Disease resistance results from foreign phytoalexin expression in a novel plant. *Nature,* **361:** 153–56.

Hamada, W. and Spanu, P.D. (1998). Co-suppression of the hydrophobin gene *Hcf–1* is correlated with antisense RNA biosynthesis in *Cladosporium fulvum. Molecular and General Genetier*, **259:** 630–38.

Hammerschmidt, R. (1999). Phytoalexins: What have we learned after 60 years? *Annual Review of Phytopathology*, **37:** 285–306.

Hammond, B.G., Campbell, K.W., Pilcher, C.D., Degooyer, T.A., Robinson, A.E. and McMillen, B.L. (2004). Lower fumonisin mycotoxin levels in the grain of Bt corn grown in the United States in 2000–2002. *Journal of Agricultural and Food Chemistry,* **52:** 1390–97.

Hammond, T.M. and Keller, N.P. (2005). RNA silencing in *Aspergillus nidulans* is independent of RNA–dependent RNA polymerase. *Genetics,* **169:** 607–17.

Hammond–Kosack, K.E. and Parker, J.E. (2003). Deciphering plant pathogen communication: fresh perspectives for molecular resistance breeding. *Current Opinion in Biotechnology*, **14:** 177–93.

Hayakawa, T., Zhu, Y., Itoh, K., Kimura, Y., Izawa, T., Shimamoto, K. and Toriyama, S. (1992). Geneticly engineered rice resistance to rice strip virus, an insect–transmitted virus. *Proceedings of the National Academy of Sciences*, **89:** 9865–69.

He, S.Y., Huang, H.C. and Collmer, A. (1993). *Pseudomonas syringae* pv. *syringae* harpin$_{pss}$: a protein that is secreted via the *Hrp* pathway and elicits the hypersensitive response in plants. *Cell,* **73:** 1255–66.

Hipskind, J.D. and Paiva, N.L. (2000). Constitutive accumulation of a resveratrol–glucoside in transgenic alfalfa increases resistance to *Phoma medicaginis*. *Molecular Plant-Microbe Interaction,* **13:** 551–62.

Honée, G., Melchers, L.S., Vleeshouwers, V.G.A.A., Van Roekel, J.S.C. and De Wit, P.J.G.M. (1995). Production of the *AVR9* elicitor from the fungal pathogen *Cladosporium fulvum* in transgenic tobacco and tomato plants. *Plant Molecular Biology,* **29:** 909–20.

Hovmøller, M.S. (2007). Source of seedling and adult plant resistance to *Puccinia strüfomis* f. sp. *tritici* in European wheats. *Plant Breeding,* **126:** 225–33.

Howie, W., Joe, L., Newbigin, E., Suslow, T. and Dunsmuir, P. (1994). Transgenic tobacco plants which express the *chiA* gene from *Serratia marcescens* have enhanced tolerance to *Rhizoctonia solani*. *Transgenic Research,* **3:** 90–98.

Hu, X., Bidney–Yalpani, D.L., Duvick, N., Crasta, J.P. and Folkerts, O. (2003). Overexpression of a gene encoding hydrogen peroxide–generating oxalate oxidase evokes defense responses in sunflower. *Plant Physiology,* **133:** 170–81.

Huang, Y. and McBeath, J.H. (1997). Plant promoters controlled expression of cecropin B gene in transgenic potatoes renders disease resistance to soft rot bacteria. *Phytopathology,* **87:** S45. (Abs.).

Hultmark, D., Engström, Å., Andersson, K., Steiner, H., Bennich, H. and Boman, H.G. (1983). Insect immunity. Attacins, a family of antibacterial proteins from *Hyalophora cecropia*. *EMBO Journal,* **2:** 571–76.

Huttner, E., Tucker, W., Vermeulen, A., Ignart, F., Sawyer, B. and Birch, R. (2001). Ribozyme genes protecting transgenic melon plants against potyviruses. *Current Issues in Molecular Biology*, **3:** 27–34.

Igawa, T., Takahashi–Ando, N., Ochiai, N., Ohsato, S., Shimizu, T. and Kudo, T. (2007). Reduced contamination by the Fusarium mycotoxin Zearalenone in maize kernels through genetic modification with a detoxification gene. *Applied and Environmental Microbiology,* **73:** 1622–29.

Islam, A. (2006). Fungus resistant transgenic plants: Strategies, progress and lessons learnt. *Plant Tissue Culture and Biotechnology,* **16(2):** 117–38.

Itoh, Y., Takahashi, K., Takizawa, H., Nikaidou, N., Tanaka, H., Nishihashi, H., Watanabe, T. and Nishizawa, Y. (2003). Family 19 chitinase of *Streptomyces griseus* HUT6037 increases plant resistance to the fungal disease. *Biosciences Biotechnology Biochem*istry, **67:** 847–55.

Jekel, P.A., Hartmann, J.B.H. and Beintema, J.J. (1991). The primary structure of hevamine, an enzyme with lysozyme/chitinase activity from *Hevea brasiliensis* latex. *European Journal Biochemistry,* **200:** 123–30.

Jha, S., Tank, H., Prasad, B. and Chattoo, B. (2009). Expression of Dm–AMP1 in rice confers resistance to *Magnaporthe oryzae* and *Rhizoctonia solani*. *Transgenic Research,* **18(1):** 59–69.

Jones, J.D.G. (1988). Expression of bacterial chitinase protein in tobacco leaves using two photosynthetic gene promoters. *Molecular and General Genetics,* **212:** 536–42.

Kachroo, A., He, Z., Patkar, R., Zhu, Q., Zhong, J., Li, D., Ronald, P., Lamb, C. and Chattoo, B.B. (2003). Induction of H_2O_2 in transgenic rice leads to cell death and enhanced resistance to both bacterial and fungal pathogens. *Transgenic Research,* **12:** 577–86.

Kadotani, N., Nakayashiki, H., Tosa, Y., and Mayama, S. (2003). RNA silencing in the pathogenic fungus *Magnaporthe oryzae*. *Molecular Plant–Microbe Interaction,* **16:** 769–76.

Kalde, M., Barth, M., Somssich, I.E. and Lippok, B. (2003). Members of the *Arabidopsis* WRKY group III transcription factors are part of different plant defense signaling pathways. *Molecular Plant–Microbe Interactions*, **16:** 295–305.

Kalpana, K., Maruthasalam, S., Rajesh, T., Poovannan, K., Kumar, K.K., Kokiladevi, E., Raja, J.A.J., Sudhakar, D., Velazhahan, R., Samiyappan, R. and Balasubramanian, P. (2006). Engineering sheath blight resistance in elite indica rice cultivars using genes encoding defense proteins. *Plant Sciences,* **170:** 203–15.

Kamoun, S., Van West, P., Vleeshouwers, V., de Groot, K. and Govers, F. (1998). Resistance of *Nicotiana benthamiana* to *Phytophthora infestans* is mediated by the recognition of the elicitor protein INF1. *Plant Cell*, **10:** 1413–25.

Kanzaki, H., Nirasawa, S., Saitoh, H., Ito, M., Nishihara, M., Terauchi, R. and Nakamura, I. (2002). Overexpression of the wasabi defensin gene confers enhanced resistance to blast fungus (*Magnaporthe grisea*) in transgenic rice. *Theoretical Applied Genetics,* **105:** 809–14.

Kashyap, P.L, Sanghera, G.S. and Kumar, A. (2010). Quorum quenching: A new hope for phytobacterial disease management. *In*: Malik, C.P. (*Ed.*). *Biotechnology Developments and Applications*, *Pointer Publications, Jaipur*, pp. 66–86.

Kato, A., Nakamura, S., Ibrahim, H., Matsumi, T., Tsumiyama, C. and Kato, M. (1998). Production of genetically modified lysozymes having extreme heat stability and antimicrobial activity against Gram negative bacteria in yeast and in plants. *Nahrung*, **42:** 128–30.

Keller, H., Pamboukdjian, N., Ponchet, M., Poupet, A., Delon, R., Verrier, J.L., Roby, D. and Ricci, P. (1999). Pathogen–induced elicitin production in transgenic tobacco generates a hypersensitive response and nonspecific disease resistance. *Plant Cell*, **11:** 223–35.

Kern, M.F., Maraschin, S.D.F., Endt, D.V., Schrank, A., Vainstein, M,A. and Pasquali, G. (2010). Expression of a chitinase gene from *Metarhizium anisopliae* in tobacco plants confers resistance against *Rhizoctonia solani*. *Applied Biochemistry and Biotechnology*, **160:** 1933–46.

Kikkert, J.R., Ali, G.S., Wallace, P.G., Reisch, B. and Reustle, G.M. (2000). Expression of a fungal chitinase in *Vitis vinifera* L. 'Merlot' and 'Chardonnay' plants produced by biolistic transformation. *Acta Horticulture*, **528:** 297–303.

Kim, J.F. and Beer, S.V. (1998). HrpW of *Erwinia amylovora*, a new harpin that contains a domain homologous to pectate lyases of a distinct class. *Journal of Bacteriology*, **180:** 5203–10.

Kim, J.K., Jang, I.C., Wu, R., Zuo, W.N., Boston, R.S., Lee, Y.H., Ahn, I.P. and Nahm, B.H. (2003). Co–expression of a modified maize ribosome–inactivating protein and a rice basic chitinase gene in transgenic rice plants confers enhanced resistance to sheath blight. *Transgenic Research*, **12:** 475–84.

Knoth, C., Ringler, J., Dangl, J.L. and Eulgem, T. (2007). *Arabidopsis* WRKY70 is required for full RPP4–mediated disease resistance and basal defense against *Hyaloperonospora parasitica*. *Molecular Plant-Microbe Interactions*, **20:** 120–28.

Ko, K. (1999). Attacin and T4 lysozyme transgenic 'Galaxy' apple: Regulation of transgene expression and plant resistance to fire blight (*Erwinia amylovora*). Ph.D dissertation, Cornell University, Ithaca, NY. p. 194.

Ko, K., Norelli, J.L., Reynoird, J.P., Boresjza–Wysocka, E., Brown, S.K. and Aldwinckle, H.S. (2000). Effect of untranslated leader sequence of AMV RNA 4 and signal peptide of pathogenesis–related protein 1b on attacin gene expression, and resistance to fire blight in transgenic apple. *Biotechnology Letters*, **22:** 373–81.

Koiwa, H., Kato, H., Nakatsu, T., Oda J. and Sato, F. (1997). Purification and characterization of tobacco pathogenesis–related protein PR–5d, an antifungal thaumatin like protein. *Plant Cell Physiol*ogy, **38:** 783–91.

Kollar, A., Dalmay, T. and Burgyan, J. (1993). Defective interfering RNA–mediated resistance against cymbidium ringspot tombusvirus in transgenic plants. *Virology,* **193:** 313–18.

Kouassi, N.K., Chen, L., Sire, C., Bangratz–Reyser, M., Beachy, R.N., Fauquet, C.M. and Brugidou, C. (2006). Expression of rice yellow mottle virus coat protein enhances virus infection in transgenic plants. *Archives of Virology,* **151:** 2111–22.

Krishnamurthy, K., Balconi, C., Sherwood, J.E. and Giroux, M.J. (2001). Wheat puroindolines enhance fungal disease resistance in transgenic rice. *Molecular Plant–Microbe Interactions,* **14:** 1255–60.

Kumar, D. and Klessig, D.F. (2003). High–affinity salicylic acid–binding protein 2 is required for plant innate immunity and has salicylic acid stimulated lipase activity. *Proceedings of the National Academy of Sciences,* **100:** 16101–106.

Leckband, G. and Lorz, H. (1998). Transformation and expression of a stilbene synthase gene of *Vitis vinifera* L. in barley and wheat for increased fungal resistance. *Theoretical Applied Genetics,* **96:** 1004–12.

Lee, J., Rudd, J.J., Macioszek, V.K. and Scheel, D. (2004). Dynamic changes in the localisation of MAPK cascade components controllong pathogenesis–related (PR) gene expression during innate immunity in parsley. *J. Biol. Chem.,* **279:** 22440–48.

Lee, T.J., Coyne, D.P., Clemente, T.E. and Mitra, A. (2002). Partial resistance to bacterial wilt in transgenic tomato plants expressing antibacterial lactoferrin gene. *Journal of the American Society for Horticultural Science.* **127:** 158–64.

Lentini, Z., Lozano, I., Tabares, E., Fory, L., Domínguez, J., Cuervo, M. and Calvert, L. (2003). Expression and inheritance of hypersensitive resistance to rice hoja blanca virus mediated by the viral nucleocapsid protein gene in transgenic rice. *Theoretical Applied Genetics,* **106:** 1018–26.

Li, R, and Fan, Y. (1999). Reduction of lesion growth rate of late blight plant disease in transgenic potato expressing harpin protein. *Science in China (Ser. C),* **42:** 96–101.

Li, X.H., Heaton, L.A. Morris, T.J. and Simon, A.E. (1989). *Turnip crinkle virus* defective interfering RNAs intensify viral symptoms and are generated *de novo. Proceedings of the National Academy of Sciences,* **86:** 9173–77.

Liang, H., Maynard, C.A., Allen, R.D. and Powell, W.A. (2004). Increased *Septoria musiva* resistance in transgenic hybrid poplar leaves expressing a wheat oxalate oxidase gene. *Plant Molecular Biology,* **45:** 619–29.

Liu, Y., Schiff, M., and Dinesh–Kumar, S.P. (2004). Involvement of MEK1 MAPKK, NTF6 MAPK, WRKY/MYB transcription vactors, COI1 and CTR1 in N-mediated resistance to tobacco mosaic virus. *Plant Journal,* **38:** 800–809.

Logemann, J., Jach, G. and Logemann, S. (1993). Expression of a ribosome inhibiting protein (RIP) or a bacterial chitinase leads to fungal resistance in transgenic plants. *In*: Mechanisms of Plant Defense Responses (Fritig, B. and Legrand, M. *Eds*). Dordrecht: Kluwer Academic Publishers, pp. 446–48.

Lorito, M. and Scala, F. (1999). Microbial genes expressed in transgenic plants to improve disease resistance. *Journal Plant Pathology,* **81:** 73–88.

Lorito, M., Woo, S.L., Fernandez, I.G., Colucci, G., Harman, G.E., Pintor–Toro, J.A., Filippone, E., Muccifora, S., Lawerence, C.B., Zoina, A., Tuzun, S. and Scala, F. (1998). Gene from mycoparasitic fungi as a source for improving plant resistance to fungal pathogens. *Proceedings of the National Academy of Sciences,* **95:** 7860–65.

Lorito, M., Muccifora, S., Woo, S.L., Bonfante, P., Bianciotto, V., Gori, P., Filippone, E. and Scala, F. (1997). Transgenic enzyme localization and mycorrhiza infection of plants expressing a *Trichoderma harzianum* endochitinase which improves plant disease resistance. *Phytopathology,* **87:** S59.

Lorito, M., Woo, S.L., Ambrosio, M.D., Harman, G.E., Hayes, C.K., Kubicek C. and Scala, F. 1996. Synergistic interaction between cell wall degrading enzymes and membrane affecting compounds. *Molcular Plant-Microbe Interactions,* **9:** 206–13.

Makandar, R., Essig, J.S., Schapaugh, M.A., Trick, H.N. and Shah, J. (2006). Genetically engineered resistance to Fusarium head blight in wheat by expression of *Arabidopsis* NPR1. *Molecular Plant–Microbe Interactions,* **19:** 123–29.

Maldonado, A.M., Doerner, P., Dixon, R.A., Lamb, C.J. and Cameron, R.K. (2002). A putative lipid transfer protein involved in systemic resistance signalling in Arabidopsis. *Nature,* **419:** 399–403.

Maleck, K., Levine, A., Eulgem, T., Morgen, A., Schmid, J., Lawton, K., Dangl, J.L. and Dietrich, R.A. (2000). The transcriptome of *Arabidopsis thaliana* during systemic acquired resistance. *Nature Genetics,* **26:** 403–10.

Malnoy, M., Venisse, J.S., Brisset, M.N. and Chevreau, E. (2003). Expression of bovine lactoferrin cDNA confers resistance to *Erwinia amylovora* in transgenic pear. *Molecular Breeding,* **12:** 231–44.

Malnoy, M., Venisse, J.S. and Chevreau, E. (2005). Expression of a bacterial effector, harpin N, causes increased resistance to fire blight in *Pyrus communis*. *Tree Genet Genomes,* **1:** 41–49.

Malnoy, M., Jin, Q., Borejsza–Wysocka, E.E., He, S.Y. and Aldwinckle, H.S. (2007). Overexpression of the apple *MpNPR1* gene confers increased disease resistance in *malus×domestica*. *Molecular Plant–Microbe Interactions,* **20:** 1568–80.

Mangrauthia, S.K, Singh, P. and Praveen, S. (2010). Genomics of helper component proteinase reveals effective strategy for papaya ringspot virus resistance. *Molecular Biotechnology,* **44:** 22–29.

Mann, S.K., Kashyap, P.L., Sanghera, G.S., Singh, G. and Singh, S. (2008). RNA Interference: An eco-friendly tool for plant disease management. *Transgenic Plant Journal,* **2:** 110–26.

Maruthasalam, S., Kalpana, K., Kumar, K.K., Loganathan, M., Poovannan, K., Raja, J.A, Kokiladevi, E., Samiyappan, R., Sudhakar, D. and Balasubramanian, P. (2007). Pyramiding transgenic resistance in elite indica rice cultivars against the sheath blight and bacterial blight. *Plant Cell Reports,* **26:** 791–804.

Mauch, F. and Staehelin, L.A. (1989). Functional implications of the subcellular localization of ethylene–induced chitinase and β–1, 3-glucanase in bean leaves. *Plant Cell,* **1:** 447–57.

McDonald, B.A. and Linde, C. (2003). The population genetics of plant pathogens and breeding strategies for durable resistance. *Euphytica,* **124:** 163–80.

McDowell, J.M. and Dangl, J.L. (2000). Signal transduction in the plant immune response. *Trends Biochemistry Science,* **25:** 79–82.

McDowell, J.M. and Woffenden, B.J. (2003). Plant disease resistance genes: recent insights and potential applications. *Trends in Biotechnology*, **21:** 178–83.

McHale, L., Tan, X.P., Koehl, P. and Michelmore, R.W. (2006). Plant NBS–LRR proteins: adaptable guards. *Genome Biology* 7: http://genomebiology.com/2006–7/4/212/ abstract.

Metraux, J.P., Burkhart, W., Moyer, M., Dincher, S., Middlesteadt, W., Williams, S., Payne, G., Carnes, M. and Ryals, J. (1989). Isolation of a complementary DNA encoding a chitinase with structural homology to a bifunctional lysozyme/ chitinase. *Proceedings of the National Academy of Sciences,* **86:** 896–900.

Miao, W., Wang, X., Li, M., Song, C., Wang, Y., Hu, D. and Wang, J. (2010). Genetic transformation of cotton with a harpin–encoding gene hpa_{Xoo} confers an enhanced defense response against different pathogens through a priming mechanism, *BMC Plant Biology,* **10:** 67.

Mitra, A. and Zhang, Z. (1994). Expression of a human lactoferrin cDNA in tobacco cells produces antibacterial protein(s). *Plant Physiology,* **106:** 977–81

Moffat, A.S. (2001). Finding new ways to fight plant diseases. *Science,* **292:** 2270–73.

Molina, A. and García–Olmedo, F. (1997). Enhanced tolerance to bacterial pathogens caused by the transgenic expression of barley lipid transfer protein LTP2. *Plant Journal,* **12:** 669–75.

Molina, A., Ahl Goy, P., Fraile, A., Sánchez–Monge, R. and García–Olmedo, F. (1993). Inhibition of bacterial and fungal pathogens by thionins of types I and II. *Plant Science,* **92:** 169–77.

Montesano, M., Brader, G. and Palva, E.T. (2003). Pathogen derived elicitors: searching for receptors in plants *Molecular Plant Pathology,* **4:** 73–79.

Montesinos, E. (2007). Antimicrobial peptides and plant disease control. *FEMS Microbiological Letters*, **270:** 1–11.

Moreno, M., Segura, A. and García–Olmedo, F. (1994). Pseudothionin–St1, a potato peptide active against potato pathogens. *European Journal of Biochemistry,* **223:** 135–39.

Mou, Z., Fan, W. and Dong, X. (2003). Inducers of plant systemic acquired resistance regulate NPR1 function through redox changes. *Cell,* **113:** 935–44.

Mourgues, F., Brisset, M. and Chevreau, E. (1998). Strategies to improve plant resistance to bacterial diseases through genetic engineering. *Trends in Biotechnology,* **16:** 203–209.

Murray, F., Llewellyn, D., McFadden, H., Last, D., Dennis E. and Peacock, W.J. (1999). Expression of the *Talaromyces flavus* glucose oxidase gene in cotton and tobacco reduces fungal infection, but is also phytotoxic. *Molecular Breeding,* **5:** 219–32.

Nakagami, H., Pitzschke, A. and Hirt, H. (2005). Emerging MAP kinase pathways in plant stress signalling. *Trends Plant Science,* **10:** 339–46.

Nakajima, H., Muranaka, T., Ishige, F., Akutsu, K. and Oeda, K. (1997). Fungal and bacterial disease resistance in transgenic plants expressing human lysozyme. *Plant Cell Reports,* **16:** 674–79.

Nakayashiki, H., Hanada, S., Nguyen, B.Q., Kadotani, N., Tosa, Y. and Mayama, S. (2005). RNA silencing as a tool for exploring gene function in ascomycete fungi. *Fungal Genetics and Biology,* **42:** 275–83.

Nandi, A., Welti, R. and Shah, J. (2004). The *Arabidopsis thaliana* dihydroxyacetone phosphate reductase gene Suppressor of Fatty Acid Desaturase Deficiency1 is required for the activation of systemic acquired resistance. *Plant Cell,* **16:** 465–77.

Napoli, C., Lemieux, C. and Jorgensen, R. (1990). Introduction of a chimeric chalcone synthase gene into petunia results in reversible co-suppression of homologous genes *in trans*. *Plant Cell,* **2:** 279–89.

Naqvi, A.R., Islam, M.N., Choudhury, N.R. and Haq, Q.M. (2009). The fascinating world of RNA interference. *International Journal of Biological Science,* **5:** 97–117.

Narayanan, N.N., Baisakh, N., Oliva, N.P., VeraCruz, C.M., Gnanamanickam, S.S., Datta, K. and Datta, S.K. (2004). Molecular breeding: marker–assisted selection combined with biolistic transformation for blast and bacterial blight resistance in indica rice (cv. CO39). *Molecular Breeding,* **14:** 61–71.

Navarro, L., Zipfel, C., Rowland, O., Keller, I., Robatzek, S., Boller, T. and Jones, J.D. (2004). The transcriptional innate immune response to flg22 interplay and overlap with Avr gene–dependent defense responses and bacterial pathogenesis. *Plant Physiology,* **135:** 1113–28.

Nguyen, T.C., Lakshman, D.K., Han, J, Galvez, L.C. and Mitra, A. (2011). Transgenic plants expressing antimicrobial lactoferrin protein are resistant to a fungal pathogen. *Journal of Plant Molecular Biology and Biotechnology,* **2(1):** 1–8.

Nicholson, R.L. and Hammerschmidt, R. (1992). Phenolic compounds and their role in disease resistance. *Annual Review of Phytopathology,* **30:** 369–89.

Niu, C., Akasaka–Kennedy, Y., Faustinelli, P., Joshi, M., Rajasekaran, K., Yang, H., Chu, Y., Cary, J., and Ozias–Akins, P. (2009). Antifungal activity in transgenic peanut (*Arachis hypogaea* L.) conferred by a nonheme chloroperoxidase gene. *Peanut Science,* **36:** 26–132.

Norelli, J.L, Mills, J.Z, Momol, M.T. and Aldwinkle, H.S. (1998). Effect of cercropintype transgenes on fire blight resistanc of apple. *Acta Horticulturae*, **489:** 273–78.

Norelli, J.L., Borejsza–Wysocka, E., Momol, M.T., Mills, J.Z., Grethel, A., Alkwinckle, H.S., Ko, K., Brown, S.K., Bauer, D.W., Beer, S.V., Abdul–Kader, A.M. and Hanke, V. (1999). Genetic transformation for fire blight resistance in apple. *Acta Horticulturae,* **489:** 295–96.

Nürnberger, T., Brunner, F., Kemmerling, B., and Piater, L. (2004). Innate immunity in plants and animals: striking similarities and obvious differences. *Immunol. Rev.* **198:** 249–66.

Osusky, M. (2004). Transgenic Research. *Spring, German,* **13:** 181–90.

Osusky, M., Zhou, G.Q., Osuska, L., Hancock, R.E., Kay, W.W. and Misra, S. (2000). Transgenic plants expressing cationic peptide chimeras exhibit broad spectrum resistance to phytopathogens. *Nature Biotechnology,* **18:** 1162–66.

Parlevliet, J.E. (2003). Durability of resistance against fungal, bacterial and viral pathogens; present situation. *Euphytica,* **124:** 147–56.

Pavli, O.I, Kelaidi, G.I, Tampakaki, A.P, and Skaracis, G.N. (2011). The *hrpZ* gene of *Pseudomonas syringae* pv. *phaseolicola* enhances resistance to rhizomania disease in transgenic *Nicotiana benthamiana* and Sugar Beet. *PLoS ONE,* **6:** e17306.

Peng, J.L., Dong, H.S., Dong, H.P., Delaney, T.P. and Bonasera, B.M. (2003) Harpin–elicited hypersensitive cell death and pathogen resistance requires the *NDR1* and *EDS1* genes. *Physiological and Molecular Plant Pathology,* **62:** 317–26.

Peterhänsel, C., Freialdenhoven, A., Kurth, J., Kolsch, R. and Schulze–Lefert, P. (1997). Interaction analyses of genes required for resistance responses to powdery mildew in barley reveal distinct pathways leading to leaf cell death. *Plant Cell,* **9:** 1397–1409.

Pieterse, C.M.J. and Van Loon, L.C. (2004). NPR1: the spider in the web of induced resistance signalling pathways. *Current Opinion in Plant Biology,* **7:** 456–64.

Powell, A.L., Van Kan, J.A.L., Ten Have, A., Visser, J., Greve L.C., Bennett, A.B. and Labavitch, J.M. (2000). Transgenic expression of pear PGIP in tomato limits fungal colonisation. *Molecular Plant-Microbe Interactions,* **13:** 942–50.

Prasad, B.D., Jha, S. and Chattoo, B.B. (2008). Transgenic indica rice expressing *Mirabilis jalapa* antimicrobial protein (Mj–AMP2) shows enhanced resistance to the rice blast fungus *Magnaporthe oryzae*. *Plant Science,* **175:** 364–71.

Prins, M., Laimler, M., Noris, E., Schubert, J., Wassenegger, M., and Tepfer, M. (2008). Strategies for antiviral resistance in transgenic plants. *Molecular Plant Pathology,* **9:** 73–83.

Prins, M., Lohuis, D., Schots, A. and Goldbach, R. (2005). Phage display selected single-chain antibodies confer high levels of resistance against *Tomato spotted wilt virus. Journal of General Virology,* **86:** 2107–13.

Punja, Z.K. (2006). Recent developments toward achieving fungal disease resistance in transgenic plants. *Canadian Journal of Plant Pathology,* **28S1:** S298–S308.

Qiu, D., Mao, J., Yang, X. and Zeng, H. (2009). Expression of an elicitor–encoding gene from *Magnaporthe grisea* enhances resistance against blast disease in transgenic rice. *Plant Cell Reports,* **28:** 925–33.

Qiu, Y. and Yu, D. (2009). Over-expression of the stress–induced *OsWRKY45* enhances disease resistance and drought tolerance in *Arabidopsis*. *Environmental Experimental Botany,* **65:** 35–47.

Quilis, J., Penas, G., Messeguer, J., Brugidou, C. and San Segundo, B. (2008). The Arabidopsis AtNPR1 inversely modulates defense responses against fungal, bacterial, or viral pathogens while conferring hypersensitivity to abiotic stresses in transgenic rice. *Molecular Plant–Microbe Interaction,* **21:** 1215–31.

Rajasekaran, K., Cary, J.W., Jacks, T.J., Stromberg, K., and Cleveland, T.E. (2000). Inhibition of fungal growth *in planta* and *in vitro* by transgenic tobacco expressing a bacterial non heme chloroperoxidase gene. *Plant Cell Reports*, **19:** 333–38.

Rajasekaran, K., Cary, J.W., Jaynes, J.M., Cleveland, T.E. (2007). Disease resistance conferred by the expression of a gene encoding a synthetic peptide in transgenic cotton (*Gossypium hirsutum* L.) plants. *Plant Biotechnology Journal,* **3:** 545–54.

Reimann–Philipp, U., Schrader, G., Martinoia, E., Barkholt, V. and Apel, K. (1989). Intracellular thionins of barley. A second group of leaf thionins closely related to but distinct from cell wall-bound thionins. *Journal of Biological Chemistry,* **264:** 8978–84.

Roberts, W.K. and Selitrennikoff, C.P. (1988). Plant and bacterial chitinases differ in antifungal activity. *Journal of General Microbiology,* **134:** 169–76.

Romeis, T. (2001). Calcium–dependent protein kinases play an essential role in a plant defence response. *EMBO Journal,* **20:** 5556–67.

Rubio, T., Borja, M., Scholthof, H.B., Feldstein, P.A., Morris, T.J. and Jackson, A.O. (1999). Broad–spectrum protection against tombusviruses elicited by defective interfering RNAs in transgenic plants. *Journal of Virology,* **73:** 5070–78.

Rudolph, C., Schreier, P.H. and Uhrig, J.F. (2003). Peptide–mediated broad-spectrum plant resistance to tospoviruses. *Proceedings of the National Academy of Sciences,* **100:** 4429–34.

Rushton, P.J., Torres, J.T., Parniske, M., Wernert, P., Hahlbrock, K. and Somssich, I.E. (1996). Interaction of elicitor–induced DNA binding proteins with elicitor response elements in the promoters of parsley PR1 genes. *EMBO Journal,* **15:** 5690–5700.

Sanghera, G.S, Kashyap, P.L, Singh, G. and Teixeirs da Silva, J.A. (2011). Trangenics: Fast track for crop amelioration. *Transgenic Plant Journal,* **5:** 1–26.

Scheel, D. (1998). Resistance response physiology and signal transduction. *Current Opinion in Plant Biology,* **1:** 305–10.

Schillberg, S., Zimmermann, S., Zhang, M.Y. and Fischer, R. (2001). Antibody–based resistance to plant pathogens. *Transgenic Research,* **10:** 1–12.

Schlaich, T., Urbabiak, B., Plissonnier, M.L., Malgras, N. and Sautter, C. (2007). Exploration and Swiss field testing of a viral gene for specific quantitative resistance against smuts and bunts in wheat. *Advances in Biochemical Engineering and Biotechnology,* **107:** 97–112.

Shao, M., Wang, J., Dean, R.A., Lin, Y., Gao, X., and Hu, S. (2008). Expression of a hairpin–encoding gene in rice confers durable non-specific resistance to *Magnaporthe grisea. Journal of Plant Biotechnology.* **6:** 73–81.

Shengji, M, Hongya, G, Lijia, Q. and Zhangliang, C. (2003). Obtaining transgenic rice resistant to rice fungal blast disease by controlled cell death strategy. *Chinese Science Bulletin,* **48:** 1753–59.

Shetty, N.P., Jørgensen, H.J.L., Sharathchandra, R.G., Collinge, D.B. and Shetty, H.S. (2008). Roles of reactive oxygen species in interactions between plants and eukaryotic pathogens. *European Journal of Plant Pathology,* **121:** 267–80.

Sitbon, F., Hennion, S., Little, C.H.A. and Sundberg, B. (1999). Enhanced ethylene production and peroxidase activity in IAA overproducing transgenic tobacco plants is associated with increased lignin content and altered lignin composition. *Plant Science,* **141:** 165–73.

Sivamani, E., Huet, H., Shen, P., Shen, P., Ong, C.A., de Kochko, A., Fauquet, C. and Beachy, R.N. (1999). Rice plant (*Oryza sativa* L.) containing rice tungro spherical virus (RTSV) coat protein transgenes are resistant to virus infection. *Molecular Breeding,* **5:** 177–85.

Sohn, S., Kim, Y., Kim, B., Lee, S., Lim, C.K., Hur, J.H. and Lee, J. (2007). Transgenic tobacco expressing the *hrpNEP* gene from *Erwinia pyrifoliae* triggers defense responses against *Botrytis cinerea. Molecular Cells,* **24:** 232–39.

Sridevi, G., Parameswari, C., Sabapathi, N., Raghupathy, V. and Veluthambi, K. (2008). Combined expression of chitinase and β–1, 3-glucanase genes in indica rice (*Oryza sativa* L.) enhances resistance against *Rhizoctonia solani. Plant Science,* **175:** 283–90.

Stanley, J., Saunders, K., Pinner, M.S. and Wong, S.M. (1997). Novel defective interfering DNAs associated with ageratum yellow vein geminivirus infection of *Ageratum conyzoides. Virology,* **239:** 87–96.

Stanley, J., Frischmuth, T. and Ellwood, S. (1990). Defective viral DNA ameliorates symptoms of geminivirus infection in transgenic plants. *Proceedings of the National Academy of Sciences,* **87:** 6291–95.

Stark–Lorenzen, P., Nelke, B., Hanssler, G., Ühlbach, M. and Thomzik, J.E. (1997). Transfer of a grapevine stilbene synthase gene to rice (*Oryzae sativa* L). *Plant Cell Reports,* **16:** 668–73.

Strittmatter, G. (1995). Inhibition of fungal disease development in plants by engineering controlled cell death. *Biotechnology,* **13:** 1085–89.

Suslow, T.V., Matsubara, D., Jones, J., Lee, R. and Dunsmuir, P. (1988). Effect of expression of bacterial chitinase on tobacco susceptibility to leaf brown spot. *Phytopathology,* **78:** 1556.

Tai, T.H., Dahlbeck, D., Clark, E.T., Gajiwala, P., Pasion, R. and Whalen, M.C. (1999). Expression of the *Bs2* pepper gene confers resistance to bacterial spot disease in tomato. *Proceedings of the National Academy of Sciences,* **96:** 14153–58.

Takakura, Y., Che, F.S., Ishida, Y., Tsutsumi, F., Kurotani, K., Usami, S., Isogai, A. and Imaseki, H. (2008). Expression of a bacterial flagellin gene triggers plant immune responses and confers disease resistance in transgenic rice plants. *Molecular Plant Pathology,* **9:** 525–29.

Takase, K., Hagiwara, K., Onodera, H., Nishizawa, Y., Ugaki, M., Omura, T., Numata, S., Akutsu, K., Kumura, H. and Shimazaki, K. (2005). Constitutive expression of human lactoferrin and its N-lobe in rice plants to confer disease resistance. *Biochemistry and Cell Biology,* **83:** 239–49.

Talbot, N.J., Kershaw, M.J., Wakley, G.E., de Vries, O.M.H., Wessels, J.G.H. and Hamer, J. E. (1996). MPG1 encodes a fungal hydrophobin involved in surface interactions during infection–related development of *Magnaporthe grisea. Plant Cell,* **8:** 985–99.

Tampakaki, A.P., Skandalis, N., Gazi, A.D., Bastaki, M.N. and Sarris, P.F. (2010). Playing the 'Harp': evolution of our understanding of *hrp/hrc* genes. *Annual Review Phytopathology,* **48:** 17.1–17.24.

Tang, X.Y., Xie, M.T., Kim, Y.J., Zhou, J.M., Klessig, D.F. and Martin, G.B. (1999). Overexpression of *Pto* activates defense responses and confers broad resistance. *The Plant Cell,* **11:** 15–29.

Tavladoraki, P., Benvenuto, E., Trinca, S., De Martinis, D., Cattaneo, A. and Galeffi, P. (1993). Transgenic plants expressing a functional single-chain Fv antibody are specifically protected from attack. *Nature,* **366:** 469–72.

Tenllado, F., Martínez–Gracia, B., Vargas, M. and Díaz–Ruíz, J.R. (2003). Crude extracts of bacterially expressed dsRNA can be used to protect plants against virus infections. *BMC Biotechnology,* **3:** 1–11.

Terakawa, T., Takaya, N., Hortuchi, H., Koike, M. and Takagi, M. (1997). A fungal chitinase gene from *Rhizopus oligosporus* confers antifungal activity to tobacco. *Plant Cell Reports,* **16:** 439–43.

Thomzik, J.E, Stenzel, K., Stocker, R., Schreier, P.H, Hain, R. and Stahl, D.J. (1997). Synthesis of a grapevine phytoalexin in transgenic tomatoes (*Lycopersicon esculentum* Mill.) conditions resistance against *Phytophthora infestans. Physiological and Molecular Plant Pathology,* **51:** 265–78.

Tör, M., Brown, D., Cooper, A., Woods–Tör, A., Sjölander, K., Jones, J.D.G. and Holub, E.B. (2004). Arabidopsis downy mildew resistance gene RPP27 encodes a receptor–like protein similar to CLAVATA2 and tomato *Cf–9. Plant Physiology,* **135:** 1100–12.

Toubart, P., Desiderio, A., Salvi, G., Cervone, F., Daroda, L. and De Lorenzo, G. (1992). Cloning and characterization of the gene encoding the endopolygalacturonase–inhibiting protein (PGIP) of *Phaseolus vulgaris* L. *Plant Journal,* **2:** 367–73.

Toyoda, H., Matsuda, Y., Yamaga, T., Ikeda, S., Morita, M., Tamai, T. and Ouchi, S. (1991). Suppression of the powdery mildew pathogen by chitinase microinjected into barley coleoptile epidermal cells. *Plant Cell Reports,* **10:** 217–20.

Tripathi, S.J., Suzuki, Y., Ferreira, S.A. and Gonsalves, D. (2008). Papaya ringspot virus–P: characteristics, pathogenicity, sequence variability and control. *Molecular Plant Pathology,* **9:** 269–80.

Trognitz, F., Manosalva, P., Gysin, R., Nino–Liu, D., Simon, R., Herrera, M., Trognitz, B., Ghislain M. and Nelson, R. (2002). Plant defense genes associated with quantitative resistance to potato late blight in *Solanum phureja* x dihaploid *S. tuberosum* hybrids. *Molecular Plant-Microbe Interactions,* **15:** 587–97.

Tsukamoto, S., Morita, S., Hirano, E., Yokoi, H. and Tanaka, K. (2005). A novel cis–element that is responsive to oxidative stress regulates three antioxidant defense genes in rice. *Plant Physiology,* **137:** 317–27.

Udvardi, M.K, Kakar, K., Wandrey, M., Montanri, O., Murray, J., Andraiankaja, A., Zhang, J.Y., Benedito, V., Hofer, J.M.I., Cheng, F. and Town, C.D. (2007). Legume transcription factors: global regulators of plant development and response to the environment. *Plant Physiology,* **144:** 538–49.

Van der Vossen, E., Sikkema, A., Hekkert, B.T.L., Gros, J., Stevens, P. and Muskens, M. (2003). An ancient R gene from the wild potato species *Solanum bulbocastanum* confers broad–spectrum resistance to *Phytophthora infestans* in cultivated potato and tomato. *Plant Journal,* **36:** 867–82.

Van Loon, L.C., Rep, M. and Pieterse, C.M.J. (2006). Significance of inducible defense-related proteins in infected plants. *Annual Review of Phytopathology,* **44:** 135–62.

Verberne, M.C., Verpoorte, R., Bol, J.F., Mercado–Blanco, J. and Linthorst, H. (2000). Overproduction of salicylic acid in plants by bacterial transgenes enhances pathogen resistance. *Nature Biotechnology,* **18:** 779–83.

Wang, G.L., Song, W.Y., Ruan, D.L., Sideris, S. and Ronald, P.C. (2007). The cloned gene, *Xa21*, confers resistance to multiple *Xanthomonas oryzae* pv. *oryzae* isolates in transgenic plants. *Molecular Plant–Microbe Interactions,* **9:** 855.

Wattad, C., Kobiler, D., Dinoor, A., and Prusky, D. (1997). Pectate lyase of *Colletotrichum gloeosporioides* attacking avocado fruits—cDNA cloning and involvement in pathogenicity. *Physiological Plant Pathology,* **50:** 197–212.

Wegener, C.B. (2001). Transgenic potatoes expressing an *Erwinia* pectate lyase gene–results of a 4-year field experiment. *Potato Research,* **44:** 4 401–410.

Wegener, C., Bartling, S., Olsen, O., Weber, J. and Vonwettstein, D. (1996). Pectate lyase in transgenic potatoes confers pre-activation of defence against *Erwinia carotovora*. *Physiological and Molecular Plant Pathology,* **49:** 359–76.

Wolter, M., Hollricher, K., Salamini, F., and Schulze–Lefert, P. (1993). The *mlo* resistance alleles to powdery mildew infection in barley trigger a developmentally controlled defense mimic phenotype. *Molecular and General Geneties,* **239:** 122–28.

Wu, K., Tian, L., Hollingworth, J., Brown, D.C. and Miki, B. (2002). Functional analysis of tomato *Pti4* in *Arabidopsis*. *Plant Physiology,* **128:** 30–37.

Wu, G., Shortt, B.J., Lawrence, E.B., Léon, J., Levine, E.B., Raskin, I. and Shah, D.M. (1997). Activation of host defense mechanisms by elevated production of H_2O_2 in transgenic plants. *Plant Physiology,* **115:** 427–35.

Xiao, X.W., Chu, P.W.G., Frenkel, M.J., Tabe, L.M., Shukla, D.D., Hanna, P.J., Higgins, T.J.V., Muller, W.J. and Ward, C.W. (2000). Antibody–mediated improved resistance to ClYVV and PVY infections in transgenic tobacco plants expressing a single–chain variable region antibody. *Molecular Breeding,* **6:** 421–31.

Yamamizo, C., Kazuo, K., Akira, K., Shinpei, K., Kazuhito, K., Jonathan, D.G., Jones, N.D. and Hirofumi, Y. (2006). Rewiring mitogen–Activated protein kinase cascade by positive feedback confers potato blight resistance. *Plant Physiology,* **140:** 681–92.

Yang, T. and Poovaiah, B.W. (2002). A calmodulin–binding/CGCG box DNA–binding protein family involved in multiple signalling pathways in plants. *Journal of Biology and Chemistry,* **277:** 45049–58.

Yao, K., De Luca, V. and Brisson, N. (1995). Creation of a metabolic sink for tryptophan alters the phenylpropanoid pathway and the susceptibility of potato to *Phytophthora infestans*. *Plant Cell,* **7:** 1787–99.

Yoshikawa, N., Saitou, Y., Kitajima, A., Chida, T., Sasaki, N. and Isogai, M. (2006). Interference of long-distance movement of grapevine berry inner necrosis virus in transgenic plants expressing a defective movement protein of apple chlorotic leaf spot virus. *Phytopathology,* **96:** 378–85.

Zhai, W.X., Wang, W.M., Zhou, Y.L., Li, X.B., Zheng, X.W. and Zhang, Q. (2002). Breeding bacterial blight–resistant hybrid rice with the cloned bacterial blight resistance gene *Xa21. Molecular Breeding,* **8:** 285–93.

Zhang, H., Li, G., Li, W. and Song, F. (2009). Transgenic strategies for improving rice disease resistance. *African Journal of Biotechnology,* **8:** 1750–57.

Zhang, J., Li, X., Jiang, G., Xu, Y. and He, Y. (2006). Pyramiding of *Xa7* and *Xa21* for the improvement of disease resistance to bacterial blight in hybrid rice. *Plant Breeding,* **125:** 600–605.

Zhang, S. and Liu, Y. (2001). Activation of salicylic acid–induced protein kinase, a mitogen–activated protein kinase, induces multiple defense responses in tobacco. *Plant Cell,* **13:** 1877–89.

Zhang, Y. and Lewis, K. (1997). Fabatins: new antimicrobial plant peptides. *FEMS Microbiology Letters,* **149:** 59–64.

Zhang, Y., Fan, W., Kinkema, M., Li, X. and Dong, X. (1999). Interaction of NPR1 with basic leucine zipper protein transcription factors that bind sequences required for salicylic acid induction of the *PR–1* gene. *Proceedings of the National Academy of Sciences,* **96:** 6523–28.

Zhang, Z., Coyne, D.P., Vidaver, A.K. and Mitra, A. (1998). Expression of human lactoferrin cDNA confers resistance to *Ralstonia solanacearum* in transgenic tobacco plants. *Phytopathology,* **88:** 730–34.

Zhao, B.Y., Lin, X.H., Poland, J., Trick, H., Leach, J. and Hulbert, S. (2005). From the cover: A maize resistance gene functions against bacterial streak disease in rice. *Proceedings of the National Academy of Sciences,* **102:** 15383–88.

Zheng, Z., Qamar, S.A., Chen, Z. and Mengiste, T. (2006). Arabidopsis WRKY33 transcription factor is required for resistance to necrotrophic fungal pathogens. *Plant Journal,* **48:** 592–605.

Zhu, H., Xu, X., Xiao, G., Yuan, L. and Li, B. (2007). Enhancing disease resistances of Super Hybrid Rice with four antifungal genes. *Sci. China C. Life Sci.,* **50:** 31–39.

Zimmerman, S., Schillberg, S., Lao, Y.C. and Fischer, R. (1998). Intracellular expression of TMV–specific single-chain Fv fragments leads to improved virus resistance in *Nicotiana tabacum. Molecular Breeding,* **4:** 369–79.

Zipfel, C., Robatzek, S., Navarro, L., Oakeley, E.J., Jones, J.D., Felix, G. and Boller, T. (2004). Bacterial disease resistance in *Arabidopsis* through flagellin perception. *Nature,* **428:** 764–67.

5

Biotechnology Driven Insect Pest and Plant Parasitic Nematode Management: Current Trends and Future Prospects

TUSHAR K. DUTTA[1], A.K. MANDAL[2] AND SOMEN ACHARYA[3]

ABSTRACT

Plant biotechnology has placed great hopes in the field of insect pest management. With the advent of genetic transformation techniques and molecular biology, it has become possible to clone and insert genes in crop plants to confer resistance against insect pests and nematodes. Since the late 1980s, biotechnology has contributed a lot, through development of transgenic plants by expressing insecticidal genes such as Bt, *trypsin inhibitors, lectins, ribosome inactivating proteins, secondary plant metabolites, vegetative insecticidal proteins, and small RNA viruses can also be used alone or in combination for pest control. This chapter is an attempt to highlight current trends and future prospects of biotechnology driven insect pest and nematode management.*

Key words: Transgenics, RNAi, *Bt*, Genetic engineering, Protease inhibitors, Lectins.

1. INTRODUCTION

Insect pests and nematodes are a major constraint to increased global production of food. There are an estimated 67000 pest species that damage

[1] Indian Agricultural Research Institute, New Delhi - 110 012.
[2] Directorate of Seed Research, Mau (U.P.) - 275 101.
[3] Defence Institute of High Altitude Research (DIHAR), DRDO, C/o 56 APO, Leh–Ladakh, India.
Corresponding author: E-mail: nemaiari@gmail.com

agricultural crops, of which approximately 9000 species are insects and mites (Ross and Lembi, 1985). Losses due to insect pests have been estimated at 14% of the total agricultural production (Oerke *et al.*, 1994). In India, the average annual losses have been estimated to be 17.5% valued at US$17.28 billion in eight major field crops (cotton, rice, maize, sugarcane, rapeseed–mustard, groundnut, pulses, coarse cereals, and wheat) (Dhaliwal *et al.*, 2010). Insects not only cause direct losses to the agricultural production but also cause losses indirectly due to impaired quality of the produce and their role as vectors of various plant pathogens (Kumar *et al.*, 2008). On the other hand, plant parasitic nematodes (PPN) accounts for 10–20% loss of world's agricultural output equivalent to $125 billion per annum in monetary value (Chitwood, 2003). PPNs are unique in their ubiquitous nature and persistence in the soil. The conflicting nature of their attack allows their presence to often pass unnoticed while crops slowly decline in vigour and yield. Rarely is any crop free from attack of these tiny and microscopic pathogens.

Problems have been encountered in obtaining resistance sources against a particular pest/pathogen and emergence of resistance breaking pathogen races or biotypes. These problems can now be overcome by using modern day cellular and molecular approaches to plant biotechnology which can facilitate the transfer of existing sources of insect/nematode resistance across conventional barrier to reproduction into other related or even unrelated crop species. Genetic manipulation of *Bacillus thuringiensis* (*Bt*) genes encoding for proteins toxic to insects offers an opportunity to produce genetically modified strains with more potent and transgenic plant expressing *Bt* toxin. Besides the *cry* genes from *B. thuringiensis*, other bacterial genes such as *choM* and *ipt* from *Actinomyces* and *Agrobacterium*, respectively, have strong insecticidal properties and have been transformed into cotton, brinjal, tobacco *etc.* in order to determine their ability to control insect infestation (Jauhar, 2006). Additionally, several proteins that are effective against certain insects such as the vegetative insecticidal proteins (VIP), α-endotoxin, a variety of secondary metabolites and proteins of plant origin are amenable to genetic manipulation (Gatehouse, 2008). There has been increasing number of candidate genes and powerful RNAi tool that are deployed for designing insect tolerant crops.

Therefore, knowledge and continuous research in the field of genetic engineering is the key to assess the potential of biotechnology for effective and sustainable management of insect pests and nematodes. This chapter addresses the potential of plant biotechnology for developing entirely new system of pest management that integrate traditional wisdom with advance knowledge of molecular biology and genetic transformation to the sustainability of agriculture.

2. TRANSGENIC PLANTS EXPRESSING *BACILLUS THURINGIENSIS* TOXINS

Designing of insect pest resistant transgenic crops have been one of the major successes of applying plant biotechnology to agriculture. Cotton (*Gossypium hirsutum*) resistant to lepidopteran larvae and maize (*Zea mays*) resistant to both lepidopteran and coleopteran larvae have become widely used in global agriculture and led to reductions in pesticide usage and lower production costs (Brookes and Barfoot, 2005; Christou *et al.*, 2006).

2.1. *Bacillus thuringiensis (Bt)* Toxins

During the past three decades, considerable research efforts have been invested to introduce insecticidal crystal protein genes from *Bacillus thuringiensis* (*Bt*) into plants by biotechnological approaches. *Bt*-genes were first introduced and expressed in tobacco (Barton *et al.*, 1987; Vaeck *et al.*, 1987) and the transgenic plants exhibited a certain level of insect resistance. Subsequently, several *Bt* genes have been expressed in transgenic plants, including potato, sugarcane, cotton, brinjal and rice, etc. Delannay *et al.* (1989) for the first time reported field performance of transgenic tomato plants expressing δ-endotoxin gene. Though Cry1Ab protein was effective against tobacco hornworm, higher level of gene expression was needed for the control of tomato fruit worm (*Helicoverpa* sp). Results of field trials of *Bt* transgenic tobacco (Hoffman *et al.*, 1992) and cotton (Wilson *et al.*, 1992) expressing truncated δ-endotoxin genes were encouraging. Since then, there have been several reports of field-tested δ-endotoxin expressing transgenic crops. The progress made in development and deployment of insect-resistant transgenic plants with different toxin genes for pest management in different crops (Table 1) is discussed in following sections.

Table 1: Insect-Resistant transgenic crops based on *Bt* toxin from *Bacillus thuringiensis*

Transgenic crop	***Bt gene***	***Target insect pest***	***Reference(s)***
Maize	*Cry1Ab*	*Heliothis zeae*	Koziel *et al.* (1993) Armstrong *et al.* (1995)
	Cry1Ac	*Pectinophora gossippiella*	Buschman *et al.* (1998)
Potato	*Cry3A*	*Leptinotarsa decemlineata*	Perlak *et al.* (1993)
	Cry1Ab	Potato tuber moth (*Phthorimaea opercullela*)	Duck and Evola (1997)

Table 1: (*Contd...*)

Table 1: (*Contd...*)

Transgenic crop	***Bt gene***	***Target insect pest***	***Reference(s)***
	Cry3	Colorado potato beetle (*Leptinotarsa decemlineata*)	Jenkins *et al.* (1999)
	Cry1Ac	Potato tuber moth (*Phthorimaea opercullela*)	Ebora *et al.* (1994)
	Cry1Ac9Cry5	Potato tuber moth (*Phthorimaea opercullela*)	Davidson *et al.* (2002)
	Cry5	Potato tuber moth (*Phthorimaea opercullela*)	Douches *et al.* (2004)
Cotton	*Cry1Ac*	Cotton boll worm	Perlak *et al.* (1990) Jenkins *et al.* (1997)
	Cry1Ab	Pink boll worm (*Pectinophora gossypiella*)	Wilson *et al.* (1992) Benedict *et al.* (1996)
	Vip3Aa	Lepidopteran insects	Artim (2003)
Tomato	*Cry1Ab*	Tobacco hornworm	Delannay *et al.* (1989)
Brinjal	*Cry1Ab*	Fruit borer	Kumar *et al.* (1998)
	Cry3B	Fruit borer	Iannacone *et al.* (1997)
Chickpea	*Cry1Ac*	Pod borer	Kar *et al.* (1997) Sanyal *et al.* (2005)
Sugarcane	*Cry1Ab*	Shoot borer	Arencibia *et al.* (1997)
Rice	*Cry1Ab*	Yellow stem borer, Striped stem borer	Datta *et al.* (1998) Shu *et al.* (2000) Ye *et al.* (2001)
	Cry1Ac	Yellow stem borer	Nayak *et al.* (1997)
	Cry2A–1Ac–gna	Yellow stem borer, Rice leaf folder	Maqbool *et al.* (1998)
	Cry1B– 1Aa	Yellow stem borer	Raina *et al.* (2002)
	Cry1Ac– 2A	Yellow stem borer, Rice leaf folder	Bashir *et al.* (2004)
Groundnut	*Cry1Ac*	Lesser corn stalk borer, (*Elasmopalpus lignosellus*)	Singsit *et al.* (1997)

European corn borer [ECB, *Ostrinia nubilalis* (Hübner)], for example, causes a loss of up to 2000 million dollars annually in the USA alone (Hyde *et al.*, 1999). Resistance breeding by conventional means is cumbersome and fraught with uncertainty. Through 12 years of breeding Syngenta, a Swiss agrochemical company, was able to produce a corn cultivar with only 10% resistance to ECB. However, *Bt* gene, when bioengineered into the corn genome, conferred almost complete resistance to ECB. It took Syngenta only 5 years to engineer the *Bt* gene into corn. Thus, Bt-corn acquired the biopesticide status. Scientists at the University of Minnesota estimated that farmers got several times greater returns on their investment by using Bt corn for insect control, compared to the chemical insecticide (Ostlie *et al.*, 1997). The Bt corn hybrids had 4 to 8% higher grain yields than standard hybrids when infested with ECB (Lauer and Wedberg, 1999). Moreover, Bt corn is beneficial to the environment and the Bt-induced insect resistance in corn is much safer to farmers and other field workers, than the chemical pesticides. Based on safety data, the U.S. Environmental Protection Agency (EPA) authorized commercial planting of *Bt* corn varieties.

Several transgenic crops with insecticidal genes have been introduced in temperate regions of the world (Sharma *et al.*, 2003). Transgenic rice varieties resistant to yellow stem borer [*Scirpophaga incertulas* (Walker)] have been developed in India (Ramesh *et al.*, 2004). Because of its higher productivity and positive health effects through reduced pesticide usage *Bt* cotton has been commercialized aggressively especially in Asian countries like China (Huang *et al.*, 2002) and India (Whitfield, 2003). Carrière *et al.* (2003) found long-term regional suppression of pink bollworm [*Pectinophora gossypiella* (Saunders)] by *Bt* cotton. *Bt* rice has the potential to eliminate yield losses caused by lepidopteran insects, estimated at 2 to 10% of Asia's annual rice yield of 523 million tonnes (High *et al.*, 2004). Field trials revealed higher tolerance of transgenic rice against yellow stem borer (Bashir *et al.*, 2004). Most recently, an insect-resistant rice variety (Xianyou 63) developed by inserting *Bt* gene, showed resistance to rice stem borer (*S. incertulas*) and leaf roller [*Cnaphalocrocis exigua* (Butler)] and is on the threshold of being released for commercial cultivation in China. This insect-resistant variety is reported to benefit small farmers due to higher crop yields and reduced use of pesticides (Huang *et al.*, 2005).

Bt strains show differing specificities of insecticidal activity toward pests, and constitute a large reservoir of genes encoding insecticidal proteins, which are accumulated in the crystalline inclusion bodies produced by the bacterium on sporulation (Cry proteins, Cyt proteins) or expressed during bacterial growth (Vip proteins). The three-domain Cry proteins have been extensively studied; their mechanism of action involves a proteolytic activation step, which occurs in the insect gut after ingestion, followed by interaction of one or both of domains II and III with receptors on the surface of cells of the insect gut epithelium. This interaction led to oligomerization of the protein, and domain I was responsible for the formation of an open channel through the cell membrane (Bravo *et al.*, 2007). The resulting ionic leakage destroys the cell, leading to breakdown of the gut, bacterial proliferation, and insect death. However, not all pests are adequately targeted by the *Bt* toxins used at present, and there is still a need to develop solutions to specific problems, *e.g.*, resistance to sap-sucking pests and pests of stored products. Various strategies to improve expression level of Bt toxins have been reported by various workers is illustrated in the following sections.

2.2. Plastid Genome Transformation

Expression of Bt toxins in transgenic plants need to be at a sufficient level to confer adequate protection against target pests. Transformation of the nuclear genome with genes encoding *Bt* genes give very low levels of expression unless extensive modifications, which include removal of AT-rich regions from the coding sequence and use of modified constitutive or tissue-specific promoters, were carried out. These methods were established within the first stage of the development of this technology and are now considered routine, although they do pose significant technical problems.

In contrast, introduction of unmodified *Bt* genes into the chloroplast genome results in high levels of toxin accumulation (McBride *et al.*, 1995), as the plastid genome is bacterial in origin. This method has not been widely adopted, due to significant technical problems in achieving stable transformation of the plastid genome and in transforming plastids in species other than tobacco (*Nicotiana tabacum*). Nevertheless, these difficulties can be overcome; various Cry proteins have been expressed in plastids of tobacco (De Cosa *et al.*, 2001), and *Cry1Ab* has been expressed in soybean (*Glycine max*) plastids (Dufourmantel *et al.*, 2005). Overexpression of the *Cry2Aa2* operon in plastids is effective in giving broad-spectrum protection against a range of pests. This technique has the potential advantage that plastid-encoded characteristics are predominantly maternally inherited in most plants, so that pollen from transgenic crops is less likely to disperse the transgene to non transgenic plant stocks. The inherent containment addresses concerns about the coexistence of transgenic and organic agricultural practices and may lead to wide adoption of this technique.

2.3. Novel *Bt* Toxins

Several novel Bt insecticidal proteins, which have no sequence similarity to three–domain Cry proteins, have been expressed in transgenic plants. Binary toxins require two components for activity and are exemplified by the Cry34/35 and Vip1/2 toxins, which are active against corn rootworm (*Diabrotica virgifera*). Cry34/35 has been expressed in transgenic maize (Moellenbeck *et al.*, 2001) that is the basis of a successful commercial product as no re-engineering of coding sequences was necessary, in contrast to an alternative approach using modified Cry3Bb1. Single-chain Vip proteins, such as Vip3, are active against lepidopteran larvae, with a broader range of toxicity than individual Cry proteins. Their range of activity can be further extended by protein engineering (Fang *et al.*, 2007).

2.4. Plants Expressing Multiple Toxins

The specificity of *Bt* Cry toxins toward target pest species was accompanied by minimized effects on non-target insects and other beneficial organisms.

However, deployment of transgenic crops expressing a single specific *Bt* toxin can lead to problems in the field, where secondary pest species are not affected, and can cause significant damage to the crop. Introduction of additional *Bt cry* genes into the crop can afford protection against a wider range of pests. Commercial use of transgenic cotton containing two *Bt* genes began in 1999, 3 years after the release of the original single Bt variety. Cotton plants expressing both Cry1Ac and Cry2Ab proteins were more toxic to bollworms (*Helicoverpa zea*) and two species of armyworms (*Spodoptera frugiperda* and *Spodoptera exigua*) than cotton expressing Cry1Ac alone in laboratory trials (Stewart *et al.*, 2001) and in greenhouse and field trials (Chitkowski *et al.*, 2003).

Expression of multiple Cry proteins can also be beneficial in prevention of resistance to toxin activity in the target pest(s). Although the approved refuge strategy has been highly successful in containing pest resistance to Bt toxins expressed in transgenic plants (Tabashnik *et al.*, 2005), targeting different receptors in the insect is more effective because multiple mutations are required to produce the loss of sensitivity to the toxins. Transgenic broccoli plants expressing either Cry1Ac or Cry1C or both proteins were exposed to a population of diamond back moth (*Plutella xylostella*), which carried *Bt* genes at a relatively low frequency (Zhao *et al.*, 2003). After 24 pest generations, plants carrying two transgenes showed a significant delay in the selection of a resistant population of insects. However, simultaneous exposure of the insects to plants carrying either one or two *Bt* genes actually increased resistance breakdown in the two gene plants (Zhao *et al.*, 2005), suggesting that the multiple-toxin approach is not a cure-all solution. Other results have also shown that pests can acquire resistance to multiple toxins, *e.g.*, a strain of the lepidopteran cotton pest *Heliothis virescens* had simultaneous resistance to Cry1Ac and Cry2Aa, with a different genetic basis of resistance to each toxin (Gahan *et al.*, 2005).

Improvements in plant transformation methods, such as extending the species range of *Agrobacterium*–mediated gene transfer methods to monocots and using plasmid vectors containing multiple gene constructs to allow introduction of multiple transgenes at a single genetic locus, have enabled the expression of multiple toxins in transgenic plant varieties. The recent announcement of a transgenic maize variety containing six insect resistance genes active against corn rootworm and lepidopteran pests (rootworm: Cry34Ab1 + Cry35Ab1, modified Cry3Bb1; Lepidoptera: Cry1F, Cry1A.105, Cry2Ab2) as a "one-stop" solution to pest problems (Gatehouse, 2008) exemplifies this gene stacking technology.

2.5. Protein Engineering in *Bt* Toxins

Mutagenesis of Cry toxins has been used extensively in studying the mechanism of action of these proteins (Bravo *et al.*, 2007) and has been exploited to produce novel recombinant toxins. There are several approaches of protein engineering in Bt toxins summarized in subsequent sections.

2.5.1. *Domain Exchange in Three-domain Cry Toxins*

The structural similarity of all members of the family of three-domain *Bt* toxins, and the separate roles of the domains in the processes of receptor binding and channel formation, suggested that combining domains from different proteins could generate active toxins with novel specificities. Transfer of the carbohydrate–binding domain III generated a Cry1Ab–Cry1C hybrid that was highly toxic to armyworm (*S. exigua*), an insect resistant to Cry1A toxins; the presence of the Cry1Ca domain III was sufficient to confer toxicity toward *Spodoptera* (de Maagd *et al*., 2000). More remarkably, a hybrid Cry protein, containing domains I and III from Cry1Ba and domain II of Cry1Ia, conferred resistance to the lepidopteran pest potato tuber moth (*Phthorimaea operculella*) and to the coleopteran Colorado potato beetle (*Leptinotarsa decemlineata*) when expressed in transgenic potato (Naimov *et al*., 2003). The parental Cry proteins in this hybrid are lepidopteran specific, with no toxicity toward coleopterans such as the potato beetle, demonstrating the creation of a novel specificity.

2.5.2. *Mutagenesis of Three-domain Cry Toxins*

Modification of *Bt* toxins by site-directed mutagenesis to increase toxicity toward target pests has been employed as an alternative to the "domain swap" approach. The key role of domain II in three-domain Cry proteins in mediating interactions with insect receptors has been exploited by mutation of amino acid residues in the loop regions of this domain. Mutation of Cry1Ab increased its toxicity toward larvae of gypsy moth (*Lymantria dispar*) by up to 40-fold (Rajamohan *et al*., 1996), and similar strategies were used to increase the toxicity of Cry3A protein toward target coleopteran pests (Wu *et al*., 2000). Liu and Dean (2006) did rational design of toxins by the engineering of toxicity of lepidopteran–specific toxin Cry1Aa against mosquito larvae. Alternatively, a directed evolution system based on phage display technology for producing toxins with improved binding to a receptor, and thus increased toxicity, has been described (Ishikawa *et al*., 2007).

A current commercial transgenic maize variety (MON863) with resistance to corn rootworm expressed a modified version of the *Bt* Cry3Bb1 toxin (Vaughn *et al*., 2005). Unmodified Cry3Bb1 is active against a number of coleopteran species, but toxicity toward Western corn rootworm was not sufficient to give adequate protection at levels of expression achievable in maize. A large number of variants of the native Cry3Bb, incorporating a series of specific mutations that aimed to improve the channel–forming ability of the toxin, were produced and screened for activity (English *et al*.,

2003). Mutations were carried out to: (a) increase hydrophobicity of the protein in domain I region containing sheets of bound water molecules and in loop regions; (b) increase the mobility of the channel–forming helices in domain I by disrupting hydrogen-bond formation; (c) increase the mobility and flexibility of loop regions in domain I; (d) alter potential ion–pair interactions and metal-binding sites; and (e) reduce or eliminate binding to carbohydrates in the insect gut by mutation of a loop region between domains I and II. The toxicity of the protein toward corn rootworm was increased approximately 8-fold, giving a product that could be expressed at levels sufficient for adequate protection against rootworm. Recent results, showing that oligomerization of Cry toxins subsequent to binding to the cadherin receptor on the insect gut surface is a necessary step in the mechanism of toxicity, have led to a strategy to engineer Cry proteins to be effective against insects that have become resistant to normal toxins by receptor mutation (Soberon *et al.*, 2007). Removal of the α–1 helix of domain I resulted in a protein that did not require binding to cadherin to oligomerize and was toxic to resistant insects. Expression of these modified toxins in plants is yet to be attempted.

2.5.3. *Fusion Proteins*

Transformation of plants with a gene construct containing a single translationally fused coding sequence encoding two Cry proteins has been used as an alternative to separate the constructs (Bohorova *et al.*, 2001), although there was no apparent advantage over simpler methods for introducing two genes. However, addition of sequences from other proteins can be used to introduce extra functionality into Cry toxins. For example, the Gal-binding lectin domain (B-chain) from the ribosome–inactivating protein ricin was fused C terminally to domain III of Cry1Ac (Mehlo *et al.*, 2005), giving a fusion protein that could bind to Gal residues in potential receptors in the insect, in addition to binding to *N*-acetyl galactosamine residues by domain III of the toxin. Expression of the fusion protein in transgenic maize and rice (*Oryza sativa*) plants gave resistance to larvae of stem borers (*Chilo suppressalis*) and leaf armyworm (*Spodoptera littoralis*), whereas plants expressing the unmodified Cry1Ac were susceptible to both insects. Plants expressing the fusion protein were also resistant to *Cicadulina mbila*; the lectin domain may cause this effect because *Bt* toxins are not effective against hemipterans.

3. ENGINEERING PLANT DEFENSIVE PROTEINS

Engineering plants to express proteins that are end-products of the wounding response, such as proteinase inhibitors and polyphenol oxidase, has generally failed to give more than partial protection against insect herbivores, due to pre-adaptation by the pests. However, several examples of exploiting plant defensive proteins have shown promise in addressing specialized insect resistance problems.

3.1. Protease Inhibitors

Many insect pests, particularly Lepidoptera, depend on serine proteases (trypsin, chymotrypsin and elastase endoproteases) as their primary protein digestive enzymes. Insects also produce SPIs (serine protease inhibitors), which are active against and presumably involved in regulating their digestive proteases. Genes encoding members of various SPIs have been cloned and introduced into transgenic plants (Wasmann *et al.*, 1994). Other species of insects rely on cysteine proteases as their primary digestive protease. These can be targeted with thiol protease inhibitors (TPIs). TPIs have been reported to be effective for controlling the maize rootworm, *Diabrotica* sp. (Edmonds *et al.*, 1996). Transgenic tobacco plants expressing trypsin inhibitor gene have resulted in increased mortality, reduced insect growth and plant damage by *H. virescens* (Hilder *et al.*, 1987). Similar results have also been shown in *H. zea* (Hoffmann *et al.*, 1992). Transgenic tobacco has also been shown to enhance protection against *Spodoptera littoralis* and *Manduca sexta*. Thomas *et al.* (1994) reported that a cDNA encoding the anti-elastase protease inhibitor (*PI*) from *M. sexta* reduced the onset of thrips predation, suggesting that this protocol can establish insect resistance in various legume crops. Sweet potato (cv. Tainong 57) trypsin inhibitor (*TI*) gene introduced into tobacco (cv. W38) retarded larval growth of *S. litura* as compared to control plants (Yeh *et al.*, 1997). Several protease inhibitor gene constructs have been introduced into different transgenic crops (Table 2).

Table 2: Insect-resistant transgenic crops expressing protease inhibitors from plants

Crop	*Transgene*	*Target insect pest*	*Reference(s)*
Tobacco	*CpTI*	*Helicoverpa armigera*	Gatehouse *et al.* (1993) Zhao *et al.* (1998) Zhang *et al.* (2004)
	SBTI	*Helicoverpa armigera* *Heliothis virescens,* *Heliothis zea, Spodoptera littoralis, Manduca sexta,* and *Helicoverpa armigera*	Hilder *et al.* (1987) Hoffman *et al.* (1992) Yeh *et al.* (1997) McManus *et al.* (1999) Charity *et al.* (1999)
	GTPI	*Helicoverpa armigera.*	Wu *et al.* (1997)
	Brassica juncea, BjTI	Tobacco cut worm (*Spodoptera litura*)	Mandal *et al.* (2002)
	PI 2 (MTI–2)	*Spodoptera littoralis*	De Leo *et al.* (1998)

Table 2: (*Contd...*)

Table 2: (*Contd...*)

Crop	*Transgene*	*Target insect pest*	*Reference(s)*
Potato	*Kti3*, *C–II*, and *PU–IV* from soybean	*Spodoptera littoralis*	Marchetti *et al.* (2000)
Cotton	*CpTI*	*Helicoverpa armigera*	Li *et al.* (1998)
Maize	*Oryzacystatin*	Southern maize rootworm (*Diabrotica undecimpunctata*)	Edmonds *et al.* (1996)
Rice	*SBTI*	Brown plant hopper (*Nilaparvata lugens*)	Lee *et al.* (1999)
	CpTI	*Chilo suppressalis* and *Sesamia inferens.*	Xu *et al.* (1996)
	WTI–1B	*Chilo suppressalis*	Mochizuki *et al.* (1999)
Sugarcane	Proteinase inhibitor II	*Antitrogus consanguineus.*	Nutt *et al.* (1999)
Cabbage	*CpTI*	*Pieris rapae*	Fang *et al.* (1997)
	Modified *CpTI* (*sck*)	*Pieris rapae*	Yang *et al.* (2002)
Potato	*OCI*	*Leptinotarsa decemlineata*	Lecardonnel *et al.* (1999)
Wheat	*TI* from barley	Angoumois grain moth, *Sitotroga cerealella*	Altpeter *et al.* (1999)

Some plants produce high levels of natural SPIs, and introduction of an additional heterologous inhibitor increases the level of resistance, *e.g.*, in sweet potato to *Euceps postfaciatus* (Golmirizaie *et al.*, 1997). Insects in general are adapted to counter the defensive mechanism of their natural hosts, but may still be susceptible to a related defense mechanism from the non-host plants. Jongsma *et al.* (1995) observed that only 18% of the gut proteinase activity from caterpillars raised on transgenic plants expressing potato inhibitor II was sensitive to it, compared to 78% in caterpillars raised on control plants. The ability of some species to compensate for protease inhibition by switching on to an alternative proteolytic activity or over-producing the existing activity would limit the application of protease inhibitors in such species. Deployment of protease inhibitors for insect control requires a detailed analysis of the particular crop-insect interactions. The range of dissociation constants (Kd) for different PIs (protease inhibitors) with specific proteases is large, and this can be used to select the most effective inhibitor for gene transfer in a particular situation (Christeller and Shaw, 1989), *e.g.*, transgenic tobacco expressing high levels of Kunitz type of trypsin inhibitor from soybean (SBTI) performs better than the tobacco plants expressing CpTI–tobacco against *H. virescens*. Proteolysis by gut extracts was 40-fold more susceptible to inhibition by SBTI than to CpTI (Gatehouse *et al.*, 1993).

Expression of potato trypsin inhibitor gene confers resistance to insects in rice (Duan *et al.*, 1996). Accumulation of soybean Kunitz trypsin inhibitor (SKTI) in rice confers resistance to brown plant hopper (*Nilaparvata lugens*) (Lee *et al.*, 1999). A gene encoding a cowpea trypsin inhibitor (CpTI) in transgenic rice significantly increased resistance to *C. suppressalis* and *Sesamia inferens* (Xu *et al.*, 1996). The results suggested that the cowpea trypsin inhibitor might be useful for the control of rice insect pests. Cowpea trypsin inhibitor gene CpTI has also been introduced into *Brassica oleracea* var. *capitata* cultivars Yingchun and Jingfeng (Fang *et al.*, 1997). The transformed plants showed resistance to *P. rapae* in laboratory tests. Vain *et al.* (1998) transformed 25 clones with genes coding for an engineered cysteine proteinase inhibitor (oryzacystatin–I delta D86, OC–I).

Adults of *Psylloides chrysocephala* fed identically on leaf discs from control or transformed plants of oilseed rape expressing constitutively oryzacystatin I (*OCI*) (Girard *et al.*, 1998). When larvae were reared on transgenic plants expressing *OCI*, they showed an increase in weight compared to control plants. Larvae fed on transgenic plants also exhibited a two-fold increase in both cysteine and serine proteolytic activity in response to the presence of *OCI*. On the contrary, transformed poplar (cv. Jean Pourtet) did not affect larval mortality, growth and pupal weights of *Lymantria dispar* and *Clostera anastomosis* (Confalonieri *et al.*, 1998).

3.2. α-Amylase Inhibitors

The α-amylase inhibitors of legume seeds expressed resistance against coleopteran seed weevils. Transgenic tobacco expressing *WAAI* (α-amylase inhibitors from wheat) gene reported to increase mortality of the lepidopteran larvae between 30 to 40% (Carbonero *et al.*, 1993). The bean (*Phaseolus vulgaris*) α-amylase inhibitor gene was expressed in seeds of transgenic garden pea (*Pisum sativum*) and other grain legumes, using a strong seed-specific promoter (Shade *et al.*, 1994). The resulting seeds contained up to 3% of the foreign protein and were resistant to larvae of the pea weevil, *Bruchus pisorum* (Morton *et al.*, 2000). This strategy was directed towards coleopteran seed herbivores (Table 3), which have a neutral or acidic gut pH, so the inhibitor is not inactivated leading to starch digestion. Deployment of transgenic crops expressing α–amylase inhibitor gene has not been exploited so far. Safety concerns have arisen as a result of the induction of systemic immunological responses in mice fed peas expressing the amylase inhibitor protein (Prescott and Hogan, 2005), which might be due to altered post-translational processing in pea compared to the natural bean host.

Table 3: Insect pest resistant transgenic crops expressing α-amylase inhibitors

Crop	*Transgene*	*Target pest*	*Reference(s)*
Pea	α–amylase inhibitors from *Phaseolus vulgaris*	Pea bruchids	Shade *et al*. (1994)
Tobacco	α–amylase inhibitors from wheat (*WAAI*)	Lepidopteran larvae	Carbonero *et al*. (1993)
	α–amylase inhibitors from bean (*BAAI*)	*Callosobruchus* sp.	Shade *et al*. (1994); Schroeder *et al*. (1995)
Adzuki beans	α–amylase	Bruchids	Ishimoto and Chrispeels (1996)
Pea	α–amylase inhibitors (alpha AI–1 and alpha AI–2)	Pea bruchid	Morton *et al*. (2000)

3.3. Lectins

Potential exploitation of lectin genes to confer insect resistance in transgenic plants (Table 4) has targeted hemipteran plant pests, which are not affected by known *Bt* toxins but were shown to be susceptible to lectin toxicity. Greater insecticidal activity was observed in chitin binding lectins from wheat germ and common bean. Transgenic maize expressing *WGA* has shown moderate activity against *O. nubilalis* and *Diabrotica* sp. (Maddock *et al*., 1991). However, mammalian toxicity of this lectin is high and it may not be therefore a good candidate for use in genetic transformation of crops. Sap sucking Hemipterans can be controlled by certain lectins (Hilder *et al*., 1995). A gene encoding the mannose specific lectin from snowdrop expressed in tobacco showed enhanced resistance to peach potato aphid (*Myzus persicae*). Transgenic potato expressing *GNA* had shown to be less susceptible to peach potato aphid (Gatehouse *et al*., 1996). However, the viability of the coccinellids was reduced when fed on aphids reared on transgenic plants.

Rice cystanin I in transgenic potato caused up to 53% mortality in larvae reared on transgenic leaves, compared to <17% mortality in control (Lecardonnel *et al*., 1999). Fitches *et al*. (1997) observed that at 21 days after hatch, mean larval biomass of the tomato moth (*Lecanobia oleracea*) was reduced by 32 and 23%, in the artificial diet containing GNA and excised leaves of transgenic tomato plants, respectively; survival decreased by nearly 40% in the glasshouse bioassay. Concanavalin A (ConA) inhibits development of tomato and peach–potato aphid when expressed in transgenic potato plants (Cao *et al*., 1999). Larvae of tomato moth fed on ConA in artificial diet exhibited decreased survival, but had only a small effect on larval weight (Gatehouse *et al*., 1999). Constitutive expression of *ConA* in transgenic potato retarded larval development and decreased the larval weight by >45%, but showed no significant effect on survival. Larvae of cotton budworm fed on transformed cotton tissues exhibited a reduction in weight, but the lectin did not result in larval mortality (Satyendra *et al.*, 1998). A 35% reduction in larval growth was observed following feeding on F_2 plants transformed with lectin A gene. Plants bearing lectin B had reduced larval weight and plants segregating for lectin C retarded larval growth.

Table 4: Insect resistant transgenic crops expressing plant lectin genes

Crop	*Transgene*	*Target insect pest*	*Reference(s)*
Tobacco	*GNA* from *Galanthus nivalis*	Peach potato aphid (*Myzus persicae*)	Boulter (1993)
Pea	Lectin	*Heliothis virescens*	Boulter (1993)
Cotton	*GNA*	Cotton bud worm, (*Heliothis virescens*)	Satyendra *et al.* (1998)
Rice	*GNA*	Brown plant hopper (*Nilaparvata lugens*) and green leafhopper (*Nephotettix virescens*)	Yang *et al.* (1998)
	GNA	Leafhopper (*Empoasca fabae*)	Habibi *et al.* (1992)
Potato	*GNA* and *ConA*	Peach potato aphid (*Myzus persicae*)	Gatehouse *et al.* (1999)
Tomato	*GNA*	Tomato moth (*Lecanobia oleracea*)	Fitches *et al.* (1997)
Maize	Wheat agglutinin (*WGA*)	*Ostrinia nubilalis* and *Diabrotica* spp.	Maddock *et al.* (1991)
Mustard	*WGA*	Mustard aphid (*Lipaphis erysimi*)	Kanrar *et al.* (2002)
Sugarcane	*GNA*	Sugarcane grubs, (*Antitrogus consanguineus*)	Nutt *et al.* (1999)

Transgenic sugarcane plants engineered to express either the potato proteinase inhibitor II (cv. UP 87) or the snowdrop lectin gene (cv. G 87) showed increased antibiosis to larvae of sugarcane grubs (*Antitrogus consanguineus*) in glasshouse trials (Nutt *et al.*, 1999). Plants transformed with a proteinase inhibitor from ornamental tobacco did not affect the weight gain of grubs. Snowdrop lectin at levels greater than 0.04% of total soluble protein decreased the fecundity, but not the survival of grain aphids (Stoger *et al.*, 1999). It is suggested that transgenic approaches using insecticidal genes such as GNA in combination with integrated pest management present promising opportunities for the control of wheat pests.

Expression of snowdrop lectin (*GNA*) in transgenic rice plants using constitutive or phloem-specific promoters showed partial resistance to brown plant hopper (*Nilaparvata lugens*) and other hemipteran pests.

Reductions of up to 50% in survival were observed, with reduced feeding, development, and fertility of survivors (Foissac *et al.*, 2000). Similar partial resistance to hemipterans has also been obtained by expression of lectin from garlic (*Allium sativum*) leaves (ASA-L) in transgenic rice (Saha *et al.*, 2006a). Transgenic rice plants expressing *ASA-L* were shown to decrease transmission of *Rice tungro virus* by its insect vector, presumably as a result of decreased feeding by the pest (Saha *et al.*, 2006b). Concerns about possible consequences to higher animals of ingesting snowdrop lectin have limited further progress, although a recent study incorporating a 90 day feeding trial found no adverse effects resulting from consumption of transgenic rice expressing *GNA* by rats (Poulsen *et al.*, 2007).

4. ENGINEERING INSECTICIDAL PROTEINS

4.1. *Photorhabdus luminescens* Insecticidal Proteins

Nematodes of *Heterorhabditis* spp. that contain symbiotic enterobacteria are widely used for small-scale biological control of insect pests. When nematodes enter an insect host, bacterial cells from the nematode gut are released into the insect circulatory system. Toxins secreted by the bacteria cause cell death in the insect host, leading to a lethal septicemia. *Photorhabdus luminescens*, the well-investigated bacterial species of this type, contains a large number of potentially insecticidal components (Ffrench–Constant, 2007). One of the orally toxic components, toxin A was selected for further study. The encoding gene *tcdA* was cloned and assembled into expression constructs, containing 5′ and 3′ un-translated region sequences from a tobacco osmotin gene to improve expression levels of mRNA and protein in transgenic plants. Expression of toxin A at levels >0.07% of total soluble protein in leaves of transgenic Arabidopsis (*Arabidopsis thaliana*) plants (Liu *et al.*, 2003) gave almost complete protection against larvae of the lepidopteran tobacco hornworm (*Manduca sexta*). Leaf extracts from these plants were also toxic to corn rootworm, showing cross-species protection. Commercial development of this technique is likely.

4.2. Cholesterol Oxidase

Bacterial cholesterol oxidase has an insecticidal activity comparable to *Bt* toxins, dependent on its enzyme activity, which is thought to promote membrane destabilization. Expression constructs containing part or all of the coding sequence of the protein, or the coding sequence fused to a chloroplast–targeting peptide, resulted in production of active enzyme in transgenic tobacco (Corbin *et al.*, 2001). However, phenotypic abnormalities were observed in transgenic plants unless the enzyme was localized in chloroplasts, possibly as a result of interference with steroidal signaling pathways. Leaf tissue from all transgenic plants was toxic to boll weevil larvae. The cholesterol oxidase gene appears to be an obvious candidate for introduction into the chloroplast genome rather than the plant nuclear genome, which would avoid potential problems caused by enzyme activity in the cytoplasm.

4.3. Avidin

Avidin has a strong insecticidal effect on many insects, although susceptibility varies widely between different insect species (apparently based on biotin requirements). Expression of avidin in transgenic maize initially aimed to produce the protein as a high-value product, but maize seed containing more than 0.1% avidin (of total protein) was fully resistant to larvae of three different coleopteran storage pests (Kramer *et al.*, 2000). The protein has also been expressed in other transgenic plants to confer resistance. Targeting of the foreign protein (*e.g.* targeting sequences of potato proteinase inhibitors) away from the cell cytoplasm (Murray *et al.*, 2002) is important to avoid developmental abnormalities in the plants. No further development of this promising method has been reported.

5. ENGINEERING PRIMARY AND SECONDARY METABOLISM ENZYMES

Transgenic expression of certain enzymes can be another alternative to *Bt* genes. One of the most important enzymes is chitinase, which is an important component of insect integument. Transgenic tobacco plants expressing chitinase have shown increased resistance to lepidopteran insects (Ding *et al.*, 1998; Gatehouse *et al.*, 1995). A chitinase from *Serratia marcescens* showed synergistic action with Bt toxin against *S. littoralis* (Rigev *et al.*, 1996). Corbin *et al.* (1994) reported the expression of cholesterol oxidase gene in bacteria and plant protoplasts. Cholesterol oxidase from *Streptomyces* is highly toxic to cotton boll weevil (*Anthonomus grandis*) (Cho *et al.*, 1995). Lipoxygenase exhibited toxic effects (Shukle and Murdock, 1983) and expressed in transgenic plants. Polyphenol oxidases and peroxidases increased the inhibitory effect of 5CQA (5-Caffeoyl quinic acid) and cholorogenic acid by oxidising the dihydroxy groups to ubiquinones that covalently bind to nucleophilic (–SH2 and NH2) groups of proteins, peptides and amino acids.

Use of bacterial isopentenyl transferase (*ipt*) gene, involved in cytokinin biosynthesis, has been fused with a promoter from the proteinase inhibitor II (*PI–IIK*) gene and introduced into *Nicotiana plumbaginifolia* (Smigocki *et al.*, 1993). *Manduca sexta* larvae consumed up to 70% less of the PI–II–

ipt leaf material on flowering plants than the larvae feeding on control plants; development of *Myzus persicae* nymphs was also delayed. Approximately half as many nymphs reached adulthood on PI–II–ipt leaves than on controls. Zeatin and zeatinriboside levels in leaves remaining on PI–II–ipt plants after hornworm feeding were elevated by about 70-fold and the chlorophyll a/b content was twice to that of controls.

Manduca sexta larvae feeding on tomato plants, constitutively expressing a prosystemin antisense gene, had approximately 3 times higher growth rates than the larvae feeding on nontransformed control plants (Orozco–Cardenas *et al.*, 1993). The levels of proteinase inhibitor I and inhibitor II proteins in leaves of tomato plants expressing the antisense prosystemin gene remained at undetectable levels until the sixth day of larval feeding and then increased throughout the plants (100 to 125 μg per g of leaf tissue after 14 days). In control plants, levels of proteinase inhibitor I and II proteins increased rapidly from the second day of larval feeding, and by the eighth day contained levels of 225 and 275 μg per g of leaf tissue, respectively, and then increased slowly thereafter. Prosystemin mRNA levels in antisense and control plants after 6 and 12 days of larval feeding correlated with levels of inhibitor I and II protein levels. These experiments demonstrate that resistance of plants toward an insect pest can be modulated by genetically engineering a gene encoding a component of the inducible systemic signalling system regulating a plant defensive response.

The availability of genes encoding the biosynthetic enzymes of secondary metabolism has made transfer of biosynthetic pathways between plants feasible. Genes encoding two Cyt P450 oxidases and a UDP-glycosyltransferase from sorghum (*Sorghum bicolor*) have been transferred to Arabidopsis (Tattersall *et al.*, 2001), resulting in production of the cyanogenic glycoside dhurrin (Kristensen *et al.*, 2005). The resulting plants produced hydrogen cyanide on tissue damage and showed enhanced resistance to attack by the flea beetle, *Phyllotreta nemorum*, a specialist feeder on crucifers. Other secondary metabolites that have been produced in transgenic plants include the alkaloid caffeine in tobacco by the introduction of three genes encoding *N*-methyl transferases (Kim *et al.*, 2006).

6. ENGINEERING OF VOLATILE COMMUNICATION COMPOUNDS

Engineering volatiles emitted by plants offers possibilities for new methods of crop protection. Volatile composition has been altered in tobacco by RNA interference (RNAi)–mediated suppression of a *cytP450* oxidase gene expressed in trichomes, and in Arabidopsis by constitutive overexpression of a plastid dual linalool/nerolidol synthase gene (Wang *et al.*, 2001). The transgenic plants deterred aphid colonization but were not wholly resistant. Volatiles can also be used as attractants for natural enemies of pests, *e.g.*, Arabidopsis plants transformed with the maize terpene synthase gene *TPS10* emitted the sesquiterpene volatiles normally produced in maize and attracted parasitoid wasps that attack maize pests (Schnee *et al.*, 2006). Volatiles used by insects to communicate with each other can also be exploited. The sesquiterpene (*E*)–β–farnesene is an alarm pheromone in aphids and attracts aphid predators and parasitoids. When Arabidopsis was transformed with an (*E*)–β–farnesene synthase gene from mint (*Mentha piperita*), the resulting plants showed significant levels of aphid deterrence in choice experiments and were attractive to the aphid parasitoid *Diaeretiella rapae* (Beale *et al.*, 2006).

7. GENE PYRAMIDING

Combining conventional host plant resistance with *Bt* genes can provide a germplasm base to achieve durable resistance to insect pests (Table 5). Transgenic cotton with *Bt* + *GNA* conferred resistance to bollworm, *H. armigera* and the cotton aphid, *A. gossypii* (Liu *et al.*, 2003). Segregation of resistant and susceptible plants in F_2 and BC_1 populations fitted the 3:1 and 1:1 ratios, respectively; indicating that resistance to bollworm was controlled by one pair of dominant genes and inherited in Mendelian manner with no cytoplasmic effects. The *Bt* gene and *OC* gene have been co–transformed to tobacco chloroplast with the particle bombardment method. Transgenic tobacco containing both genes had enhanced toxicity to the larvae of corn earworm, *H. zea*, as compared to plants containing *Bt* or *OC* alone. The *Bt* and *OC* genes were inherited maternally (Su *et al.*, 2002). Sugarcane plants transformed with both the GNA and a proteinase inhibitor from *Nicotiana alata* Link and Oho, caused a significant negative effect on the weight gain of *Dermolepida albohirtum* larvae (Nutt *et al.*, 2001). Gahakwa *et al.* (2000) studied the stability of transgene expression in 40 independent rice plants representing 11 diverse cultivated varieties. Each line contained three or four different transgenes delivered by particle bombardment, either by co-transformation, or in the form of a co-integrate vector. Approximately 75% of the lines showed Mendelian inheritance of all transgenes, suggesting integration at a single locus. The levels of transgene expression varied among different lines, but primary transformants showing high-level expression of the *GNA*, *gusA*, *hpt*, and *bar* genes were transmitted to the progeny. Six transgenes (three markers and three insect resistance genes) were stably expressed over four generations of transgenic rice plants. Transgene expression was stable, and this represented a step toward genetic engineering from model varieties to elite breeding lines grown in different parts of the world.

Table 5: Gene pyramiding to increase the effectiveness of transgenic plants for pest management

Transgene(s)	***Result***	***Reference(s)***
Serine PIs + *Bt*	Enhances activity of *Bt* genes	Macintosh *et al.* (1990)
Tannic acid + *Bt*	Increases activity of *Bt* genes	Gibson *et al.* (1995)
cry1Ac + *CpTi*	More effective than plants expressing *Bt* gene alone	Zhao *et al.* (1998)
cry1Ac + *cry1C*	Plants producing both toxins caused rapid mortality	Cao *et al.* (1999)
cry1Ac + *cry1Ab*	Showed greater insecticidal activity	Stewart *et al.* (2001)

8. RNA INTERFERENCE (RNAI)

RNA interference already proved its usefulness in functional genomic research on insects having considerable potential in crop protection (Huvenne and Smagghe, 2010). Disrupting gene function by the use of RNAi is a well-established technique in insect genetics based on delivery by injection into insect cells or tissues (Table 6). The key to the success of this approach is: (i) identification of a suitable insect target and (ii) dsRNA delivery, which includes *in planta* expression of dsRNA and delivery of sufficient amounts of intact dsRNA for uptake by the insect (Turner *et al.*, 2006). Although different approaches were used for the generation of insect-resistant plants, careful target selection was common to both (Fig. 1).

Table 6: Development of RNAi in insects and RNAi mediated insect pest management

Target insect-pest	***Gene(s)***	***Methods***	***Effects***	***Reference(s)***
Egyptian cotton leafworm (*Spodoptera littoralis*)	β–*actin*	Injection	Disruption of sperm release	Gvakharia *et al.*, 2003
	Circadian clock gene *per*	Injection	Delayed sperm release	Kotwica *et al.*, 2009
Light brown apple moth (*Epiphyas postvittana*)	*EposCXE1* (Carboxy-lesterase) and *EposPBP1* (Pheromone binding protein)	Feeding	Inhibition of gene expression	Turner *et al.*, 2006
Cotton bollworm (*Helicoverpa armigera*)	Cytochrome P450 gene *CYP6AE14*	Feeding	Inhibition of larval growth	Mao *et al.*, 2007

Table 6: (*Contd...*)

Table 6: (*Contd...*)

Target insect-pest	***Gene(s)***	***Methods***	***Effects***	***Reference(s)***
	Glutathione-S-transferase gene *GST1*	Feeding	Inhibition of gene expression	Mao *et al.*, 2007
Beet armyworm (*Spodoptera exigua*)	Chitin synthase gene	Injection	Disorder in the insect cuticle, no expansion of the larval tracheal epithelial wall, and other larval abnormalities	Chen *et al.*, 2008
Japanese pine sawyer (*Monochamus alternates*)	Laccase gene *MaLac2*	Injection	Pupal and adult cuticle sclerotisation, death at a high dose	Niu *et al.*, 2008
Red flour beetle (*Tribolium Castaneum*)	Chitin synthase genes *TcCHS1* and *TcCHS2*	Injection	Disruption in all types of moulting (larva-larva, larva-pupa, and pupa-adult), cessation of ingestion, decrease in larval size, and reduction of chitin content in the midgut	Arakane *et al.*, 2005
	Chitinase–like proteins *TcCHT5*, *TcCHT10*, *TcCHT7*, and *TcIDGF4*	Injection	Effects on pupal-adult moulting, egg hatching, larval moulting, pupation, and adult metamorphosis, abdominal contraction and wing/elytra extension and adult eclosion	Zhu *et al.*, 2008
Western corn rootworm (*Diabrotica virgifera* LeConte)	Vacuolar ATPase (*v–ATP*)	Feeding	Delayed larval development and increased mortality	Baum *et al.*, 2007

Table 6: (*Contd...*)

Table 6: (*Contd...*)

Target insect-pest	***Gene(s)***	***Methods***	***Effects***	***Reference(s)***
Striped flea beetle (*Phyllotreta striolata*)	Arginine kinase gene *AK*	Feeding	Delayed development, increased mortality, and reduced fertility	Zhao *et al.*, 2008
Mediterranean field cricket (*Gryllus bimaculatus*)	Circadian clock gene *per*	Injection	Complete loss of circadian control of locomotor activity and electrical activity in the optic lobe	Moriyama *et al.*, 2008
	Nitric oxide synthase gene *NOS*	Injection	Destruction of long-term memory	Takahashi *et al.*, 2009
German cockroach (*Blattella germanica*)	*BgRXR* gene	Injection	Inhibition of pupal eclosion	Martin *et al.*, 2006
	Pigment-dispersing factor gene *pdf*	Injection	Effects on insect night activity	Lee *et al.*, 2009
American grasshopper (*Schistocerca americana*)	Eye colour gene *vermilion*	Injection	Suppression of ommochrome formation and systematic expression	Dong and Friedrich, 2005
Brown planthopper (*Nilaparvata lugens*)	Trehalose phosphate synthase (*TPS*) (*NlTPS* mRNA)	Feeding	Disturbed development through disruption in the TPS enzymatic activity, reduction of insect survival rate	Chen *et al.*, 2010
Turnip sawfly (*Athalia rosae*)	*Ar white* gene	Injection	White phenocopy inembryonic eyepigmentation	Sumitani *et al.*, 2005
European honey bee (*Apis mellifera*)	Transcription factor gene *Relish*	Injection	Inhibition of *Relish* gene expression and reduction in the expression of two other immune genes, *abaecin* and *hymenoptaecin*	Schlüns and Crozier, 2007

Table 6: (*Contd...*)

Table 6: (*Contd...*)

Target insect-pest	*Gene(s)*	*Methods*	*Effects*	*Reference(s)*
Triatomid bug (*Rhodnius prolixus*)	Salivary nitrophorin 2 gene *NP2*	Injection and feeding	Shortened plasma coagulationtime	Araujo *et al.*, 2006
Savannah tsetsefly (*Glossina morsitan smorsitans*)	*TsetseEP* gene and transferrin gene *2A192*	Feeding	Inhibition of *TsetseEP* gene expression, but no inhibition of *2A192* gene expression	Walshe *et al.*, 2009
Eastern subterranean Termite (*Reticulitermes flavipes*)	Cellulase enzyme gene *Cell–1* and casteregulatory Hexamerin storage protein gene *Hex–2*	Feeding	Reduction in group fitness and Increased mortality	Zhou *et al.*, 2008

Transgenic maize producing dsRNA directed against V-type ATPase of corn rootworm showed suppression of mRNA maturation in the insect and reduction in feeding damage compared to controls (Baum *et al.*, 2007). Similarly, transgenic tobacco and Arabidopsis expressing dsRNA directed against a detoxification enzyme (Cyt P450 gene *CYP6AE14*) for gossypol in cotton bollworm caused the insect to become more sensitive to gossypol in the diet (Mao *et al.*, 2007). When the striped flea beetle (*Phyllotreta striolata*) was fed with dsRNA, the invertebrate specific phosphotransferase arginine kinase was successfully silenced in the gut, and due to disruption of cellular energy homeostasis, the development of the beetle was severely impaired (Zhao *et al.*, 2008). Bautista *et al.* (2009) studied the influence of silencing the cytochrome P450 gene *CYP6BG1* that is over-expressed in a permethrin-resistant diamondback moth (*Plutella xylostella*) strain. When the gene was silenced after consumption of a droplet of dsRNA solution, the moths became significantly more sensitive to the pyrethroid insecticide. When eastern subterranean termites (*Reticulitermes flavipes*) were fed with cellulose discs supplemented with dsRNA, silencing of the digestive cellulose enzyme and caste regulatory hexamerin storage protein led to reduced termite fitness and increased mortality (Zhou *et al.*, 2008). Along with it transgenic insects have also been tested as a mechanism for pest control (Scolari *et al.*, 2011). For both transgenic plants and insects, very few species have been investigated; therefore, further research is essential to explore the potential use of RNAi as an effective means for pest control.

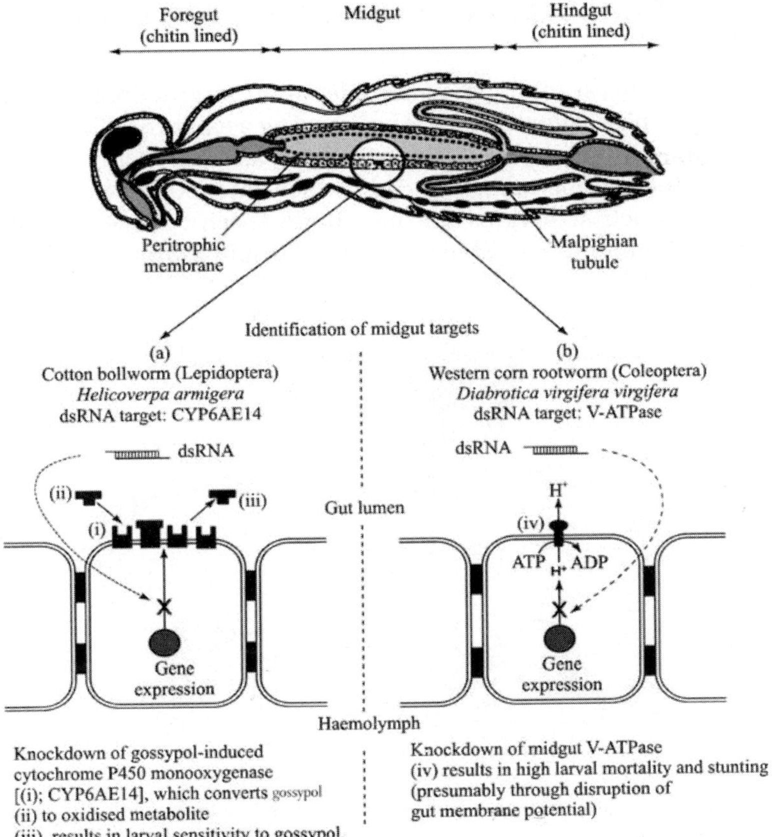

Fig. 1: *RNAi based approaches for generating insect-resistant transgenic plants.* Double-stranded RNA (dsRNA) produced *in planta* can lead to targeted gene silencing in Lepidoptera and Coleoptera pest species. dsRNAs corresponding to specific insect targets are expressed *in planta* and are cleaved by endogenous plant Dicer enzymes to produce short interfering RNAs (siRNAs) of around 21 nucleotides. Large dsRNA and siRNA cleavage products are expressed throughout plant tissues and are orally delivered to insect herbivores feeding on transgenic plant material. For gene–silencing to initiate in targeted insect pests, large dsRNAs and siRNAs must persist in the insect gut, and sufficient quantities must be present for uptake into cells in contact with RNAs. Approach (a): a gut–specific cytochrome monooxygenase, *CYP6AE14*, has been identified (i) whose expression correlates with larval growth on diets containing gossypol (ii) a cotton secondary metabolite *CYP6AE14* is presumably involved in detoxification of gossypol (iii) because specific knockdown of this gene product by dsRNAs delivered in artificial diet and by transgenic plant material increases larval sensitivity to gossypol. Approach (b): a screening approach to identify a lethal phenotype in *Diabrotica virgifera* when midgut V-type ATPase A (V-ATPase) (iv) was downregulated by dsRNAs delivered in artificial diet feeding trials and transgenic corn which results in disruption of electrochemical gradient across the gut epithelia and ultimately high larval mortality (*Source*: Price and Gatehouse, 2008).

9. GENETIC ENGINEERING FOR RESISTANCE AGAINST NEMATODES

With the new understanding of nematode–plant interactions at molecular level (Fig. 2), new avenues for engineering resistance have opened. Nematode offers a variety of targets for such an approach, including the exoskeleton, the feeding tube, digestion, essential metabolic processes and reproduction. Such strategies are nematode specific, pose lesser risk to man and environment, can show resistance against broader spectrum of PPN and durable.

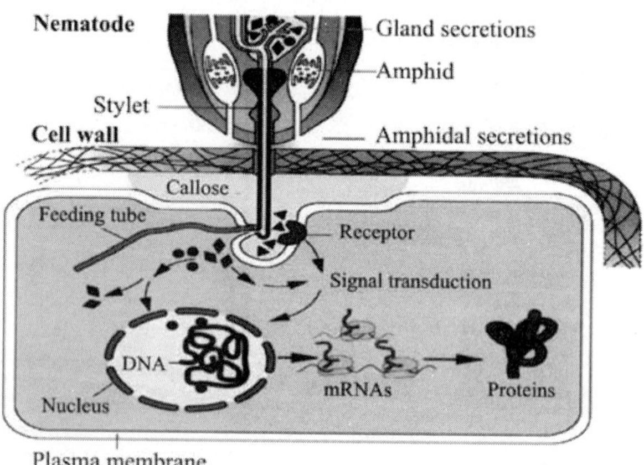

Fig. 2: Cell and gene responses of plant root due to esophageal gland secretions through nematode stylet. Effectors secreted by the stylet leads to the successful establishment of feeding cells (giant cells), essential for root-knot nematode development and reproduction. Giant cells are metabolically hyperactive and form a nutrient sink for the nematode around which root galls are formed.

9.1. Cloning of Resistance Genes

Resistance in nematology is considered to be the ability of the host plant to prevent multiplication of the parasite. The resistance genes prevent nematodes to establish, feed or develop and reproduce in crop plants. In many cases the host-nematode interaction is mediated by single resistance genes (R genes) in the host and single avirulence (*Avr*) genes in the nematode. Recognition initiates a chain of defense reactions, often including a hypersensitive response (HR) consisting of localized cell necrosis at the infection site. New sources of resistance that either arise spontaneously in cell culture or are specifically engineered can be exploited in a wide range of crops (Table 7). Tables 8 and 9 depict mapping of all the nematode resistance genes in different crop plants reported so far using RFLP and RAPD technique.

Table 7: Somatic hybridization in different crops to develop resistance against various PPNs

Crops	*Species*	*Nematodes*	*Reference(s)*
Brassica	*Sinapsis alba* and *Brassica napus*	*Heterodera schachtii*	Lelivelt *et al.*, 1993a
	Raphanus sativus and *B. napus*	*H. schachtii*	Lelivelt *et al.*, 1993b
Potato	*Solanum tuberosum* and *S. bulbocastanum*	*Globodera rostochiensis*	Austin *et al.*, 1993; Mollers and Wenzel, 1991
		Meloidogyne chitwoodi race1	Brown *et al.*, 1996
	S. tuberosum PDH 40 and PDH 417	*Globodera pallida* Pa2 and Pa3	Cooper–Bland *et al.*, 1993

The first resistance (R) gene to be cloned against nematode was the *Hs1*$^{pro-1}$ from sugar beet against *H. schachtii* (Cai *et al.*, 1997). This R gene was isolated from *Beta procumbens*, a wild species of beet. The deduced amino acid sequence of *Hs1*$^{pro-1}$ predicts a small protein with a possible amino terminal signal sequence and a possible membrane spanning region. The predicted mature protein of 282 amino acids contains a leucine rich region at its amino terminus, but this does not fit the pattern of leucine rich repeat (LRR) (Ellis and Jones, 1998). The second nematode resistance gene to be cloned was the *Mi* gene from tomato against the root-knot nematode species, *M. incognita*, *M. javanica* and *M. arenaria* (Williamson, 1998; Milligan *et al.*, 1998). Transgenic plants with *Mi* also show resistance against the potato aphid, *Macrosiphum euphorbiae* (Rossi *et al.*, 1998) and the white fly, *Bemisia tabaci*. *Mi* encodes a protein of 1257 amino acids that is a member of the plant resistance protein family characterized by the presence of nucleotide binding site (NBS) and LRR domains (Williamson, 1999). The commercial potential of this gene was studied by *Agrobacterium* mediated transformation and disease tests of susceptible lines of tomato, tobacco, eggplant and lettuce. Segregation for resistance against both *M. incognita* and *M. euphorbiae* was demonstrated in progeny of transformed tomato line, while transformed cultivars of eggplant showed resistance only against *M. incognita*. Stable inheritance of resistance was demonstrated through three generations for each of the four transgenic lines of eggplant.

Another nematode resistance gene cloned is the *Hero* gene of tomato which confers wide spectrum resistance to potato cyst nematodes, over 95% against *Globodera rostochiensis* and over 85% against *G. pallida*. It could be very valuable for incorporating PCN resistance in potato cultivars. Although these R-genes are used in many breeding programs, their effectiveness is limited. For example, the *Gro-1*; formerly *H1* resistance gene from *Solanum tuberosum* spp. *andigena* is present in more than 50 European potato cultivars to obtain resistance against *Globodera rostochiensis*, but is effective against only two of the eight 'pathotypes' of the potato cyst nematode. Gene *Gro1–4* confers resistance to *G. rostochiensis* pathotype Ro1. The deduced *Gro1–4* protein differed by 29 amino acid changes from susceptible members of the *Gro1* gene family.

Table 8: RFLP for mapping nematode resistance gene in different crop species

Crop	*R–gene*	*Nematode*	*RFLP marker*	*Reference(s)*
Potato	H1	*G. rostochiensis* pathotypes Ro1 and Ro4	2.7 cM from H1	Pineda *et al.*, 1993
	Gro1	Ro1 and Ro5	Closely linked	Ballvora *et al.*, 1995
	Gro VI	Ro1	Mapped on chromosome 5	Jacobs *et al.*, 1996
	Gpa2	*G. pallida* pathotyes Pa2 and Pa3	Mapped on chromosome 10 and 11	Kreike *et al.*, 1994
	Gpa2 + Rx	*G. pallida* + Potato virus X	115 kb apart	Van der Vossen *et al.*, 2000
	Rmc1	*M. chitwoodi*	Chromosome 10	Brown *et al.*, 1996
Tomato	Mi–1	*M. incognita*, *M. javanica*, *M. arenaria*	Chromosome 6	Ho *et al.*, 1992
	Mi–3	*M. incognita*, *M. javanica*	Linkage with NR14 and Mi–3	Yaghoobi *et al.*, 1995
	Mi–1	*M. incognita*	550 kb apart from Mi–1 along with AFLP	Kaloshian *et al.*, 1998
Soybean	Rhg4	*Heterodera glycines* race 3	1 cM apart from Rhg4 within linkage group A	Webb *et al.*, 1995
	Rhg1	*H. glycines* race 3 and 5	Detected polymorphism	Skorupska *et al.*, 1994
Wheat	Cre1	*H. avenae*	Mapped the gene along with STS	Williams *et al.*, 1994

Gpa2, a gene derived from *S. tuberosum* ssp. *andigena* conferring resistance to some isolate of potato cyst nematode, *Globodera pallida*, has been cloned (Rouppe *et al.*, 1999). It is physically tightly linked to the virus resistance gene *Rx* on chromosome 12 of potato. Although *Rx* and *Gpa2* confer resistance to unrelated pathogens, they seem to share a common ancestral gene as they are members of the same cluster and share a high degree of homology. This indicates they share components of the same signal transduction pathway.

Table 9: RAPD for mapping nematode resistance gene in different crop species

Crop	*R–gene*	*Nematode*	*RAPD marker*	*Reference(s)*
Sugarbeet	$Hs1^{pat1}$	*H. schachtii*	Closely linked with $Hs1^{pat1}$	Salentijn *et al.*, 1995
Citrus		*Tylenchulus semipenetrans*	Along with SCAR identified linkage with R–gene of *Citrus tristeza* virus	Ling *et al.*, 2000
Carrot	Mj–1	*M. javanica*	Minimum 0.8 cM to maximum 5.7 cM apart from Mj–1	Boiteux *et al.*, 2000
Cotton	rkn1 and Mi–2	*M. incognita*	along with STS closely linked with rkn1 and Mi–2	Niu *et al.*, 2007

In resistant tomato no giant cell was initiated, though the nematode enters and migrates to feeding site. A localized necrosis, suggesting a HR, was visible near the anterior end of the nematode within a few days of infection. The resistance mediated by $Hs1^{pro-1}$, did not appear to involve a hypersensitive response (Jung, 1998). The nematodes died in the late J2 stage because of degradation of the feeding structure (syncytium) in the host. Cloned genes offer a potential for transfer of resistance to other related crop plants which may be expected to harbour suitable intracellular signalling pathways and therefore be competent to mount a resistance response.

9.2. Gene Silencing

The molecular cloning of host genes and promoters involved in the formation of feeding sites (mediates effective transfer of nutrients from the plants to the feeding nematodes) could facilitate the production of antisense constructs to inhibit expression of key genes. The antisense RNA is thought to base pair with its target RNA forming double stranded RNA. This duplex formation may inhibit mRNA maturation and/or translation or it may lead to rapid mRNA degradation.

With the recent advent of gene expression control via small interfering RNA (siRNA) and micro RNA (miRNA) molecules, biologists are exploring genes and development from a new perspective (Baulcombe, 2007). Understanding of this ubiquitous phenomenon has revealed RNA interference (RNAi) as a powerful tool to manipulate gene expression and analyse gene function (Fire *et al.*, 1998). Induction of RNAi by delivering double-stranded RNA (dsRNA) has been very successful in the model non-parasitic nematode *Caenorhabditis elegans*, and the majority of its genes have been characterized using this technique (Kamath *et al.*, 2003). Delivering dsRNA to PPNs can be accomplished by soaking the nematodes with dsRNA solution mixed with the neurotransmitters like resorcinol, octopamine, DMT, serotonin etc. Downregulation of target genes in plant-parasitic nematodes was first accomplished by octopamine-stimulated oral ingestion of dsRNA by pre-parasitic J2s of the cyst nematodes *Globodera pallida* and *H. glycines* (Urwin *et al.*, 2002). Using similar *in Vitro* dsRNA delivery approaches, downregulation of various housekeeping genes leading to reduced parasitic ability, delayed egg hatching, impaired motility, and ability to locate and invade roots has been demonstrated in both root-knot and cyst nematode (Urwin *et al.*, 2002; Bakhetia *et al.*, 2005, 2007; Chen *et al.*, 2005; Fanelli *et al.*, 2005; Lilley *et al.*, 2005; Rosso *et al.*, 2005; Shingles *et al.*, 2007).

The success of the above *in vitro* dsRNA ingestion and down-regulation of the target genes inspired the *in planta* delivery of dsRNA to the feeding nematodes. Although both root-knot and cyst nematodes were found to be responsive to *in vitro* dsRNA uptake and down-regulation of the target gene, the most convincing success of *in planta* delivery of dsRNA to feeding nematodes came from root-knot nematodes (Huang *et al.*, 2006; Yadav *et al.*, 2006). Huang *et al.* (2006) reported significant reduction of gall formation by different root-knot nematode species on transgenic *Arabidopsis* plants expressing dsRNA for the parasitism gene 16D10, which is unique to root-knot nematodes. Similarly, tobacco RNAi lines expressing dsRNA for two root-knot nematode housekeeping genes (a splicing factor and an integrase gene) induced down-regulation of the target genes and reduced parasitic success (Yadav *et al.*, 2006). More recently, a virus-induced gene silencing approach (VIGS) has been used to attempt delivery of siRNA to the feeding cyst nematodes, *H. schachtii* (Valentine *et al.*, 2007). The study did demonstrate the selective uptake of syncytium-expressed monomeric red fluorescent protein (mRFP) and down-regulation of the target gene, glyceraldehyde–3–phosphate dehydrogenase (GADPH), in the feeding cyst nematodes effectively reducing nematode parasitic success.

9.3. Plantibodies

Immunomodulation is a molecular technique that allows the interference with cellular metabolism or pathogen infectivity by the ectopic expression of genes encoding antibodies or antibody fragments. Thus plants can be engineered to produce functional antibodies or antibody fragments known as plantibodies to protect it from nematode infection. Sedentary endoparasites alter plant cell cycle and cell differentiation to form feeding cells, and the chemicals that redirect the fate of the plant cells are thought to originate from the nematode pharyngeal glands. Therefore, inactivation of the active components in these secretions would result in the nematode failing to initiate a suitable feeding cell. The heavy and light chains of a murine monoclonal antibody specific to stylet secretions of *M. incognita* were expressed in tobacco (Baum *et al.*, 1996).

Plantibodies have also been generated that recognize secretory-excretory proteins on the cuticle surface and the amphids of *Globodera pallida* and *G. rostochiensis*. Plantibodies affected nematode movement and had resulted delayed root penetration (Fioretti *et al.*, 2002). The effectiveness of plantibodies against nematode parasitism is yet to be demonstrated although it has been used successfully to combat other plant pathogens. Optimization of immunomodulation technique can make it a powerful and attractive tool for gene inactivation, complementary to the classical antisense and co-suppression approaches.

9.4. Protease Inhibitors

Proteases/proteinases/proteolytic enzymes are a group of enzymes whose catalytic function is to hydrolyze peptide bonds of proteins. There are four classes of proteases (cysteine, serine, aspartyl and metallo) which differ in their ability to hydrolyze various peptide bonds. Each type of proteases, *e.g.*, fungal protease, pepsin, trysin, chymotrypsin, papain, bromelain, and subtilisin differs in their ability to break specific kind of peptide bonds (Burrows and de Waele, 1997).

Nematodes need proteinases to digest protein in food to amino acid for protein synthesis. A strategy to block the genes or produce protease inhibitors (PI) in host plants can starve the nematodes to death. Protease inhibitors of all classes occur in plants and have been constituent of human diet, thus a safe alternative. The potential of this strategy was first demonstrated for serine PI, cowpea trypsin inhibitor (*CpTI*). *CpTI* expressed in transgenic potato impaired early growth of *G. pallida* and influenced its sexual fate towards male juveniles. A modified rice cysteine PI, Oc–IΔD86 (cystatin) expressed as a transgene in *Arabidopsis thaliana*, had a profound effect on the size and fecundity of females of both *Heterodera schachtii* and *M. incognita*. No females of either species achieved minimum size required for egg production. The rice genes, *Oc-I* and *Oc-II* were the first plant cystatins to be cloned. Thus use of PI as nematode antifeedant approach offers prospects for novel plant resistance and reduced use of environmentally damaging nematicides (Atkinson *et al.*, 2003; Lilley *et al.*, 2011).

Constructs based on a cystatin from sunflower and a protein engineered variant of a rice cystatin (Oc–IΔD86) conferred similar level of resistance on potato plants to *Globodera* spp. as chicken egg white cystatin (CEWC) under the control of CaMV35S. In containment trials, the reproductive success of *H. schachtii*, *M. incognita* and *Rotylenchulus reniformis* was reduced by approximately 70% in *Arabidopsis* expressing the modified cystatin. A comparable degree of resistance was also observed in transgenic potatoes, rice and banana (Atkinson, 2003; Lilley *et al.*, 2004).

9.5. Targeting the Nematode Feeding Site

An alternative strategy to target the nematode directly is to disrupt the feeding sites that sedentary endoparasites initiate within the plant. Once the J2 sets up a feeding site it becomes sessile and its continued existence and development are entirely dependent on the feeding cells. Therefore, killing the feeding cell would result in the death of the nematode and thus confer resistance to host plant. Theoretically, it can be achieved by expressing a cell death gene specifically within the feeding site upon its induction. Primarily this approach requires a promoter that drives the cell death system specifically in the nematode feeding cell. The concept of using promoters of genes upregulated in nematode feeding sites and linking them to effectors that could target either the nematode or the feeding site itself was clearly stated by Gurr *et al.* (1991).

The cytotoxic barnase (an extracellular ribonuclease produced by *Bacillus amyloliquefaciens*) is expressed using a feeding cell induced promoter, whereas the neutralizing gene encoding barstar (intracellular inhibitor of barnase) is expressed using a constitutive promoter that is down-regulated in the feeding cell. Utility of barnase was first demonstrated in plant cell death system for emasculation in maize from tapetum specific promoter, but its potency led to death of non-target cells through leaky expression. To overcome this inhibitor protein barstar was applied using a constitutive promoter, CaMV35S. The restorer barstar has been expressed constitutively throughout the plant to avoid non-specific barnase damage in non-target cells of the plant. Some resistance to both *M. incognita* and *H. schachtii* has been reported using this approach. Further research on this aspect is required with respect to specificity of gene function at the feeding cell level and also to check the damage of non-target cells as promoters like CaMV35S are often not truly constitutive, but are down-regulated in the older regions of roots (Mariani *et al.*, 1992; Ohl *et al.*, 1997).

9.6. Ribosome Inactivating Proteins

Ribosome-inactivating proteins (RIPs) are N-glycosidases that depurinate the conserved α–sarcin loop of large rRNAs, resulted in inactivation of the ribosome thus blocking protein synthesis. The RIPs have great potential as cell-killing agents as their ability to inhibit translation. Type 2 RIPs like ricin and abrin are highly toxic heterodimeric proteins as the enzymatic chain is linked to a lectin, facilitating uptake into the cell. In transgenic tobacco plants both the recombinant RIP and the two component RIP system

(RIPα and RIPβ), under the control of promoters predominantly active in the root or nematode feeding site, had inhibitory effect on the development of *M. javanica* females (Thomas and Cottage, 2006).

9.7. Protein Engineering for Expressing Nematode Toxins

Exotoxins and endotoxins produced by the soil bacterium, *Bacillus thuringiensis* have the ability to kill insects. Endotoxins are encoded by *Cry* group of genes, used to develop transgenics conferring resistance against Lepidoptera and Dipteran insects. Cry proteins are bigger in size and as a result get excluded from the oral uptake by the feeding tubes of cyst, root-knot and presumably other PPNs. Deployment of protein engineering technique to reduce the size of the toxin molecule and its receptor specificity can unravel better management options of PPNs. Feeding tubes formed in the feeding cells act as molecular sieves or ultra-filtration devices in species specific manner and limit the maximum size of antinematode effectors that can be delivered orally to PPNs. *H. schachtii* excludes dextrans of 40kDa but not 22kDa, proteins of 22kDa and 28kDa, but not 11kDa, are also excluded by this nematode(Burrows and de Waele, 1997).

9.8. Activation of Signal Transduction Cascade

Peroxidase activity associated with nematode invasion produce bursts of the reactive oxygen intermediate hydrogen peroxide affecting both the nematode and plant cells. The transient increase in H_2O_2 may act as an antimicrobial agent and could have some effects on the invading nematodes. This transient increase initiates other defensive responses through lipid peroxidation releasing phytoalexins. Transgenic tomato plants expressing annexin like cDNA of *Arabidopsis thaliana* have showed reduced penetration of *M. incognita* (Ohl *et al*., 1997).

9.9. Lectins

Lectins are proteins or glycoprotein substances, usually of plant origin, of non-immunoglobulin nature, capable of specific recognition of and reversible binding to, carbohydrate moieties of complex glycoconjugates without altering the covalent structure of any of the recognized glycosyl ligands. Lectins interaction with certain carbohydrates is as specific as the enzyme-substrate or antigen-antibody interactions. Lectins may bind with free sugar or with sugar residues of polysaccharides, glycoproteins, or glycolipids which can be free or bound. Expression of snowdrop (*Galanthus nivalis*) lectin (GNA) in potato and oil-seed rape confers partial resistance to cyst nematodes and in *Arabidopsis* to root-knot nematodes (Ripoll *et al*., 2003). Lectin may bind glycoproteins localized on the surface of the nematodes, on the chemoreceptors from the amphids, or in the amphidial secretions and interfere with the nematode sensory perception and its ability to establish feeding cells.

10. PLANT BIOTECHNOLOGY: SUCCESS OR FAILURE

The production of transgenic insect-resistant plants, and their continuing development, has been a major scientific success, mirrored by the practical success of a limited number of pest-resistant genetically modified crops in India and some other countries like USA and China. Bt transgenic crops have been used successfully to provide resistance against a number of insect-pests for a decade or so. The year 1996 marked a milestone in agricultural biotechnology, when for the first time *Bt* transgenic varieties of potato, cotton and corn were released for commercial cultivation. Since then the global area of transgenic crops continued to grow at a sustained double-digit growth rate. In 2006, out of 102 million ha of transgenic crops grown by 10.3 million farmers in 22 countries, 19 million ha (19%) were planted with Bt crops and 13.1 million ha (13%) with stacked genes for insect resistance (James, 2008). Insect resistance has been the second most widely used trait in transgenic crops after herbicide resistance in global agriculture (Gomez–Barbero and Rodriguez–Cerezo, 2007).

11. CONCLUSIONS AND FUTURE PROSPECTS

The development and deployment of transgenic plants with insect resistance will continue to expand in the future. This approach of controlling insects would offer the advantage of allowing some degree of selection for specificity effects, so that pests but not the beneficial organisms are targeted. Incorporation of *Bt* genes will have a tremendous effect on pest management. We need to pursue the management strategy that reflects the pest biology, insect plant interactions and their influence on the natural enemies to prolong the life span of the transgenics. Refugia can play an important role in resistance management and should take into account the pest complex, the insect hosts and the environment. Expression of more than one gene (gene pyramiding) and single chain antibody genes, which would be compatible with the likely trends in pesticide discovery using biology derived target based methods. Emphasis should be placed on combining exotic genes with conventional host plant resistance, and also with traits conferring resistance to other insect pests and diseases of importance in the target region. Several genes conferring resistance to insects can also be deployed as multilines or synthetics.

Recently emerged RNAi technology should be wisely harnessed to attain broad spectrum and sustained level of insect protection. As, the effects of transgenic plants on the activity and abundance of natural enemies vary across crops and the insect species involved, therefore, studies on the adverse effects of transgenic crops on natural enemies are largely through poor quality of the host or early mortality of the insect host, rather than through direct toxicity to the natural enemies should be strengthened. Wherever the transgenic crops have shown adverse effects on natural enemies, these effects may still be far lower than those of the broad-spectrum pesticides. Deployment of transgenic plants would also lead to a major reduction in the number of insecticide sprays, resulting in increased activity of natural enemies and a safer environment in which to live. Thus, there is need to adopt biosafety protocol and resistance management strategies, organize public awareness, and present the benefits of Bt transgenic crops to improve social attitude for their rational deployment. If deployed appropriately plant biotechnology could be one of the best environment friendly insect–pest management strategies.

Host-plant resistance has been prioritized over chemical, biological, physical, cultural and regulatory control components for nematode management as it is cost-effective in nature, they don't require any additional cost input because resistant crop is "self-protected" and should yield well on infested land. But abiotic factors play a major role in genetic expression of resistance as increase in soil temperature may reduce the level of resistance. Existence of pathotypes or races in nematode species is other factor contributing significantly in failure of resistant varieties which have been developed after incorporating resistance source of that particular variety resistant to that particular pathotype or race. Genetic manipulation of crop plants through gene pyramiding where several anti-nematode genes of different modes of action are expressed together in the same plant may be a possible remedy. But, genetic modification could lead to crops with enhanced invasiveness and persistence. These new genotypes could invade natural ecosystems and cause undesirable change, either through spread of the crops themselves or through hybridization with wild relatives. Thus new technologies for incorporating nematode resistance in plants hold exciting prospects for plant breeders but they will complement, rather than replace the current range of techniques that have worked so well.

ACKNOWLEDGEMENTS

We acknowledge Mr. Achintya Pramanik, Ph. D. Student, Division of Entomology, IARI, New Delhi and Dr. Premlal Kashyap, Directorate of Seed Research, Mau, U.P. for providing some valuable study materials related to entomological part of this chapter.

REFERENCES

Altpeter, F., Diaz, I., McAuslane, H.., Gaddour, K., Carbonero, P. and Vasil, I.K. (1999). Increased insect resistance in transgenic wheat stably expressing trypsin inhibitor *CMe. Molecular Breeding,* **5:** 53–63.

Arakane, Y., Muthukrishnan, S., Kramer, K.J., Specht, C.A., Tomoyasu, Y., Lorenzen, M.D., Kanost, M. and Beeman, R.W. (2005). The *Tribolium* chitin synthase genes *TcCHS1* and *TcCHS2* are specialized for synthesis of epidermal cuticle and midgut peritrophic matrix. *Insect Mol. Biol.,* **14(5):** 453–63.

Araujo, R.N., Santos, A., Pinto, F.S., Gontijo, N.F., Lehane, M.J. and Pereira, M.H. (2006). RNA interference of the salivary gland nitrophorin 2 in the triatomine bug *Rhodnius prolixus* (Hemiptera: Reduviidae) by dsRNA ingestion or injection. *Insect Biochem. Mol. Biol.,* **36(9):** 683–93.

Arencibia, A., Vazquez, R.I., Prieto, D., Tellez, P., Carmona, E.R., Coego, A., Hernandez, L., de la Riva, G.A. and Selman–Housein, G. (1997). Transgenic sugarcane plants resistant to stem borer attack. *Molecular Breeding,* **3:** 247–55.

Armstrong, C.L., Parker, G.B. and Pershing, J.C. (1995). Field evaluation of European corn borer control in progeny of 173 transgenic corn events expressing an insecticidal protein from *Bacillus thuringiensis*. *Crop Sci.,* **35:** 550–57.

Artim, L. (2003). Application for determination of non–regulated status for lepidopteran insect protected VIP3A cotton transformation event COT102. Submitted by Syngenta Seeds, Inc., Research Triangle Park, NC 27709–2257 to the Biotechnology Regulatory Services, Riverdale, MD.

Atkinson, H.J. (2003). Strategies for resistance to nematodes in *Musa* spp. *In*: Genetic transformation strategies to address the major constraints to banana and plantain production in Africa (*Ed.*: Atkinson, H.J. *et al.*). INIBAP, France. p. 130.

Atkinson, H.J., Urwin, P.E. and McPherson, M.J. (2003). Engineering plants for nematode resistance. *Annual Review of Phytopathology,* **41:** 615–39.

Austin, S., Pohlman, J.D., Brown, C.R. and Mojtahedi, H. (1993). Interspecific somatic hybridization between *Solanum tuberosum* L. and *S. bulbocastanum* dun. as a means of transferring nematode resistance. *American Potato Journal,* **70(6):** 485.

Bakhetia, M., Charlton, W., Atkinson, H.J. and McPherson, M.J. (2005). RNA interference of dual oxidase in the plant nematode *Meloidogyne incognita. Molecular Plant Microbe Interactions,* **10:** 1099–1106.

Bakhetia, M., Urwin, P.E. and Atkinson, H.J. (2007). qPCR analysis and RNAi define pharyngeal gland cell-expressed genes of *Heterodera glycines* required for initial interactions with the host. *Mol. Plant–Microbe. Interact.* **20:** 306–12.

Ballvora, A., Hesselbach, J., Niewöhner, J., Leister, D., Salamini, F. and Gebhardt, C. (1995). Marker enrichment and high-resolution map of the segment of potato chromosome VII harbouring the nematode resistance gene *Gro1*. *Molecular and General Genetics*. **249(1):** 82–90.

Barton, K., Whitely, H. and Yang, N.S. (1987). *Bacillus thuringiensis* δ-endotoxin in transgenic *Nicotiana tabacum* provides resistance to lepidopteran insects. *Plant Physiol.* **85:** 1103–1109.

Bashir, K., Husnain, T., Fatima, T., Latif, Z., Mehdi, S.A. and Riazuddin, S. (2004). Field evaluation and risk assessment of transgenic Indica basmati rice. *Mol. Breed,,* **13:** 301–12.

Baulcombe, D. (2007). Amplified Silencing. *Science,* **315:** 199–200.

Baum, J.A., Bogaert, T., Clinton, W., Heck, G.R., Feldmann, P., Ilagan, O., Johnson, S., Plaetinck, G., Munyikwa, T. and Pleau, M. *et al.* (2007). Control of coleopteran insect pests through RNA interference. *Nat Biotechnol,* **25:** 1322–26.

Baum, T.J., Hyatt, A., Parrott, W.A., Pratt, L.H. and Hussey, R.S. (1996). Expression in tobacco of a monoclonal antibody specific to stylet secretions of the root-knot nematode. *Mol. Plant–Microbe Int.* **9:** 382–87.

Bautista, M.A.M., Miyata, T., Miura, K. and Tanaka, T. (2009). RNA interference-mediated knockdown of a cytochrome P450, CYP6BG1, from the diamondback moth, *Plutella xylostella*, reduces larval resistance to permethrin. *Insect Biochemistry and Molecular Biology* **39:** 38–46.

Beale M.H., Birkett, M.A., Bruce, T.J.A., Chamberlain, K., Field, L.M., Huttly, A.K., Martin, J.L., Parker, R., Phillips, A.L. and Pickett, J.A. *et al.* (2006). Aphid alarm pheromone produced by transgenic plants affects aphid and parasitoid behavior. *Proc Natl Acad Sci USA,* **103:** 10509–13.

Benedict, J.H., Sachs, E.S., Altman, D.W., Deaton, D.R., Kohel, R.J., Ring, D.R. and Berberich, B.A. (1996). Field performance of cotton expressing CryIA insecticidal crystal protein for resistance to *Heliothis virescens* and *Helicoverpa zea* (Lepidoptera: Noctiudae). *Journal of Economic Entomology,* **89:** 230–38.

Bohorova N., Frutos, R., Royer, M., Estanol, P., Pacheco, M., Rascon, Q., McLean, S. and Hoisington, D. (2001). Novel synthetic *Bacillus thuringiensis* cry1B gene and the cry1B–cry1Ab translational fusion confer resistance to southwestern corn borer, sugarcane borer and fall armyworm in transgenic tropical maize. *Theor Appl Genet,* **103:** 817–26.

Boiteux, L.S., Belter, J.G., Roberts, P.A. and Simon, P.W. (2000). RAPD linkage map of the genomic region encompassing the root-knot nematode (*Meloidogyne javanica*) resistance locus in carrot. *Theoretical and Applied Genetics,* **100(3):** 439–46.

Boulter, D. (1993). Insect pest control by copying nature using genetically engineered crops. *Phytochemistry* **34:** 1453–66.

Bravo, A., Gill, S.S. and Soberón, M. (2007). Mode of action of *Bacillus thuringiensis* Cry and Cyt toxins and their potential for insect control. *Toxicon,* **49:** 423–35.

Brookes, G. and Barfoot, P. (2005). GM crops: the global economic and environmental impact: the first nine years 1996–2004. *Ag Bio Forum,* **8:** 15.

Brown, C.R., Yang, C.P., Mojtahedi, H., Santo, G. S. and Masuelli R. (1996). RFLP analysis of resistance to Columbia root–knot nematode derived from *Solanum bulbocastanum* in a BC_2 population. *Theoretical and Applied Genetics,* **92(5):** 572–76.

Burrows, P.R. and de Waele, D. (1997). Engineering resistance against plant parasitic nematodes using anti-nematode genes. *In*: Cellular and Molecular Aspects of Plant-Nematode Interactions (*Eds*.: Fenoll, C., Grundler, F.M.W. and Ohl, S.A.). Kluwer Academic Publishers, Dordrecht, Netherlands. p. 286.

Buschman, L., Sloderbeck, P., Guo, Y., Higgins, R. and Witt, M. (1998). Corn borer resistance and grain yield of *Bt* and non-*Bt* corn hybrids at Garden city, Kansas, in 1997. *In*: Progress Report-814, agricultural experiment station and co-operative extension service, Kansas State University. pp. 34–38.

Cai, D., Kleine, M., Kifle, S., Harloff, H.J., Sandal, N.N., Marcker, K.A., Klein–Lankhorst, R.M., Salentijn, E.M.J., Lange, W., Stiekema, W.J., Wyss, U., Grundler, F.M.W. and Jung, C. (1997). Positional cloning of a gene for nematode resistance in sugar beet. *Science,* **275:** 832–34.

Cao, J., Tang, J.D., Strizhov, N., Shelton, A.M. and Earle, E.D. (1999). Transgenic broccoli with high levels of *Bacillus thuringiensis* Cry1C protein control diamondback moth larvae resistant to Cry1A or Cry1C. *Molecular Breeding,* **5:** 131–41.

Carbonero, P., Royo, J., Diaz, I., Garcia–Maroto, F., Gonzalez–Hidalgo, E., Gutierez, C. and Casanera, P. (1993). Cereal inhibitors of insect hydrolases (α-amylases and trypsin): Genetic control, transgenic expression and insect pests. *In*: Bruening, G.J., Garcia–Olmedo, F. and Ponz, F.J. (*Eds.*), *Workshop on Engineering Plants Against Pests and Pathogens*, 1–13 January, 1993. Madrid, Spain: Instituto Juan March de Estudios Investigacions.

Carrière, Y., Ellers–Kirk, C., Sisterson, M., Antilla, L., Whitlow, M., Dennehy T.J. and Tabashnik, B.E. (2003). Long-term regional suppression of pink bollworm by *Bacillus thuringiensis* cotton. *PNAS USA,* **100:** 1519–23.

Charity, J.A., Anderson, M.A., Bittisnich, D.J., Whitecross, M. and Higgins, T.J.V. (1999). Transgenic tobacco and peas expressing a proteinase inhibitor from *Nicotiana alata* have increased resistance. *Molecular Breeding,* **5:** 357–65.

Chen, J., Zhang, D., Yao, Q., Zhang, J., Dong, X., Tian, H., Chen, J. and Zhang, W. (2010). Feeding-based RNA interference of a trehalose phosphate synthase gene in the brown planthopper, *Nilaparvata lugens. Insect Molecular Biology,* **19(6):** 777–86.

Chen, Q., Rehman, S., Smant, G. and Jones, J.T. (2005). Functional analysis of pathogenicity proteins of the potato cyst nematode *Globodera rostochiensis* using RNAi. *Molecular Plant Microbe Interactions,* **7:** 621–25.

Chen, X., Tian, H., Zou, L., Tang, B., Hu, J. and Zhang, W. (2008). Disruption of *Spodoptera exigua* larval development by silencing chitin synthase gene A with RNA interference. *Bulletin of Entomological Research,* **98(06):** 613–19.

Chitkowski, R.L., Turnipseed, S.G., Sullivan, M.J. and Bridges, W.C. (2003). Field and laboratory evaluations of transgenic cottons expressing one or two *Bacillus thuringiensis* var. *kurstaki* Berliner proteins for management of noctuid (*Lepidoptera*) pests. *J. Econ. Entomol,* **96:** 755–62.

Chitwood, D.J. (2003). Research on plant-parasitic nematode biology conducted by the United States Department of Agriculture–Agricultural Research Service. *Pest Manag Sci.,* **59:** 748–53.

Cho, H., Choi, K., Yamashita, M., Morikawa, H. and Murooka, Y. (1995). Introduction and expression of the *Streptomyces* cholesterol oxidase gene (*choA*), a potent insecticidal protein active against boll–weevil larvae, into tobacco cells. *Applied Microbiology and Biotechnology,* **44:** 133–38.

Christeller, J.T. and Shaw, B.D. (1989). The interaction of a range of serine proteinase inhibitors with bovine trypsin and *Costelytra zealandica* trypsin. *Insect Biochemistry,* **19:** 233–41.

Christou, P., Capell, T., Kohli, A., Gatehouse, J.A. and Gatehouse, A.M.R. (2006). Recent developments and future prospects in insect pest control in transgenic crops. *Trends in Plant Science,* **11:** 302–308.

Confalonieri, M., Allegro, G., Balestrazzi, A., Fogher, C. and Delledonne, M. (1998). Regeneration of *Populus nigra* transgenic plants expressing a Kunitz proteinase inhibitor (KTi3) gene. *Molecular Breeding,* **4:** 137–45.

Cooper–Bland, S., Maine, M.J.D., Fleming, M.L.M.H., Phillips, M.S., Powell, W. and Kumar, A. (1993). Synthesis of intraspecific somatic hybrids of *Solanum tuberosum*: assessments of morphological, biochemical and nematode (*Globodera pallida*) resistance characteristics. *Journal of Experimental Botany,* **45(9):** 1319.

Corbin, D.R., Grebenok, R.J., Ohnmeiss, T.E., Greenplate, J.T. and Purcell, J.P. (2001). Expression and chloroplast targeting of cholesterol oxidase in transgenic tobacco plants. *Plant Physiology,* **126(3):** 1116–28.

Corbin, D.R., Greenplate, J.T., Wong, E.Y. and Purcell, J.P. (1994). Cloning of an insecticidal cholesterol oxidase gene and its expression in bacteria and plant protoplasts. *Applied Environmental Microbiology,* **60:** 4239–44.

Datta, K., Vasquez, A., Tu, J., Torrizo, L., Alam, M.F., Oliva, N., Abrigo, E., Khush, G.S. and Datta, S.K. (1998). Constitutive and tissue-specific differential expression of the cryIA(b) gene in transgenic rice plants conferring resistance to rice insect pests. *Theoretical and Applied Genetics,* **97:** 20–30.

Davidson, M.M., Jacobs, J.M.E., Reader, J.K., Butler, R.C., Frater, C.M., Markwick, N.P., Wratten, S.D. and Conner, A.J. (2002). Development and evaluation of potatoes transgenic for a cry1Ac9 gene conferring resistance to potato tuber moth. *Journal of American Society of Horticulture Science,* **127:** 590–96.

De Cosa, B., Moar, W., Lee, S.B., Miller, M. and Daniell, H. (2001). Overexpression of the Bt cry2Aa2 operon in chloroplasts leads to formation of insecticidal crystals. *Nat Biotechnol,* **19:** 71–74.

De Leo, F., Bonade–Bottino, M.A., Ceci, L.R., Gallerani, R. and Jouanin, L. (1998). Opposite effects on *Spodoptera littoralis* larvae of high expression level of a trypsin inhibitor in transgenic plants. *Plant Physiology,* **118:** 997–1004.

de Maagd, R.A.,Weemen–Hendriks, M., Stiekema, W. and Bosch, D. (2000). *Bacillus thuringiensis* delta-endotoxin Cry1C domain III can function as a specificity determinant for *Spodoptera exigua* in different, but not all, Cry1–Cry1C hybrids. *Appl Environ Microbiol,* **66:** 1559–63.

Delannay, X., LaVallee, B.J., Proksch, R.K., Fuchs, R.L., Sims, S.K., Greenplate, J.T., Marrone, P.G., Dodson, R.B., Augustine, J.J., Layton, J.G. and Fischhoff, D.A. (1989). Field performance of transgenic tomato plants expressing *Bacillus thuringiensis* var *kurstaki* insect control protein. *Bio. Technol.,* **7:** 1265–69.

Dhaliwal, G.S., Jindal, V. and Dhawan, A.K. (2010). Insect pest problems and crop losses: Changing trends. *Indian Journal of Ecology,* **37:** 1–7.

Ding, X., Gopalakrishnan, B., Johnson, L.B., White, F.F., Wang, X., Morgan, T.D., Kramer, K.J. and Muthukrishnan, S. (1998). Insect resistance of transgenic tobacco expressing an insect chitinase gene. *Transgenic Research,* **7:** 77–84.

Dong, Y. and Friedrich, M. (2005). Nymphal RNAi: Systemic RNAi mediated gene knockdown in juvenile grasshopper. *BMC Biotechnol,.* **5:** 25–32.

Douches, D.S., Pett, W., Santos, F., Coombs, J., Grafius, E., Li, W., Metry, E.A., NASR E.–din, T. and Madkour, M. (2004). Field and storage testing Bt potatoes for resistance to potato tuberworm (Lepidoptera: Gelichiidae). *J. Econ. Entomol.* **97:** 1425–31.

Duan, X., Li, X., Xue, Q., Abo el Saad, M., Xu, D. and Wu, R. (1996). Transgenic rice plants harboring an introduced potato proteinase inhibitor II gene are insect resistant. *Nature Biotechnology,* **14:** 494–98.

Duck, N.B. and Evola, S.V. (1997). Use of transgenes to increase host plant resistance to insects: Opportunities and challenges. *In*: Advances in insect control: The role of transgenic plants (*Eds*., Carozzi, N.B. and Koziel, M.G.). Taylor and Francis Ltd., London. pp. 1–20.

Dufourmantel, N., Tissot, G., Goutorbe, F., Garcon, F., Muhr, C., Jansens, S., Pelissier, B., Peltier, G. and Dubald, M. (2005). Generation and analysis of soybean plastid transformants expressing *Bacillus thuringiensis* Cry1Ab protoxin. *Plant Mol. Biol.,* **58:** 659–68.

Ebora, R.V., Ebora, M.M. and Sticklen, M.B. (1994). Transgenic potato expressing the *Bacillus thuringiensis* cryIA(c) gene effects on the survival and food consumption of *Phthorimaea operculella* (Lepidoptera: Gelechiidae) and *Ostrinia nubilalis* (Lepidoptera: Noctuidae). *J. Econ. Entomol.*, **87:** 1122–27.

Edmonds, H.S., Gatehouse, L.N., Hilder, V.A. and Gatehouse, J.A. (1996). The inhibitory effects of the cysteine protease inhibitor, oryzacystatin, on digestive proteases and on larval survival and development of the southern corn rootworm (*Diabrotica undecimpunctata* Howard). *Entomologia Experimentalis et Applicata,* **78:** 83–94.

Ellis, J. and Jones, D. (1998). Structure and function of proteins controlling strain specific pathogen resistance in plants. *Currents Opinion in Plant Biology,* **1:** 288–93.

English, L.H., Brussock, S.M., Malvar, T.M., Bryson, J.W., Kulesza, C.A., Walters, F.S., Slatin, S.L., Von Tersch, M.A. and Romano, C. (2003). Coleopteran–resistant transgenic plants and methods of their production. US Patent No. 7,227,056.

Fanelli, E., Di Vito, M., Jones, J.T. and De Giorgi, C. (2005). Analysis of chitin synthase in a plant parasitic nematode, *Meloidogyne artiellia*, using RNAi. *Gene,* **349:** 87–95.

Fang, H.J., Li, D.L., Wang, G.L. and Li, Y.H. (1997). An insect-resistant transgenic cabbage plant with the cowpea trypsin inhibitor (CpTi) gene. *Acta Botanica Sinica* **39:** 940–45.

Fang, J., Xu, X., Wang, P., Zhao, J.Z., Shelton, A.M., Cheng, J., Feng, M.G. and Shen, Z. (2007). Characterization of chimeric *Bacillus thuringiensis* Vip3 toxins. *Appl Environ Microbiol,* **73:** 956–61.

Ffrench–Constant R.H., Dowling, A. and Waterfield, N.R. (2007). Insecticidal toxins from Photorhabdus bacteria and their potential use in agriculture. *Toxicon,* **49:** 436–51.

Fioretti, L., Porter, A., Haydock, P.J. and Curtis, R. (2002). Monoclonal antibodies reactive with secreted–excreted products from the amphids and the cuticle surface of *Globodera pallida* affect nematode movement and delay invasion of potato roots. *International Journal for Parasitology,* **32:** 1709–18.

Fire, A., Xu, S.Q., Montgomery, M.K., Kostas, S.A., Driver, S.E. and Mello, C.C. (1998). Potent and specific genetic interference by double-stranded RNA in *Caenorhabditis elegans. Nature,* **391:** 806–11.

Fitches, E., Gatehouse, A.M.R. and Gatehouse, J.A. (1997). Effects of snowdrop lectin (GNA) delivered via artificial diet and transgenic plants on the development of tomato moth (*Lecanobia oleracea*) larvae in laboratory and glasshouse trials. *Journal of Insect Physiology,* **43:** 727–39.

Foissac, X., Loc, N.T., Christou, P., Gatehouse, A.M.R. and Gatehouse, J.A. (2000). Resistance to green leafhopper (*Nephotettix virescens*) and brown plant hopper (*Nilaparvata lugens*) in transgenic rice expressing snowdrop lectin (*Galanthus nivalis agglutinin*; GNA). *J. Insect Physiol,* **46:** 573–83.

Gahakwa, D., Maqbool, S.B., Fu, X., Sudhakar, D., Christou, P. and Kohli, A. (2000). Transgenic rice as a system to study the stability of transgene expression: Multiple heterologous transgenes show similar behaviour in diverse genetic backgrounds, *Theoretical and Applied Genetics,* **101:** 388–99.

Gahan, L.J., Ma, Y.T., Coble, M.L.M., Gould, F., Moar, W.J. and Heckel, D.G. (2005). Genetic basis of resistance to Cry1Ac and Cry2Aa in *Heliothis virescens* (Lepidoptera: Noctuidae). *J. Econ. Entomol.,* **98:** 1357–68.

Gatehouse, A.M.R., Davison, G.M., Stewart, J.N., Gatehouse, L.N., Kumar, A., Geoghegan, *I.E.*, Birch, A.N.E. and Gatehouse, J.A. (1999). Concanavalin A inhibits development of tomato moth (*Lecanobia oleracea*) and peach potato aphid (*Myzus persicae*) when expressed in transgenic potato plants. *Molecular Breeding,* **5:** 153–65.

Gatehouse, A.M.R., Down, R.E., Powell, K.S., Sauvion, N., Bahbe, Y., Newell, C.A., Merryweather, A., Hamilton, W.D.O. and Gatehouse, J.A. (1996). Transgenic potato plants with enhanced resistance to the peach–potato aphid *Myzus persicae. Entomologia Experimentalis et Applicata,* **79:** 295–307.

Gatehouse, A.M.R., Powell, K.S., Van Damme, E.J.M. and Gatehouse, J.A. (1995). Insecticidal properties of plant lectins. *In*: Lectins, Biomedical Perspectives (*Eds.* Pusztai, A. and Bardocz, S.). London, UK: Taylor and Francis.

Gatehouse, A.M.R., Shi, Y., Powell, K.S., Brough, C., Hilder, V.A., Hamilton, W.D.O., Newell, C., Merryweather, A., Boutler, D. and Gatehouse, J.A. (1993). Approaches to insect resistance using transgenic plants. *Philosophical Transactions of the Royal Society of London, Biological Sciences* (B), **342:** 279–86.

Gatehouse, J.A. (2008). Biotechnological prospects for engineering insect-resistant plants. *Plant Physiology,* **146:** 881–87.

Gibson, D.M., Gallo, L.G., Krasnoff, S.B. and Ketchum, R.E.B. (1995). Increased efficiency of *Bacillus thuringiensis* subsp. *kurstaki* in combination with tannic acid. *Journal of Economic Entomology,* **88:** 270–77.

Girard, C., Le–Metayer, M., Zaccomer, B., Bartlet, E., Williams, I., Bonade–Bottino, M., Pham–Delegue, M.H. and Ouanin, L. (1998). Growth stimulation of beetle larvae reared on a transgenic oilseed rape expressing a cysteine proteinase inhibitor. *Journal of Insect Physiology,* **44:** 263–70.

Golmirizaie, A., Zhang, D.P., Nopo, L., Newell, C.A., Vera, A. and Cisneros, F. (1997). Enhanced resistance to West Indian sweet potato weevil (*Euscepes postfaciatus*) in transgenic 'Jewel' sweet potato with cowpea trypsin inhibitor and snowdrop lectin. *Hortscience,* **32:** 435.

Gómez–Barbero, M. and Rodríguez–Cerezo, E. (2007). GM Crops in EU agriculture–A case study for the Bio4EU project (http://bio4eu.jrc.ec.europa.eu/documents/FINALGMcropsintheEUBIO4EU.pdf).

Gurr, S.J., McPherson, M.J., Scollan, C., Atkinson, H.J. and Bowles, D.J. (1991). Gene expression in nematode–infected plant roots. *Molecular and General Genetics,* **1226:** 361–66.

Gvakharia, B.O., Bebas, P., Cymborowski, B. and Giebultowicz, J.M. (2003). Disruption of sperm release from insect testes by cytochalasin and beta-actin mRNA mediated interference. *Cell Mol. Life Sci.,* **60(8):** 1744–51.

Habibi, J., Backus, E.A. and Czapla, T.H. (1992). Effect of plant lectins on survival of potato leafhopper. *Proceedings, XIX International Congress of Entomology*, Beijing, China, p. 373.

High, S.M., Cohen, M.B., Shu, Q.Y. and Altosaar, I. (2004). Achieving successful deployment of *Bt* rice. *Trends Plant Sci,* **9:** 286–92.

Hilder, V.A., Gatehouse, A.M.R., Sheerman, S.E., Baker, R.F. and Boulter, D. (1987). A novel mechanism of insect resistance engineered into tobacco. *Nature,* **330:** 160–63.

Hilder, V.A., Powell, K.S., Gatehouse, A.M.R., Gatehouse, J.A., Gatehouse, L.N., Shi, Y., Hamilton, W.D.O., Merryweather, A., Newell, C.A., Timans, J.C., Peumans, W.J., Van Damme, E. and Boulter, D. (1995). Expression of snowdrop lectin in transgenic tobacco plants results in added protection against aphids. *Transgenic Research,* **4:** 18–25.

Ho, J.Y., Weide, R., Ma, H.M., Van Wordragen, M.F., Lambert, K.N., Koornneef, M., Zabel, P. and Williamson, V.M. (1992). The root-knot nematode resistance gene (Mi) in tomato: construction of a molecular linkage map and identification of dominant cDNA markers in resistant genotypes. *The Plant Journal,* **3(6):** 971–82.

Hoffmann, M.P., Zalom, F.G., Wilson, L.T., Smilanick, J.M., Malyj, L.D., Kisen, J., Hilder, V.A. and Barnes, W.M. (1992). Field evaluation of transgenic tobacco containing genes encoding *Bacillus thuringiensis* δ-endotoxin or cowpea trypsin inhibitor: Efficacy against *Helicoverpa zea* (Lepidoptera: Noctuidae). *Journal of Economic Entomology,* **85:** 2516–22.

Huang, G., Dong, R., Allen, R., Davis, E.L., Baum, T.J. and Hussey, R.S. (2006). A root–knot nematode secretory peptide functions as a ligand for a plant transcription factor. *Molecular plant–Microbe Interactions,* **19:** 463–70.

Huang, J., Hu, R., Rozelle, S. and Pray, C. (2005). Insect–resistant GM rice in farmers' fields: Assessing productivity and health effects in China. *Science,* **308:** 688–90.

Huang, J., Rozelle, S., Pray, C. and Wang, Q. (2002). Plant biotechnology in China. *Science,* **295:** 674–77.

Huvenne, H. and Smagghe, G. (2010). Mechanisms of dsRNA uptake in insects and potential of RNAi for pest control: A review. *Journal of Insect Physiology,* **56:** 227–35.

Hyde, J., Martin, M.A., Preckel, P.V. and Edwards, C.R. (1999). The economics of *Bt* corn: Valuing protection from the European corn borer. *Review of Agricultural Economics,* **21:** 442–54.

Iannacone, R., Grieco, P.D. and Cellini, F. (1997). Specific sequence modifications of a cry3B endotoxin gene result in high levels of expression and insect resistance. *Plant Mol. Biol.* **34:** 485–96.

Ishikawa, H., Hoshino, Y., Motoki, Y., Kawahara, T., Kitajima, M., Kitami, M., Watanabe, A., Bravo, A., Soberon, M. and Honda, A. *et al.* (2007). A system for the directed evolution of the insecticidal protein from *Bacillus thuringiensis. Mol Biotechnol,* **36:** 90–101.

Ishimoto, M. and Chrispeels, M.J. (1996). Protective mechanism of the Mexican bean weevil against high levels of alpha-amylase inhibitor in the common bean. *Plant Physiology,* **111(2):** 393–401.

Jacobs, J.M.E., Van, E.H.J., Horsman, K., Arens, P.F.P., Verkerk–Bakker, B., Jacobsen, E., Periera, A. and Stiekema, W.J. (1996). Mapping of resistance to the potato cyst nematode *Globodera rostochiensis* from the wild potato species *Solanum vernei. Molecular Breeding,* **2(1):** 51–60.

James, C. (2008). Global Status of Commercialized Biotech/GM Crops: 2008. ISAAA Brief No.39. ISAAA: Ithaca, NY (http://www.isaaa.org/resources/publications/briefs/39/executivesummary/pdf/Brief%2039%20–%20Executive%20Summary%20–%20English.pdf)

Jauhar, P. (2006). Modern biotechnology as an integral supplement to conventional plant breeding: the prospects and challenges. *Crop Science,* **46:** 1841–59.

Jenkins, J.N. (1999). Transgenic plants expressing toxins from *Bacillus thuringiensis. In*: Biopesticides: Use and Delivery (*Eds.*: Hall, F.R. and Menn, J.J.). Totowa, New Jersey: Humana Press, p. 232.

Jenkins, J.N., McCarty, J.C.Jr., Buehler, R.E., Kiser, J., Williams, C. and Wofford, T. (1997). Resistance of cotton with delta-endotoxin genes from *Bacillus thuringiensis* var. *kurstaki* on selected Lepidopteran insects. *Agron. J.,* **89:** 768–80.

Jongsma, M.A., Bakker, P.L., Peters, J., Bosch, D. and Stiekema, W.J. (1995). Adaptation of *Spodoptera exigua* larvae to plant proteinase inhibitors by induction of proteinase activity insensitive to inhibition. *PNAS USA.* **92:** 8041–45.

Jung, C. (1998). A singular gene doubles up pest resistance. *Nature Biotechnology* **16:** 1315–16.

Kaloshian, I., Yaghoobi, J., Liharska, T., Hontelez, J., Hanson, D., Hogan, P., Jesse, T., Wijbrandi, J., Simons, G., Vos, P., Zabel, P. and Williamson, V.M. (1998). Genetic and physical localization of the root–knot nematode resistance locus Mi in tomato. *Molecular and General Genetics,* **257(3):** 376–85.

Kamath, R.S., Frase, A.G., Dong, Y., Poulin, G., Durbin, R., Gotta, M., Kanapin, A., Le Bot, N., Moreno, S., Sohrmann, M., Welchman, D.P., Zipperlen, P. and Ahringer, J. (2003). Systematic functional analysis of the *Caenorhabditis elegans* genome using RNAi. *Nature,* **16:** 231–37.

Kanrar, S., Venkateswari, J., Kirti, P.B. and Chopra, V.L. (2002). Transgenic Indian mustard (*Brassica juncea*) with resistance to the mustard aphid (*Lipaphis erysimi* Kalt.). *Plant Cell Reports,* **20:** 976–81.

Kar, S., Basu, D., Das, S., Ramkrishnan, N.A., Mukherjee, P., Nayak, P. and Sen, S.K. (1997). Expression of CryIA(c) gene of *Bacillus thuringiensis* in transgenic chickpea plants inhibits development of pod borer (*Heliothis armigera*) larvae. *Transgenic Research,* **6:** 177–85.

Kim, Y.S., Uefuji, H., Ogita, S. and Sano, H. (2006). Transgenic tobacco plants producing caffeine: a potential new strategy for insect pest control. *Transgenic Res,* **15:** 667–72.

Kotwica, J., Bebas, P., Gvakharia, B.O. and Giebultowicz, J.M. (2009). RNA interference of the period gene affects the rhythm of sperm release in moths. *J. Biol. Rhythms.* **24(1):** 25–34.

Koziel, M.G., Beland, G.L., Bowman, C., Carozzi, N.B., Crenshaw, R., Crossland, L., Dawson, J., Desai, N., Hill, M., Kadwell, S., Launis, K., Lewis, K., Maddox, D., McPherson, K., Meghji, M.R., Merlin, E., Rhodes, R., Warren, G.W., Wright, M. and Evola, S.V. (1993). Field performance of elite transgenic maize plants expressing an insecticidal protein derived from *Bacillus thuringiensis. Biotechnol,* **11:** 194–200.

Kramer, K.J., Morgan, T.D., Throne, J.E., Dowell, F.E., Bailey, M. and Howard, J.A. (2000). Transgenic avidin maize is resistant to storage insect pests. *Nat Biotechnol,* **18:** 670–74.

Kreike, C.M., de Koning, J.R.A., Vinke, J.H., Van Ooijen, J.W., Gebhardt, C. and Stiekema, W.J. (1994). Mapping of loci involved in quantitatively inherited resistance to the potato cyst–nematode *Globodera rostochiensis* pathotype Ro1. *Theoretical and Applied Genetics,* **87(4):** 464–70.

Kristensen, C., Morant, M., Olsen, C.E., Ekstrom, C.T., Galbraith, D.W., Moller, B.L. and Bak, S. (2005). Metabolic engineering of dhurrin in transgenic Arabidopsis plants with marginal inadvertent effects on themetabolome and transcriptome. *Proc Natl Acad Sci USA,* **102:** 1779–84.

Kumar, P.A., Mandaokar, A., Sreenivasu, K., Chakrabarti, S.K., Bisaria, S., Sharma, S.R., Kaur, S. and Sharma, R.P. (1998). Insect-resistant transgenic brinjal plants. *Mol. Breed.* **4:** 33–37.

Kumar, S., Chandra, A. and Pandey, K.C. (2008). *Bacillus thuringiensis* (*Bt*) transgenic crop: An environment friendly insect-pest management strategy. *Journal of Environmental Biology,* **29(5):** 641–53.

Lauer, J. and Wedberg, J. (1999). Grain yield of initial *Bt* corn hybrid introductions to farmers in the Northern corn belt. *J. Prod. Agric.,* **12:** 373–76.

Lecardonnel, A., Chauvin, L., Jouanin, L., Beaujean, A., Prevost, G. and Sangwan Norreel, B. (1999). Effects of rice cystatin I expression in transgenic potato on Colorado potato beetle larvae. *Plant Science,* **140:** 71–79.

Lee, C.M., Su, M.T. and Lee, H.J. (2009). Pigment dispersing factor: an output regulator of the circadian clock in the German cockroach. *J. Biol. Rhythms.* **24(1):** 35–43.

Lee, S.I., Lee, S.H., Koo, J.C., Chun, H.J., Lim, C.O., Mun, J.H., Song, Y.H. and Cho, M.J. (1999). Soybean Kunitz trypsin inhibitor (SKTI) confers resistance to the brown planthopper (*Nilaparvata lugens* Stal) in transgenic rice. *Molecular Breeding,* **5:** 1–9.

Lelivelt, C.L.C., Lange, W. and Dolstra, O. (1993b). Intergeneric crosses for the transfer of resistance to the beet cyst nematode from *Raphanus sativus* to *Brassica napus. Euphytica,* **68(1):** 111–20.

Lelivelt, C.L.C., Leunissen, E.H.M., Frederiks, H.J., Helsper, J.P.F.G. and Krens, F.A. (1993a). Transfer of resistance to the beet cyst nematode (*Heterodera schachtii* Schm.) from *Sinapis alba* L. (white mustard) to the *Brassica napus* L. gene pool by means of sexual and somatic hybridization. *Theoretical and Applied Genetics,* **85(6):** 688–96.

Li, Y.E., Zhu, Z., Chen, Z.X., Wu, X., Wang, W. and Li, S.J. (1998). Obtaining transgenic cotton plants with cowpea trypsin inhibitor. *Acta Gossypii Sinica,* **10:** 237–43.

Lilley, C.J., Goodchild, S.A., Atkinson, H.J. and Urwin, P.E. (2005). Cloning and characterization of *Heterodera glycines* aminopeptidase cDNA. *International Journal of Parasitology,* **35:** 1577–85.

Lilley, C.J., Kyndt, T. and Gheysen, G. (2011). Nematode resistant GM crops in Industrialized and Developing countries. *In: Genomics and Molecular Genetics of Plant Nematode Interctions (Eds.: Jones, J., Ghesen, G. and Fenoll, C.). Springerlink.* p. 517.

Lilley, C.J., Urwin, P.E., Johnston, K.A. and Atkinson, H.J. (2004). Preferential expression of a plant cystatin at nematode feeding sites confers resistance to *Meloidogyne incognita* and *Globodera pallida. Plant Biotechnology Journal,* **2(1):** 3–12.

Ling, P., Duncan, L.W., Deng, Z., Dunn, D., Hu, X., Huang, S. and Gmitter Jr, F.G. (2000). Inheritance of citrus nematode resistance and its linkage with molecular markers. *Theoretical and Applied Genetics,* **100(7):** 1010–17.

Liu, D., Burton, S., Glancy, T., Li, Z.S., Hampton, R., Meade, T. and Merlo, D.J. (2003). Insect resistance conferred by 283-kDa *Photorhabdus luminescens* protein TcdA in *Arabidopsis thaliana. Nature Biotechnology,* **21:** 1222–28.

Liu, X.S. and Dean, D.H. (2006). Redesigning *Bacillus thuringiensis* Cry1Aa toxin into a mosquito toxin. *Protein Eng. Des. Sel.,* **19:** 107–111.

Macintosh, S.C., Kishore, G.M., Perlack, F., Marrone, P.G., Stone, T.B., Sims, S.R. and Fuchs, R.L. (1990). Potentiation of *Bacillus thruringiensis* insecticidal activity by serine protease inhibitors. *J. Agric. Food Chem.,* **38:** 1145–52.

Maddock, S.E., Hufman, G., Isenhour, D.J., Roth, B.A., Raikhel, N.V., Howard, J.A. and Czapla, T.H. (1991. Expression in maize plants of wheat germ agglutinin, a novel source of insect resistance. In *3rd International Congress of Plant Molecular Biology.* Tucson, Arizona, USA. p. 372. Abstract.

Mandal, S., Kundu, P., Roy, B. and Mandal, R.K. (2002). Precursor of the inactive 2S seed storage protein from the Indian mustard *Brassica juncea* is a novel trypsin inhibitor: Characterization, post-translational processing studies, and transgenic expression to develop insect-resistant plants. *Journal of Biological Chemistry,* **277:** 37161–68.

Mao, Y.B., Cai, W.J., Wang, J.W., Hong, G.J., Tao, X.Y., Wang, L.J., Huang, Y.P. and Chen, X.Y. (2007). Silencing a cotton bollworm P450 monoxygenase gene by plant-mediated RNAi impairs larval tolerance of gossypol. *Nat Biotechnol,* **25:** 1307–13.

Maqbool, S.B., Husnain, T., Raizuddin, S. and Christou, P. (1998). Effective control of yellow rice stem borer and rice leaf folder in transgenic rice indica varieties Basmati 370 and M 7 using novel δ-endotoxin Cry2A *Bacillus thuringiensis* gene. *Molecular Breeding,* **4:** 501–507.

Marchetti, S., Delledonne, M., Fogher, C., Chiaba, C., Chiesa, F., Savazzini, F. and Giordano, A. (2000). Soybean Kunitz, *C-II* and *PI-IV* inhibitor genes confer different levels of insect resistance to tobacco and potato transgenic plants. *Theoretical and Applied Genetics,* **101:** 519–26.

Mariani, C., Gossele, V., De Beuckeleer, M., De Block, M., Goldberg, R.B., De Greef, W. and Leemans, J. (1992). A chimaeric ribonuclease–inhibitor gene restores fertility to male sterile plants. *Nature,* **357:** 384–87.

Martin, D., Maestro, O., Cruz, J., Mane–Padros, D. and Belles X. (2006). RNAi studies reveal a conserved role for RXR in molting in the cockroach *Blattella germanica. J. Insect Physiol.*, **52(4):** 410–16.

McBride, K.E., Svab, Z., Schael, D.J., Hogan, P.S., Stalker, K.M. and Maliga, P. (1995). Amplification of a chimeric *Bacillus* gene in chloroplasts leads to an extraordinary level of an insecticidal protein in tobacco. *Bio / Technology,* **13:** 362–65.

McManus, M.T., Burgess, E.P.J., Philip, B., Watson, L.M., Laing, W.A., Voisey, C.R. and White, D.W.R. (1999). Expression of the soybean (Kunitz) trypsin inhibitor in transgenic tobacco: effects on larval development of *Spodoptera litura. Transgenic Research,* **8:** 383–95.

Mehlo, L., Gahakwa, D., Nghia, P.T., Loc, N.T., Capell, T., Gatehouse, J.A., Gatehouse, A.M.R. and Christou, P. (2005). An alternative strategy for sustainable pest resistance in genetically enhanced crops. *Proc Natl Acad Sci USA,* **102:** 7812–16.

Milligan, S., Bodeau, J., Yaghoobi, J., Kaloshian, I., Zabel, P. and Williamson, V.M. 1998. The root-knot resistance gene Mi from tomato is a member of the leucin zipper, nucleotide binding leucine–rich repeat family of plant genes. *Molecular and General Genetics,* **257:** 376–85.

Mochizuki, A., Nishizawa, Y., Onodera, H., Tabei, Y., Toki, S., Habu, Y., Ugaki, M. and Ohashi, Y. (1999). Transgenic rice plants expressing a trypsin inhibitor are resistant against rice stem borers, *Chilo suppressalis. Entomologia Experimentalis et Applicata* **93:** 173–78.

Moellenbeck, D.J., Peters, M.L., Bing, J.W., Rouse, J.R., Higgins, L.S., Sims, L., Nevshemal, T., Marshall, L., Ellis, R.T. and Bystrak, P.G. *et al.* (2001). Insecticidal proteins from *Bacillus thuringiensis* protect corn from corn rootworms. *Nat Biotechnol,* **19:** 668–72.

Mollers, C. and Wenzel, G. (1991). Somatic hybridization dihaploid potato protoplasts as a tool for potato breeding. *Botanica acta,* **105:** 133–39.

Moriyama, Y., Sakamoto, T., Karpova, S.G., Matsumoto, A., Noji, S. and Tomioka, K. (2008). RNA interference of the clock gene period disrupts circadian rhythms in the cricket *Gryllus bimaculatus. J. Biol. Rhythms*, **23(4):** 308–18.

Morton, R.L., Schroeder, H.E., Bateman, K.S., Chrispeels, M.J., Armstrong, E. and Higgins, T.J.V. (2000). Bean alpha-amylase inhibitor 1 in transgenic peas (*Pisum sativum*) provides complete protection from pea weevil (*Bruchus pisorum*) under field conditions. *Proceedings National Academy of Sciences USA,* **97:** 3820–25.

Murray, C., Sutherland, P.W., Phung, M.M., Lester, M.T., Marshall, R.K. and Christeller, J.T. (2002). Expression of biotin-binding proteins, avidin and streptavidin, in plant tissues using plant vacuolar targeting sequences. *Transgenic Res,* **11:** 199–214.

Naimov, S., Dukiandjiev, S. and de Maagd, R.A. (2003). A hybrid *Bacillus thuringiensis* delta-endotoxin gives resistance against a coleopteran and a lepidopteran pest in transgenic potato. *Plant Biotechnol J.,* **1:** 51–57.

Nayak, P., Basu, D., Das, S., Basu, A., Ghosh, D., Ramakrishnan, N.A., Ghosh, M. and Sen, S.K. (1997). Transgenic elite indica rice plants expressing CryIAc d-endotoxin of *Bacillus thuringiensis* are resistant against yellow stem borer (*Scirpophaga incertulas*). *Proc. Natl. Acad. Sci. USA.* **94:** 2111–16.

Niu, B.L., Shen, W.F., Liu, Y., Weng, H.B., He, L.H., Mu, J.J., Wu, Z.L., Jiang, P., Tao, Y.Z. and Meng, Z.Q. (2008). Cloning and RNAi–mediated functional characterization of *MaLac2* of the pine sawyer, *Monochamus alternatus. Insect Mol. Biol.,* **17(3):** 303–312.

Niu, C., Hinchliffe, D.J., Cantrell, R.G., Wang, C., Roberts, P.A. and Zhang, J. (2007). Identification of Molecular markers associated with root-knot nemarode resistance in Upland Cotton. *Crop Science,* **47:** 951–60.

Nutt, K.A., Allsopp, P.G., Geijskes, R.J., McKeon, M.G., Smith, G.R. and Hogarth, D.M. (2001). Canegrub resistant sugarcane. *Proceedings of the XXIV Congress*, Brisbane, Australia, 17–21 September, 2001. *International Society of Sugarcane Technologists,* **2:** 582–84.

Nutt, K.A., Allsopp, P.G., McGhie, T.K., Shepherd, K.M., Joyce, P.A., Taylo, G.O., McQualter, R.B., Smith, G.R. and Hogarth, D.M. (1999). Transgenic sugarcane with increased resistance to canegrubs. *In*: *Proceedings of the 1999 Conference of the Australian Society of Sugarcane Technologists*, 27–30 April, 1999. Townsville, Brisbane, Australia, pp. 171–76.

Oerke, E.C., Dehne, H.W., Schonbeck, F. and Weber A. (1994). Crop production and crop protection: Estimated losses in major food and cash crops. Elsevier, Amsterdam, The Netherlands.

Ohl, S.A., Van der Lee, F.M. and Sijmons, P.C. (1997). Anti-feeding structure approaches to nematode resistance. *In*: Cellular and Molecular Aspects of Plant–Nematode Interactions (*Eds.*: Fenoll, C., Grundler, F.M.W. and Ohl, S.A.). Kluwer Academic Publishers, Dordrecht, Netherlands. p. 286.

Orozco–Cardenas, M., McGurl, B. and Ryan, C.A. (1993). Expression of an antisense prosystemin gene in tomato plants reduces resistance toward *Manduca sexta* larvae. *PNAS USA,* **90:** 8273–76.

Ostlie, K.R., Hutchinson, W.D. and Hellmich R.L. (1997). *Bt* corn and European corn borer. North Central Regional publication 602. University of Minnesota St. Paul.

Perlak, F.J., Deaton, R.W., Armstrong, T.A., Fuchs, R.L., Sims, S.R., Greenplate, J.T. and Fischhoff, D.A. (1990). Insect resistant cotton plants. *Biol. Technol.,* **8:** 939–43.

Perlak, F.J., Stone, T.B., Muskopf, Y.M., Petersen, L.J., Parker, G.B., McPherson, S.A., Wyman, J., Love, S., Reed, G., Biever, D. and Fischhoff, D.A. (1993). Genetically improved potatoes: Protection from damage by Colorado potato beetles. *Plant Mol. Biol.,* **22:** 313–21.

Pineda, O., Bonierbale, M.W., Plaisted, R.L., Brodie, B.B. and Tanksley, S.D. (1993). Identification of RFLP markers linked to the H1 gene conferring resistance to the potato cyst nematode *Globodera rostochiensis*. *Genome,* **36(1):** 152–56.

Poulsen, M., Kroghsbo, S., Schroder, M., Wilcks, A., Jacobsen, H., Miller, A., Frenzel, T., Danier, J., Rychlik, M., Shu,Y., Emami, K., Sudhakar, D., Gatehouse, A., Engel, K.H. and Knudsen, I.B. (2007). A 90-day safety study in Wistar rats fed genetically modified rice expressing snowdrop lectin *Galanthus nivalis* (GNA). *Food Chem Toxicol,* **45:** 350–63.

Prescott, V.E. and Hogan, S.P. (2005). Genetically modified plants and food hypersensitivity diseases: Usage and implications of experimental models for risk assessment. *Pharmacology and Therapeutics,* **111(2):** 374–83.

Price, D.R.G. and Gatehouse, J.A. (2008). RNAi–mediated crop protection against insects. *Trends in Biotechnology,* **26(7):** 393–400.

Raina, S.K., Talwar, D., Nayak, N.R., Khanna, H.K. and Grover, M. (2002). Field evaluation and generation of two-gene *Bt* transgenics of indica rice. *In*: Abstract Intl. Rice Cong., Bejing, China, p. 287.

Rajamohan, F., Alzate, O., Cotrill, J.A., Curtis, A. and Dean, D.H. (1996). Protein engineering of *Bacillus thuringiensis* delta-endotoxin: mutations at domain II of CryIAb enhance receptor affinity and toxicity toward gypsy moth larvae. *Proc Natl Acad Sci USA,* **93:** 14338–43.

Ramesh, S., Nagadhara, D., Reddy, V.D. and Rao, K.V. (2004). Production of transgenic indica rice resistant to yellow stem borer and sap–sucking insects, using super-binary vectors of *Agrobacterium tumefaciens*. *Plant Sci,* **166:** 1077–85.

Rigev, A., Keller, M., Strizhov, M., Sneh, B., Prudovsky, E., Chet, I., Ginzberg, I., Koncz, Z., Schell, J. and Zilberstein, A. (1996). Synergistic activity of a *Bacillus thuringinensis* δ-endotoxin and bacterial endochitinase against *Spodoptera littoralis* larvae. *Applied Environmental Microbiology,* **62:** 3581–86.

Ripoll, C., Favery, B., Lecomte, P., Van Damme, E., Peumans, W., Abad, P. and Jouanin, L. (2003). Evaluation of the ability of lectin from snowdrop (*Galanthus nivalis*) to protect plants against root-knot nematodes. *Plant Science,* **164:** 517–23.

Ross, M.A. and Lembi, C.A. (1985). Applied weed science. Burgess Publishing Co., Minneapolis, p. 340.

Rossi, M., Goggin, F.L., Milligan, S.B., Kaloshian, I., Ullman, D.E. and Williamson, V.M. (1998). The nematode resistance gene Mi of tomato confers resistance against the potato aphid. *PNAS USA,* **95:** 9750–54.

Rosso, M.N., Dubrana, M.P., Cimbolini, N., Jaubert, S. and Abad, P. (2005). Application of RNA interference to root–knot nematode genes encoding esophageal gland proteins. *Molecular Plant Microbe Interactions,* **18:** 615–20.

Rouppe van der Voort, J.R., Kanyuka, K., Van der Vossen, E., Bendahmane, A., Mooijiman, P., Klein Lankhorst, R., Stiekman, W., Baulcombe, D. and Bakker, J. (1999). Tight physical linkage of the nematode resistance gene *Gpa2* and the virus resistance gene *Rx* on single segment introgressed from wild species *Solanum tuberosum* subsp. *andigena* CPC1673 into cultivated potato. *Molecular Plant Microbe Interactions,* **12:** 197–206.

Saha, P., Dasgupta, I. and Das, S. (2006b). A novel approach for developing resistance in rice against phloem limited viruses by antagonizing the phloem feeding hemipteran vectors. *Plant Mol Biol,* **62:** 735–52.

Saha, P., Majumder, P., Dutta, I., Ray, T., Roy, S.C. and Das, S. (2006a). Transgenic rice expressing *Allium sativum* leaf lectin with enhanced resistance against sap-sucking insect pests. *Planta,* **223:** 1329–43.

Salentjin, E.M.J., Arens De–Reuver, M.J.B., Lange, W., Bock, T.S.M., Stiekema, W.J. and Klein–Lankhorst, R.M. (1995). Isolation and characterization of RAPD–based markers linked to the beet cyst nematode resistance locus (*Hs1^{pat1}*) on chromosome 1 of *B. patellaris*. *Theor. Appl. Genet.,* **90:** 885–91.

Sanyal, I., Singh, A.K., Meetu, K. and Devindra, A.V. (2005). Agrobacterium–mediated transformation of chickpea (*Cicer arietinum* L.) with *Bacillus thuringiensis* cry1Ac gene for resistance against pod borer insect *Helicoverpa armigera*. *Plant Science,* **168:** 1135–46.

Satyendra, R., Stewart, J.M. and Wilkins, T. (1998). Assessment of resistance of cotton transformed with lectin genes to tobacco budworm. *In:* Special Report. Fayetteville: Arkansas Agricultural Experiment Station, University of Arkansas, pp. 95–98.

Schlüns H. and Crozier R.H. (2007). Relish regulates expression of antimicrobial peptide genes in the honeybee, *Apis mellifera*, shown by RNA interference. *Insect Mol. Biol.* **16(6):** 753–59.

Schnee, C., Kollner, T.G., Held, M., Turlings, T.C.J., Gershenzon, J. and Degenhardt, J. (2006). The products of a single maize sesquiterpene synthase form a volatile defense signal that attracts natural enemies of maize herbivores. *Proc. Natl. Acad. Sci. USA* **103:** 1129–34.

Schroeder, H.E., Gollasch, S., Moore, A., Tabe, L.M., Craig, S., Hardie, D.C., Chrispeels, M.J., Spencer, D. and Higgins, T.J.V. (1995). Bean alpha-amylase inhibitor confers resistance to pea weevil (*Bruchus pisorum*) in transgenic peas (*Pisum sativum* L.). *Plant Physiology,* **107:** 1233–39.

Scolari, F., Siciliano, P., Gabrieli, P., Gomulski, L.M., Bonomi, A., Gasperi, G. and Malacrida, A.R. (2011). Safe and fit genetically modified insects for pest control: from lab to field applications. *Genetica,* **139:** 41–52.

Shade, R.E., Schroeder, H.E., Pueyo, J.J., Tabe, L.M., Murdock, L.L., Higgins, T.J.V. and Chrispeels, M.J. (1994). Transgenic pea seeds expressing alpha-amylase inhibitor of the common bean are resistant to bruchid beetles. *Bio/Technology,* **12:** 793–96.

Sharma, H.C., Sharma, K.K., Seetharama, N. and Crouch, J.H. (2003). The utility and management of transgenic plants with *Bacillus thuringiensis* genes for protection from pests. *New Seeds Journal,* **5:** 53–76.

Shingles, J., Lilley, C.J., Atkinson, H.J. and Urwin, P.E. (2007). *Meloidogyne incognita:* Molecular and biochemical characterization of a cathepsin L cysteine proteinase and the effect on parasitism following RNAi. *Experimental Parasitology,* **115:** 114–20.

Shu, Q., Ye, G., Cui, H., Cheng, X., Xiang, Y., Wu, D., Gao, M., Xia, Y., Hu, C., Sardana, R. and Altosaar, I. (2000). Transgenic rice plants with a synthetic *cry1Ab* gene from *Bacillus thuringiensis* were highly resistant to eight lepidopteran rice pest species. *Mol. Breed.*, **6:** 433–39.

Shukle, R.H. and Murdock, L.L. (1983). Lipoxygenease, trypsin inhibitor, and lectin from soybeans: effect on larval growth of *Manduca sexta* (Lepidoptera: Sphingidae). *Environmental Entomology,* **12:** 787–91.

Singsit, C., Adang, M.J., Lynch, R.E., Anderson, W.F., Wang, A., Cardineau, G. and Ozias–Akins, P. (1997). Expression of a *Bacillus thuringiensis cryIA(c)* gene in transgenic peanut plants and its efficacy against lesser cornstalk borer. *Transgenic Research,* **6:** 169–76.

Skorupska, H.T., Choi, I.S., Rao–Arelli, A.P. and Bridges, W.C. (1994). Resistance to soybean cyst nematode and molecular polymorphism in various sources of Peking soybean. *Euphytica,* **75(1):** 63–70.

Smigocki, A., Neal, J.W.Jr., McCanna, I. and Douglass, L. (1993). Cytokinin–mediated insect resistance in *Nicotiana* plants transformed with the *ipt* gene. *Plant Molecular Biology,* **23:** 325–35.

Soberon, M., Pardo–Lopez, L., Lopez, I., Gomez, I., Tabashnik, B.E. and Bravo, A. (2007). Engineering modified Bt toxins to counter insect resistance. *Science,* **318:** 1640–42.

Stewart, S.D., Adamczyk, J.J., Knighten, K.S. and Davis, F.M. (2001). Impact of *Bt* cottons expressing one or two insecticidal proteins of *Bacillus thuringiensis* Berliner on growth and survival of noctuid (*Lepidoptera*) larvae. *J. Econ. Entomol.,* **94:** 752–60.

Stoger, E., Williams, S., Christou, P., Down, R.E. and Gatehouse, J.A. (1999). Expression of the insecticidal lectin from snowdrop (*Galanthus nivalis* agglutinin; GNA) in transgenic wheat plants: Effects on predation by the grain aphid *Sitobion avenae. Molecular Breeding,* **5:** 65–73.

Su, N., Sun, M., Yang, B., Meng, K., Liu, C.Y., Ni, P.C. and Shen, G.F. (2002). The insect resistance of OC and *Bt* transplastomic plants and the phenotype of their progenies. *Hereditas Beijing,* **24:** 288–92.

Sumitani, M., Yamamoto, D.S., Lee, J.M. and Hatakeyama, M. (2005). Isolation of white gene orthologue of the sawfly, *Athalia rosae* (Hymenoptera) and its functional analysis using RNA interference. *Insect Biochem. Mol. Biol.,* **35(3):** 231–40.

Tabashnik, B.E., Dennehy, T.J. and Carrie're, Y. (2005). Delayed resistance to transgenic cotton in pink bollworm. *Proc. Natl. Acad. Sci. USA,* **102:** 15389–93.

Takahashi, T., Hamada, A., Miyawaki, K., Matsumoto, Y., Mito, T., Noji, S. and Mizunami, M. (2009). Systemic RNA interference for the study of learning and memory in an insect. *Journal of Neuroscience Methods,* **179(1):** 9–15.

Tattersall, D.B., Bak, S., Jones, P.R., Olsen, C.E., Nielsen, J.K., Hansen, M.L., Hoj, P.B. and Moller, B.L. (2001). Resistance to an herbivore through engineered cyanogenic glucoside synthesis. *Science,* **293:** 1826–28.

Thomas, C. and Cottage, A. (2006). Genetic Engineering for Resistance. *In*: Plant Nematology (*Eds.*: Perry, R.N. and Moens, M.). CABI Publishers, UK. p. 447.

Thomas, J.C., Wasmann, C.C., Echt, C., Dunn, R.L., Bohnert, H.J. and McCoy, T.J. (1994). Introduction and expression of an insect proteinase inhibitor in alfalfa (*Medicago sativa* L.). *Plant Cell Reports,* **14:** 31–36.

Turner, C.T., Davy, M.W., MacDiarmid, R.M., Plummer, K.M., Birch, N.P. and Newcomb, R.D. (2006). RNA interference in the light brown apple moth, *Epiphyas postvittana* (Walker) induced by double-stranded RNA feeding. *Insect Mol Biol,* **15:** 383–91.

Urwin, P.E., Lilley, C.J. and Atkinson, H.J. (2002). Ingestion of double stranded RNA by preparasitic juvenile cyst nematodes leads to RNA interference. *Molecular Plant Microbe Interactions,* **15:** 747–52.

Vaeck, M., Reynaerts, A., Hoftey, H., Jansens, S., DeBeuckleer, M., Dean, C., Zabeau, M., Van Montagu, M. and Leemans, J. (1987). Transgenic plants protected from insect attack. *Nature,* **327:** 33–37.

Vain, P., Worland, B., Clarke, M.C., Richard, G., Beavis, M., Liu, H., Kohli, A., Leech, M., Snape, J. and Christou, P. (1998). Expression of an engineered cysteine proteinase inhibitor (Oryzacystatin–I delta D86) for nematode resistance in transgenic rice plants. *Theoretical and Applied Genetics,* **96:** 266–71.

Valentine, T.A., Randall, E., Wypijewski, K., Chapman, S., Jones, J. and Oparka, K.J. (2007). Delivery of macromolecules to plant parasitic nematodes using a tobacco rattle virus vector. *Plant Biotechnology Journal,* **5:** 827–34.

Van der Vossen, E.A.G., Van der Voort, J.N.A.M.R., Kanyuka, K., Bendahmane, A., Sandbrink, H., baulcombe, D.C., Bakker, J., Stiekema, W.J. and Klein–Lankhorst,

R.M. (2000). Homologues of a single resistance-gene cluster in potato confer resistance to distinct pathogens: a virus and a nematode. *The Plant Journal,* **23(5):** 567–76.

Vaughn, T., Cavato, T., Brar, G., Coombe, T., DeGooyer, T., Ford, S., Groth, M., Howe, A., Johnson, S. and Kolacz, K. *et al.* (2005). A method of controlling corn rootworm feeding using a *Bacillus thuringiensis* protein expressed in transgenic maize. *Crop Sci.* **45:** 931–38.

Walshe, D.P., Lehane, S.M., Lehane, M.J. and Haines, L.R. (2009). Prolonged gene knockdown in the tsetse fly *Glossina* by feeding double stranded RNA. *Insect Mol. Biol.,* **18(1):** 11–19.

Wang, E., Wang, R., DeParasis, J., Loughrin, J.H., Gan, S. and Wagner, G.J. (2001). Suppression of a P450 hydroxylase gene in plant trichome glands enhances natural–product–based aphid resistance. *Nat Biotechnol,* **19:** 371–74.

Wasmann, C.C., Echt, C., Dunn, R.L., Bohnert, H.J. and McCoy, T.J. (1994). Introduction and expression of an insect proteinase inhibitor in alfalfa (*Medicago sativa* L.). *Plant Cell Reports,* **14:** 31–36.

Webb, D.M., Baltazar, B.M., Rao–Arelli, A.P., Schupp, J., Clayton, K., Keim P. and Beavis, W.D. (1995). Genetic mapping of soybean cyst nematode race-3 resistance loci in the soybean PI 437.654. *Theoretical and Applied Genetics,* **91(4):** 574–81.

Whitfield, J. (2003). Transgenic cotton a winner in India. Nature News Service, Macmillan Magazines Ltd. Feb. 7, 2003.

Williams, K.J., Fisher, J.M. and Langridge, P. (1994). Identification of RFLP markers linked to the cereal cyst nematode resistance gene (cre) in wheat. *Theoretical and applied Genetics,* **89:** 927–30.

Williamson, V.M. (1998). Root-knot resistance genes in tomato and their potential for future use. *Annual Review of Phytopathology,* **36:** 277–93.

Williamson, V.M. (1999). Plant nematode resistance genes. *Currents Opinion in Plant Biology* **2:** 327–31.

Wilson, F.D., Flint, H.M., Deaton, W.R., Fischhoff, D.A., Perlak, F.J., Armstrong, T.A., Fuchs, R.L., Berberich, S.A., Parks, N.J. and Stapp, B.R. (1992). Resistance of cotton lines containing a *Bacillus thuringiensis* toxin to pink bollworm (Lepidoptera: Gelechiidae) and other insects. *J. Eco. Entomol.,* **85:** 1516–21.

Wu, S.J., Koller, C.N., Miller, D.L., Bauer, L.S. and Dean, D.H. (2000). Enhanced toxicity of *Bacillus thuringiensis* Cry3A delta-endotoxin in coleopterans by mutagenesis in a receptor binding loop. *FEBS Lett,* **473:** 227–32.

Wu, Y.R., Llewellyn, D., Mathews, A. and Dennis, E.S. (1997). Adaptation of *Helicoverpa armigera* (Lepidoptera: Noctuidae) to a proteinase inhibitor expressed in transgenic tobacco. *Molecular Breeding,* **3:** 371–80.

Xu, D.P., Xue, Q.Z., McElroy, D., Mawal, Y., Hilder, V.A. and Wu, R. (1996). Constitutive expression of a cowpea trypsin inhibitor gene, *CpTi,* in transgenic rice plants confers resistance to two major rice insect pests. *Molecular Breeding,* **2:** 167–73.

Yadav, B.C., Veluthambi, K and Subramaniam, K. (2006). Host-generated double stranded RNA induces RNAi plant-parasitic nematodes and protects the host from infection. *Mol. Biochem. Parasitol.,* **148:** 219–22.

Yaghoobi, J., Kaloshian, I., Wen, Y. and Williamson, V.M. (1995). Mapping a new nematode resistance locus in *Lycopersicon peruvianum*. *Theoretical and Applied Genetics,* **91(3):** 457–64.

Yang, C.D., Tang, K.X., Wu, L.B., Li, Y., Zhao, C.Z., Liu, G.J. and Shen, D.L. (1998). Transformation of haploid rice shoots with snowdrop lectin gene (*GNA*) by *Agrobacterium*–mediated transformation. *Chinese Journal of Rice Science,* **12:** 129–33.

Yang, G.D., Zhu, Z., Li, Y., Zhu, Z.J., Xiao, G.F. and Wei, X.L. (2002). Obtaining transgenic plants of Chinese cabbage resistant to *Pieris rapae* L. with modified CpTI gene (*sck*). *Acta Horticulturae Sinica,* **29:** 224–28.

Ye, G., Yao, H., Cui, H., Cheng, X., Hu, C. and Xia, Y. (2001). Field evaluation of resistance of transgenic rice containing a synthetic cry1Ab gene from *Bacillus thuringiensis* Berliner to two stem borers. *J. Econ. Entomol.*, **94:** 271–76.

Yeh, K.W., Lin, M.L., Tuan, S.J., Chen, Y.M., Lin, C.Y. and Kao, S.S. (1997). Sweet potato (*Ipomoea batatas*) trypsin inhibitors expressed in transgenic tobacco plants confer resistance against *Spodoptera litura. Plant Cell Reports,* **16:** 696–99.

Zhang, J. H., Wang, C.Z., Qin, J.D. and Guo, S.D. (2004). Feeding behaviour of *Helicoverpa armigera* larvae on insect-resistant transgenic cotton and non-transgenic cotton. *Journal of Applied Entomology*, **128(3):** 218–25.

Zhao, J.Z., Cao, J., Collins, H.L., Bates, S.L., Roush, R.T., Earle, E.D. and Shelton, A.M. (2005). Concurrent use of transgenic plants expressing a single and two *Bacillus thuringiensis* genes speeds insect adaptation to pyramided plants. *Proc Natl Acad Sci USA,* **102:** 8426–30.

Zhao, J.Z., Cao, J., Li, Y.X., Collins, H.L., Roush, R.T., Earle, E.D. and Shelton, A.M. (2003). Transgenic plants expressing two *Bacillus thuringiensis* toxins delay insect resistance evolution. *Nat Biotechnol,* **21:** 1493–97.

Zhao, J.Z., Shi, X.P., Fan, X.L., Zhang, C.Y., Zhao, R.M. and Fan, Y.L. (1998). Insecticidal activity of transgenic tobacco co-expressing *Bt* and *CpTI* genes on *Helicoverpa armigera* and its role in delaying the development of pest resistance. *Rice Biotechnology Quarterly,* **34:** 9–10.

Zhao, Y.Y., Yang, G., Wang–Pruski, G. and You, M.S. (2008). *Phyllotreta striolata* (Coleoptera: Chrysomelidae): arginine kinase cloning and RNAi–based pest control. *European Journal of Entomology,* **105:** 815–22.

Zhou, X.G., Wheeler, M.M., Oi, F.M. and Scharf, M.E. (2008). RNA interference in the termite, *Reticulitermes flavipes* through ingestion of double-stranded RNA. *Insect Biochemistry and Molecular Biology,* **38:** 805–15.

Zhu, Q., Arakane, Y., Beeman, R.W., Kramer, K.J. and Muthukrishnan, S. (2008). Functional specialization among insect chitinase family genes revealed by RNA interference. *PNAS USA,* **105(18):** 6650–55.

6

Recent Advances in Cultural Management of Crop Pests

R. Gopi[1], S. Ramesh Babu[2] and Chandan Kapoor[3]

ABSTRACT

Cultural control is a method of crop protection using careful timing and a combination of agronomic practices to make the environment less favorable for the increase of certain pests or diseases. For centuries, farmers used only nonchemical methods of pest control including cultural method. Growing concern of environmental pollution and pesticide resistance has made the scientists, environmentalists, policy makers and even farmers to look upto conventional methods like cultural methods. Cultural methods can be used alone or as components of pest management programmes. Cultural control methods are cheap, simple, effective farmer– and eco–friendly, safe and easy to use, but require detailed knowledge of pest and plant biology and ecology and it may be necessary to integrate several practices for maximum pest control. The various cultural control measures discussed in this chapter are site selection, organic amendments, tillage, sanitation, sowing and planting, spacing, irrigation, manuring, mulching, pruning, cover crops, trap crops, inter crop and mixed crops, which are effective not only in managing the pest population but also effective in maintaining natural enemies and antagonistic microorganisms that parasitize the pest population.

Key words: Cultural management, Crop pest, Plant diseases.

1 Plant Pathology, ICAR Research Complex for NEH Region, Sikkim Centre, Tadong, Gangtok - 737 102, India.
2 Dept of Entomology, Agricultural Research Station, MPUAT, Banswara, Rajasthan - 327 001, India.
3 Plant Breeding, ICAR Research Complex, Sikkim Centre, Tadong, Gangtok - 737 102, India.
Corresponding author: E-mail: ramaraj.muthu.gopi2@gmail.com

1. INTRODUCTION

Cultural control methods are the oldest methods of pest control. With the introduction of synthetic chemicals using more sophisticated technology and their quick action, cultural control methods were given less importance. In recent years cultural control is one of the specific areas of research which has generated interest among the scientific community in order to lessen the risk of pesticides. Certainly chemical protection is the most effective method of pest control; however healthy cultural operations improve the degree of control achieved through pesticide application. Sometime pesticide spray in both horticultural and arable crops is not economically feasible and environmentally viable. Pesticides spray is sometimes ineffective or erratic in their performance on the pest population.

Cultural control refers to that broad set of management technological options which may be manipulated by agricultural procedures to achieve their crop production goals. Cultural control measures such as field hygiene or prevention of pest immigration are fundamental for the effective control of pests (Yano, 2004). Begum *et al.* (1995) reported that high management practices reduced leaf spotting of *Alternaria* by 59% and also 55% higher seed yield was obtained. Similarly cultural practices reported to be very ineffective in controlling rust in Asparagus (Mullen and Viss, 1996), *Thielaviopsis basicola* in Citrus (Graham, 1994), *Meloidogyne incognita* population in Cowpea (McSorley, 1994), *Maruca vitrata* pod borer in Cowpea (Sharma, 1998).

Cultural control measures are important in Integrated Pest Management (IPM), because of their preventive nature, yet cultural practices have received little attention, possibly due to the difficulty of testing by conventional methods. Growers may also be reluctant to adopt as they require significant changes in conventional cropping practices. The cultural practices (CP) are classified into three categories according to Katan (2000). In this regard there are three categories of CP: (a) CP for regular purposes which can also be used for disease control, *e.g.*, irrigation; (b) CP which is used solely or mainly for pest control, *e.g.*, sanitation; (c) CP which can be used for both agricultural purposes and pest control, *e.g.*, crop rotation. CP for pest control can be used before, at or after planting. They include crop rotation, fallow, flooding, deep ploughing, flaming, soil solarization–which involves a combination of physical and biological processes, adjusting planting date, irrigation, fertilization, compost, weed control, herbicide application sanitation, tillage and others.

2. CULTURAL PRACTICES

2.1. Site Selection

Site selection plays a key role in pest and disease management. Site should be ideal for the crop and the natural enemies of the pest but unfavorable for the pest itself. A site should have good soil drainage. Heavy, poorly drained fields should be avoided as poor drainage can stress crop plants and increase the risk of root rot and damping off. A site with good air drainage is very important where the plants are exposed to direct sunlight. Planting with rows parallel to prevailing winds will promote fast drying of the foliage and fruit which will help prevent fruit rots and leaf diseases such as leaf spot, leaf blight and leaf scorch.

2.2. Organic Amendments

Organic soil amendments (Fig. 1), including animal and green manures and wastes from processed animal products such as blood meal, bone meal, have been used for centuries as soil supplements. Organic amendments increase the resistance of plants to pests and diseases. It also induces activity of various antagonistic microorganisms. Some release insecticidal, nematicidal and fungicidal compounds during decomposition. In the chemical era the impact of amendments on plant diseases was ignored, although in the past, the use of amendments was a recognized means for disease control. Thus organic amendments are good candidates as a replacement for methyl bromide used for fumigation for soil borne pest and disease control. The effect of organic treatments persists for several years. Organic amendments, unlike fumigants, reduce populations of pathogens while increasing those of most other soil microorganisms. The suppression of the disease appeared to be related to an increased microbial activity in the soil amended with organic matter (Osunlaja, 1989). The mechanism of disease control for high-nitrogen–containing amendments is the generation of ammonia and/or nitrous acid following degradation of the amendments by microorganisms. McSorley and Gallaher (1995) studied the effects of organic amendment and crop establishment on nematode densities of *Paratrichodorus minor, Pratylenchus* spp. and *Criconemella* spp. in various vegetable crops. In compost treated plants of okra and yellow squash final densities of *Paratrichodorus* and *Pratylenchus* spp. were low. Whereas final densities of *Meloidogyne incognita* was high and not affected by compost treatment. Integrated treatment with phytosanitary cultural practices and organic soil amendment suppressed the root-knot nematode in brinjal (Chakraborti, 2001) and Ufra disease in rice (Chakraborti, 1999). The fruit damage by American boll worm was significantly lower in compost-treated early-maturing and small-fruited cultivars of tomato Punjab Kesri (Kaur and Singh, 2000). Survival of *Fusarium oxysporum* f. sp. *asparagi*, *Rhizoctonia solani*, and *Verticillium dahliae* in inoculum samples buried 15 cm deep was strongly reduced in soil amended with broccoli and grass and covered with polythene sheet (Blok *et al.*, 2000). Soil amended with stalks of sunflower, alfalfa and Hungarian vetch reduced significantly onion bulb rot caused by *Fasarium oxysporum* f.sp. *cepae* and *Aspergillus niger* in a soil naturally infested with the pathogens under field conditions.

Various tradeoffs suggest that farmers should alternate between conventional and minimum tillage with frequent additions of organic matter to enhance several aspects of soil quality and reduce pest, disease and yield problem that can occur with continuous minimum tillage (Jackson *et al.*, 2004). The root wilt disease intensity was reduced by crop diversification and recycling of organic residues from home-shed farms in coconut (Anithakumari, 2007).

Fig. 1: Organic amendment (FYM)

2.3. Tillage

Two primary tillage systems currently being utilized in modern agriculture are conventional tillage (Fig. 2.) and conservation tillage. In recent years conservation tillage has increased significantly throughout the United States over the past decade (Dill Macky and Jones, 2000). No-tillage cultivation of soybean results in incidence of some diseases and a reduction in that of others (Vallone, 1998). While the risk of foliar diseases may increase over time, soil-borne diseases may decrease in no-till conservation tillage systems because of increased biological activity and beneficial microorganisms. Dispersal of soil inocula of Phytophthora blight of bell pepper was suppressed and final disease incidence was 2.5 to 43% when stubble from a fall-sown, no-till, wheat cover crop was present (Ristaino *et al.*, 1997). The conventional tillage after harvest under furrow irrigation decreased the degree of aggregation of sclerotia of *Sclerotinia minor* in lettuce after each season but the distribution pattern of sclerotia changed little by the associated sub-surface drip irrigation (Wu and Subbarao, 2003). Brown *et al.* (1996) reported that tillage systems affect population of thrips vector on ground nut. The mean number of larvae and percentage tiller infestation were significantly less in the combination of cultural practices having conventional tillage as a component having no tillage. Incidence and severity of Fusarium head blight was lower in mold board-plowed plots than in either chisel plow or no-till plots. In mold board-plowed plot burying residues has been reported to favor the decline of residue-associated pathogens in residue of wheat including *Fusarium graminearum* (Sutton and Vyn, 1990). Tillage practices that destroy residue such as burning and ploughing are effective against diseases such as tan spot, but the trend is towards more reduced or no till farming. Many factors are responsible for the increasing incidence of the pathogens in no-tillage soybean system. The most important being high soil fertility and soil water conservation which promote vegetative growth and in some cases lodging. Root disease severity of wheat was less under zero tillage than conventional tillage (Bailey *et al.*, 2000). Traditional tillage reduced sheath blight incidence in rice (Rodriguez *et al.*, 1999). No-till appeared to reduce densities of *Phyllotreta cruciferae* in crucifers compared to conventional tillage. The greater structural diversity of the no-till plots might be interfering with host plant location (Milbrath *et al.*, 1995).

Fig. 2: Tillage operation

2.4. Sanitation

Field sanitation (Fig. 3) is the removal or destruction of diseased plant residues. Sanitation involves eradication of weed hosts, alternate and collateral hosts, and infested plant debris including fruits, flowers, plant debris, and cleaning of field borders. Sanitation breaks pest or disease cycle. Sanitation works because the part of the plant that contains the pathogen (or inoculum) are removed before it spreads and infects another plant or another part of the same plant. The removal of some diseases whenever they appear helps to reduce the chance of diseases spreading during the growing season. Field sanitation is recommended for the damping off diseases (Gutierrez, *et al.,* 1997) and in combination with fungicides for fruit rots in strawberry (Legard *et al.,* 1997). Sanitation forms an integral part of the management of *Botrytis* in greenhouses (Hausbeck and Moorman, 1996). Cutting of wild grapevines is a cultural control strategy for grape berry moth *Paralobesia viteana* Clemens (Jenkins and Isaacs, 2007). Gaur and Singh (1995) reported stubble burning before chick pea sowing reduced *Ascochyta* infection. Information on cultural practices obtained from interviews and field visits indicates that practices such as late harvesting and the leaving behind of residues of previous sorghum, millet and maize crops increased termite infestation. However the cutting down and burning of the vegetation in a farm before planting significantly reduced the percentage of plants attacked by termites (Umeh *et al.*, 2001). Rice hull burning (RHB) is a traditional cultural practice of many onion growers in San Hose city, Nueava Dciha, Philippines. Burning of 15 cm deep rice hulls significantly reduced the *Meloidogyne graminicola* in the soil and increased onion yield. Moreover increasing the thickness of rice hull burned to 30 cm deep resulted in a yield increase to 44.2% over no RHB and 11.9% over 15 cm deep rice hulls (Gergon *et al.*, 2001). Brown *et al.* (1996) reported that volunteer ground nuts may play an important role as fall and spring reservoirs for *Tomato spotted wilt virus* (TSWV) and reduction of volunteers may be an important component of an integrated disease management programme. Maintaining weed free plots until 70 days after sowing was effective in controlling pod boring insects including *Euborellia stali* and *Dolylus* sp. in ground nut (Senguttavan and Dhanakodi, 1997). Populations of *Helicotylenchus dihystera* increased and crop yield decreased as weeding and hoeing interval increased in French bean (Sharma and Bhatia, 2002).

Fig. 3: Sanitation

2.5. Sowing and Planting

Sowing and planting are the very important operations that decide the appearance and spread of pest and diseases in crop plants. Sharma and Bhatia (2002) reported that sowing time had significant effect on nematode population and the crop yield in French bean. Knodel *et al.* (2008) studied the effect of two spring planting dates of *Brassica napus* and different insecticide based management strategies on the feeding injury caused by flea beetles. Adult beetle emergence coincided with the emergence of the early planted canola, and this resulted in greater feeding injury in the early planted canola than late planted one. Sunflower beetle adult, *Zygogramma exclamationis* and larval populations decreased as planting date was delayed (Charlet and Knodel, 2003). Later transplanting of paddy induced higher population of *Nilaparvata lugens, Sogetella furcifera, Laodelphax striatellus* and *Nephotettix cincticeps*; and *Cnaphalocrocis medinalis* caused greater damage in earlier transplanted paddy fields (Lee and Ma, 1997). Later sowing resulted in severe infestations of leaf miner *Hydrellia prosternalis* of rice (Sherif *et al.*, 1997). Decrease in seed rates, line sowing in blond psyllium decreased the downy mildew (Rathore, 2002). Late sowing of maize cv. Partap resulted in lowest dead heart formation by the stem borer *Chilo partellus* (Singh and Kanta, 1999).

2.6. Spacing

Close spacing and dense canopy will encourage the faster rate of multiplication of various pests and diseases. Hence, optimum spacing and density are to be maintained to manage pest and disease problems in crop plants. Sherif *et al.* (1997) reported plots with smaller spacing after transplanting (10 × 10 and 10 × 15 cm) were less infested by rice leaf miner *Hydrellia prosternalis* than wider spaced plots two weeks after transplanting. Tripathy and Rathi (2000) reported wide row spacing of 60 cm along with other cultural practices reduced the Botrytis grey mold of chick pea. A progressive increase in row spacing corresponds to a decrease in the severity of Septoria leaf spot in tomato (Kumar and Sugha, 2000). Population build up of hoppers was higher in closer plant spacing of 10 × 10 cm compared to wider spacing of 30 × 30 cm in rice. On the contrary predatory mirid bug and spider multiplied faster in plots with sparse than high plant density. Stalk girdling began earlier in high-density plots and a larger proportion of plants were girdled compared with low-density plots on all sampling dates in both years. Certain cultural tactics, in particular reduced plant spacing, have potential to delay the onset of girdling behavior by *Dectes texanus* larvae in sunflower and thus mitigate losses that otherwise result from the lodging of girdled plants (Michaud *et al.*, 2009). Population of the mustard aphid, *Lipaphis erysimi* (Kaltenbach), increased significantly as the inter-row spacing decreased in canola (Sarwar, 2008).

2.7. Irrigation

Water is an elixir of life. Water can be used directly to suffocate insects and; indirectly by changing the overall health of the plants. Moisture is an important limiting factor that affects the survival of some pests. Irrigation management is clearly an important criterion for pest and disease control. Both water stress and water excess favors the pest and disease development. Over watering greatly favors soil-borne pathogenic fungi. Where sufficient water is available, flooding is sometimes used for insect and nematode control. Flooding is used in controlling Panama wilt in banana and leaf folder in rice. Plant bacterial pathogen also depends upon rain and sprinkler irrigation for their spread and survival. Overhead sprinkler irrigation can enhance dissemination and infectivity of foliar fungal pathogens. Permanent flooded rice plots encouraged infestation and longer irrigation intervals decreased infestation rates (Sherif *et al.*, 1997). The number of sclerotia and degree of aggregations of sclerotia within a season or year to year under subsurface drip irrigation was not significant (Wu and Subbarao, 2003). Both plastic mulch and drip irrigation was effective for controlling pod and stem blight and purple blotch without significant reduction in soybean yield (Oh and Kim, 2001). Campos and Civantos (2000) reported irrigation influences pests by modifying tree vegetative status and soil micro climatic conditions in olive. There was less foliar blight disease development with the double irrigation treatment than in the single irrigation treatment in wheat (Mahto, 2001).

2.8. Fertilizer and Manuring

The maintenance of balanced nutrition increases the resistance in crop plants to various crop diseases. Some diseases such as certain cankers are, more prevalent on plants that are deficient in mineral nutrition. Application of N and especially NPK resulted in the decrease of pest resistance in plants and increase of the damage degree. A clear indication of plant nutrition and disease appearance is given by Nagaraja *et al.* (2006) that brown spot and blast in finger millet appears in severe form where adequate fertilizer is not given. Perez–Marco (2000), reported spring nitrogen fertilization increased the number of plants affected by *Fusarium* sp. in wheat. Manuring with NPK and $ZnSO_4$ along with other cultural and chemical practices was found to be the most effective among the treatments given in controlling Fusarium wilt in guava (Hamiduzzaman *et al.*, 1999). Plant nutrition can have direct and indirect effect on pest suppression for example nitrogen will increase the vegetative growth of crop plants and also pest and disease susceptibility. On the other hand P and K are known to reduce the incidence of pests. Godika *et al.* (2001) reported that application of 40 Kg K/ha reduced the infection of Alternaria blight and white rust in Indian mustard. Pathak and Godika (2010) also reported 100%

recommended dose of fertilizer along with FYM or 100% recommended dose of fertilizer with one spray of 0.1% thiourea at 50% flowering stage lowers the incidence of Alternaria blight and white rust in Indian mustard. High nitrogenous fertilizer predisposed grapevines to infestation by *Botrytis cinerea* and increased the disease severity (Hasan, 1998). Krishnakumar and Maheswarappa (2010) reported that the yield of root wilt-affected coconut palms can be improved over the years or maintained without further deterioration by integrated nutrient management involving chemical fertilizers and organic manures.

2.9. Mulching

Mulching (Fig. 4) offers great agro-ecological potential. It enhances beneficial microbes, conserve water and soil and increases fertility and control weed. All commercial tomato growers in USA rely upon a raised bed poly ethylene mulch production system to grow their crops (Rich *et al.*, 2009). Straw spread on the soil surface suppressed final cucumber mosaic virus incidence by 25–40% (Bwye *et al.*, 1999). Rhainds *et al.* (2001) reported that reflective mulch may suppress the incidence of damage by tarnished plant bug both directly by reducing number of nymphs per flower cluster, and indirectly by enhancing productivity of strawberry plants. In plastic mulching seed infestation of soybean remarkably reduced by drip irrigation compared to overhead sprinkling but not reduced in no-mulching cultivation. Ampofo and Massamo (1998) reported that mulching with bananas and bracken fern leaves increased plant tolerance to the bean stem maggots on beans. Sometimes mulching has negative effect on crop pest management. The use of straw mulch as a pest management option for *Microtheca ochroloma* in crucifer crops on organic farms is not recommended as it increases the *M. ochroloma* in crucifers (Manrique *et al.*, 2010).

Fig. 4: Mulching

2.10. Cutting and Pruning

Removal and destruction of dead or infected branches can greatly reduce over-wintering stages of pest population. Cutting down and pruning of the vegetation in a farm before planting significantly reduced the percentage of plant attacked by termites (Umeh *et al.*, 2001). Pruning of trees enables management of tree shapes, increased growth of fruiting spurs improved fruit color and management of disease by the removal of diseased stems or dead wood that can harbor pathogens. Hasan (1998) reported removal of leaves around clusters, when practiced two or three times during the season and thinning of berries reduced significantly Botrytis bunch rot development in grape. Houma *et al.* (1998) also demonstrated that Botrytis bunch rot of grape can be suppressed by the removal of leaves from around flower clusters. Uniform nipping of tender branches of chick pea after 45 or 60 days of planting reduced the disease by Botrytis grey mold. Kumar and Sugha (2000) reported tomato plants which were staked with lower leaf defoliated developed less leaf spot caused by *Septoria lycopersici*. Summer pruning resulted in a small change in the apple canopy micro climate decreasing the hours of RH and increasing the evaporative potential and also improved spray deposition and thereby controlling fly speck disease in apple. Holb (2005) reported that strong pruning significantly decreased leaf scab on the susceptible and the moderately susceptible cultivars at all sites compared with un-pruned one in apple. Uddin and Stevenson (1998) reported reduction of Phomopsis shoot blight disease incidence by selectively removing infected shoots from the peach trees.

2.11. Cover Crops

Cover crops are those crops that are cultivated between the growing seasons of the main cash crop. Cover crops are normally grown to manage weeds, to conserve the soil moisture and nutrients, organic matter and also to manage soil-borne pests and diseases. Further, cover crop plant surfaces can support healthy populations of beneficial microorganisms, including various types of yeasts that can migrate onto a cash crop after planting or transplanting. By including cover crops in crop rotations and not spraying insecticides, beneficials are protected for further crops. They include annual grains such as millet, rye and sorghum–sudan hybrids and legumes such as vetch, clover, and cowpea. Wheat cover crop suppressed the dispersal of soil inocula of Phytophthora blight of bell pepper and final disease incidence was 2.5%–4.3% (Ristaino *et al.*, 1997). Cover crops like brassicas and many grasses can be used to manage nematodes. Cover crops with documented nematicidal properties against at least one nematode species include sorghum–sudangrass hybrids (*Sorghum bicolor X S. bicolor* var. *sudanese*), marigold (*Tagetes patula*), hairy indigo (*Indigofera hirsuta*), showy crotalaria (*Crotalaria spectabilis*), sunnhemp (*Crotalaria juncea*), velvetbean (*Mucuna deeringiana*), rapeseed (*Brassica rapa*), mustards and radish (*Raphanus satiuus*). Rye has been shown to be effective in minimizing the damage caused by root-knot nematodes (*Meloidogyne incognita*) when grown as a cover crop and then incorporated into the soil prior to planting (McBride *et al.*, 2000).

2.12. Trap Crops

In recent times crop diversification is growing interest among the farmers for insect pest management. One method of diversification is trap cropping. Trap crops (Fig. 5) are composed of one or more plant species that are grown to attract insects in order to protect the cash crop from the pest. Trap crops can be manipulated in time or space so that they attract insects at a critical period in the pests and/or the crops life cycle. Trap crops are often grown earlier than the main cash crop and usually be destroyed before the insects reproduce. The trap crop can be a different plant species, variety or just a different growth stage of the same species as the main crop, as long as it is more attractive to the pest when they are present. It necessitates that the trap crop be more attractive to the pest as either a food source or oviposition site than the main crop. Indeed, the relative attractiveness and size of the trap crop in a landscape are important factors in the arresting of the pest and consequent success of a trap cropping system. Trap cropping varies according to factors such as plant characteristics, the basis of deployment, and the use of combined approaches (Shelton and Bannes–Perez, 2006). Studies conducted for four years in Idaho, USA to evaluate the efficacy of Adagio oil radish and Metex white mustard (*Sinapis alba*) on *Heterodera schachtii* indicated that trap crops reduced the nematode population significantly with an increase in sugar beet yield (Hafez and Sundararaj, 2000). Trap crops should have some chemical (olfactory/ gustatory) or physical (tactile/visual) stimuli to attract the pests over main crops. Diamond back moth a specialist folivore on cruciferous vegetable crops seem to be attracted to the phytochemicals like glucosinolates and their volatile hydrolytic products for its selective feeding and oviposition (Renwick, 2002). Several trap crops like Indian mustard (*Brassica juncea*), collards (*Brassica oleracea* var. *acephala*) and yellow rocket (*Barberea vulgaris*) have been found to be effective in attracting *Plutella xylostella* adults over the cabbage (Mitchell *et al.*, 2000; Shelton and Nault, 2004; Badenes–Perez *et al.*, 2004). Chinese cabbage will be an ideal alternative to mustard for use as a trap crop in the management of diamond back moth of cabbage (Satpathy *et al.*, 2010). The tropical army worm, *Spodoptera litura* is an important pest of tobacco *Nicotiana tabacum*. Zhou *et al.* (2010) reported that castor and taro hosted significantly more *S. litura* than peanut, sweet potato or tobacco in a green house trial, and tobacco field plots with taro rows hosted significantly fewer *S. litura* than those with rows of other trap crops or without trap crops, provided the taro was in a fast growing stage. When these crops were grown along with egg plant and soybean, in separate plots tobacco plots hosted more *S. litura* than the other crop plants in the season but late in the season, taro plots hosted significantly more *S. litura* than other crops.

Fig. 5: Trap cropping: Bhendi (okra) in cotton field to trap sucking pests

2.13. Inter Crops and Mixed Crop

Mixed or inter-cropping of plant species and cultivars is a common cultural practice in many countries. Intercropping (Fig. 6) is one form of mixed cropping. Besides attracting natural enemies, intercropping also controls weeds and keep the soil healthy. In inter-cropping weeds or secondary crops will interfere directly with pests in a bottom-up manner. This approach is reflected in the resource concentration hypothesis, which proposes that concentrated areas of host plant are easier for herbivores to find and colonize (Root, 1973). The presence of plants distantly related to the crop plant can visually or chemically interfere with specialist herbivores, making the habitat less favorable. Non crop plants, however, can encourage generalist herbivores that feed on both non-crop and crop plants (Schellhorn and Sork, 1997). Another approach for managing pests involves mixed cropping. It is a type of cropping in which more than one crop is grown on the same piece of land. This reduces insects and other pests by encouraging natural enemies due to:

1. Greater temporal and spatial distribution of nectar and pollen sources
2. Increased ground cover
3. Increased prey, affecting alternative food sources when the pest species are scarce or at appropriate time in the predator's life cycle.

Fig. 6: Chick pea and mustard inter-cropping to control gram pod borer

Intercropping tomato with cowpea reduced bacterial wilt (*Ralstonia solanacearum*) on tomato (Michel *et al.*, 1997). Disease incidence of bacterial wilt (*R. solanacearum*) in potato was significantly lower in treatment containing healthy tubers + full earthing up at planting and intercropping with cabbage. Tripathi and Rathi (2000) reported that wider spaced inter cropping with dwarf or semi dwarf wheat varieties reduced the disease severity of Botrytis grey mold in chick pea. Intercropping with gobbi sarson and taking a higher number of cuttings resulted in a significant reduction in tiller infection of Sclerotinia rot in Egyptian clover. On maize, intercropping with cassava reduced egg and immature numbers of *Sesamia calamistis* by 67 and 83%, respectively, as a result of reduced host finding by the ovipositing adult moth and of higher egg parasitism by *Telenomus* spp. (Schulthess *et al.,* 2004).

2.14. Crop Rotation

An effective crop rotation is one in which a crop of one plant family is followed by one from a different family that is not a host crop of the pest to be controlled. Crop rotation interrupts normal life cycle of pests and diseases. Dill–Macky and Jones (2000) reported that wheat grown after soybeans had reduced incidence and severity of Fusarium Head Blight (FHB). The infestation of termites in wheat was significantly less under rice–wheat rotation compared with a wheat fallow system (Chander and Garg, 1999). Reduced tillage also should be coupled with crop rotation, which negatively affects many wheat pests including tan spot, take-all disease (Bockus, 1998). Prestes *et al.* (2002) reported that leaf blotch incidence was higher under monoculture and no-till than under crop rotation. The use of perennial grass in rotation with tomatoes has largely been overlooked but has the potential to reduce soil-borne pest and weed problems (Rich *et al.*, 2009). Lemaga (2001) reported planting two different crops in two consecutive seasons resulted in superior wilt control, compared to planting the same crop in potato field. Recent research has shown that some plants like crucifer crops like broccoli have a suppressive effect on diseases. After harvest broccoli and other crucifer crops, the plant residue is plowed into the soil; the decomposition of the broccoli stems and leaves releases natural chemical that can significantly reduce the number of *Verticillium dahliae* micro sclerotia.

3. ADVANTAGES

1. Cultural control practices are not costly.
2. It utilizes resources available to farmers, such as labor or indigenous materials.
3. Cultural control methods are generally eco- and farmer-friendly.
4. They are compatible with other pest management practices.
5. It reduces resistance and resurgence development in pest population.
6. It does not have undesirable residues in food, feed and killing of natural enemies of pests.
7. It is more profitable for low value crops.

4. DISADVANTAGES

1. Most of the cultural methods suppress populations of some pests but can increase some others for example clipping of leaves in paddy is practiced to control stem borer attack whereas it increases the incidence of bacterial leaf blight in paddy.
2. Some cultural practices decreases pests but also decreases yield.
3. Many practices require community–wide adoption, which may be difficult to achieve *e.g.*, crop rotation, sowing etc.
4. It requires thorough knowledge of pest or pathogen biology and ecology.

CONCLUSIONS

Current pest management practices heavily rely on chemical control. Control of pests using chemicals like pesticides and fungicides causes various health hazards besides increasing the cost of cultivation to the growers. To change this situation cultural control is an effective alternative. However to reap the real benefit of cultural control participation of people from all level is required. Government food and agricultural policies should be modified to give importance to cultural control. Government should come up with a new legislation to empower the government department to make regulations for the cultural measures which require the involvement and participation of the community *e.g.*, crop rotation and timely sowing. Research and development activities with respect to sustainable agriculture including cultural control measures should be strengthened by providing enough financial support. Awareness on cultural control measures also should be created by public education campaign.

REFERENCES

Ampofo, J.K.O. and Massomo S.M. (1998). Some cultural strategies of bean stem maggots (Diptera: Agromyzidae) on beans in Tanzania. *African Crop Science Journal,* **6(4):** 351–56.

Anithakumari, P. (2007). Integrated farming of coconut based homesteads in root (wilt) affected area—the impact of extension interventions. *Journal of Plantation Crops,* **35(3):** 152–57.

Badenes–Perez, F.R., Shelton, A.M. and Nault Brain, A. (2004). Evaluating trap crops for diamondback moth. *Plutella xylostella* (Lepidoptera: Plutellidae), *J. Econ. Entomol.,* **97(4):** 1365–72.

Bailey, K.L., Gossen, B.D., Derksen, D.A. and Watson, P.R. (2000). Impact of agronomic practices and environment on diseases of wheat and lentil in southeastern Saskatchewan. *Can. J. Plant Sci.,* **80(4):** 917–27.

Begum, H.A., Meah, M.B., Howlider, M.A.R. and Islam, N. (1995). Development of alternaria blight on *Brassica campestris* in relation to crop management practices. *Bangladesh Journal of Plant Pathology,* **11(1–2):** 29–32.

Blok, W.J., Lamers, J.G., Termorshuizen, A.J. and Bollen, G.J. (2000). Control of soilborne plant pathogens by incorporating fresh organic amendments followed by tarping. *Phytopathology,* **90:** 253–59.

Bockus, W.W. (1998). Control strategies for stubble borne pathogens of wheat. *Can. J. Plant Pathol.,* **20(4):** 371–75.

Brown, S.L., Todd J.W and Culbreath A.K. (1996). Effect of selected cultural practices on incidence of tomato spotted wilt virus and populations of thrips vectors in peanuts. *Acta–Horticulture,* **431:** 491–98.

Bwye, A.M., Jones R.A.C. and Proudlove W. (1999). Effects of different cultural practices on spread of cucumber mosaic virus in narrow–leafed lupins (*Lupinus angustifolius*). *Australian Journal of Agricultural Research,* **50(6):** 985–96.

Campos, M. and Civantos, M. (2000). Cultural practices in Olive and impact on pests and diseases. *Olivae,* **84:** 40–46.

Chakraborti, S. (2001). Integrated management of root nematode in brinjal. *Indian J. Nematol.,* **31(1):** 80–83.

Chakraborti, S. (1999). Non-synthetic chemical based management approaches for rice stem nematode. *Pest Management and Economic Zoology,* **7(2):** 161–65.

Chander, S. and Garg, R.N. (1999). Effect of nitrogen and cultural practices on pests of rice and wheat under rice–wheat cropping system. *Annals of Plant Protection Sciences,* **17(2):** 159–62.

Charlet, L.D. and Knodel J.J. (2003). Impact of planting date on sunflower beetle (Coleoptera: Chrysomelidae) infestation, damage, and parasitism in cultivated sunflower. *J. Econ. Entomol.,* **96(3):** 706–13.

Dill–Macky, R. and Jones, R.K. (2000). The effect of previous crop residues and tillage head blight of wheat. *Plant Dis.,* **84(1):** 71–76.

Yano, E. (2004). Recent Development of Biological Control and IPM in Greenhouses in Japan. *Journal of Asia–Pacific Entomology,* **7(1):** 5–11.

Gaur, R.B. and Singh, R.D. (1995). Effect of some cultural practices on perenation of Ascochyta fungus of chickpea. *Advances in Agricultural Research in India,* **4:** 36–46.

Gergon, E.B., Miller, S.A., Davide, R.G. Opina, O.S. and Obien S.R. (2001). Evaluation of cultural practices (surface burning, deep ploughing, organic amendments) for management of rice root-knot nematode in rice–onion cropping system and their effect on onion (*Allium cepa* L.) yield. *International Journal of Pest Management,* **47(4):** 265–72.

Godika, S., Pathak A.K. and Jain J.P. (2001). Integrated management of alternaria blight and white rust diseases of mustard. *Indian J. Agri. Sci.,* **7(11):** 733–35.

Graham, J.H. (1994). Control of black root rot on citrus seedlings in peat–based media. *Proceedings of the Florida State Horticultural Society,* **107:** 21–26.

Gutierrez, W.A., Shews, H.D. and Melton, T.A. (1997). Sources of inoculum and management of *Rhizoctonia solani* damping off on tobacco transplants under greenhouse conditions. *Plant Disease,* **81:** 604–606.

Hamiduzzaman, M.M., Meah M.B. and Doula, M.A.U. (1999). An Integrated approach for guava wilt control. *Pakistan Journal of Scientific and Industrial Research,* **42(5):** 244–47.

Hasan, S.A. (1998). Effect of cultural practices, fungicides and biologicalcontrol agents on Botrytis bunch rot of grapes, *Bulletin–OILB / SROP,* **21(6):** 41–51.

Hafez, S.L. and Sundararaj, P. (2000). Impact of agronomic and cultural practices of green manure crops for the management of *Heterodera schachtii* in sugarbeet. *International Journal of Nematology,* **10(2):** 177–82.

Hausbeck, M.K. and Moorman, G.W. (1996). Managing *Botrytis* in greenhouse–grown flower crops. *Plant Disease,* **80:** 1212–19.

Holb, I.J. (2005). Effect of pruning on apple scab in organic apple production. *Plant Dis.*, **89:** 611–18.

Houma, A.R. Cherif, M. and Boubaker, A. (1998). Effect of nitrogen fertilizing, green pruning and fungicide treatments on Botrytis bunch rot of grapes. *J. Plant. Pathol.*, **80(2):** 115–24.

Jackson, L.E., Ramirez, I., Yokita, R., Fennimore, S.A., Koike, D.M., Henderson, W.E., Calderon Chaney, F.Y. and Klonsky, K. (2004). On–farm assessment of organic matter and tillage management on vegetable yield, soil, weeds, pests, and economics in California, *Agriculture, Ecosystems and Environment,* **103(3):** 443–63.

Jenkins, P.E. and Isaacs, R. (2007). Cutting wild grapewines as a cultural control strategy for Grape berry moth (Lepidoptera: Tortricidae). *Environ. Entomol,* **36(1):** 187–94.

Katan, J. (2000). Physical and cultural methods for the management of soil-borne pathogens. *Crop Protection,* **19(8–10):** 725–31.

Kaur, S. and Singh, D. (2000). Economics of controlling *Helicoverpa armigera* (Hubner) through suitable varieties and cultural practices in tomato. *Indian Phytopathology,* **27(2):** 185–88.

Knodel, J.J., Denise, L., Olson Bryan, K., Robert, H. and Henson, A. (2008). Impact of planting dates and insecticide strategies for managing crucifer flea beetles

(Coleoptera: Chrysomelidae) in spring–planted canola. *J. Econ. Entomol.,* **101(3):** 810–21.

Krishnakumar, V. and Maheswarappa, H.P. (2010). Integrated nutrient management for root (wilt) diseased coconut (*Cocos nucefera*) palms. *Indian J. Agri. Sci.,* **80(5):** 394–98.

Kumar, S. and Sugha, S.K. (2000). Role of cultural practices in the management of Septoria leaf spot of tomato. *Indian Phytopathology,* **53(1):** 105–106.

Lee, S. and Ma, K. (1997). Occurrence of major insect pests in machine transplanted and direct seeded rice paddy field. *Kor. J. App. Entomol.,* **36(2):** 141–44.

Legard, D.E., Chandler, C.K. and Bartz, J.A. (1997). The control of strawberry disease by sanitation. *Proceedings of the third international strawberry symposium* Veldhoven, Netherlands, 29 April to 4 May 1996. *Acta Horticulturae,* **439:** 917–22.

Lemaga, B. (2001). Integrated control of potato bacterial wilt in Kabale District, Southwestern Uganda. *Scientist and farmer partners in research for the 21st Century Program report* 1999–2000.

Mahto, B.N. (2001). Effect of cultural practices on foliar blight incidence in wheat. *Annals of Agricultural Research,* **22(4):** 452–56.

Manrique, V., Cecil, O., Montemayor, R.D., Skvarch Cave, E.A. and Smith. B.W. (2010). Effect of straw mulch on populations of *Microtheca ochroloma* (Coleoptera: Chrysomelidae) and ground predators in turnip *Brassica rapa* in Florida. *Florida Entomologist,* **93(3):** 407–11.

McBride, R.G., Mikkelsen, R.L. and Barker, K.R. (2000). The role of low molecular weight organic acids from decomposing rye in inhibiting root-knot nematode populations in soil, *Applied Soil Ecology,* **15:** 243–51.

McSorley, R. (1994). Nematode management in sustainable agriculture. Environmentally Sound Agriculture:–*Proceedings of the Second Conference,* pp. 517–22.

Mcsorley, R. and Gallaher, N. (1995). Cultural practices improve crop tolerance to nematodes. *Nematropeca,* **25(1):** 53–60.

Michaud, J.P., Stahlman, P.W., Jyoti, J.L. and Grant, A.K. (2009). Plant spacing and weed control affect sunflower stalk insects and the girdling behavior of *Dectes texanus* (Coleoptera: Cerambycidae). *J. Econ. Entomol.,* **102(3):** 1044–53.

Michel, V.V., Wang, J.F., Midmore, D.J. and Hartman, G.L. (1997). Effects of intercropping and soil amendment with urea and calcium oxide on the incidence of bacterial wilt of tomato and survival of soil-borne *Pseudomonas solanacearum* in Taiwan. *Plant Pathology,* **46:** 600–10.

Milbrath, L.R., Weiss, M.J. and Schatz, B.G. (1995). Influence of tillage system, planting date, and oilseed crucifers on flea beetle populations (Coleoptera: Chrysomelidae). *Can. Entomol.,* **127(3):** 289–93.

Mitchell, E.R., Hu, G.Y. and Johanowicz, D. (2000). Management of diamondback moth (Lepidoptera: Plutellidae) in cabbage using collard as a trap crop. *HortScience,* **35:** 875–79.

Mullen, R.J. and Viss, T.C. (1996). Control of asparagus rust in the Sacramento–San Joaquin Delta Region of California. *Acta Horticulturae,* **415:** 297–300.

Nagaraja, A., Jagdesh, K., Krishna, Sudir. and Gowde, K.T. (2006). Effect of organic and inorganic fertilizers on incidence of finger millet blast and brown spot. *J. Mycol. Pl. Pathol,* **33(2):** 465.

Oh, J. and Kim, D. (2001). Effect of cultural practices on the occurrence of pod and stem blight and purple blotch, and on soybean growth. *Res. Plant Dis.*, **7(2):** 107–111.

Osunlaja, S.O. (1989). Effect of organic soil amendments on the incidence of brown spot disease in maize caused by *Physoderma maydis*. *Journal of Basic Microbiology,* **29(8):** 501–505.

Pathak, A.K. and Godika, S. (2010). Effect of organic fertilizers, Biofertilizers, Antagonist and nutritional supplements on yield and disease incidence in Indian mustard in arid soil. *Indian J. Agri. Sci.,* **80(7):** 652–54.

Perez–Marco, P. (2000). Cultural practices affecting the rate of Fusarium damage in durum wheats. *Options Mediterraneennes Serie A,–Seminaires–Mediterraneens,* **40:** 403–405.

Prestes, A.M., Santos Dos, H.P. and Reis E.M. (2002). Effect of cultural practices on the incidence of leaf blotches of wheat. *Pesquisa Agropecuaria Brasileira,* **37(6):** 791–97.

Rathore, B.S. (2002). Management of downy mildew of blond psyllium through cultural practices. *Plant Dis. Res.,* **17(1):** 83–85.

Renwick, J.A.A. (2002). The chemical world of crucifers voles: lures treats and traps. *Entomologia Experimentia et Applicata,* **104:** 35–42.

Rhainds, M., Kovach, J., Dosa, E.L., and English–Loeb, G. (2001). Impact of reflective mulch on yield of strawberry plants and incidence of damage by tarnished plant bug (Heteroptera: Miridae). *J. Econ. Entomol.,* **94(6):** 1477–84.

Rich. J.R., Rhoads, F.M. and Olson, S.M. (2009). Sustainable tomato production system using perennial grass rotation and disease resistant varieties. *Acta Hort.*, **809:** 227–34.

Ristaino J.B., Parra, G. and Campbell, C.L. (1997). Suppression of Phytophthora blight in bell pepper by a no-till wheat cover crop. *Phytopathology,* **87(3):** 242–49.

Rodriquez , H.A., Nass, H., Cardona, R. and Aleman, L. (1999). Alternatives to control sheath blight caused by *Rhizoctonia solani* in rice. *Fitopatologia Venezolana,* **12(1):** 18–21.

Root, R.B. (1973). Organization of a plant–arthropod association in simple and diverse habitats: the fauna of collards (*Brassicae oleracea*), *Ecol. Monogr.*, **43:** 95–124.

Sarwar, M. (2008). Plant spacing–a non polluting tool for aphid (Hemiptera: Aphididae) management in canola, *Brassica napus*. *Journal of Economic Society of Iran,* **27(2):** 13– 22.

Satpathy, S., Shivalingaswamy, T.M., Kumar, A., Rai, A.B. and Rai, M. (2010). Potentiality of Chinese cabbage (*Brassica rapa* subsp. *pekinensis*) as a trap crop for diamondback moth (*Plutella xylostella*) management in cabbage. *Indian J. Agri. Sci.,* **80(3):** 238–41.

Schellhorn, N.A. and Sork, V.L. (1997). The impact of weed diversity on insect population dynamics and crop yield in collards, *Brassica oleraceae* (Brassicaceae), *Oecologia,* **111:** 233–40.

Schulthess, F., Chabi–Olaye, A. and Gounou, S. (2004). Multi-trophic level interactions in a cassava–maize mixed cropping system in the humid tropics of West Africa. *Bull Entomol Res.*, **94(3):** 261–72.

Senguttavan, T. and Dhanakodi, C.V. (1997). Management of pod borer in groundnut through manipulation of cultural practices in alfisols. *Journal of Oilseeds Research,* **14(2):** 269–73.

Sharma, H.C. (1998). Bionomics, host plant resistance, and management of the legume pod borer, *Maruca vitrata*—a review. *Crop Protection,* **17(5):** 373–86.

Sharma, G.C. and Bhatia, M. (2002). Effect of cultural practices on nematode populations and yield of Frence beans. *Nematologia Mediterranea,* **30(1):** 23–25.

Shelton, A.M. and Nault, B.A. (2004). Dead-end trap cropping, a technique to improve management of the diamondback moth, *Plutella xylostella* (Lepidoptera: Plutellidae). *Crop Protection,* **23:** 497–503.

Shelton, A.M. and Badenes–Perez, F.R. (2006). Concepts and applications of trap cropping in pest management. *Annu. Rev. Entomol.*, **51:** 285–308.

Sherif, M.R., Khodair, I. and El–Habashy, M. (1997). Cultural practices to manage the rice leaf miner, *Hydrella prosternalis* (Diptera: Ephydridae) in Egypt. *Egyptian Journal of Agricultural Research,* **75(3):** 611–22.

Singh, L. and Kanta, U. (1999). Manipulation of cultural practices in the management of *Chilo partellus* (Swinhoe) in *kharif* maize. *Journal of Insect Science,* **12(1):** 54–57.

Sutton, J.C. and Vyn, T.J. (1990). Crop sequence and tillage practices in relation to diseases of winter wheat in Ontario. *Can. J. Plant Pathol.,* **12:** 358–68.

Tripathi, R.S. and Rathi, Y.P.S. (2000). Effect of certain cultural practices on severity of *Botrytis* grey mould of chickpea. *Indian Phytopathology,* **53(2):** 172–74.

Uddin, W. and Stevenson, K.L. (1998). Seasonal development of *Phomopsis* shoot blight of peach and effects of selective pruning and shoot debris management on disease incidence. *Plant Dis.*, **82:** 565–68.

Umeh, V.C., Waliyar, F., Traore, S., Chaibou, I.M., Omar, B. and Detognon, J. (2001). Farmers' opinions and influence of cultural practices on soil pest damage to groundnut in West Africa. *Insect Science and Its Application,* **21(3):** 257–65.

Vallone, S. (1998). Disease management in no-tillage soybean systems. *JIRCAS–Working–Report,* **13:** 35–42.

Wu, B.M. and Subbarao, K.V. (2003). Effect of irrigation and tillage on the temporal and spatial dynamics of *Sclerotinia minor* sclerotia and lettuce incidence. *Phytopathology,* **93(12):** 1572–80.

Zhou, Z., Chen, Z. and Xu, Z. (2010). Potential of trap crops for integrated management of the tropical armyworm, *Spodoptera litura* in tobacco. *Journal of Insect Science,* **117:** 1–11.

7

Legal Approaches in Pest Management—An Imperative in a Globalized World

M. Suresh[1], C.V. Sameer Kumar[1] and C. Sudhakar[1]

ABSTRACT

The rapid increase in demographic pressure during the past few decades coupled with unprecedented losses due to various biotic and abiotic stresses is forcing the agricultural researchers to develop strategies that ensure food security particularly in the third world countries. An important facet of food production is the control of pests that otherwise damage crops making them unsuitable for human and livestock consumption. Food hygiene and plant health are the key factors governing the market access of agricultural commodities in view of liberalized world trade in agriculture. The basic strategies and techniques of plant protection that are already in practice viz., *exclusion, eradication protection and immunization play an important role in augmenting and stabilizing crop yields. However, legal or regulatory procedures which are based on exclusion principle are necessary to restrict the immigration of foreign pests and prevent the dispersal of established pests. Sanitary and phytosanitary measures ensure environmental protection, safeguard consumer interests and thus promote the trade between various countries leading to economic welfare of the countries. Sanitary or phytosanitary measures include all relevant laws, decrees, regulations, requirements and procedures, testing, inspection, certification and approval procedures. Regulatory approaches,* viz., *quarantine (both domestic and international) as well as precautionary approaches, prevent the introduction of invasive pests thus protecting the plant health and environment. Several organizations at international, national and at regional level monitoring*

1 Agricultural Research Station, Acharya N. G. Ranga Agricultural University, Tandur, Rangareddy District, Andhra Pradesh - 501 141, India.
Corresponding author: E-mail: sureshiari@yahoo.co.in

and coordinating sanitary and phytosanitary related issues. IPPC is a multilateral treaty that makes the provision for the application of measures by governments to protect their plant resources from harmful pests which may be introduced through international trade. While at regional level, nine regional plant protection organizations participate in activities to achieve the objectives of IPPC. They also cooperate with the commission on phytosanitary measures and the IPPC in the development of international standards. At national level NPPOs and Directorates established by respective governments fulfill the functions specified in the IPPC treaty. With rapid globalization and liberalized economic policies, international trade is touching the new heights. Consequently, the risk of spread of invasive species is also ringing the warning bells. The number and diversity of potentially harmful organisms being moved across the globe is steadily increasing. With changing face of world agricultural scenario, it has become imperative to thoroughly understand and adopt the legal approaches for effective pest management at global level.

***Key words*:** Pest management, Exclusion, Quarantine, Sanitary and Phytosanitary measure, Invasive pests, IPPC, ISPMs, Precautionary approach, Pest risk analysis, Pest reports, DIP act, Inspection, and Certification.

---

1. INTRODUCTION

Liberalization of world trade in agriculture with the advent of the WTO has thrown up many challenges to developing countries in gaining market access to developed countries particularly in compliance with international standards on food hygiene and plant health.

Plant disease or pest management has the key role in meeting the international standards on plant health. Plant disease/pest management practices rely on anticipating occurrence/or forecasting of disease and attacking vulnerable points in the disease cycle (*i.e.,* weak links in the infection chain). Therefore, correct diagnosis of a disease is necessary to identify the pathogen, which is the real target of any disease management programme. Particularly for effective disease management, a thorough understanding of the disease cycle, including climatic and other environmental factors that influence the cycle, and cultural requirements of the host plant are essential (Arneson, 2001 and Jacobsen, 2001).

The many strategies, tactics and techniques used in disease management can be grouped under one or more very broad principles of action. Differences between these principles often are not clear. The simplest system consists of two principles, prevention (prophylaxis in some early writings) and therapy (treatment or cure) (Fry, 1982).

Regarding, plant disease management, one early proposal by Whetzel (1929) included four general disease control principles *viz.*, exclusion, eradication, protection and immunization.

Exclusion measures are aimed at preventing new pathogens and disease from reaching an uninfected area and avoiding contact between the pathogen and the crop field. Legal or regulatory approaches are the most effective means of prevention of any pest which works on this exclusion principle (Maloy, 2005).

This principle is defined as any measure that prevents the introduction of a disease-causing agent (pathogen) into a region, farm, or planting (Maloy, 1993). The basic strategy assumes that most pathogens can travel only short distances without the aid of some other agent such as humans or other vector, and that natural barriers like oceans, deserts, and mountains create obstacles to their natural spread. In many cases pathogens are moved with their host plants or even on non-host material such as soil, packing material or shipping containers. Karnal bunt of wheat has A1 quarantine pest status under EPPO and not present in EU/UK earlier, but in the present globalized world, severe threat exists for EU/UK along with raised imports from US. Soybean rust (caused by *Phakopsora pachyrhizi*) has been found recently in the southeastern U.S. and precautions have been undertaken to prevent further spread. Due to its destructiveness, South American leaf blight (SALB) (caused by *Microcyclus ulei*) is a feared disease in the major rubber producing region of Indonesia, and contingency plans have been proposed to chemically defoliate rubber trees by aerial application of herbicides if the pathogen is detected. It is hoped that this would prevent establishment of the pathogen in the region.

An important and practical strategy for excluding pathogens is to produce pathogen-free seed or planting stock through certification programs for seeds and vegetatively propagated plant materials such as potatoes, grapes, tree fruits, etc. These programs utilize technologies that include isolation of production areas, field inspections, and removal of suspect plants to produce and maintain pathogen-free stocks. Planting stock that is freed of pathogens can be increased by tissue culture and micro propagation techniques as well as be maintained in protective enclosures such as screen houses to exclude pathogens and their vectors. Exclusion may be accomplished by something as simple as cleaning farming equipment to remove contaminated debris and soil that can harbor pathogens such as *Verticillium*, nematodes or other soil borne organisms and prevent their introduction into non-infested fields (Maloy and Baudoin, 2001).

Quarantine, both of export and import has gained considerable attention so that Indian agriculture is protected from the ingress of exotic pests and diseases. Identification and declaration of pest free areas for export may also be gaining importance in promotion of exports.

In this context, regulatory or legal approaches are a global imperative and a lively area of policy discussion as countries seek strong protection to safeguard their environmental and economic interests particularly from the weed, pest and/or pathogens.

2. LEGAL APPROACHES OR REGULATORY ACTIONS

Regulatory actions are often employed to prevent immigration of foreign pests or to prevent the dispersal of established pests. Such actions are termed as legal methods of pest management.

In the United States, the Animal and Plant Health Inspection Service (APHIS), a division of the U.S. Department of Agriculture, is responsible for promulgating and enforcing plant quarantine measures. There are also state agencies that deal with local quarantines. Internationally, eight regional plant protection organizations (PPOs) were established in 1951 by the International Plant Protection Convention sponsored by the Food and Agricultural Organization of the United Nations. This was revised in 1997 and now includes nine regional PPOs. The European and Mediterranean Plant Protection Organization (EPPO) is the oldest of the regional PPOs. The regional PPOs have no regulatory authority such as APHIS or other governmental agency, but function to develop strategies against the introduction and spread of pests and to coordinate the use of phytosanitary regulations to ensure agreement among the different member countries.

The IPPC's International Standard for Phytosanitary Measures, Guidelines on Lists of Regulated Pests, defines the term "pest" to potentially include an imported plant itself to the extent that, as a weed, it may be injurious to other plants and plant products. This has been interpreted broadly by the IPPC parties to include environmental effects, such as threats to a country's native plants. Other definitions therein help the reader understand IPPC terminology and that the convention focuses only on "regulated pests," that is, those plants, animals, or pathogens that may be prohibited, quarantined, or otherwise subjected to "pest risk management" by the country to which they are being considered for import.

Legal approaches are very important to address the global phenomenon of harmful invading of non-native species of all kinds in to both human-managed and relatively wild areas.

2.1. SPS Agreement and Plant Health Management

The decision to start the Uruguay Round trade negotiations was made after years of public debate, including debate in national governments. The decision to negotiate an agreement on the application of sanitary and phytosanitary measures was made in1986 when the Round was launched. The SPS negotiations were open to all of the 124 governments which participated in the Uruguay Round. Many governments were represented by their food safety or animal and plant health protection officials. The negotiators also drew on the expertise of technical international organizations such as the FAO, the Codex and the OIE. Developing countries participated in all aspects of the Uruguay Round negotiations to an unprecedented extent. In the negotiations on sanitary and phytosanitary measures, developing countries were active participants, often represented by their national food safety or animal and plant health experts. Both before and during the Uruguay Round negotiations, the GATT Secretariat assisted developing countries to establish effective negotiating positions. The SPS Agreement calls for assistance to developing countries to enable them to strengthen their food safety and animal and plant health protection systems. FAO and other international organizations already operate programmes for developing countries in these areas.

2.1.1. *Sanitary and Phytosanitary Measures* (*SPM*)

For the purposes of the SPS Agreement, Sanitary and Phytosanitary measures are defined as any measures applied:

- To protect human or animal life from risks arising from additives, contaminants, toxins or disease-causing organisms in their food;
- To protect human life from plant- or animal-carried diseases;
- To protect animal or plant life from pests, diseases, or disease-causing organisms;
- To prevent or limit other damage to a country from the entry, establishment or spread of pests.

These include sanitary and phytosanitary measures taken to protect the health of fish and wild fauna, as well as of forests and wild flora. Measures for environmental protection (other than as defined above), to protect consumer interests, or for the welfare of animals are not covered by the SPS Agreement (Jagadeeshwar *et al.*, 2009). These concerns, however, are addressed by other WTO agreements (*i.e.*, the TBT Agreement, 1994).

Sanitary or phytosanitary measures include all relevant laws, decrees, regulations, requirements and procedures including, *inter alia*, end product criteria; processes and production methods; testing, inspection, certification and approval procedures; quarantine treatments including relevant requirements associated with the transport of animals or plants, or with the materials necessary for their survival during transport; provisions on relevant statistical methods, sampling procedures and methods of risk assessment; and packaging and labeling requirements directly related to food safety.

2.1.2. *International Standards, Guidelines and Recommendations*

(a) For food safety, the standards, guidelines and recommendations established by the Codex Alimentarius Commission relating to food additives, veterinary drug and pesticide residues, contaminants, methods of analysis and sampling, and codes and guidelines of hygienic practice;

(b) For animal health and zoonoses, the standards, guidelines and recommendations developed under the auspices of the International Office of Epizootics;

(c) For plant health, the international standards, guidelines and recommendations developed under the auspices of the Secretariat of the International Plant Protection Convention in cooperation with regional organizations operating within the framework of the International Plant Protection Convention; and

(d) For matters not covered by the above organizations, appropriate standards, guidelines and recommendations promulgated by other relevant international organizations open for membership to all Members, as identified by the Committee.

2.1.3. *Important Definitions*

(a) Pest:	Any species, strain or biotype of plant, animal or pathogenic agent injurious to plants or plant products [FAO, 1990; revised FAO, 1995; IPPC, 1997]
(b) Pest categorization:	The process for determining whether a pest has or has not the characteristics of a quarantine pest or those of a regulated on quarantine pest [ISPM Pub. No. 11, 2001]
(c) Pest Free Area:	An area in which a specific pest does not occur as demonstrated by scientific evidence and in which, where appropriate, this condition is being officially maintained [FAO, 1995].

(or)

An area, whether all of a country, part of a country, or all or parts of several countries, as identified by the competent authorities, in which a specific pest or disease does not occur.

Note: A pest- or disease-free area may surround, be surrounded by, or be adjacent to an area-whether within part of a country or in a geographic region which includes parts of or all of several countries- in which a specific pest or disease is known to occur but is subject to regional control measures such as the establishment of protection, surveillance and buffer zones which will confine or eradicate the pest or disease in question.

(d) Pest Risk Analysis: The process of evaluating biological or other scientific and economic evidence to determine whether a pest should be regulated and the strength of measures to be taken against it [FAO, 1995; revised IPPC, 1997].

(e) Pest status (in an area): Presence or absence, at the present time, of a pest in an area, including where appropriate its distribution, as officially determined using expert judgement on the basis of current and historical pest records and other information [CEPM[@], 1997; revised ICPM[#], 1998].

(f) Phytosanitary action: An official operation such as inspection, testing, surveillance or treatment, undertaken to implement phytosanitary regulations or procedures [ICPM, 2001].

(g) Phytosanitary certification: Use of phytosanitary procedures leading to the issue of a phytosanitary certificate [FAO, 1990].

(h) Phytosanitary measure: Any legislation, regulation or official procedure having the purpose to prevent the introduction and/or spread of quarantine pests, or to limit the economic impact of regulated non quarantine pests [FAO, 1995; revised IPPC, 1997; ISC, 2001].

(i) Quarantine pest: A pest of potential economic importance to the area endangered thereby and not yet present there, or present but not widely distributed and being officially controlled [FAO,1990; revised FAO, 1995; IPPC, 1997].

(j) Regulated article: Any plant, plant product, storage place, packaging, conveyance, container, soil and any other organism, object or material capable of harboring or spreading pests, deemed to require phytosanitary measures, particularly where international transportation is involved [FAO, 1990; revised FAO, 1995; IPPC, 1997].

(k) Regulated non-quarantine pest: A non-quarantine pest whose presence in plants for planting affects the intended use of those plants with an economically unacceptable impact and which is therefore regulated within the territory of the importing contracting party [IPPC, 1997].

(l) Regulated pest: A quarantine pest or a regulated non-quarantine pest [IPPC, 1997].

(m) Risk assessment: The evaluation of the likelihood of entry, establishment or spread of a pest or disease within the territory of an importing Member according to the sanitary or phytosanitary measures which might be applied, and of the associated potential biological and economic consequences; or the evaluation of the potential for adverse effects on human or animal health arising from the presence of additives, contaminants, toxins or disease-causing organisms in food, beverages or feedstuffs.

(n) Appropriate level of sanitary or phyto sanitary protection: The level of protection deemed appropriate by the Member establishing sanitary or phytosanitary measure to protect human, animal or plant life or health within its territory.

Note: Many Members otherwise refer to this concept as the "Acceptable Level of Risk".

(o) Area of low pest or disease prevalence: An area, whether all of a country, part of a country, or all or parts of several countries, as identified by the competent authorities, in which a specific pest or disease occurs at low levels and which is subject to effective surveillance, control or eradication measures.

Note: CEPM[@]: Committee of Experts on Phytosanitary Measures

ICPM[#]: Interim Commission on Phytosanitary Measures

3. REGULATORY APPROACHES

Legal or regulatory approaches can be broadly categorized in to two categories *viz*

1. Plant quarantine system, and
2. Precautionary system.

3.1. Quarantine

The word quarantine has its roots in the Latin word for forty. It originally referred to the period of detention which was imposed on ships' passengers to allow latent cases of disease to develop before passengers were permitted to land. The earliest record of such restrictions for human disease goes back to the latter half of the 14th century. Subsequently, as governments became more concerned with the spread of pests destructive to agricultural and forest crops, new controls were gradually introduced under the name Plant Quarantine.

Plant Quarantine thus can be defined as *"a legal restriction on the movement of agricultural commodities for the purpose of exclusion, prevention or delay in the spread of plant pests and diseases in un-infected areas"*.

3.1.1. *Quarantine Pest*

A quarantine pest is defined by the IPPC as "a pest of potential economic importance to the area endangered thereby and not yet present there, or present but not widely distributed and being officially controlled" (IPPC, 1997). In other words, it is any organism that is injurious or potentially injurious, directly or indirectly, to plants or plant products or by-products of plants and includes bacteria, fungi, insects, mites, molluscs, nematodes, other plants and viruses, not present in a specified area at risk or, if present, being controlled by an NPPO.

Early detection is essential to success of quarantine. Once established, some species spread so quickly that quarantine measures are merely futile. *e.g.*, cereal leaf beetle in South West Michigan states (Horn, 1988).

3.1.2. *Regulated Non–quarantine Pest (RNQP)*

A regulated non-quarantine pest (RNQP) is defined by the IPPC as "a non-quarantine pest whose presence in plants for planting affects the intended use of those plants with an economically unacceptable impact and which is therefore regulated within the territory of the importing contracting party" (IPPC, 1997). RNQPs are usually widely distributed in the country where they are regulated.

3.1.3. *Quarantine Measures*

Plant quarantine measures are of three types:

3.1.3.1. *Domestic quarantine*

It is aimed at preventing spread of a pest/disease of a particular area to the other parts of the country *i.e.*, to restrict the inter-state movement of invasive pests *e.g.*, presently in India domestic quarantine measures exist for nine invasive pests *viz.* two insect pests, Fluted scale (*Iceria purchesi*), San Jose scale (*Quadraspidiotus perniciosus*), coffee berry borer and codling moth,and four diseases, wart (*Synchytrium endobioticum*) of potato, bunchy top (virus), mosaic (virus) of banana, apple scab and potato cyst nematode with a view to preventing the spread of these pests (Chattopadhyay, 1991; Reddy and Joshi, 1992).

3.1.3.2. *International quarantine*

It is aimed preventing the entry of the pest from one country to the importing country or *vice-versa*.

3.1.3.3. *Total embargos*

It refers to the complete restriction on the movement of the plant or plant material or complete prohibition on the import of the material that possesses the risk. *e.g.*, golden nematode of potato.

3.1.4. *Types of Plant Quarantine*

- Pre-entry plant quarantine:
- Post-entry plant quarantine:

3.1.4.1 *Pre-entry plant quarantine*

These are the quarantine measures adopted at the point of entry to prevent the entry of the pathogen in to the country or area where it is not present earlier. The broad categorization of pre entry quarantine requirements for different materials (for India) was given in Table 1 (Plant Quarantine (PQ) Order, 2003).

Table 1: Pre-entry plant quarantine regulations—India: (Adopted from PQ order, 2003)

S. No.	*Plant Material for import*	*Pre-entry quarantine requirement*
1.	Seeds/plants/plant material for sowing/planting/propagation and for consumption	• Import permit issued by authorized plant quarantine officer in India • Phytosanitary Certificate issued at the country of origin of the consignment
2.	Soil, earth, clay and similar material for any microbiological, soil-mechanics, or mineralogical investigations and peat for horticultural purposes.	Special permit issued by the Plant Protection Adviser (PPA)
3.	Insects or microbial cultures including mushroom, algae and other bio-control agents	Special permit issued by the Plant Protection Adviser
4.	Germplasm/transgenic/ genetically modified organisms for research/experimental purpose by public/private sector institutions	Special permit issued by the Director, National Bureau of Plant Genetic Resources (NBPGR), New Delhi, India.
5.	Transgenic/genetically modified organisms	Approval of Review Committee on Genetic Manipulation (RCGM) under Department of Biotechnology
6	Articles packed with raw/solid wood packing material	The treatment of packing material include fumigation with methyl bromide @48 g/m^3 for 16 hrs at 21°C and above (or) any equivalent thereof or Heat (dry heat/ hot water) Treatment at 56°C for 30 mins. (or) Kiln Drying or Chemical Impregnation (or) any other treatment that meet Heat Treatment specifications of ISPM along with a Phytosanitary Certificate stating that packing material has been appropriately treated.

3.1.4.2 *Post-entry quarantine (PEQ)*

Post-entry quarantine measures (PEQ) are the restrictions imposed on the material once entered in the country and which meets the pre-entry quarantine requirements. The PEQ measures are imposed under specialized facilities which varies based on the material on which the PEQ measures are to be imposed (Post-entry quarantine manual, 2010).

In accordance with provisions of Plant Quarantine (Regulation of Import into India) Order, 2003, the importer shall be required to establish the post-entry quarantine facilities such as an isolated field/nursery/glass house/screen house/poly house etc., that are duly certified by the Inspection Authorities (IAs) in accordance with guidelines prescribed by Plant Protection Advisor (PPA) for importing plants and plant materials that require post-entry quarantine. The importer shall establish these facilities sufficiently in advance so that the same may be ready for use at the time of arrival of consignment but not latter.

The importer shall inform in advance the concerned IA about the time of planting of imported plant material in approved facility and permit complete access to the post-entry quarantine facility for the inspection of plants and also abide by his instructions concerning plants growing under post-entry quarantine. The IA may permit the release of plants grown under post-entry quarantine, if they are found free from pests for the period specified in the permit. Where the plants are found affected by quarantine pests, the IA shall order the destruction or return to the country of origin, of the affected consignment of plants of whole or part of.

If the pest is not of quarantine importance IA may advise necessary control measures and release the plant material if satisfied that the pest is controlled or issue order for destruction. The importer shall destroy the affected plant material when ordered in a prescribed manner under the supervision of IA. The requirements for establishing different types of quarantine facilities were given below (PQ Order, 2003).

3.1.4.3 *Requirements of open field quarantine facility*

1. The farm/nursery site shall be fenced all round up to a height of 1.25 m with a single entry point with a lockable gate. It shall be isolated from similar cropped area by at least 1 km and weedfree.
2. The nursery or cropped area shall be bordered with 3–4 rows of densely populated crops such as *dhaincha* or spiny Sesbania (*Sesbania bispinosa*), sunnhemp (*Crotalaria juncea*) and *subabul* (*Leucaena leucocephala*) etc., to serve as insect barrier.
3. An isolation distance of 100 m shall be adopted to segregate different varieties of the same crop or different sources of the same crop species or alternatively each population is isolated by a polythene barrier up to a minimum height of 1.25 m.
4. A suitable signboard must be displayed at the gate indicating restricted area and entry to the area require the approval of the owner of PEQ facility.

5. The farm/nursery shall have the facilities for soil pasteurization or solarization or fumigation, incineration, spraying and watering facilities.
6. A buffer zone of not less than 3 m in width must be maintained in vegetation-free condition around the inside perimeter of the approved area by regular weeding.

3.1.4.4. *Requirements of closed quarantine facility (glass / screen / poly house)*

1. The glass/screen/poly house facility shall have double door entry with entrance porch with inner door fitted with automatic spring door closure and outer sliding door provided with external lock.
2. The vents/openings of glass/screen/poly house facility shall be screened with insect-proof mesh of stainless steel/phosphor bronze, 40 meshes to a linear inch and woven of 30 gauge wire. If nylon mesh is used it should be of good quality and evenly woven.
3. The drainage must be through a gully trap with a provision for insect-proofing in case of concrete flooring.
4. The facility shall have small isolated cubicle for growing indicator plants for virus detection and a head house for storage of pots, green house tools/equipment and storage bins for storing sterile soil or potting mixture, soil sterilization or solarization/incineration, spraying equipment/watering facilities etc.
5. The entrance shall be provided with disinfectant pad for sterilization of feet and basin for disinfecting hands.
6. The facility shall have a minimum floor space area of 10 sq.m. However, larger houses shall have insect-proof screened partitions for isolation of different kind of plant species/varieties.
7. Suitable yellow sticky traps shall be hanged at various intervals at crop canopy level to monitor insect population.

3.1.4.5. *Requirements of tissue culture quarantine facility*

1. The tissue culture facility shall have small growth chamber or separate incubation place for holding imported tissue culture during PEQ.
2. The facility shall have an isolated poly/screen house facility for tissue culture hardening. However where import of tissue cultures is made for *in vitro* propagation, such facility is not required.
3. The facility shall have Enzyme Linked Immunosorbent Assay (ELISA) facility such as virus screening of stock material.

3.1.5. *Examples of Phytosanitary Measures Which May be Applied to Consignments Include (PQ Order, 2003)*

- Inspections or testing for freedom from a pest—this is a practical measure for visible pests or for pests which produce visible symptoms on plants and plant products, for example, inspection of fresh fruit in the packing house at the time of packing for the presence of mites.
- Inspection and certification at points of origin—the exporting country may be asked to inspect the shipment and officially document (certify) that the shipment is free from regulated pests before export.
- Prohibition of parts of the host—an importing country may prohibit specific parts of a host plant that could be a pathway for movement of a quarantine pest, and allow importation of other parts of the same host because the pest would not be present on that part, for example, a pest that is present on fruits and leaves of a deciduous flowering shrub is not necessarily present on dormant planting stock.
- Pre-entry or post-entry quarantine—the importing country may define certain control conditions, inspection and possible treatment of shipments upon their entry into the country. Often this involves isolating the shipments from other material capable of harbouring regulated pests until such time that it can be determined that the imported material is free from such pests, for example, fruit tree planting stock may be grown in isolation from hosts or vectors of specified virus diseases until such time that it can be tested and proven to be virus-free.
- Specified conditions for preparation of the consignment—the importing country may specify steps which must be followed in order to prepare the consignment for shipment. These conditions can include things such as packaging requirements, clean-up requirements and movement of the commodity under quarantine, *e.g.*, a requirement for shipments of fresh fruit to be packaged in new containers only, instead of re-used containers, may be effective in preventing the inadvertent movement of hitchhiker insects or soil which may contaminate used containers.
- Removal of the pest from the consignment by treatment or other methods—the importing country may specify chemical or physical treatments which must be applied to the consignment before it may be imported. For example, chemical treatments such as fumigants, insecticides, fungicides and herbicides, or physical treatments such as cold storage or heat treatment may be effective in eliminating specified quarantine pests from the shipment. Regulations and international obligations pertaining to the use of chemical pest control

products must be respected while establishing treatment requirements of this nature, *e.g.*, the use of methyl bromide.

- Prevention of establishment by limiting the use, distribution or timing of the consignment—the importing country may restrict the end use of the product (*e.g.*, for processing only) or the distribution of the product following importation in order to reduce the risk posed by certain pests to an acceptable level.
- Permitting importation during periods of the year when climatic conditions are not conducive to successful introduction, or when the pest is not present on the consignment, may be an effective measure for some pests or commodities. For example, potatoes for consumption may sometimes be imported from higher risk areas, if treated with a sprout inhibitor and packaged into small bags and sold in urban areas, where they are much less likely to be planted.

3.2. Precautionary Systems

Precautionary approach is the advancement–of regulatory approach for addressing this problem of invasive species. Invasive species often have devastating effects on natural areas, human health, agricultural, human-built structures and industry. The number and diversity of potentially harmful organisms being moved around the world is steadily increasing. Comparing these changes in land use and climate are rendering some habitat more susceptible to damaging invasions.

A major reason to regulate the intentional importation of a new plant species is the possibility that the plant itself may become a weed (Jenkins, 2005). In many countries, the majority of harmful non-native weeds have resulted from escaped plants that were intentionally imported in the past for ornamental, agricultural, forestry, hobby, and other uses. They have caused extensive economic losses, necessitated the spraying of expensive and damaging herbicides, and altered landscapes and bodies of water, causing impacts that cascade through ecosystems. A few examples: invading water hyacinth in Lake Victoria and elsewhere across Africa and Asia has blanketed thousands of kilometers of shorelines, blocked human access for fishing and other uses, damaged key wildlife areas, and promoted the growth of disease vectors. Introduction of the Australian Melaleuca tree in south Florida in the United States has converted thousands of hectares of the unique "Everglades" marsh to monospecific forest islands that are inhospitable to native wildlife.

In addition to resulting in many weed invasions, imported plants have carried plant pests such as the wood–devouring Asian long-horned beetle and plant pathogens such as the microbes that caused Dutch Elm disease and chestnut blight, with devastating environmental and economic consequences. While precise figures are not available, global crop losses and other costs and damages from all introduced weeds, plant pests, and plant pathogens are estimated to amount to many tens of billions–perhaps hundreds of billions–of dollars (US) annually.

At the same time that the problems of freely imported plants are receiving greater recognition, most major importing and exporting countries of plants are now parties to trade-facilitating agreements that impose international standards on national regulatory measures.

Several pre cautionary systems for regulating plant import have become in to effect in the last ten years in the form of clean/dirty/grey list approaches, also known as precautionary three list approach, in which species proposed for import are classified as either.

1. ***Clean-list*:** After assessment, the material to be imported, meets the Country's acceptable Level of Risk and is *allowed* to be imported.
2. ***Dirty list*:** After assessment, the material in question, fails to meet the country's acceptable level of risk and is placed in *Non-Provisionally prohibited.*
3. ***Grey-list*:** When significant information about the risk if any associated with the imported material is missing are placed under *Provisionally prohibited pending further information* and needs a full risk assessment.

The importer should ensure that the consignments are free from quarantine weeds. The importer should ensure that no prohibited/restricted plant species are imported except those by authorized institutes and that every consignment of plant species imported shall be accompanied by a phytosanitary certificate issued by an authorized officer at the country of origin of consignment, containing additional declarations for freedom from specified pests as indicated in the import permit. Also the importer should ensure that the special conditions prescribed there under are complied with. The importers should familiarize themselves with the list of plant species covered under Convention on International Trade in Endangered Species (CITES) and it shall be the responsibility of importer to ensure that the provisions of CITES are complied. The importers should also familiarize themselves with the list of plant species covered under prohibited list of plants/plant materials under Export and Import Policy (EXIM Policy), for restricted list of plants/plant materials covered under Export and Import Policy.

Several precautionary systems for regulating plant imports have come into effect in the last ten years in the form of clean/dirty/gray list approaches.

Thus, for example, all proposed live plant imports of new species for New Zealand or Western Australia as a general matter require a risk toward adopting stricter import controls is the emergence of better science on invasive species. That is, general scientific recognition now exists of the environmental risk associated with unregulated plant imports. Also, better, more reliable, and faster predictive tools now exist for assessing weed and disease risks based on models, decision trees, international databases, data sharing networks, and so on, allowing more precautionary approaches to be implemented without unduly restricting trade.

4. GLOBAL ORGANIZATION OF LEGAL BODIES–PLANT HEALTH MANAGEMENT

4.1. FAO—International Plant Protection Convention (IPPC)

The introduction and spread of plants and pests of plants from one geographical area to another is an issue of worldwide concern, and is addressed at the international level by several agreements. The principal agreement aimed at preventing the spread and introduction of pests of plants and plant products is the International Plant Protection Convention (IPPC).

The International Plant Protection Convention (IPPC) is a multilateral treaty for international cooperation in plant protection. The Convention makes provision for the application of measures by governments to protect their plant resources from harmful pests (phytosanitary measures) which may be introduced through international trade.

The Convention also aims to protect plant health while limiting interference with international trade. The IPPC applies to cultivated plants, natural flora and plant products, and includes both direct and indirect damage by pests (thus including plants as pests of plants *e.g.*, weeds). In addition to plants and plant products, the IPPC also extends to storage places, packaging, conveyances, containers, soil and any other organism, object or material capable of harbouring or spreading pests of plants. Countries that have ratified the IPPC are referred to as contracting parties. Over 80% of the countries in the world are contracting parties to the IPPC. Contracting parties agree to cooperate with one another in their attempts to prevent the international spread of pests of plants. This includes exchanging information on pests of plants, providing technical and biological information necessary for pest risk analysis, and participation in any special campaigns for combating pests (NAAS policy paper 12, 2001).

Contracting parties have the right to use phytosanitary measures to regulate imports, but have an obligation to do so only where necessary and technically justified. Some other obligations of contracting parties include the following:

- Establishing a National Plant Protection Organization (NPPO),
- Publishing and transmitting their phytosanitary requirements, restrictions and prohibitions,
- Conducting surveillance for pests,
- Notifying trading partners of non-compliance with import requirements and emergency actions taken, and
- Exchanging information on pests of plants (including pest reporting).

The IPPC is deposited with the Director-General of the FAO and is administered through the IPPC Secretariat located in FAO's Plant Protection Service. The IPPC was first adopted in 1951 and has been amended twice, most recently in 1997.The revision of the IPPC agreed in 1997 and which entered into legal force on 2nd October, 2005 represents an updating of the Convention to reflect contemporary phytosanitary concepts and the role of the IPPC in relation to the Uruguay Round Agreements of the WTO, particularly the SPS Agreement. The Secretariat of the IPPC is located at the FAO headquarters in Rome. More details about IPPC can be obtained from its official website www.iipc.int.

4.2. CPM and International Standards for Phytosanitary Measures (ISPMs)

The Commission on Phytosanitary Measures (CPM) is the governing body of the IPPC and it meets on an annual basis. The CPM has adopted a number of International Standards for Phytosanitary Measures (ISPMs) that provide guidance to countries and assist contracting parties in meeting the aims of the Convention. Some of the areas covered by ISPMs include: surveillance, pest risk analysis, establishment of pest free areas, export certification, phytosanitary certificates and pest reporting.

International Standards for Phytosanitary Measures (ISPMs) are adopted by contracting parties to the IPPC through the Commission on Phytosanitary Measures. ISPMs are the standards, guidelines and recommendations recognized as the basis for phytosanitary measures applied by Members of the World Trade Organization under the Agreement on the Application of Sanitary and Phytosanitary Measures. Non-contracting parties to the IPPC are encouraged to observe these standards. There are 24 ISPM's under these guidelines (Table 2).

Table 2: International Standards for Phytosanitary Measures (ISPMs)Adopted from ISPM, 2009 edition by IPPC

S. No.	*ISPM*	*Year*	*Particulars*
1.	ISPM No. 1	1993	Principles of plant quarantine as related to international trade
2.	ISPM No. 2	1995	Guidelines for pest risk analysis.
3.	ISPM No. 3	2005	Guidelines for the export, shipment, import and release of biological control agents and other beneficial organisms.
4.	ISPM No. 4	1995	Requirements for the establishment of pest free areas.
5.	ISPM No. 5	2005	Glossary of phytosanitary terms.
6.	ISPM No. 6	1997	Guidelines for surveillance.
7.	ISPM No. 7	1997	Export certification system.
8.	ISPM No. 8	1998	Determination of pest status in an area.
9.	ISPM No. 9	1998	Guidelines for pest eradication programmes.
10.	ISPM No. 10	1999	Requirements for the establishment of pest free places of production and pest free production sites.
11.	ISPM No. 11	2004	Pest risk analysis for quarantine pests, including analysis of environmental risks and living modified organisms.
12.	ISPM No. 12	2001	Guidelines for phytosanitary certificates.
13.	ISPM No. 13	2001	Guidelines for the notification of non-compliance and emergency action.
14.	ISPM No. 14	2002	The use of integrated measures in a systems approach for pest risk management.
15.	ISPM No. 15	2002	Guidelines for regulating wood packaging material in international trade.
16.	ISPM No. 16	2002	Regulated non-quarantine pests: concept and application.
17.	ISPM No. 17	2002	Pest reporting.
18.	ISPM No. 18	2003	Guidelines for the use of irradiation as a phytosanitary measure.
19.	ISPM No. 19	2003	Guidelines on lists of regulated pests.
20.	ISPM No. 20	2004	Guidelines for a phytosanitary import regulatory system.
21.	ISPM No. 21	2004	Pest risk analysis for regulated non-quarantine pests.
22.	ISPM No. 22	2005	Requirements for the establishment of areas of low pest prevalence.
23.	ISPM No. 23	2005	Guidelines for inspection.
24.	ISPM No. 24	2005	Guidelines for the determination and recognition of equivalence of phytosanitary measures.
25.	ISPM No. 25	2006	Consignments in transit

Table 2: (*Contd...*)

Table 2: *(Contd...)*

S. No.	*ISPM*	*Year*	*Particulars*
26.	ISPM No. 26	2006	Diagnostic protocols for regulated pests
27.	ISPM No. 27	2006	Diagnostic protocols for regulated pests
28.	ISPM No. 28	2009	Phytosanitary treatments for regulated pests
29.	ISPM No. 29	2007	Recognition of pest free areas and areas of low pest prevalence
30.	ISPM No. 30	2008	Establishment of areas of low pest prevalence for fruit flies (Tephritidae)
31.	ISPM No. 31	2008	Methodologies for sampling consignments
32.	ISPM No. 32	2009	Categorization of commodities according to their pest risk

4.3. National Plant Protection Organizations

A National Plant Protection Organization (NPPO) is an official service established by a government to fulfill the functions specified in the IPPC (IPPC website). NPPOs implement the phytosanitary laws and/or regulations issued by their governments. Responsibilities of NPPOs under the IPPC include:

- Issuance of phytosanitary certificates,
- Surveillance and inspection,
- Controlling pests (for example, administering treatments, preventing spread, disinfection or disinfestation),
- Protecting endangered areas,
- Conducting pest risk analyses,
- Ensuring phytosanitary security of consignments from certification until export, and
- Designation, maintenance and surveillance of pest free areas and areas of low pest prevalence.

To facilitate information exchange between the IPPC and contracting parties, each country has a designated official contact point, which is most often the NPPO.

4.4. Regional Plant Protection Organizations

Contracting parties also cooperate with each other within their regions through Regional Plant Protection Organizations (RPPOs). The functions of RPPOs include:

- Participating in activities to achieve the objectives of the IPPC,

- Disseminating information relating to the IPPC, and
- Cooperating with the CPM and the IPPC secretariat in the development of international standards.

There are currently nine RPPOs which are as follows (IPPC website):

- Asia and Pacific Plant Protection Commission (APPPC)—Southeast Asia, Indian subcontinent, Australia and New Zealand. Website: https://www.ippc.int/id/13497
- Comunidad Andina de Naciones (CAN)—Andean community. Website:www.comunidadandina.org
- Comité de Sanidad Vegetal del Cono Sur (COSAVE)—Southern cone of South America. Website: www.cosave.org
- Caribbean Plant Protection Commission (CPPC)—Caribbean Islands and Central America. Website: https://www.ippc.int/id/13470
- European and Mediterranean Plant Protection Organization (EPPO)—Europe and Mediterranean. Website: www.eppo.org
- Inter–African Phytosanitary Council (IAPSC)—Africa. Website: www.auappo.org
- North American Plant Protection Organization (NAPPO)—North America. Website: www.nappo.org
- Organismo Internacional Regional de Sanidad Agropecuaria (OIRSA)—Central America. Website: ns1.oirsa.org.sv
- Pacific Plant Protection Organization (PPPO)—Southwest Pacific Islands. Website: www.spc.int/pps

RPPOs each have their own independent statutes and conduct their own regional programmes. They may also create regional standards for their member countries. RPPOs cooperate with each other and with FAO and meet annually at the Technical Consultation, a meeting coordinated by the IPPC Secretariat. Lists of member countries of each RPPO can be found on the IPPC website: https://www.ippc.int.

5. PEST RISK ANALYSIS (PRA)

Pest risk analysis (PRA) is a science-based process that provides the rationale for determining appropriate phytosanitary measures for a specified PRA area. It is a process that evaluates technical, scientific and economic evidence to determine whether an organism is a potential pest of plants and, if so, how it should be managed.

Pest risk analysis is a process consisting of three stages (Fig. 1.)

1. Initiation of the PRA through identification of a pest or pathway, or review or revision of an existing phytosanitary policy,
2. Pest risk assessment, and
3. Pest risk management.

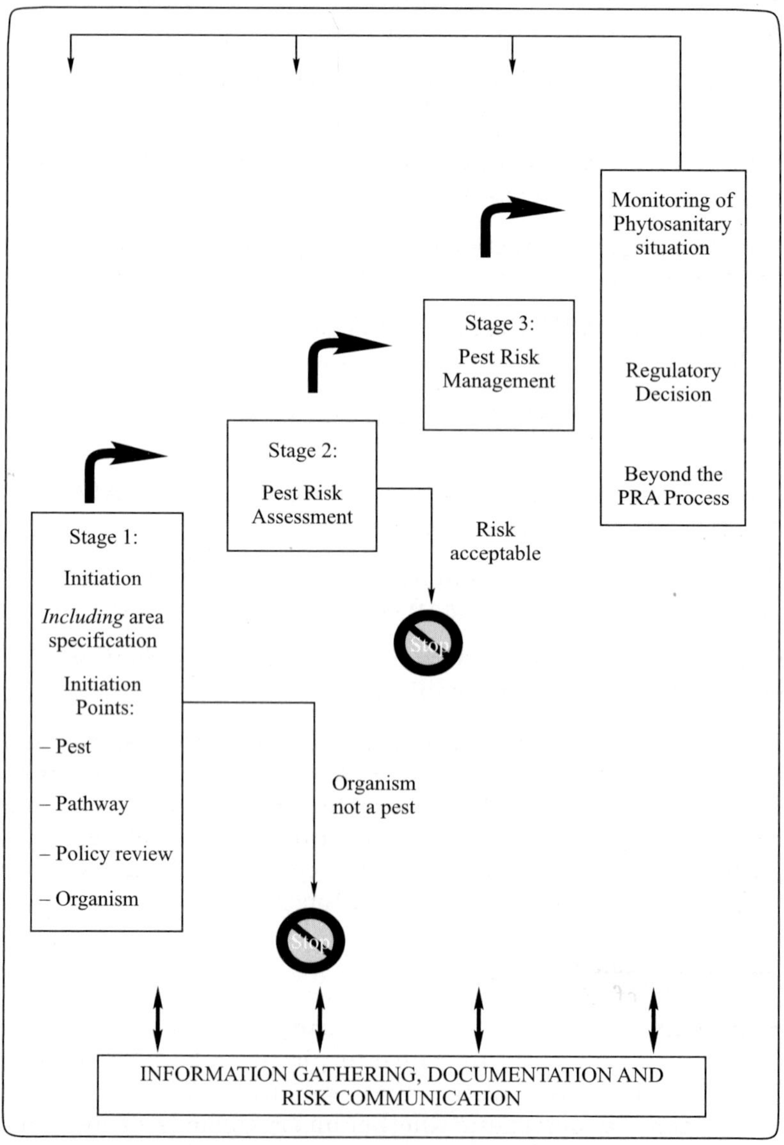

Fig. 1: Pest Risk Analysis Adpted from ISPM No. 2 (2009)

Risk communication is an integral component that occurs throughout each step.

Under the IPPC, the term plant pest refers to all organisms harmful to plants or plant products including other plants, bacteria, fungi, insects and other animals, mites, molluscs, nematodes, and viruses. Pests can be either regulated or not, and the IPPC recognizes and defines two categories of regulated pests of plants: quarantine pests and regulated, non-quarantine pests. PRA assists with determining whether a pest fits either of these two categories and the strength of phytosanitary measures, if any, that should be taken in response to it (Jagadeeshwar *et al.*, 2009).

If it is determined that the organism is a potential quarantine pest of plants, the probability of introduction and spread and the magnitude of potential consequences is evaluated using scientific, technical and economic evidence. If the pest risk is deemed unacceptable, the analysis may continue by suggesting management options that will reduce the pest risk to an acceptable level. These pest risk management options may be used to establish phytosanitary regulations.

A PRA may also consider the pest risks posed by the introduction of organisms associated with a particular pathway, such as a traded commodity. In most cases the commodity itself does not pose a pest risk but it might carry organisms that are pests of plants.

6. PEST REPORTING

Pest reporting allows countries to adjust as necessary their phytosanitary requirements and actions to take into account any changes in risk. It provides useful current and historical information for operation of phytosanitary systems.

The main purpose of pest reporting is to communicate immediate or potential danger. Immediate or potential danger normally arises from the occurrence, outbreak or spread of a pest that is a quarantine pest in the country in which it is detected, or a quarantine pest for neighboring countries and trading partners. The provision of reliable and prompt pest reports confirms the operation of effective surveillance and reporting systems within countries (ISPM, publication No. 17, 2002). List of recently reported pests of different groups from different parts of the world was given in Table 3.

Accurate information on pest status facilitates technical justification of measures and helps to minimize unjustified interference with trade. Every country needs pest reports for these purposes, and can only obtain them by the cooperation of other countries. Phytosanitary actions taken by importing countries based on pest reports should be commensurate with the risk and technically justified. NPPO should make provisions for collection, verification and analysis of domestic pest reports by the way of surveillance–either by general/specific means and by sources of information and finally conducts verification and analysis–by identification of pest concerned–thus establishing pest status in the country. NPPO also conducts PRA–for getting new/unexpected pest situations.

Table 3: Recent Examples of various Pests Reported—EPPO alert list: *Source*: http://www.eppo.org/QUARANTINE/Alert_List/alert_list.htm.Date of visit: 11.09.2011.

Pest name	***Main host plant/ habitats***	***Entry year***
Insects and mites		
Agrilus anxius (Coleoptera: Buprestidae)	*Betula*	2010–02
Chrysophtharta bimaculata (Coleoptera: Chrysomelidae)	*Eucalyptus*	2010–05
Drosophila suzukii (Diptera: Drosophilidae)	Polyphagous (fruit crops)	2010–01
Halyomorpha halys (Hemiptera: Pentatomidae)	Polyphagous	2008–10
Keiferia lycopersicella (Lepidoptera: Gelechiidae	*Lycopersicon esculentum*	2010–11
Leucinodes orbonalis (Lepidoptera: Pyralidae)	Solanaceae	2008–01
Oemona hirta (Coleoptera: Cerambycidae)	Polyphagous	2010–10
Psacothea hilaris (Coleoptera: Cerambycidae)	Moraceae (*Ficus, Morus*)	2008–10
Strauzia longipennis (Diptera: Tephritidae)	*Helianthus annuus*	2011–02
Thaumatotibia leucotreta (Lepidoptera: Tortricidae)	Polygaphagous (citrus, cotton, maize, prunus)	2011–05
Nematodes		
Meloidogyne ethiopica	Polyphagous	2011–01
Fungi		
Chalara fraxinea	*Fraxinus*	2007–09
Fusarium oxysporum f.sp. *lactucae*	*Lactuca sativa, Valerianella locusta*	2009–09
Melampsora euphorbiae	*Euphorbia pulcherrima* and wild *Euphorbia* spp.	2008–03
Phytophthora pinifolia	*Pinus radiate*	2009–01
Phytophthora ramorum	*Lithocarpus, Quercus, Rhododendron, Viburnum* and other woody ornamentals	2001–01

Table 3: (*Contd...*)

Table 3: (*Contd...*)

Pest name	***Main host plant/ habitats***	***Entry year***
Bacteria		
Acidovorax citrulli	*Citrullus lanatus, Cucumis melo.*	2009–07
Candidatus *Liberibacter solanacearum*	*Solanum tuberosum, Lycopersicon esculentum, Capsicum* spp.	2009–05
Pseudomonas syringae pv. *aesculi*	*Aesculus hippocastanum*	2009–06
Pseudomonas syringae pv. *actinidiae*	*Actinidia* spp. (kiwifruit)	2009–11
Virus/viroids		
Spiroplasma kunkelii	*Zea mays*	2008–01
Pepino mosaic virus	*Lycopersicon esculentum*	2000–01
Tomato apical stunt pospi viroid	*Lycopersicon esculentum*	2003–07
Tomato torrado virus	*Lycopersicon esculentum*	2009–02
Invasive plants		
Araujia sericifera	River banks, forests, crops, urban areas, wastelands	2008–03
Delairea odorata	Humid habitats, river banks, forests, grassland, road and rail, wastelands	2009–02
Eriochloa villosa	River banks, road and rail networks, wastelands, arable crops, permanent crops	2008–09
Hydrilla verticillata	Inland wetlands (marshes, peat bogs) and continental waters	2009–03
Hygrophila polysperma	Continental waters, banks of continental water, river banks/canal sides (dry river beds)	2010–01
Miscanthus sinensis	Riverbanks, pastures, road and rail networks, wastelands	2011–04
Myriophyllum heterophyllum	Water bodies, slow flowing waters	2009–01
Parthenium hysterophorus	Arable crops, orchards, river banks, road and rail networks, wastelands	2011–03
Pennisetum setaceum	Pastures, natural grasslands, deserts, road and rail networks, wastelands	2009–03

Table 3: (*Contd...*)

Table 3: (*Contd...*)

Pest name	***Main host plant/ habitats***	***Entry year***
Stipa trichotoma, Stipa neesiana and *Stipa tenuissima*	Arable lands, pastures, natural grasslands, road and rail networks, wastelands	2009–06
Verbesina encelioides	Arable land, pastures, road and rail networks, other artificial surfaces (*e.g.*, wastelands)	2008–10

6.1. A Pest Report Should Clearly Indicate

- The identity of the pest with scientific name (where possible, to the species level, and below species level, if known and relevant
- The date of the report
- Host(s) or articles concerned (as appropriate)
- The status of the pest.
- Geographical distribution of the pest (including a map, if appropriate)
- The nature of the immediate or potential danger, or other reason for reporting.
- It may also indicate the phytosanitary measures applied or required, their purpose, and
- Any other information as indicated for pest records

If all the information is not available on the pest situation then a preliminary report should be made and updates made, as further information becomes available (Table 4).

7. NATIONAL PHYTOSANITARY CONTROL SYSTEM–INDIA

The plant quarantine (PQ) regulations are operative in India through the *Destructive Insects and Pests (DIP) Act, 1914* and amendments issued there under. The DIP act was passed in 1914 and has been revised time to time with several amendments. There are a number of lacunae under DIP Act which already resulted in introduction of serious pests into the country *viz.* paddy blast caused by *Pyricularia oryzae.* By keeping these points under consideration, a new Quarantine Bill was proposed (Prem Narain and Aparna Sawhney, 2005).

Fig. 2: Important insect pests/disease/invasive plants reported in recent times. 1A—*Strauzia longipennis* and 1B—*Thaumatotibia leucotreta*—Insect pests. 1C—*Meloidogyne ethiopica* (Nematode) 2A—*Acidovorax citrulli*, 2B—*Melampsora euphorbiae* and 2C—*Pepino mosaic virus*–Diseases. 3A—*Araujia sericifera* , 3B—*Verbesina encelioides,* 3C—*Eriochloa villosa* (wooly cup grass), 3D–*Parthenium hysterophorus* (Congress weed), 3E–*Delairea odorata*, 3F–*Stipa tenuissima* and 3G–*Pennisetum setaceum*–Invasive plants.

Table 4: Typical Pest Report format (*Source*: IPPC website)

Country	***Identity of pest***	***Host(s) or article(s) concerned***	***Status of pest (under ISPM no. 8)***	***Title***	***Date***	***Last updated***	***Report number***
Brazil [BRA]	*Puccinia kuehnii* (W. Krüger) E.J. Butler	Sugar cane	Transient: actionable, under surveillance	Pest Notification–Sugar Cane Orange Rust–January 5th 2010	11.01.2010	11.01.2010	BRA–01/1
Australia [AUS]	*Phakopsora euvitis*	Vitis spp; and native *Ampelocissus* species under experimental conditions	Absent: pest eradicated	Eradication of grapevine leaf rust (GVLR) from Australia	11.03.2008	11.03.2008	AUS–10/1
Canada [CAN]	*Tomicus piniperda* (Linnaeus)		Present: subject to official control	Pine Shoot Beetle–Update of the Regulated Areas–Canada	15.12.2009	15.12.2009	CAN–01/2
Barbados [BRB]	*Fusarium udum* Butler. (telomorph: *Gibberella indica*).	Pigeon pea (*Cajanus cajan*)		First report of *Fusarium udum*	29.06.2010	29.06.2010	BRB–01/1
Jamaica [JAM]	*Ralstonia solanacearum* Bacterium	Banana and Plantain (*Musa* spp.)	Present: under eradication	Occurrence of the Moko Disease in Jamaica	04.05.2006	04.05.2006	JAM–01/1
South Africa [ZAF]	*Scirtothrips dorsalis*		Absent: pest no longer present	SA no longer has *Scirtothrips dorsalis*	26–01–2007	26–01–2007	ZAF–01/1
United Kingdom [GBR]	*Scirtothrips dorsalis*	Various plants growing in a glasshouse	Present: only in protected cultivation	*Scirtothrips dorsalis*	28–01–2009	28–01–2009	GBR–13/1

7.1. Plant Quarantine Bill

A new draft Plant Quarantine Bill is under consideration by the Ministry of Agriculture for replacing the existing DIP Act of 1914 with the following mandates *viz.*

- To prevent the introduction of quarantine pests in India by regulating the importation of plants, plant products and other objects;
- To prevent the spread of quarantine pests outside India by regulating the exportation of plants, plant products and other objects;
- To prevent the spread of quarantine pests from one State to another;
- To regulate the introduction in India of new or beneficial organisms and soil;
- To give effect to international agreements to which India is a party and in particular to the International Plant Protection Convention, the Agreement on the Application of Sanitary and Phytosanitary Measures, and the Agreement on Technical Barriers to Trade; and
- To provide for the constitution of the Plant Quarantine Authority of India and to ensure efficiency and accountability in the implementation of the above objectives.

7.2. Plants, Fruits and Seeds (PFS) Order

The import of seeds/plants/plant materials both for sowing/planting/propagation and consumption, and soil/peat–moss, etc., is regulated by the *Plants, Fruits and Seeds (PFS) (Regulation of Import into India) Order, 1989* (PFS Order) and amendments issued there under. Both import permits issued by the Plant Protection Adviser or any authorized officer under the above–said notification and the Phytosanitary Certificate issued at the country of origin are mandatory for import of plants/plant material for sowing/planting/propagation as well as consumption.

A special permit issued by the Plant Protection Adviser regulates the import of soil or peat/sphagnum moss.

Further, the import of cotton is regulated by 'Import of Cotton into India Regulations, 1972. As per the above notifications, cotton shall not be imported into India by sea except through the ports of Bhavnagar, Kolkata [Calcutta], Chennai [Madras], Cochin, Mumbai [Bombay] and Tuticorin. The fumigation is mandatory for the import of cotton bales into India at the port of entry to prevent the entry of cotton boll weevil.

The import of live insects is regulated by Notification No. 193/40A dated 3 February 1941 and the import of live fungi is regulated by Notification

No. 16–5(1)/43A dated 10th May 1943. Some items do not require any import permit and the details were furnished in Table 5.

Table 5: Items do not require clearance certificate: (adopted from FPS Order, 1989)

Sl. No.	*Name of the item not requiring Plant Quarantine Clearance*	*Reasons for exempting from Plant Quarantine Clearance*
1.	Asafoetida (Hing)	• No Phytosanitary risk is likely to be associated with it being tree exudate
2.	Medium Density Fiberboard from Malaysia	• Heat treatment • Physical processing • Chemical Processing
3.	Manufactured/Processed Tea	• Processed • Hygienically packed in aluminum foil
4.	Processed plywood, particle board, oriented strand board or veneer	• Chemical processing • Heat treatment • Physical processing
5.	Dead yeast	• Being a processed material unlikely to be infested by pests
6.	Paper bags	• Being a processed/manufactured material unlikely to be infested by pests
7.	Dietary food health supplement in capsule form	• Being a processed/manufactured material unlikely to be infested by pests
8.	Waxed/bleached coir bristle fibre	• Waxed • Bleached
9.	Gambier extract	• Being a processed/manufactured material unlikely to be infested by pests
10.	Marigold pellets undergone processing	• Heat treatment • Physical processing
11.	Processed Jute products	• Heat treatment • Chemical processing • Proper packing in storage
12.	Processed oat flakes	• Heat treatment • Physical processing
13.	Tea saponin extract	• Steam distillation • Drying and extraction processing
14.	Natural waxed coconut fibre	• Heating • Waxing
15.	Agar agar extract, ingredient for dehydrated culture media	• Alkalizing and washing • Bleaching, acidification and extraction

Table 5: (*Contd...*)

Table 5: *(Contd...)*

Sl. No.	*Name of the item not requiring Plant Quarantine Clearance*	*Reasons for exempting from Plant Quarantine Clearance*
		• Dehydration, drying and packing
16.	Wood chips of less than 6 mm thickness	• Not regulated by Plant Quarantine (Regulation of Import into India) Order, 2003, unless they are found to be harbouring any regulated pests specified in this order
17.	Processed Cotton linters pulp	• Mechanical cleaning and washing • Bleaching with Sodium hypochlorite • Acid treatment with sulphuric acid • Drying
18.	Vacuum dried carrot flakes	• Mechanical cleaning and washing • Steam peeling at 100°C • Vacuum drying and packing
19.	Soya protein isolate	• Cleaning and extraction at 80 °C • Drying at 120°C, centrifuging, acid washing, sterilization at 140°C, deodorization and homogenization at 80°C and spray drying at 145°C and packing • Non-Genetically Modified Organism (Non-GMO) product
20.	Exudates (gum) of various trees	• No Phytosanitary risk is likely to be associated with tree exudes.

The import of germplasm is permitted by the Director, NBPGR, New Delhi and for research purpose by the institutes or organizations under Indian Council of Agricultural Research (ICAR)/State Agricultural Universities. The import of transgenic seeds for research/trial purpose is permitted under a special permit issued by the Director, NBPGR, New Delhi after its import clearance by the Recombination Committee on Genetic Engineering and Modification of the Department of Biotechnology (DBT) constituted under the chairmanship of the Adviser, DBT.

A draft Plant Quarantine Order is in the process of notification to replace existing PFS (Regulation of Import into India) Order, 1989 and amendments issued there under to provide comprehensive regulations for importation of plants and plant products both for consumption and propagation; germplasm and transgenic; bio-control agents; live insects and microbial cultures including algae; and genetically modified organisms (GMOs), soil, etc.

7.3. Organizational Structure of Regulatory Bodies in India

The Directorate of Plant Protection, Quarantine and Storage established under the Ministry of Agriculture (Department of Agriculture and Cooperation) is responsible for the administration of PQ regulations issued under the DIP Act, 1914. The Directorate is headed by the Plant Protection Adviser and assisted by Joint Director (PQ) in PQ matters. A total of 57 points of entry are notified for import of plants and plant material, of which 29 PQ stations are established (Table 6). However, the import of seed and propagating plant material is restricted through five major stations, *viz.*, Amritsar, Chennai, Kolkata, Mumbai and New Delhi. The above regional stations are headed by Deputy Directors (Entomology/Plant Pathology) and assisted by Assistant Directors specialized in Nematology, Bacteriology, Virology and weed science and the minor stations are headed by Plant Protection Officers specialized in entomology or plant pathology.

Table 6: Points of Entry for Import of plants/plant materials and other Articles Adopted from Schedule I of Annexure–5 of PQ Order, 2003

Seaports	*Airports*	*Land Frontier Stations*
1. Alleppey (Kerala)*	1. Amritsar (Punjab)@	1. Agartala (Tripura)
2. Bhavnagar (Gujarat)	2. Bangalore (Karnataka)	2. Amritsar Rly. Stn. (Punjab)
3. Kolkata (West Bengal)@	3. Kolkata (West Bengal)	3. Attari Rly. Stn.(Punjab)
4. Calicut (Kerala)	4. Chennai (Tamil Nadu)	4. Attari Wagha Border Check post (Punjab)
5. Chennai (Tamil Nadu)@	5. Hyderabad (Andhra Pradesh)	5. Bongaon (West Bengal)
6. Cochin (Kerala)	6. Mumbai (Maharashtra)	6. Gede Road Rly. Stn. (West Bengal)
7. Cuddalore (Tamil Nadu)*	7. New Delhi (Delhi) @	7. Jogbani (Bihar)
8. Goa* (Goa)	8. Patna (Bihar)	8. Moreh (Manipur)
9. Gopalpur (Orissa)*	9. Tiruchirapalli (Tamil Nadu)	9. Panitanki (West Bengal)
10. Haldia (West Bengal)*	10. Trivandrum (Kerala)	10. Raxual (Bihar)
11. Jamnagar (Gujarat)	11. Varanasi (Uttar Pradesh)	11. Rupadiha (Uttar Pradesh)
12. Beypore (Kerala)*	12. Guwahati (Assam)	12. Sonauli (Uttar Pradesh)
13. Kakinada (Andhra Pradesh)		13. Banbasa (Uttaranchal)
14. Kandla (Gujarat)		
15. Karwar (Karnataka)*		
16. Krishnapatnam (Andhra Pradesh)*		

Table 6: (*Contd...*)

Table 6: (*Contd...*)

Seaports	*Airports*	*Land Frontier Stations*
17. Machlipatnam (Andhra Pradesh)*		
18. Mandvi (Gujarat)		
19. Mangalore (Karnataka)		
20. Mumbai (Maharashtra)@		
21. Mundra (Gujarat)		
22. Nagapatinam (Tamil Nadu)*		
23. Nava Shiva (Maharashtra)		
24. Navlakhi (Gujarat)*		
25. Okha (Gujarat)*		
26. Paradeep (Orissa)*		
27. Pondicherry		
28. Porbander (Gujarat)		
29. Rameshwaram (Tamil Nadu)		
30. Tiruvananthapuram (Kerala)		
31. Tuticorin (Tamil Nadu)		
32. Veraval (Gujarat)*		
33. Visakhapatnam (Andhra Pradesh)		
34. Vizhinjam (Kerala)		

* For import of food grains and timber only

@ Import of Propagating material is restricted to these stations.

Keeping in view the significant role played by the phytosanitary services in safe conduct of global trade in agriculture, the Government of India (Ministry of Agriculture) has established modern pest diagnostic laboratory facilities with high-tech scientific equipment at four regional centers of Amritsar, Chennai, Kolkata and New Delhi. The project was aimed at developing and strengthening of PQ facilities at major ports through capacity building and human resource development. Under the FAO/UNDP Project training fellowship visits, study tour programs were organized in developed countries to acquire the knowledge of quarantine procedures and guidelines, rapid diagnostic techniques and to acquaint with quarantine policies and regulations in the implementation of phytosanitary measures. Besides this, local training programs were also held to upgrade the skills of in-service PQ personnel working at those ports.

A PQ website entitled http://www.plantquarantineindia.org was designed and hosted under the above–said Project .The PQ website provides information about contact points; PQ set–up; PQ act and regulations; New Seed Policy guidelines; quarantine procedures for issuance of permit, import clearance, post entry quarantine inspection and export inspection and certification of agriculture commodities.

A software package entitled '*Quarantina 2k*' was developed for computerized issuance of permits, import release orders and the phytosanitary certificates and was installed at four major PQ stations at Amritsar/Kolkata (Calcutta)/Chennai (Madras)/New Delhi.

A National Phytosanitary Database was developed providing the data related to import inspection and export certification of agricultural commodities and issuance of permits. Further a suitable software package entitled '*Phytopest*' was developed for creating endemic pest database of prioritized commodities (Prem Narain and Aparna Sawhney, 2005).

7.4. Inspection Authorities In India

7.4.1. *Import of Seeds/plant Materials for Sowing, Planting and Propagation*

The Importer or his agent shall apply for permit for importing seeds/plant materials for sowing, planting and propagation at least seven days in advance to the permit issuing authority of the designated port of entry as per the Plant Quarantine (Regulation of Import into India) Order, 2003. All consignments of seeds and plants for propagation shall be imported only through Regional Plant Quarantine Stations of Amritsar, Chennai, Kolkata, Mumbai or New Delhi.

7.4.2. *Inspection Authorities–Post Entry Quarantine*

For post entry quarantine purpose a total of 40 inspection authorities were notified for all the states and Union territories (PQ Order, 2003). For the states, Head, Division of Plant Pathology of the respective state Agricultural Universities (or ICAR Institute) will be designated as the Inspection Authority (IAs) and for the Union Territories Head, Division of Plant Pathology of the closest Agricultural University or ICAR Institute acts as the IAs. For other specified purposes separate Inspection Authorities were identified (Table 7).

7.5. Pest Interception–India

Plant pest interceptions from imported commodities provide documented evidence of the value of plant quarantine activities. Historical records of interceptions are the best evidence of how pests enter into a country. Interception records provide a basis for decision making in Plant Protection and Quarantine (PPQ). The interception records are used in conducting pest risk assessment and in determining the personnel and equipment needs at ports of entry (Plant Quarantine Manual, 2010).

Table 7: Inspection authorities for specified purposes Adopted from Part II of Schedule XI–PQ Order, 2003

S. No.	*Name of the Inspection Authority*	*Jurisdiction*	*Purpose*
1.	Head, Advance Centre for Plant Virology, Indian Agricultural Research Institute, PUSA, New Delhi	Entire country raised plants	Tissue culture
2.	Head, Indian Institute of Horticulture Research, Hesarragatta, Bangalore	Entire country	Tissue culture raised plants
3.	Head, Institute of Himalayan Bio-Resources Technology, Palmpur, Himachal Pradesh	Entire country	Tissue culture raised plants

A list of pests/races, not known to present earlier, that were intercepted before their spread thus aiming at a successful preventive measure was given in Table 8.

Table 8: Examples of different categories of pests intercepted in quarantine in India. (*Source*: Mandal, 2011)

Categories of pest intercepted	*Pest*	*Host/race*	*Source country*
Unknown in India	*Peronospora manshurica*	Soybean	USA
	Uromyces betae	Sugarbeet	USA and Italy
	Fusarium nivale	Wheat	UK
	Cowpea mottle virus	Cowpea	Philippines
	Tomato black ring virus	French bean	CIAT (Columbia)
	Heterodera schachtii	Sugarbeet	Denmark
	Anthonomus grandis	Cotton	USA
	Quadrastichodella eucalypti	Eucalyptus	Australia
Known to occur but the race/ biotype/strain intercepted is not known to occur	*Helminthosporium maydis,*	Sorghum/race	USA
	Pea seed-borne mosaic virus,	TBroadbean	—
	Burkholderia solanacearum	Groundnut/ biovar 3	Australia

CONCLUSIONS

Agriculture is the life line of the country as nearly two-third of the population depends on farming and India made rapid strides towards meeting the food demand to equally as well as globally. With the increase in world food trade especially over the few decades, the implementation of food safety and plant health along with quality standards have become imperative.

Fruit and other perishable products being carried in to the USA by travellers are confiscated to prevent entry of pests. California prohibits transport of apples from northwest counties into the remainder of the state by posting quarantine signs to prevent the spread of apple maggot. New York prohibits movement of gravel from counties bordering the eastern shore of Lake Ontario to prevent the spread of the alfalfa snout beetle. Imported animals are often held in quarantine for a period of time to allow inspection for pests and diseases. Such legal actions are not always effective, but they enable interceptions of many pests that could add to the pest problems already present in the county.

A need exists to reconcile those aspects of international legal regime that promote a relatively unrestricted plant trade with the aspects that allow trade restrictions in order to accommodate a country's acceptable level of risk from imported weeds, pests and pathogens in the countries appropriate level of protection.

Consequent to the adoption of the SPS Agreement, there is a growing concern among the developing countries that stringent SPS measures imposed by the developed country trading partners may impede the export of agriculture produce from developing countries, as the latter are not well-equipped to address various issues related to SPS Agreements due to lack of information on measures that affect exports; lack of infrastructure facilities and scientific support for inspection, testing, diagnosis, and treatment; lack of risk assessment/surveillance capabilities; and lack of preparedness to associate with setting of international standards and participation in dispute settlement. On the other hand, trade by developed countries has increased many folds in the absence of countervailing measures by developing countries leading to widening the gap of the trade balance. Many of the developing countries including India have to face stiff challenges of product disqualification on account of pesticide residues/aflatoxins exceeding the limits set by the national standards of various EEC countries, which are higher than those prescribed by Codex Alimentarius Commission.

The size, geographical complexity, diversified agro-climatic conditions and cropping patterns and huge population are present major challenges to India with regard to control of trans-boundary plant pests and animal diseases. Particularly, India has to share a long porous border with Bangladesh, Bhutan, Nepal, Pakistan and Myanmar and thus there is an urgent need to promote strong regional cooperation and linkages in developing effective mechanism for harmonizing SPS measures among the countries within the region. Under rapidly changing external environment, the need to develop an integrated national PQ service with world-class infrastructure facilities and various systems approach such as information management, pest risk analysis, pest surveillance, phytosanitary-border control and export inspection and certification, was considered to be a high priority by the Government of India to protect the country from the inadvertent introduction of economically important exotic pests and to facilitate market access by having credible and technically competent inspection and certification systems.

A plan of action has been prepared and implemented during the Tenth Plan period (2002–07). During the plan period, a total of 17 amendments were made to the Plant Quarantine order, the latest being the 8th amendment to the Draft Plant Quarantine (regulation of import in to India) order 2010 (Department of Agriculture and Cooperation, 2011). Accordingly, the National Standards for Phytosanitary Measures for some of the important activities have already been developed and adopted including the Guidelines for Development of National Standards for Phytosanitary Measures and six draft National Standards are under the process of approval. Stanard Operating Procedures (SOP) for export/import, inspection, certification for plant and plant products were developed by Directorate of and are now under operation. The Standards which are critical for our exports have been prioritised. Further, a National Integrated Fruit Fly Surveillance Programme has also been prepared with a view to establish pest-free areas against fruit flies. To streamline the Plant Quarantine activities, efforts are being made to computerise the Plant Quarantine Stations for speedy and transparent functioning.

Further, a software was developed, by name Plant Quarantine Information System (PQIS, version 1.0), by the Ministry of Agriculture, Department of Agriculture and Cooperation in January 2011 (http://plantquarantineindia.nic.in). The objective of PQIS is to provide an efficient and effective service to importers, exporters, individuals and the Government. Plant Quarantine Information System facilitates Importers to apply online for Import Permit, Import Release Order and Exporters to apply online for Phytosanitary Certificate. Exporters and Importers can view the status of their application online and access history of their application during entire life cycle of the application.

Substantial efforts have also been made towards the international cooperation on the phytosanitary matters including participation in IPPC Standard Committee meetings for development of International Standards for Phytosanitary Measures, Training of fumigation service providers and regulators in methyl bromide fumigation under the Australian Fumigation Accreditation Scheme, finalisation of SPS Protocol for export of mango to China, harmonisation of phytosanitary measures with reference to Seed trade through the Workshops organised by Asia Pacific Seed Association (APSA), negotiations on free trade agreement with Thailand and Singapore, Memorandum of Understanding and work plan with Chile for cooperation on Phytosanitary matters and Indo–Nepal Joint trade and economic development negotiations.

REFERENCES

Arneson, P.A. (2001). Plant Disease Epidemiology: Temporal Aspects. *The Plant Health Instructor*. DOI: 10.1094/PHI–A–2001–0524–01. (Revised 2006).

Chattopadhyay, S.B. (1991). Principles and procedures of plant protection. Oxford and IBH Publishing Co. Pvt. Ltd.., New Delhi, India. pp. 377–86.

Fry, W.E. (1982). Principles of Plant Disease Management. Academic Press, New York. p. 378.

Gazettes of Department of Agriculture and Cooperation (2011). Available on http://www.agricoop.nic.in/gazette.htm. Date of visit: 22.09.2011.

Horn, D.J. (1988). Ecological approach to pest management. The Gulford Press, New York, pp: 204–206.

http://plantquarantineindia.nic.in. Plant Quarantine Information System (PQIS) default page. Date of visit: 2.09.2011.

ISPM—International Standards for Phytosanitary Measures 1–32 (2009 edition) (2009). By the secretariat of international plant protection convention. FAO, Rome. p. 432.

Jacobsen, B. (2001). Disease Management. *In:* Maloy, O.C. and Murray, T.D. *Eds.,* Encyclopedia of Plant Pathology, Wiley, New York. pp. 351–56.

Jagadeeshwar, R., Sreenivasarao, C. and Narayana Reddy, P. (2009). *In:* Reading manual for trainees–Training programme on sanitary and phytosanitary (SPS) Measures and Risk Analysis to promote trade in India. July 27–30, 2009, ANGRAU, Hyderabad, India.

Jenkins P.T. (2005). International law related to precautionary approaches to national regulation of plant imports, *Journal of World Trade,* **39:** 895–906.

Maloy, O.C. (1993). Plant Disease Control: Principles and Practice. Wiley, New York.

Maloy, O.C. (2005). Plant Disease Management. The Plant Health Instructor DOI: 10.1094/PHI–I–/2005–0202–01.

Maloy, O.C. and A. Baudoin. 2001. Disease Control Principles. *In:* Maloy, O.C. and Murray, T.D. *Eds.,* Encyclopedia of Plant Pathology. Wiley, New York. pp. 330–32.

Mandal, F.B. (2011). The management of alien species in India. *International Journal of Biodiversity and Conservation,* **3(9):** 467–73.

NAAS policy paper 12 on sanitary and phytosanitary agreement of the WTO–Advantage India, September 2001.

Post-Entry Quarantine Manual, (2010). The US Department of Agriculture (USDA). p. 322.

Plant Quarantine (Regulation of Import into India) Order, (2003) (PQ Order). The Gazette of India, Ministry of Agriculture and Cooperation. Dtd. 18th November, 2003. p. 105.

Narain, Prem. and Sawhney, Aparna. (2005). India 1 and 2. *In:* Cornelis sonneveld *Eds.,* country papers in report of the APO seminar on sanitary and phytosanitary measures held in Japan, 4–11 December, 2002. pp. 96–108.

Reddy, D.B. and Joshi, N.C. (1992). Plant Protection in India. Allied Publishers Ltd., New Delhi, India. pp. 411–20.

The Plant Fruits and Seeds (Regulation of Import into India) Order, (1989) (PFS Order). The Gazette of India, Ministry of Agriculture and Cooperation. Dtd. 27th October, 1989. p. 23.

Whetzel, H.H. (1929). The Terminology of Plant Pathology. Proc. Int. Cong. *Plant Science, Ithaca,* NY, **1926:** 1204–15.

www.eppo.org. Official website of The European and Mediterranean Plant Protection Organization (EPPO). Date of visit: 11.09.2011.

www.eppo.org/QUARANTINE/Alert_List/alert_list.html. Date of visit: 11.09.2011.

www.fao.org. Official website of Food and Agricultural Organisation (FAO), Rome, Italy. Date of visit: 04.06.2011.

www.ippc.int. Official website of International Plant Protection Convention (IPPC), Rome. Date of visit: 21.02.2011.

www.plantquarantineinida.org. Official website of Plant Quarantine Organization of India. (PQOI). Date of visit: 11.09.2011.

8

Plant Protection in Protected Culture

MOUSA NAJAFINIA[1]* AND MEHDI AZADEHVAR

ABSTRACT

Diseases are a fact of life in most greenhouse vegetable operations. Implementing an effective disease management program may well determine the success or failure of a greenhouse business. Diagnosis is the process of determining the cause or causes of a disease through examination and analysis. An accurate diagnosis is often needed before effective remedial or preventative measures can be taken. It is not unusual to encounter more than one disease on a given plant, so care should be taken to examine all of the signs and symptoms at hand. Environmental stresses can predispose plants to attack by pathogens, and infectious diseases can weaken plants so they are more susceptible to non-infectious diseases. Diagnosis is both art and a science, and diagnostic skills can be improved through experience, education and practice. Disease management strategies aim to prevent the establishment of diseases in greenhouse plantings, as well as to minimize the development and spread of any diseases that may become established in a crop. The development of a successful disease management program requires a good working knowledge of crop production requirements and an ability to recognize the key diagnostic symptoms and signs of diseases and pests. The main components of such a program are regular crop monitoring and the integration of cultural, chemical and biological control practices.

Key words: Greenhouse, Plant protection, Disease, Management and protected culture.

[1] Agricultural Research, Education and Extension Organization (AREEO), Department of Plant Protection, Agricultural Research Center of Jiroft and Kahnouj, Jiroft, Iran.

* *Corresponding author*: E-mail: mnajafinia10@yahoo.com

1. INTRODUCTION

Limitation in water, soil and increasing of world population led to concentrate on the methods to produce more and more. Production of different crops in low and high tunnels and greenhouses increased in popularity significantly in the last decade around the world. They provide the option of off-season production and expansion of markets over traditional outdoor field systems. Cultivated crops are including vegetable crops, cut flowers, ornamental plants and fruits. Among the vegetables, tomato, cucumber, watermelon (particularly under low plastic tunnels), pepper, eggplant, melon, lettuce, and other vegetable crops are grown in greenhouse. The greenhouses mostly are covered with plastic films and glass. Plastic film for covering is generally polyethylene. It lasts for one to two years (Tuzel and Ozcelike, 2004).

Greenhouse and high tunnel systems pose unique pest management challenges. The protected culture of greenhouse and high tunnel production may result in lower incidence of diseases exacerbated by rainfall. Diseases that are uncommon in open fields may appear under protected culture. Like Botrytis blight/gray mold, downy and powdery mildew, leaf spots, white mold are among the most important of these diseases.

Greenhouses have special conditions namely good environment for all diseases, moderate (warm) temperature, high humidity/moisture, and air movement to spread inoculum, overcrowded conditions, and rapid growth of cultivated plants in greenhouse which is very suitable for incidence and quick spreading of disease. The incidence of disease in protected cultivation depend on many factor such as, cultivated plant type, the layout of cultured row, greenhouse structure, plant canopy, cultivation pad, varieties, ventilation, covering material, sanitation and so on. For having a free disease greenhouse, it needs to provide all mentioned agents suitable to plant growth and unfavorable conditions for pathogens. Successful managing greenhouse diseases needs to: Accurate Diagnosis (the use of molecular techniques to detect and diagnosis of plant pathogens is highly valuable), understanding of pathogen sources, understanding of pathogen biology (life cycle, environmental requirements, mode of spread) and finally develop appropriate and effective management strategies.

2. GREENHOUSE TYPE STRUCTURES AND THEIR IMPACT ON CROP PROTECTION

(i) ***Low Tunnels (row-covers)*:** These are small structures that provide temporary protection to crops. Their height is generally 1 m or less, with no aisle for walking, so that cultural practices must be performed from the outside. Their use enhances early yields and yield volume; they also protect against unfavorable weather (Fig. 1).

(ii) ***High Tunnels (walk-in tunnels)*:** Such structures use the same cover materials as low tunnels and are high enough to perform cropping practices inside. Moderately tall crops are grown (Fig. 1).

Fig 1: Different types of structure of protected culture: Low tunnel (upper left), single high tunnel (upper right), wooden pole block-type house (lower left) and galvanized steel pipe frame house (lower right)

(iii) ***Greenhouses*:** These differ from other protection structures in that they are sufficiently high and large to permit a person to conveniently stand upright and work within (Nelson, 1985). Greenhouses appeared when glass became available for covering. Later, the introduction of plastic films permitted world-wide expansion of the greenhouse industry. There are many design variations in greenhouse structures, but only two basic types, a wooden pole block-type house or a galvanized steel pipe frame house are most common (Fig. 1). The selection of material and frame structure mostly depend on climate and financial ability of farmers. Our investigation in Iran showed that woody structure are cheaper but due to lower high and low resistance against wind and storm, high humidity and other disadvantages they are not suitable in tropical and humid regions. The incidence rate of disease in woody greenhouse is more and the use of pesticides for control of pests and disease are higher.

In terms of area, the basic greenhouse used for vegetable crops is a structure 3 to 3.4 m high, 9m wide and 56 m long, covering about 560 m^2 (Choukr–Allah, 2004). Our experience in south of Iran with tropic to sub tropic climate showed that the best structure is 4 to 4.5 m high, 7 m wide and 40 m long. They may be made as single tunnel, double and multi tunnel. The covering material used in most of these structures is polyethylene film.

Insufficient ventilation is still one of the most important problems particularly in the plastic houses. The ratio of roof ventilation to the floor space should be between 15 and 20% but it ranges between 1–4% on an average, but in recent years, roof ventilation ratio has increased (Tuzel and Ozcelike, 2004). Insufficient ventilation causes high humidity during the wintertime and makes it difficult to control the temperature particularly at the beginning and at the end of growing season. Generally plastics are opened and closed with pipes placed onto the roof and sidewalls to control temperature and humidity (Fig. 2a, b).

Fig. 2: Adjusting the temperature and relative humidity for controlling plant diseases by natural ventilation from side wall (left) and from roof (right)

3. SOIL-LESS CULTURE

One of the main reasons for development of soil-less crops is pathological. Several new substrates have been tested including rockwool, perlite, and polymer of urea, compost and organic substrate. However, research should be carried on, particularly in view of the expected future expansion of this technology in order to be able to propose precise and environmentally sound policies that certainly can be applied to soil-less culture. Once again, it is evident that a strategy of avoiding or minimizing the use of pesticides in soil-less culture will be most appropriate, using the two substrate (perlite and polymer of urea) since they could be recycled and do not present any environment risk when they are incorporated in to the soil as amendment (Choukr–Allah, 2004).

4. FERTIGATION AND CHEMIGATION IN PROTECTED CULTURES

Fertigation and chemigation are regular and widely accepted practices for fertilization and protection under protected culture. The application of plant protectants through irrigation system such as herbicides, fungicides, insecticides, nematicides, growth regulators and biocontrol agents has been expanded rapidly during the last decades (Papadopoulos, 1992). Fumigation is an established means of applying fungicides to crops. Fumigation has proven to be quite effective for preventing or controlling a wide range of foliar and soil-borne diseases on a variety of crops (Papadopoulos, 1992). Fumigation has also proven to be effective for the soil application of the water soluble soil fumigants. The effectiveness of fertigation and fungigation through drip/trickle irrigation systems has been widely established in the greenhouses of Iran.

The application *via.*, nemagation of non-fumigant and fumigant nematicides through sprinkler irrigation systems has been proven effective for management of several major nematodes. Application of nematicides via nemagation appears to offer the possibility of salvaging a crop in which nematodes are detected after the crop has been planted. This salvaging alternative is not usually practical with the application of nematicides by conventional methods. Our experience in Iran showed the best and simple way for application of nematocide is along with soil solarization operations before planting.

One of the advantages of chemigation is timeliness of application, that is, pest control agents can be applied at any time the irrigation system can be operated in wet or dry weather and at any time of the day or night. Also, sprinkler irrigation systems can be used to practice total chemigation that is the application of all chemicals required to produce a crop. Therefore sprinkler irrigation systems are very well suited to integration into IPM programme. Chemigation with drip/trickle irrigation systems can also be conveniently integrated into an IPM programme for applying those chemicals which can be practically distributed through a drip/trickle irrigation system. Chemigation with surface irrigation systems would seem to have very limited application in an IPM programme. The current interest in the development of biosensors which could be used to monitor pest populations in the field or greenhouses offer an exciting futuristic potential for further integration of chemigation into an IPM programme.

5. MANAGING DISEASES IN PROTECTED CULTURES

Good disease control is closely linked with good husbandry. Any extra measures for control, such as time table fungicide sprays, add considerably to the cost of production. Therefore, carefully observe all the basic precautions in crop hygiene from seeding, to harvesting, to crop clearing.

These principles are emphasized throughout the disease descriptions. Pesticides are coming to be regarded as disease controls of last resort. Using disease-resistant cultivars and clean seed from a reputable source, in addition to observing good sanitation practices in the seedling and transplant stages of crop production, can reduce or eliminate the need for pesticides. There has been a revolution in greenhouse production technology that demands critical reappraisal of the research and development approaches to phytopathology. At the same time, this revolution presents an exciting challenge and opportunity for the grower to control plant disease relatively cheaply and without constant recourse to pesticides.

6. PESTS AND DISEASES MONITORING

More than 90% of greenhouse farmers in Iran and 82% of tomato greenhouses in Morocco (Hanafi and Schnitzler, 2004) use insect traps for the purpose of monitoring or mass trapping. Yellow sticky traps provide good information on pest population evolution, as well as insecticide efficacy/ resistance. For non-flying pests as well as diseases, plants are checked on a daily basis by specialized workers who are trained to identify insects and diseases of importance to greenhouse crops (Hanafi and Schnitzler, 2004).

7. MONITORING AND CONTROLLING THE ENVIRONMENT

Monitoring the environment and controlling it to keep plants disease free has become a fundamental part of greenhouse management. Once the environmental parameters for the various activities of pathogens and biocontrol organisms are understood, those activities can be regulated by the relatively precise control, growers can exercise over the environment. Precision is greatly enhanced by computer monitoring and controlling of temperature, energy input and conservation, light, shade, air movement and ventilation, water vapor density, relative humidity, dew point, crop nutrition and carbon dioxide enrichment. Consideration must also be given to such external factors as site orientation, radiation, rain, wind velocity and direction.

Managing the environment for disease control has arrived relatively recently by the juxtaposition of computerized climate controls and better understanding of the autecology of fungi and the bacteria and the autecology of biocontrol organisms. Fungi have specific and often different optimum environmental requirements for sporulation, dispersal, spore germination, and infection. For example, grey mold pathogen, *Botrytis cinerea*, and other groups of fungi, such as *Alternaria, Fusarium, Phytophthora, Pythium,* and *Verticillium* spp. and the downy mildew each has its own set of ecological requirements that demand a different rationale in the design of environmental control measures. In south part of Iran, during wintertime the relative humidity inside and outside of greenhouse is moderately high and because of this the incidence of downy mildew, Alternaria leaf spot, grey mold and Sclerotinia rot diseases on cucumber are common. Once conditions for infection are recognized and their environmental parameters are defined, however, infection can be prevented simply by avoiding those conditions. In the case of fungi such as *B. cinerea,* which are dependent on a water film for spore germination and infection, preventing temperatures from reaching the dew point is an effective mechanism of disease escape. Our experiences in such conditions showed that efficiency of fungicides application without good ventilation will be very less. Pruning of infected leaves is useful for reducing source of inoculum. Bacterial diseases tend to be more serious in humid conditions, in succulent soft plants and in certain areas of the greenhouse where such adverse conditions, poor drainage and roof drips occur. To reduce roof drips, using heating systems 3–5 hr before sunrise will decrease dew formation (Figs. 3, 4, 5). Using of extra layers of polyethylene sheet about 1 m below greenhouse roof is very critical for collection of water drips and control of fungal and bacterial diseases.

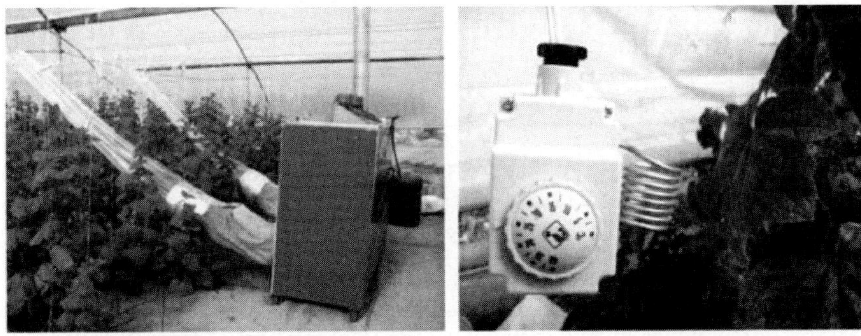

Fig. 3: Adjustable heating system for desirable temperature

The epidemiology of virus disease depends largely on the mode of transmission, whether on pruning tools or fingers or by insects. Thus, in order to devise rational management measures, the autecology of the vector becomes a subject for in-depth study. A simple crop management practice can similarly often save plants with basal stem and root rot from total destruction. Several of these diseases, such as Pythium root rot and Fusarium crown and root rot of tomatoes, are the diseases of cool and wet soils. Altered greenhouse and bench design can improve air movement, which not only reduces the risk of disease but also induces ethylene stress and hardier plants, bottom heat, a traditional means of avoiding Pythium and Rhizoctonia root rot, is enhanced in seedling and cutting trays that provide for upward air movement between the young plants.

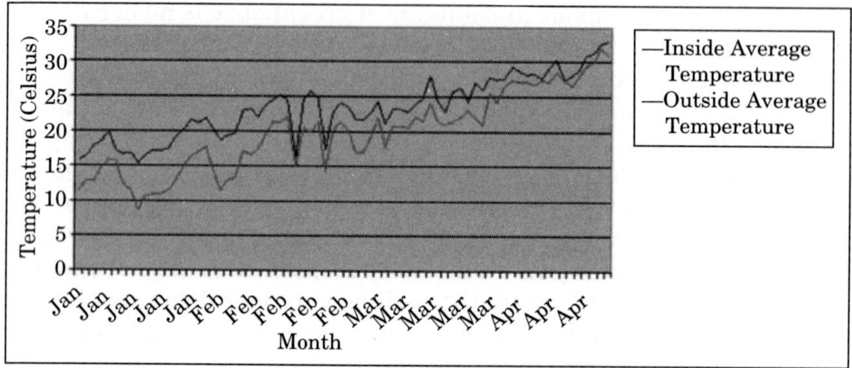

Fig. 4: Comparison of temperature changes inside and outside of greenhouse during growing season in south part of Iran (Najafinia *et al.*, 2009)

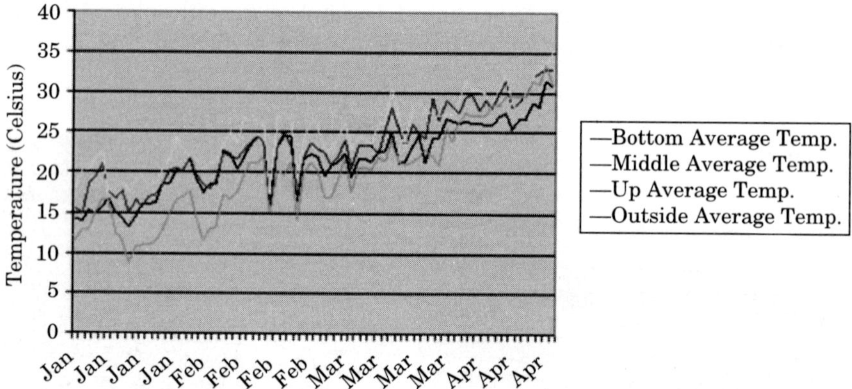

Fig. 5: Average temperature in different layers (height) of inside and outside of greenhouse during growing season in south part of Iran (Najafinia *et al.*, 2009)

8. CULTURAL CONTROL

There are several important cultural practices for disease control which should be integrated into any greenhouse management program. These serve both preventative and eradicative functions. Similar cultural control methods may be applied to specific diseases which are common to several crops. Greenhouse producers must strive to provide environmental conditions which are a compromise between those which favor plant growth on one hand and hinder disease development on the other. Attention to such details helps to prevent the onset and spread of diseases as well as to reduce the need to implement an expensive eradicative program.

8.1. Damping off, Root and Stem Rots

- Plant in a light, well drained, well prepared, pasteurized soil or rooting medium such as sand, soil, vermiculite, perlite or sphagnum moss. If pasteurized rooting media are not available, apply fungicides and mix well into the media.
- Where possible, keep soil on the dry side.
- Avoid over watering, overcrowding and planting too deeply.
- Provide good ventilation and air circulation to reduce humidity.
- Water in the morning to allow plants and soil surface to dry before evening.
- Avoid over fertilization, especially with nitrogen.
- Irrigate with clean water.
- Use raised bed in clay soil instead of flat bed in sandy soil (Fig. 6).
- Do not sow seeds too thickly.

Fig. 6: Preparing of nearly flat seed cultivation bed in sandy loam soil (right) in comparison with raised bed (left) in moderate clay soil for natural drainage is very important for control of damping off

8.2. Grey Mold, Powdery Mildew, and Leaf Spots (Bacterial and Fungal)

- Avoid placing the plants in damp, shady locations.
- Do not sprinkle foliage, particularly in late afternoon or evening.
- Take cuttings from healthy plants and disinfect tools between cuts.
- Provide good air circulation and raise night temperature to reduce humidity.

- Avoid over fertilizing, especially with nitrogen.
- Take cuttings off dry plants only.
- Do not use wet mulches.
- Space plants, especially "mother plants" to eliminate foliage contact.

Viruses

- Propagate cuttings from healthy plants.
- Disinfect cutting tools between stock plants from which cuttings are taken.
- Avoid mixing old and new plants or plants from different sources.
- Keep insects under control, especially aphids.
- Install insect net to control vectors.

8.3. Nematodes (Root and Foliar)

- Avoid introducing soil-borne nematodes from gardens and other greenhouses by strict sanitation procedures, including foot baths or overshoes for visitors.
- Avoid spreading foliar nematodes by not allowing leaves and stems to stay wet for long periods of time.
- Obtain root and foliar nematode–free stock.
- If possible, take soil samples and have them checked for plant parasitic nematodes.
- Careful representative sampling is essential if harmful nematodes are to be detected.
- Soil solarization along with application of fumigant before planting (while using soil solarization regularly each year, it suggest use fumigation once every two years).

8.4. Sanitation

Sanitation should be an integral part of a disease control program. If a source of infection is constantly present, control measures may be expensive and ineffective.

Proper sanitation is critical in the greenhouse environment. Weeds, which may harbor insect pests and some pathogens, and also reduce air movement, should be removed from inside and outside the structure.

Diseased tissue should be removed and disposed of and waste tissue should be composted or buried. The composting should be located as far away from the greenhouse as possible. Workers should wash hands often at least at the end of each row to minimize spread of pathogens. This control method includes all actions designed to eliminate or reduce the inoculum present in a plant or a plot, and so to prevent the spread of the pest or disease. Thus, removal and proper disposal of infected plant debris, that may harbor the pest or pathogen, reduces the amount of inoculum. Eliminating weeds from the edges of greenhouses, at least two weeks before planting is suggested, it delays infestation by insects and mites.

Many plant pathogens are seed-transmitted so the use of pathogen-free seeds should, therefore, be an important component of an integrated disease management.

Proper fertilization, good drainage, proper planting density, weeds control, etc. improve the plants' growth and may have a direct or indirect effect on the control of a particular pest. Soil and water salinity increase the susceptibility of tomato plants to many diseases and, particularly, to *Fusarium* and *Verticillium* wilts. Resistant varieties become susceptible when the irrigation water has a high salt content. Reducing the water salt content by mixing non-salty water from dams with salty water pumped from wells, decreases the incidence and severity of these two pathogens (Hanafi and Schnitzler, 2004). Tomato plants are pruned to remove auxiliary buds and leaves. Plant pruning, not only creates a drier microclimate in the lower plant but also provides numerous points of entry for many pathogens such as *B. cinerea*. Pruning wounds on tomato plants are less likely to become infected by these pathogens if the leaves are cut close to the stem than if a fragment of petiole is left on the stem (Hanafi and Schnitzler, 2004).

9. MECHANICAL EXCLUSION

During the last three years in Iran, we have tested various insect nets in managing different greenhouse pest and control of viral disease vectors. The good result came while using the fine mesh screens. Our results in south of Iran (The weather is very hot in summer and mild in winter), showed that the use of insect nets in early season (autumn) up to end of winter is effective but in spring time because of dusty conditions coupled with warm and humid weather (Fig. 7) have a negative effect on greenhouse climate, creating favorable conditions for foliar diseases. This problem can be solved by using pad and fan system for ventilation (Fig. 8), but it is costly and many farmers don't have financial ability to follow this system.

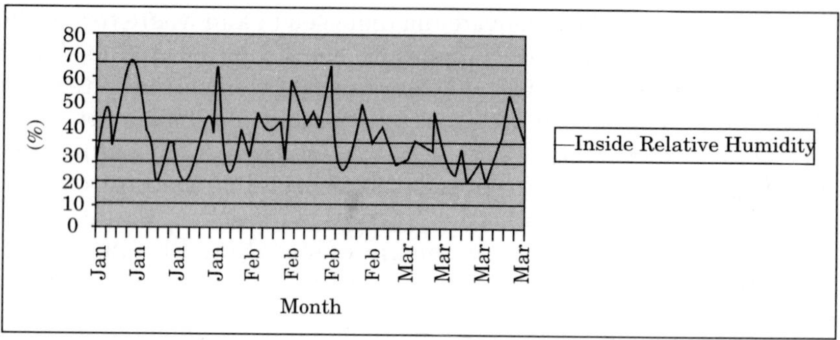

Fig. 7: Relative humidity variation trend inside the greenhouse in south part of Iran (Najafinia *et al.*, 2009)

Fig. 8: Pad (right) and fan (left) system for ventilation to reduce foliar diseases

10. BIOLOGICAL CONTROL

Diseases in greenhouse vegetables and floriculture crops can be managed effectively with biological fungicides (biofungicides). A biofungicide is composed of beneficial microorganisms, such as specialized fungi and bacteria that attack and control plant pathogens and the diseases they cause. These specialized fungi and bacteria are microorganisms that normally inhabit most soils. Biofungicides can be a viable alternative to chemical fungicides and can be used as part of an integrated disease management program to reduce the risk of pathogens developing resistance to traditional chemical based fungicides. An example of a widely used commercial biofungicide in the greenhouse industry is *Trichoderma harzianum* (TH). It protects plant roots from pathogens such as *Pythium, Rhizoctonia, Fusarium, Sclerotinia*, and *Thielaviopsis* (Sharma *et al.*, 2006; Sharma and Najafinia, 2006). It will also suppress foliar diseases such as *Botrytis* and powdery mildew (Thomas, 2002).

To optimize the effectiveness of TH or any other biofungicide, apply before the onset of disease development (preventative treatment) since they will not "cure" pre-existing pathogens. Early application of the biofungicide protects the roots against bad fungi, allowing for better development of root hairs. Always use biofungicides in conjunction with standard cultural controls including sanitation, and weekly scouting (Thomas, 2002). Because of the precision with which the greenhouse environment can be controlled, the chance of success with biological controls is grater in greenhouse crops than in field crops. Biological control may provide an additional tool to these chemical approaches for bacterial disease management. Bacterial biological control agents are now commercially available for the control of crown gall, fire blight of pear and several other diseases (Ji *et al.,* 2006). The effects of *Pseudomonas aeruginosa*, *Bacillus subtilis* and antagonistic fungus *Paecilomyces lilacinus* (provitan), on the growth and gall development of lettuce infected by root-knot nematodes *Meloidogyne* spp. was studied both in greenhouse and field environments. The results showed that the weight of lettuce planted in nematode infested soil, containing these three tested organisms, was higher than those cultivated in nematode infested soil with no control agents (Prakob *et al.,* 2009).

11. SOIL SOLARIZATION TECHNIQUE

Soil solarization is the use of clear plastic mulch placed over fallow soil during the hot months of the year in order to trap the energy from the sun and heat the soil to temperatures that are lethal to the pathogen. It has been used effectively in some locations around the world; however, it is costly and laborious. Soil solarization in south part of Iran is limited to the greenhouse industry. Up to this time, the technique has been widely accepted among growers in Iran. The area which is using soil solarization technique in south part of Iran is more than 1200 ha. In the present report, the author describes the solarization technique for the plastic greenhouse. Figure 9 represents simplified steps of the technique. The procedure of the technique is as follows: removing the plant debris's, dividing the farm into small plots, deep irrigation, deep tillage, then, the soil surface is covered with clear plastic film, and the soil is thoroughly moistened by temporary irrigation through the furrows or drip irrigation. Then the soil surface is kept closed for a period 30 to 45 (the time and period of the treatment are restricted by both meteorological conditions and cropping system) days during the summer season. By the procedure described above, the maximum soil temperatures ranging from 50 to 55°C at a depth of 10 cm. The technique in general has proved acceptable in the greenhouse industry for many areas. Soil temperature is a fundamental factor in the solarization effect. Calcium cyanamide was at first introduced to the solarization technique for the purpose of rapid decomposition of organic amendments, thereby improving soil fertility. Calcium cyanamide, however, proved to have a marked effect for suppressing many soil pathogens, when combined with heat treatment or solarization of the soil. Calcium cyanamide is employed by many growers, at a rate of 1000 kg per hectare, frequently combined with the organic materials. Solarization appears to cause a limited disease-controlling effect when a crop has a deeper root system and a pathogen inhabits deeper soil layers. Soil solarization is not always sufficiently acceptable for farm management. One limitation is the inevitable problem due to meteorological restriction. The effectiveness of solarization is greatly influenced by the weather during the treating period and locality of the field. Solarization requires high soil moisture content, and the treatment is believed to be less effective in a field with poor water retentivity. Soil temperatures are lower in soil near the sides and near the pillars of the greenhouse. The effect is relatively low for deep-rooted crops and for pathogens inhabiting deeper soil layers.

Fig. 9: Soil solarization technique for plastic greenhouses, deep tillage (a), normal disk (b), making plot and irrigation (c) and soil covering with polyethylene sheet (d) in south part of Iran (Najafinia *et al.*, 2009)

12. SAFETY PRECAUTIONS DURING APPLICATION OF CHEMICAL PESTICIDES

Many different pesticides may be applied during the many different stages of production. When designing programs, the supervisor should choose the least toxic pesticide available. Only where no other pesticide will control the pest should the most toxic product be used. Some crops require more labour-intensive management than others. When handling, there is always the potential for exposure to airborne residues still present after application and surface residues left on plants and equipment in the treated area. Employees should be aware that exposure through dermal contact can be just as dangerous as exposure through inhalation or ingestion if significant amounts are absorbed. Employees should shower after work and regularly wash work clothes. It is your responsibility to inform all employees of the potential hazard and the importance of protecting themselves.

Pesticides used in a greenhouse are available in a variety of formulations. The formulation selected directly influences the application method and therefore the type of application equipment to be used. Proper application of a pesticide depends not only on the correct method and equipment; it also involves the correct diagnosis of the pest problem, the correct selection of the pesticide to use, (assuming that the use of pesticides is the control measure to use), and finally, the correct application rate.

Application methods specific to greenhouse operations can be divided into the following types:

- High volume applications
- Low volume applications
- Smoke fumigation
- Granular applications
- Dust applications

Table 1: Tips for managing infectious and non-infectious diseases of greenhouse vegetable crops. (Howard, 2005; Sharma and Najafinia, 2006; Sharma *et al.*, 2006)

Category	*Diseases and pathogens*	*Management practices*
Seedling diseases	Seed decay (*Pythium*) • Pre-emergence damping-off (*Pythium, Rhizoctonia*)	Start plants in a clean greenhouse • Use a fungicidal seed treatment
	• Post-emergence damping-off (*Pythium, Rhizoctonia, Fusarium*)	• Use well-drained, pathogen-free growing media
	• Root rots (*Pythium, Rhizoctonia, Fusarium*)	• Use clean seedling trays • Avoid overcrowding and overwatering • Irrigate with clean, tempered water

Table 1: (*Contd...*)

Table 1: (*Contd...*)

Category	*Diseases and pathogens*	*Management practices*
		• Allow plant surfaces to dry after watering and before evening • Don't overfertilize • Control fungus gnats and shore flies • Apply drenches or sprays of chemical or biological fungicides if needed
Root, stem and vascular diseases	Root rots (*Pythium, Rhizoctonia, Fusarium*) • Crown rots (*Pythium, Fusarium, Rhizopus,* soft rot bacteria) • Stem rots (*Botrytis* [gray mold]), *Sclerotinia* [white mold], *Didymella* [gummy stem blight], *Penicillium* [blue mold], *Clavibacter* [bacterial canker]) • Wilts (*Verticillium, Fusarium*)	Choose resistant varieties • Use pathogen-free growing media • Promote good air circulation in the crop • Irrigate with clean water • Avoid unnecessary injuries to plants • Use clean tools for pruning • Remove badly infected and deadplants • Control fungus gnats and shore flies • Apply drenches or sprays of chemical or biological fungicides if needed
Foliar and fruit diseases	Leaf spots (*Alternaria*) • Scab, leaf mold (*Cladosporium*) • Powdery mildew (*Erysiphe*) • Sooty molds (*Alternaria, Cladosporium*) • Fruit rots (*Sclerotinia* [white mold], *Botrytis* [gray mold],	Choose resistant varieties • Promote good air circulation in the crop • Reduce relative humidity • Avoid wetting the foliage and fruit • Use clean tools for pruning • Remove badly infected and deadplants • Control aphids, thrips and whiteflies • Apply drenches or sprays of chemical or biological fungicides if needed

13. PHYSIOLOGICAL DISORDERS

13.1. Heat Damage

If cucumbers and other greenhouse plants are seeded beneath the plastic mulch, they may well be scorch-damaged by undue heat reflected from the plastic.

13.2. High-Temperature Wilting

Temperatures over 32°C, especially in bright sunlight, can also cause temporary wilt, both in the field and in the greenhouse. Ventilation in the greenhouse should begin at 24–27°C because if high temperatures persist, the margins of the lower leaves may die. A temporary wilt often occurs in the early part of the season, before an adequate root system has developed, and when warm, sunny days follow prolonged dull periods. Most plants recover, but some that lose lower leaves may die. In this case, mounding the base of the plant with peaty soil and watering with overhead sprinklers help recovery.

13.3. Chilling Damage

Chilled fruit has delicate scratch–like longitudinal pale fawn scarring (Fig. 10) and is often malformed.

Fig. 10: Chilling damage on fruit and plant cucumber (Najafinia, 2010)

13.4. Low-Temperature Wilting

In the greenhouse, cucumbers require a minimum temperature of 21°C. If the temperature drops suddenly, alarming wilt symptoms may appear, but the plants usually recover. If cool conditions persist, for example, near the wall of an uninsulated house, plants will be permanently stunted.

13.5. Irregular Watering

Wilting, yellowing, injured roots, and slow growth can result from poor drainage, overwatering of new transplants, or watering on cold dull days.

These conditions are more likely to occur if soils are cold or contain excess amounts of poor-quality manure. Underwatering also causes temporary wilting and plant death if the watering program is not corrected soon enough.

13.6. Nutrient Deficiencies

Balancing nutrients in the soil in amounts available to plants is a delicate matter. Any nutrient deficiency or imbalance represents a potential loss in crop production. Nutrient levels in soils, especially sterilized soils, and in crops should be properly assessed at regular intervals by your provincial agricultural representative. Although certain latitude in nutrient imbalances in soil can be tolerated, their effects in hydroponic systems are immediate and often catastrophic.

14. CHEMICAL INJURY

14.1. Nutrients and Pesticides

If incorrectly applied, most pesticides can cause injury and even death to cucumber plants, especially in the greenhouse. Before applying herbicides, soil fertilizers, foliar nutrients, fungicides, or insecticides, it is essential to know whether the plant's health depends on a particular chemical. Insurance sprays applied according to the calendar are not always justified, and many pesticides actually reduce yield. It is also essential to know which chemicals are compatible when mixed together and whether temperature, time of day, available soil moisture, water on the foliage, and condition of the crop are compatible with a particular chemical treatment or method of application.

A high proportion of disease–like symptoms, such as the residual effects of pesticides, soil fumigants, or spray drift such as herbicide effect (Fig. 11), are caused by the misuse of chemicals. Simple arithmetical errors often result in the wrong dilution of chemicals, which can be either fatal or completely ineffective. Double-check all calculations, and if in the slightest doubt, seek expert advice.

14.2. High Salts

Cucumbers planted too soon into freshly steamed greenhouse soil or unleached soil may suffer from high concentrations of inorganic salts. These salts damage the roots and result in poor growth and small blossoms. Furthermore, the plant is unable to respond to subsequent fertilizer treatments. Before planting, send soil samples to an analytical laboratory for salt assessments, and then leach and fertilize according to provincial recommendations. In general, avoid potassium nitrate, calcium nitrate, ammonium phosphate, high analysis complete fertilizers, and early fertilizers that build up high salt levels. Ammonium fertilizers are unnecessary for two months after planting. In the greenhouse, have two or three soil samples analyzed during the growing season to ensure that fertilizers are being correctly applied.

Fig. 11: Chemical injury caused by herbicide drift (Najafinia, 2007)

15. GREENHOUSE INSECT PESTS AND THEIR MANAGEMENT

Insects damage plants in a variety of ways. They attack the seed, cut off young plants, chew foliage, suck the sap, bore and tunnel into the stems and branches and transmit diseases. Insects must be correctly identified before they can be properly controlled. In addition, the following information is needed to effectively control insects in the greenhouse:

- The insect's life cycle
- The insect's feeding habits (which plants, where on the plant, typical signs of damage)
- Where the insects hides when it is not feeding

Insects do the most damage when they are feeding on plants in the greenhouse. Understanding the feeding habits of the insect may help to apply effective control measures. Some insects pierce the surface of plants and suck liquids from the tissue below. They have a proboscis that looks like a slender tubular beak; inside this proboscis are several stylets which actually do the piercing and sucking. Thrips are sucking insects that rasp the surface with the stylets until the sap flows out and then they retract the stylets and suck up the sap.

16. INSECT CONTROL

The measures taken to control insects will be affected by environmental conditions, the specific insect to be controlled and its stages of development, the possibility of insect resistance to some insecticides, the susceptibility of plants to insecticides, the presence of non-target insects, the presence of insect predators, and the choice of insecticides.

(a) ***Environmental Conditions*:** Control measures will be affected by environmental factors influencing the chemical itself, the insect pest and the plant that is being attacked.

(b) ***Temperature*:** Chemical reactions take place more readily as temperature increases. Phytotoxicity can be a greater problem if increased temperature results in greater chemical absorption into the plant or if increased temperature results in the rapid evaporation of a spray solution concentrating chemical on leaf surfaces. At elevated temperatures, and especially under dry conditions, plant activity may be slowed which could affect the rate at which a systemic insecticide will be translocated.

- The warmer the weather, the more active the insect will be. If an insect is feeding more actively, it will ingest more insecticide. Conversely, pests tend to be more dispersed in warmer weather and this may hamper control efforts. Under cool conditions, insects may congregate together allowing you to direct an insecticide application for maximum effect.

(c) ***Moisture*:** Insecticides usually work better under conditions of high relative humidity (RH). Diseases affecting insects are usually better adapted to high relative humidity. Some insects (*e.g.*, aphids) function best when the humidity is low; liquid wastes are eliminated more efficiently. Low RH when accompanied by high temperatures and ventilation could put plants under stress and slow the movement of translocated insecticides. Low RH will also encourage the evaporation of spray solution on foliage and could result in foliage burning.

(d) ***Sun/Cloud*:** Some chemicals are more subject to breakdown in the sun (photochemical breakdown). This could affect efficacy if the chemical had to remain on a leaf surface for a period of time. The activity of some insects is affected by sunny or cloudy conditions. This could influence the probability of a spray application successfully intercepting a pest. The major effect of sun/cloud on plants is related to food production and translocation. Photosynthesis and translocation will be greatest under sunny conditions (provided the plant is not under stress).

CONCLUSIONS

Because greenhouses create an ideal habitat for a variety of diseases, insects, and weeds, pest management activities are a primary factor in the cost of production and affect both yield and quality of the product. The variety of crops grown in a typical greenhouse operation makes pesticide selection complex, especially if there are food and ornamental crops in the same greenhouse, as is frequently true in bedding plant production. Limited pesticide options for some pest/crop combinations make pesticide rotation difficult, and insecticide resistance has developed in some insect populations. The appearance of new pests, and increasing importance of some old pests, leads to a continued need for education in pest identification. Integrated pest management (IPM) has been widely adopted, although the level of adoption, and sometimes the understanding of IPM, varies. Cultural methods of pest management are common but there is a desire for greater information on environmental methods of control and sanitation options. Biological control is becoming more common as a method of pest management, and there is a great need for both research and training in current and new biological control options.

Successful disease control relies on proper disease prevention practices and plant disease diagnosis (Najafinia and Sharma, 2009; 2007). Once the major diseases and pest problems have been identified, an integrated pest management (IPM) program can be followed. Several important cultural practices to control diseases should be integrated into all greenhouse IPM programs to control diseases. These serve to both prevent and minimize/eradicate diseases. Analogous cultural control methods may be applied to specific diseases that are common to several crops. Greenhouse producers must strive to provide environmental conditions that are more favorable to plant growth and development than to disease development. Careful attention to such details helps to prevent the onset and spread of diseases and may reduce the need for an expensive eradicative program.

Good disease control is closely linked with good husbandry. Any extra measures for control, such as fungicide sprays, add considerably to the cost of production. Therefore, carefully observe all the basic precautions in crop hygiene from seeding, to harvesting, to crop clearing. These principles are emphasized throughout the disease descriptions that follow. Pesticides are coming to be regarded as disease controls of last resort. Using disease-resistant cultivars and clean seed from a reputable source, in addition to observing good sanitation practices in the seedling and transplant stages of crop production, can reduce or eliminate the need for pesticides. Many greenhouse growers rely solely on biological control for insects and on hygiene and sound environment management for disease control.

In most provinces pesticides can be applied to commercial crops only by licensed operators trained in their safe use. In commercial crops and in home gardens, mix and apply pesticides strictly in accordance with the instructions on the label, and dispose of containers in ways set out

In provincial and local regulations, never deviate from the rates given on the label; twice as much will usually do twice the amount of damage. Pesticides applied as insurance sprays, in anticipation of disease outbreaks, are seldom justified. Your provincial agricultural representative is in the best position to give specific advice if epidemics threaten. With increasing availability of information and understanding on how pathogens and pests cause damages, new strategies are being devised to enhance protection that is possible. Plant breeding and biotechnology tools in combination are already providing new materials for better plant management. The pest management tools that have been deployed have had a positive impact on the environment by reducing that amount of chemical pesticides that are applied to these crops.

REFERENCES

Choukr–Allah, R. (2004). Protected culture in Morocco: new trends and developments. Proceedings of international workshop on: "The production in the greenhouse after the Era of the methyl bromide". Comiso, Italy, 1–3 April, 2004. pp. 187–94.

Hanafi, A. and Schnitzler–Wilfried, H. (2004). Integrated production and protection in greenhouse tomato in Morocco. *In*: Cantliffe, D.J., Stoffella, P.J. and Shaw, N. (*Eds*.). Proc. VII IS on Prot. Cult. Mild winter climates. *Acta Hort*., **659:** 259–300. ISHS 2004.

Howard, R. (2005). Management of major freenhouse vegetable diseases. 27th annual canadian greenhouse conference, mississauga, Oct. 5, 2005. pp. 1–6.

Jarvis, W.R. (1989). Managing diseases in greenhouse crops. *Plant Disease*, **73(3):** 190–94.

Ji, P., Campbell, H.L., Kloepper, J.W., Jones, J.B., Suslow, T.V. and Wilson, M. (2006). Integrated biological control of bacterial speck and spot of tomato under field conditions using foliar biological control agents and plant growth–promoting rhizobacteria. *Biological Control,* **36:** 358–67.

Najafinia, M. and Sharma, P. (2009). Cross pathogenicity among isolates of *Fusarium oxysporum* causing wilt in cucumber and muskmelon. *Indian Phytopathology* **62(1):** 9–13.

Najafinia, M. and Sharma, P. (2006). Vegetative compatibility groups and pathogenic variability among isolates of *Fusarium oxysporum* f.sp. *cucumerinum* causing wilt in India. International meeting on biotic and abiotic stress responses in plants, 11–13 December, ICGEB, New Delhi.

Najafinia, M. and Sharma, P. (2007). Characterization of *Fusarium oxysporum* isolates from cucumber in India using pathogenicity, vegetative compatibility (VCGs) and RAPD assay. 59th annual meeting, indian phytopathological society and national symposium on "Plant Pathogens: Exploitation and Management" January 16–18, 2007, Jabalpur (Madhya Pradesh), India.

Najafinia, M., Azadvar, M., Ghaffarinejad, S. and Moemeni, D. (2009). Developing early warning system for managing greenhouse disease. International conference on plant patholoIgy in globalized Era, IARI, New Delhi, India, November 10–13, 2009.

Nelson, P.V. (1985). *Greenhouse Operation and Management,* Prentice–Hall, New Brunswick, NJ.

Papadopoulos, I. (1992). Fertigation–chemigation in protected culture. *Cahiers Options Mediterraneenns,* **31:** 275–91.

Prakob, W., Nguen–Hom, J., Jaimasit, P., Silapapongpri, S., Thanunchai, J. and Chaisuk, P. (2009). Biological control of lettuce root knot disease by use of *Pseudomonas eruginosa*, *Bacillus subtilis* and *Paecilomyces lilacinus*. *Journal of Agricultural Technology,* **5(1):** 179–91.

Sharma, P. and Najafinia, M. (2006). Biological control of *Sclerotium rolfsi* causing rots in vegetables. National seminar on new frontiers in plant pathology, 28–30 September 2006, Shimoga, Karnataka.

Sharma, P., Sharma, N. and Najafinia, M. (2006). Fungicidal properties of some plant extracts against *Fusarium oxysporum* f. sp. *cucumerinum*. National seminar on new frontiers in plant pathology, 28–30 September 2006, Shimoga, Karnataka.

Thomas, C. (2002). Managing Plant Diseases with Biofungicides. http://hortweb.cas.psu.edu/extension/vegcrops/newsletterlist.html.

Tuzel, Y. and Ozcelike, A. (2004). Recent trends and developments of protected cultivation in turkey. Proceedings of international workshop on: "The production in the greenhouse after the era of the methyl bromide" Comiso, Italy, 1–3 April, 2004, pp. 167–75.

9

Site Specific Crop Protection

PRASHANT P. JAMBHULKAR[1] AND NARESH M. MESHRAM[2]

ABSTRACT

Site specific crop protection or precision crop protection is an integral component of precision farming which basically depends on measurement and understanding of variability. Site specific crop protection is an enabled technology and based on information and focused decision. The components include Remote Sensing (RS), Geographical Information System (GIS), Global Positioning System (GPS), soil testing, yield monitors and Variable Rate Technology (VRT). The available technologies enable us in understanding the variability and by giving site specific crop protection recommendations we can manage the variability that makes site specific crop protection viable. The disease or insect affected crop and normal crops give different tonal variation in imageries. After image analysis the great variability in tonal variation on the imagery is found. Normal crops give red, bright red and dark red colour with smooth texture but pests affected area gives pink, yellow and yellow pinkish red colour with irregular shape and rough texture. This helps farmers to pay immediate attention towards pest infestation and manage crop by spraying chemicals. Thermography allows the quantitative analysis of spatial and dynamic physiological information on the plant status. Infrared thermography is used to study spatial variability of stomatal conductance, to schedule irrigation, for monitoring of temperature stress in plants, to screen for mutants with altered stomatal control and for assessment of plant-pathogen interaction by monitoring patterns of surface leaf temperature. Site specific crop protection is a system that is designed to strengthen farming management, providing farmers improved means to spot weak productivity zones and react before they become loses, increasing yields for the farmer and reducing annual production costs and adverse environmental impacts.

[1] Agriculture Research Station, Borwat Farm, MPUAT (Udaipur) Banswara, Rajasthan - 327 001.

[2] Indian Agriculture Research Institute, Pusa Campus, New Delhi - 110 012.
Corresponding author: E-mail: prashant_pj@rediffmail.com

***Key words*:** Site specific crop protection, Remote Sensing, Spectral imaging, Hyperspectral imaging, GPS and GIS, Spatial variability, Tonal variation, Patch spraying, Thermography, Acoustic detection.

1. INTRODUCTION

Agricultural production strategies have changed dramatically over the past decade. Many of these changes have been driven by economic decisions to reduce input and maximize profits and by environmental guidelines mandating more efficient and safer use of agricultural chemicals. However, growers now have a heightened sensitivity to concern over the quality, nutritional value, and safety against various diseases and insect pest invasion. They are selecting cultivars and adjusting planting dates to accommodate anticipated patterns in weather *e.g.*, El Nino or La Nina events. They are also relying on biotechnological innovations for suppressing pests *e.g.*, insects protected (Bt) and roundup ready crops. The possibility of selling carbon credits to industry is breathing new life into on-farm conservation tillage practices that enhance carbon sequestration (Robert, 2001).

Perhaps the most significant change in agriculture during the past ten years is shift towards precision or site specific crop management. Insect, diseases and weed are the major pests that a farmer faces during cultivation. There are different methods of pest management *viz*. cultural, mechanical, chemical, and biological control. But the farmers heavily depend upon chemical control for its greater efficiency and easy handling. But the over application of pesticides leads to problem of chemical residues in soil and also in produce, where as application of sub-lethal doses may lead to development of resistance and resurgences in pests. Thus the optimising the pesticide application is very essential. Now, at the beginning of the 21st century, growers are seeking new ways to exploit the variability of pest's resurgence. In the process they need more precise information on plant condition and pest situation than was required a decade ago. Not only does this information need to be accurate and consistent across their farm and from year to year, it must also be available at temporal and spatial scales that match rapidly evolving capabilities to vary cultural procedures, irrigations and agrochemical inputs. Application of agricultural inputs at uniform rates across the field without due regard to in-field variation in pest population and its infestation and crop condition, does not yield desirable results in terms of crop yield. The management of infield variability of the disease and insect pest and weed infestation for improving crop production and minimizing the environmental impact is the crux of site specific crop protection or precision crop protection. Site specific or precision crop protection is a practice of precision agriculture which is the application of technologies and principles to manage spatial and temporal variability associated with all aspect of agricultural production for improving production and environmental quality. The success of site specific or precision crop protection depends on accurate assessment of normal crop and pest infested crop, its management and evaluation in space-time continuum in crop protection. The agronomic feasibility of site specific crop protection has been intuitive, depending largely on application of traditional arrangement recommendation at finer scales.

Site-specific agriculture takes into account the spatial variability of biotic factors, such as weeds and pathogens, and of non-biotic factors, such as nutrients or water content, and uses diverse technologies to apply at variable rates fertilizers, pesticides or other inputs, fitted to the needs of each small area defined (Blackmore, 1996). Patchy distribution of pathogens/weeds and non-biotic factors is visually observed in the fields and are well documented (Krohmann *et al.*, 2006; López–Granados *et al.*, 2006). However, pesticides/fertilizers are usually applied at a single rate over the entire agricultural field. To reduce the total amount of inputs applied and to apply pesticide/fertilizers only where biotic/non-biotic patches demand them, site-specific management (SSM) techniques are being developed to treat only where weed/pathogens/nutrient levels densities exceed (or are lower than) the economic threshold, and to reduce application rates in patches where densities remain with low infestation levels. Potential economic and environmental benefits of SSM include reduced spray volume, application time and non-target spraying (Medlin *et al.*, 2000).

A very large body of research spanning almost four decades has demonstrated that much of this information is available remotely, via aircraft and satellite based sensor system. When combined with remarkable advances in global positioning system (GPS) receivers, microcomputer, geographical information system (GIS), yield monitors and enhanced crop simulation models, remote sensing technology has the potential to transform the ways that growers manage or protect their crops through precision or site specific crop protection techniques. The different type of information can be extracted from the same satellite data. Remote sensing as a tool will give information regularly and at a low cost to enable timely reclamation. In India, production forecasting of certain crops, crops yield modelling and crop stress detection are done using remote sensing data. The first satellite remote sensed data became available in 1972. Currently a number of spacecraft imaging systems are operating using remote sensing sensors. Some of the current image systems from spacecraft platform include Indian remote sensing satellite (IRS), SPOT and IKONOS, etc. In India, application of remote sensing in agriculture had been started through detection of coconut wilt disease in Kerala coast by Dr. C. Dakshinamurti, the then Head of Agricultural Physics Division, IARI, New Delhi, mapped that area during 1968–69 using ektachrome infrared photography from a helicopter (Dakshinamurty, 1971). Remote sensing had hardly been used in crop protection perspective due to some inherent problems like, (i) remote sensing can detect surface phenomenon and insect infestation or disease infection that occurs mainly within crop canopy, (ii) biotic stress symptoms are generally confusing with abiotic stress symptoms, (iii) damaged symptoms only appear on the canopy, but sporadic in nature and too small to be detected by sensors (spatial resolution), (iv) imageries with high spatial resolution (IKONOS, QuickBird) are available now-a-days which can be used in farm unit level, but high cost is a deterrent for using over a large area, and (v) Return period of polar orbital satellites are 14–21 days, but a pest infested area requires frequent monitoring (2–3days interval) for any action plan.

Insect and disease incidence by their nature are non-recurrent and usually vary in intensity from field to field. For remote sensing applications, these two characteristics have implications for regularity and scales of both remotely sensed data from individual fields are recorded by sensors. Furthermore reference data collection must be sufficiently precise with respect to field location. Malingreau (1980) found that locating individual fields was difficult in extensive crop areas where distinct field reference points are not common and all fields "look alike". For crop protection problems with these two characteristics, Landsat data are not adequate. A remote sensing and analytical system that is more regular and timely in its coverage and of larger scale appears to be necessary. Aircraft borne multispectral scanners of multispectral photographic systems (with photos digitised by a microdensitometer if a computer is to be used for analyses of the scenes) are likely to be useful for monitoring of disease and insect problems. This condition appears to be true if remote sensing is to be used in a program to detect early outbreak of insect and disease problems.

2. THE BASIC COMPONENTS OF SITE SPECIFIC CROP PROTECTION

Site specific crop protection or precision crop protection is an integral component of precision farming which basically depends on measurement and understanding of variability. The main component of site specific crop protection system must address variability. Site specific crop protection is an enabled technology and based on information and focused decision. The components include (the enabling technologies) Remote Sensing (RS), Geographical Information System (GIS), Global Positioning System (GPS), soil testing, yield monitors and Variable Rate Technology (VRT).

Site specific crop protection requires the requisition, management, analysis and output of large amount of spatial and temporal data. Mobile computing systems were needed to function on the go in farming operations because desktop systems in the farm office were not sufficient. In site specific crop protection system, information technologies are so essential that mobile mapping system (MMS) is very important. Equipped with mapping sensors, MMS can collect field data with a GIS or combine GIS software in its mobile terminals directly. Thus MMS can monitor disease progress and insect flights process all along. Moreover it can contrast multitemporal data collected or stored in database and find what changes occurred and then give efficient crop protection plan. MMS can finish the monitor task of all the plants. Using the scanners, the peculiar plant can be found quickly, and its position also can be recorded. In fact satellite remote sensing data also can provide such distinct information. Compared with mobile mapping system, its spatial range is much larger but space resolution is lower and constrained by satellite calendar. Site specific crop protection is concerned with spatial and temporal variability and it is information based and decision focused. It is the spatial analysis capabilities of GIS, DGPS and GPS that greatly enabled farming system; particularly for guidance and digital evaluation and modelling position accuracies at the centimetre level are possible in DGPS receivers. Accurate guidance and navigation system will allow for farming operations at height and under unfavourable weather conditions.

(a) ***Remote Sensing*:** Remote Sensing (RS) is the science of obtaining and interpreting information from a distance, using sensors that are not in physical contact with the object being observed (Jensen, 1996). Today, RS is potentially a practical management tool for site-specific crop management (Casady and Palm, 2002). Remotely sensed images can be used to identify nutrient deficiencies, diseases, water deficiency or surplus, weed infestations, insect damage, hail damage, wind damage, herbicide damage, and plant populations. Information from remote sensing can be used as base maps in variable rate applications of fertilizers and pesticides. Information from remotely sensed images allows farmers to treat only affected areas of a field. Problems within a field may be identified remotely before they can be visually identified. The basic principles of remote sensing with satellites and aircraft are similar to visual observations. Energy in the form of light waves travels from the sun to Earth. Light waves travel similarly to waves traveling across a lake. The distance from the peak of one wave to the peak of the next wave is the wavelength. Energy from sunlight is called the electromagnetic spectrum. The wavelengths used in most agricultural remote sensing applications cover only a small region of the electromagnetic spectrum. Wavelengths are measured in micrometers (µm) or nanometers (nm). One µm is about .00003937 inch and 1 µm equals 1,000 nm. The visible region of the electromagnetic spectrum is from about 400 nm to about 700 nm. The green color associated with plant vigor has a wavelength that centers near 500 nm (Fig. 1).

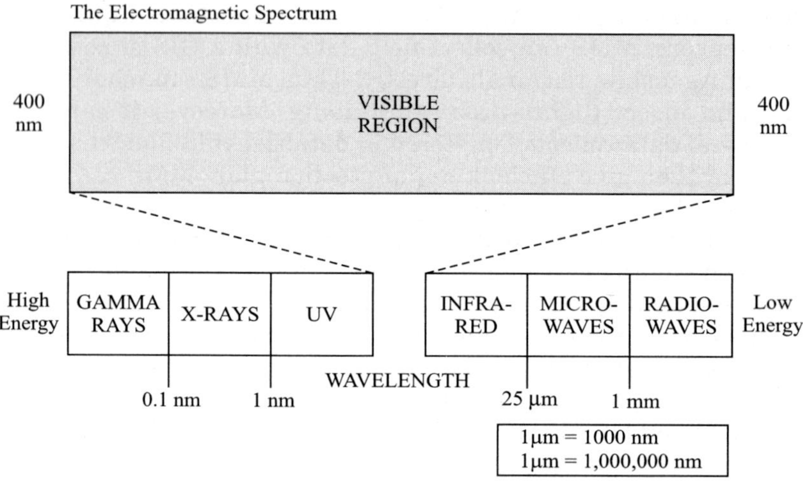

Fig. 1: The visible region of the spectrum ranges from about 0.4 µm to 0.7 µm (Liaghat and Balasundram, 2010)

- Wavelengths longer than those in the visible region and up to about 25 µm are in the infrared region. The infrared region nearest to that of the visible region is the near infrared (NIR) region. Both the visible and infrared regions are used in agricultural remote sensing. When electromagnetic energy from the sun strikes plants, three things can happen. Depending upon the wavelength of the energy and characteristics of individual plants, the energy will be reflected, absorbed, or transmitted. Reflected energy bounces off leaves and is readily identified by human eyes as the green color of plants. A plant looks green because the chlorophyll in the leaves absorbs much of the energy in the visible wavelengths and the green color is reflected. Sunlight that is not reflected or absorbed is transmitted through the leaves to the ground. Interactions between reflected, absorbed, and transmitted energy can be detected by remote sensing. The differences in leaf colors, textures, shapes, or even how the leaves are attached to plants, determine how much energy will be reflected, absorbed or transmitted. The relationship between reflected, absorbed, and transmitted energy is used to determine spectral signatures of individual plants. Spectral signatures are unique to plant species. Remote sensing is used to identify stressed areas in fields by first establishing the spectral signatures of healthy plants. The spectral signatures of stressed plants appear altered from those of healthy plants.

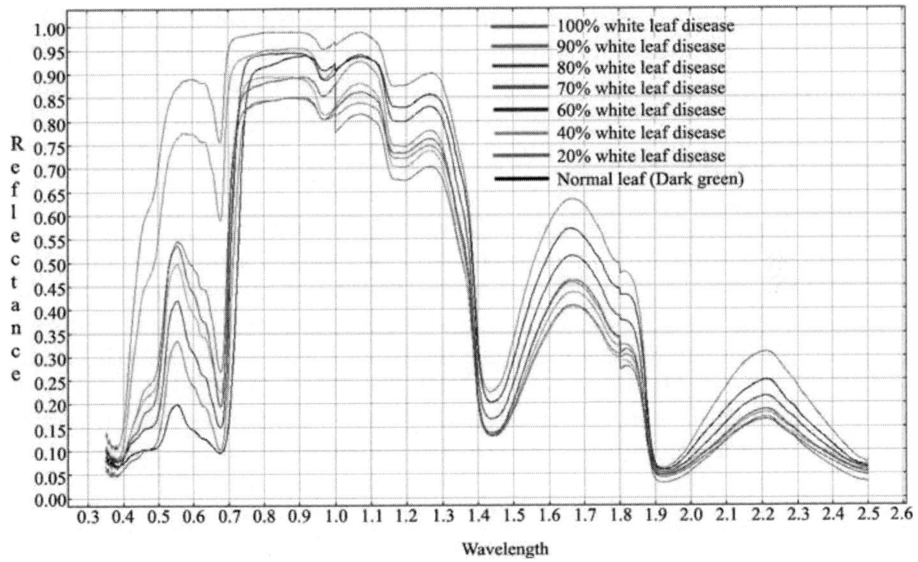

Fig. 2: Spectral signature of healthy and normal sugarcane plant

- The above Fig. 2 shows severe damage due to white leaf disease in sugarcane; revealed the highly reflectance or low absorption in the visible band of the spectrum from 400 to 700 nm. The normal green leaves, on the contrary, showed the low reflectance or highly absorption in the nonvisible range from about 750 to 1,200 nm. Interpreting the reflectance values at various wavelengths of energy can be used to assess crop health.
- The comparison of the reflectance values at different wavelengths, called a vegetative index, is commonly used to determine plant vigor. The most common vegetative index is the normalized difference vegetative index (NDVI). NDVI compares the reflectance values of the red and NIR regions of the electromagnetic spectrum. The NDVI value of each area on an image helps identify areas of varying levels of plant vigor within fields.
- There are several types of remote sensing systems used in agriculture but the most common is a passive system that senses the electromagnetic energy reflected from plants (Fig. 3). The sun is the most common source of energy for passive systems. Passive system sensors can be mounted on satellites, manned or unmanned aircraft, or directly on farm equipment. There are several factors to consider when choosing a remote sensing system for a particular application, including spatial resolution, spectral resolution, radiometric resolution, and temporal resolution.

(*i*) ***Spatial resolution*** refers to the size of the smallest object that can be located in fields or detected in an image. The basic unit in an image is called a pixel. One-meter spatial resolution means each pixel image represents an area of one square meter. The smaller an area represented by one pixel, the higher the resolution of the image.

(ii) ***Spectral resolution*** refers to the number of bands and the wavelength width of each band. A band is a narrow portion of the electromagnetic spectrum. Shorter wavelength widths can be distinguished in higher spectral resolution images. Multispectral imagery can measure several wavelength bands, such as visible green or NIR. Landsat, Quickbird, and Spot satellites use multispectral sensors. Hyperspectral imagery measures energy in narrower and more numerous bands than multispectral imagery. The narrow bands of hyperspectral imagery are more sensitive to variations in energy wavelengths and, therefore, have a greater potential to detect crop stress than multispectral imagery. Multispectral and hyperspectral imagery are used together to provide a more complete picture of crop conditions.

(iii) ***Radiometric resolution*** refers to the sensitivity of a remote sensor to variations in the reflectance levels. The higher the radiometric resolution of a remote sensor, the more sensitive it is to detecting small differences in reflectance values. Higher radiometric resolution allows a remote sensor to provide a more precise picture of a specific portion of the electromagnetic spectrum.

(iv) ***Temporal resolution*** refers to how often a remote sensing platform can provide coverage of an area. Geostationary satellites can provide continuous sensing while normal orbiting satellites can only provide data each time they pass over an area. Remote sensing taken from cameras mounted on airplanes is often used to provide data for applications requiring more frequent sensing. Cloud cover can interfere with the data from a scheduled remotely sensed data system. Remote sensors located in fields or attached to agricultural equipment can provide the most frequent temporal resolution.

(b) ***GPS and GIS:*** GPS is a referencing device capable of identifying different attributes and pest affected sites within the field and GIS maps these attributes and analyse using simple browsers or complex models. The use of GPS in agriculture is limited but it is fair to expect wide spread use of GPS in future. GIS is an ideal tool for managing data about the nature, location and spread of weeds, disease or insect pests. Allowing storage of vast amounts of data about the type of these pests, including reports on where the pest has been spotted and when, provides capabilities for tracking and predictive analysis. This crucial information allows for timely preventative measures to be deployed. GIS can also be used to track the success of such actions, such as monitoring the results of an aerial spraying program. However, to successfully use GIS in pest control, accurate information about the nature and location of the problem must first be collected.

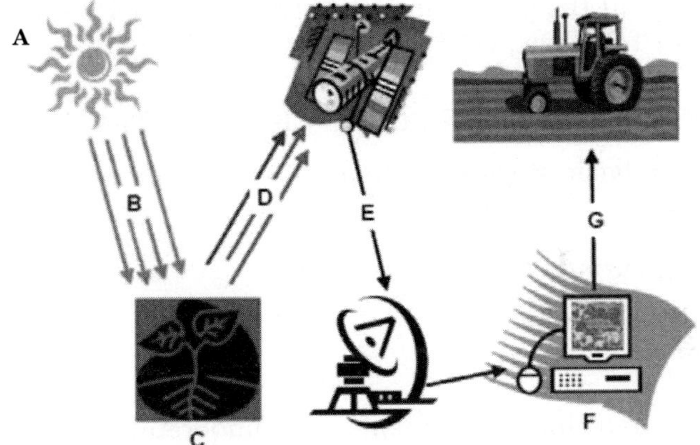

Fig. 3: Illustration of a satellite remote sensing process as applied to agricultural monitoring processes. The sun (A) emits electromagnetic energy (B) to plants (C). A portion of the electromagnetic energy is transmitted through the leaves. The sensor on the satellite detects the reflected energy (D). The data are then transmitted to the ground station (E). The data are analyzed (F) and displayed on field maps (G).

GPS systems provide an ideal solution for collecting such data, as they allow accurate position information to be collected along with necessary descriptive information. GPS-based data collection systems typically allow a data dictionary, or data structure, to be pre-defined. This ensures that the correct information is collected, in a format that is compatible with the GIS database. Some GPS-based systems also allow upload of existing GIS data. This enables the user to accurately navigate back to treatment areas for monitoring and further analysis. Attribute information can be easily updated in the field, with changes automatically logged, to allow for accurate charting of the results of a treatment or preventative program. GPS systems that are specifically designed for use in GIS data collection provide extensive functionality for recording and monitoring pest activity. Capturing more than just an accurate position, such systems provide the ability to record and store digitally exactly what is going on in the field.

Recently a GPS based crop duster to contain heavy pest population mounted on helicopter, which can spray an area as small as 4X4 meter is becoming boon for farmers. One can record observations for weed growth, pest population, unusual plant stress which can then be mapped with GIS programme. GPS guided tractors sense the exact location in the field and sends signals to computer fixed on tractor which has GIS, storing the pesticide requirement map/data in it. The GIS, in consultation with a decision support system would decide what the exact requirement of pesticides for particular location is. It then commands a variable rate agrochemical applicator, which is again attached with the tractor to spray the exact dosage at the precise location of farm. The GPS satellite used in agriculture are TOPCON, GARMIN and differential GPS.

3. BASIC STEPS IN SITE SPECIFIC CROP PROTECTION

There are three basic steps in site specific crop protection system, these are:

(a) Assessing variation

(b) Managing variations

(c) Evaluation

The available technologies enable us in understanding the variability and by giving site specific crop protection recommendations we can manage the variability that make site specific crop protection viable. And finally evaluation must be an integral part of this system. The detailed steps involved in each process are described below:

(a) ***Assessing Variability*:** Assessing variability is the critical first step in site specific crop protection system. Since it is clear that one cannot manage what one does not know. Factors and the processes that regulate or control the crop performance in terms of yield and crop protection vary in space and time. Quantifying the variability of these factors and processes and determining when and where different combinations are responsible for the spatial and temporal variation in crop yield is the challenge for system of site specific crop protection system. Techniques for assessing spatial variability are readily available and need to be precisely used in accordance with pesticide application on existing disease or insect pests. Therefore, assessing the infield variability in pest composition and population is very crucial and is the first step. The spatial variability of required determinants should be well recognised, adequately quantified and precisely located. Construction of the condition maps on the basis of variability is a crucial component. Condition maps can be generated through surveys, point sampling and interpolation, remote sensing and modelling.

- Techniques for assessing temporal variability also exist but the simultaneous reporting a spatial and temporal variation is rare. We need both the spatial and temporal statistics. We can observe the variability in susceptibility or infectivity of crop due to disease or insect pests but we cannot predict the reasons for variability. It needs the observations of crop at every growth stage of disease or every movements or flight of insect pest over the growing season, which is nothing but a temporal variation. Hence we need both the space and time statistics to apply the system of site specific crop protection. But this is not common to all variability factors that dictate crop production. Some variables are more pronounced in space rather than time, making them more conducive to current forms of precision management.

(b) ***Managing Variability*:** Once the variations of pests are adequately assessed, farmers must match agrochemical inputs to known conditions employing site specific crop protection and Variable Rate Technology (VRT). VRT is the technology of applying the farm inputs in varying rates at places and in amounts they are required to produce uniform yields throughout the entire field. In site specific variability management we can use GPS instrument, so that site specificity is pronounced and management will be easy and economical. While taking observations or plant samples, we have to note the sample site coordinates and further we can use the same for management. This results in effective use of inputs and avoids any wastage and this is what we are looking for. Thus, along with other farm machineries, variable rate sprayers are commercially available. Sprayers equipped with VRT generally possess a DGPS receiver to locate the spatial variability in the field and automatically regulate the rate of application. Due to high cost of integrated control system, farmers prefer to rely on custom hiring for using VRT.

(c) ***Evaluation*:** There are three important issues regarding precision agriculture evaluation.

(i) Economics

(ii) Environment

(iii) Technology transfer

The most important fact regarding the analysis of profitability of precision crop protection is that the value comes from the application of the data and not from the use of technology. Potential improvement of environmental quality is often cited as reason for using precision component in crop protection. Reduced agrochemical use, higher nutrient use efficiencies, increased efficiency of managed inputs and increased production of soils from degradation are frequently cited as potential benefits to the environment. Enabling technologies can make site specific or precision crop protection feasible, agronomic principles and decision rules can make it applicable and enhanced production efficiency or other forms of value can make it profitable.

4. EVENTS OF SITE SPECIFIC CROP PROTECTION

(a) ***Spatial Distribution:*** The spatial distribution of the individual pest population or infestation has a high ecological significance and a high importance in agriculture. It affects the design of the sampling program, the method of statistical analyses of the data and the pest control strategy. In agricultural ecosystems, the spatial distribution of pests and natural economies can be envisaged at the plot level (a farm or orchard) or at a wide area level (a district or state). The knowledge of spatial distribution at a plot level allows optimising the decision making on the control measures to be applied, at an area level; it also facilitates planning of monitoring.

- The study of spatial distribution of insects has been carried out traditionally by means of dispersion indices of Taylor, Morisita, Lloyd and Iwao (Taylor, 1984). More recently, the development and analysis of big sets has opened new possibilities to describe and analyse the spatial distribution on population: Geographical Information System (GIS) and geostatistics (Trematerra *et al.,* 2004). While the GIS only permits to compile and to manipulate spatially referenced data, geostatistics permits to describe correlations through space and time and to interpolate between and to extrapolate beyond sample points. The geostatistical descriptions of the spatial distribution of a population of pests leads to a map of isolines (Ribes–Dasi *et al.*, 2005). The use of pheromone traps to monitor pest population is highly extended. The variable to be used is the number of adults caught in pheromone traps and the spatial position of each trap is georeferenced by means of GPS. The maps that results from geostatistical analysis are studied by pest control advisers and they advice the appropriate control measures to farmers.

- Spatial pattern is one of the most fundamental properties of disease dynamics because it reflects the environmental forces acting on the dispersal and life cycle of a pathogen (Ristaino and Gumpertz, 2000). For this reason, researchers of plant disease epidemics are increasingly using landscape approaches to quantify and model spatial pattern of disease spread in order to understand the basic factors that influence pathogen dispersal and infection processes. Despite the strength of a landscape approach, relatively few studies have developed spatial models of disease pattern in natural systems due to the challenge of integrating numerous, spatially referenced samples of disease incidence with digital maps describing spatial variations in environmental factors and plant community structures (Lundquist and Klopfenstein, 2001).

(b) ***Crop Scouting*:** Crop scouting is being advocated as another process that can benefit from precision crop protection technologies. A grower may randomly spot problems in a field and with an appropriate DGPS data recording system on board, can easily record that point so subsequent control measures can be taken (*e.g*., spot spraying for weed, disease or insect pests occurrence). By recording the point, the grower can easily navigate back to the problem spot at different time, using navigational devices as used for soil sampling to control the problem.

- Remotely sensed imagery is also being applied to mid season crop management. Aerial photography and/or satellite based imagery are being collected to provide growers with images and information on their crop health during crop growth season. Different types of imagery can provide different information about the crop. For example, infrared imagery can be used to identify plant health and nutrition problems, while other spectral bands provide better information about disease, insect pests and weed infestations. Advances in satellites for remote sensing will provide agriculture with an enhanced set of information (Peterson *et al*., 1995). A new generation of satellites for imagery will offer spatial and temporal resolutions that have never been available to the public (Hanley, 1997). EarthWatch, a private company, built a constellation of commercial imaging satellites. The initial configuration consists of two EarlyBird satellites, the first of which was launched in the first quarter of 1997 and next two QuickBird satellites were launched by the same organisation in 1998. EarlyBird has a 3 metre resolution panchromatic sensor and a 15 metre resolution multispectral sensor. QuickBird features a 0.82 metre resolution panchromatic sensor and 3.28 metre resolution multispectral sensor. This level of spatial imagery can provide useful information to identify crop health or pest problems. Problem areas are marked so these points can be navigated using a DGPS receiver (Buick, 1997).

(c) ***Variable Rate Application of Agrochemicals*:** After collecting huge quantity of spatial and temporal information for a field, one of the final steps in precision crop protection is to automatically vary the pesticide inputs on the field as the sprayer truck drives across the field. To perform variable rate application, there is an important GIS based process where one or more spatial layers for a field are used to determine the prescription map which prescribes the optimum or suitable rates of application. Many precision agricultural GIS software packages allow the user to generate prescription maps based on earlier observations of disease or insect pests attack. This helps GIS user to formulate their own 'optimum' equations for agrochemicals/pesticides spray based on known spatio–temporal variables in the field. The resultant map would be loaded in an on-board field computer that drives the variable rate applicator. A great deal of research and a number of crop growth models are being developed which will help to automate this prescription generation process *e.g.*, research using decision support system that growers can access (*e.g.*, over the internet) or regional models can be integrated into the user's own precision crop protection GIS software.

- Variable rate applications are done by setting up a DGPS system on a spraying unit along with an onboard processor capable of importing and interpreting the prescription map. The operator can drive the vehicle about the field with the system automatically regulating application rate based on the vehicle location. Variable rate application control devices are also involved in the process since the onboard computers sends the prescribed application rate for each location, usually set to occur at a distance or time based interval, to the variable rate controller system which in turn regulates the output rate of the spray boom or the applications channels on the sprayer.

(d) ***Swath Guidance*:** Another significant advantage of DGPS technology is the ability to guide the field operator along parallel swathing patterns or along particular patterns of swathing that the operator desires to follow. Typically this is used with pesticide applications where it can be difficult to see where previous passes (swaths) were made. Although foam marker systems have been used for some time to provide guidance to applicator, a number of problems exist with these such as dissipation of the foam before the operator can use it for the next pass, the operator still needs to estimate by eye where to turn or line up alongside the foam from the previous swath which can be difficult with a 30 metre or longer spray boom. Wind can make foam drift causing gaps or overlaps in applications and the application must be done in day light (Kohls, 1996). Differential GPS for parallel swath guidance that removes the foam drift problem can make it possible for the operator to apply for longer hours, operate at night, operate in foggy or snow conditions. It is important to note how this can improve pesticide efficacy under conditions of high humidity where plant chemical surface contact and plant uptake of chemicals can be significantly improved. The primary purpose of using DGPS for using parallel swath guidance is to improve the applications or coverage of agrochemical or pesticide on to a crop by being able to easily identify the skips or overlaps in applications to a field. Some precision agricultural systems used for parallel swath guidance will display a graphical map in the field, with the areas already applied as shaded or coloured swaths so the operator can quickly identify areas which have not been applied to. This allows operator to return to sites that have been missed and rectify the problem while they are still in the field.

(e) ***Spectral Imaging*:** Assessment of health of crop, as well as early detection of crop infestations, is critical ensuring good agricultural productivity. Stress associated with, for example, moisture deficiencies, insect, fungal disease or weed infestations, must be detected early enough to provide an opportunity for the farmer to mitigate. This process requires remote sensing imagery be provided on a frequent basis (at a minimum, weekly). Optical remote sensing can see beyond the visible wavelengths in to the infrared, where wavelengths are highly sensitive to crop vigour as well as crop stress and crop damage. Remote sensing imagery allows a farmer to observe images of his infected field and make timely decisions about managing the crop. When the plants are affected by stress, such as that cause by disease or insect damage, changes occurs in spectral reflectance characteristics of the foliage. Colwell (1956) reported previsual detection of wheat rust using colour infrared film, provided the photos were obtained under certain conditions of development of the disease, illumination and film exposure. Manzer and Cooper (1967) showed that colour infrared film can be effective tool for detecting late blight of potatoes. The enhancement capabilities of colour–infrared film clearly make it very useful tool for monitoring plant diseases and insect infestations. Such conditions cause a difference in tone that makes the stressed vegetation distinguishable from the normal red tone of healthy surrounding vegetation. The disease or insect affected crop and normal crops give different tonal variation in imageries. During a study of pest damage detection in cropping systems of IndoGangetic plains, GPS gave exact location and position on the imagery. The image interpretation keys shape, size, tone, texture and association help to detect imagery and confirmation during ground survey. After image analysis found the great variability in tonal variation on the imagery. Normal crops give red, bright red and dark red colour with smooth texture but pests affected area give pink, yellow and yellow pinkish red colour with irregular shape and rough texture. This helps farmers to pay immediate attention towards pest infestation and manage crop by spraying chemicals. Early detection of disease and insect pests or weeds when their spatial extent is small, reduces the cost of control and increases the possibility of successful eradication (Rejmanek and Pitcairn, 2002). Now-a-days detection capabilities have improved because sensor technology and classification techniques have become more sophisticated. Crop stress study now depends on multispectral satellite image (QuickBird) and on a hyperspectral image of an air borne sensor (AVIS). Using digital camera, the images of plant disease or insect pests are captured before and the professional botanist can give the consultation based on the transmitted image. The database of spectral reflectance of plant disease is needed for spectral identification in the diagnosis of plant disease or insect pests. It becomes also possible to acquire the quantitative data with regard to the colour of plant disease or insect pests. Acquisition of such data will give quantitative information useful for diagnosis. And the accuracy of colour reproduction using multispectral camera is superior to the conventional RGB camera. Multispectral imaging technique combines space imaging and spectral detecting. It can obtain the spectral information and image information of object at the same time. Multispectral analysis done by high resolution satellite 'QuickBird' which collect reflectance data 450 km above ground. The very high spatial resolution of 0.7 m in PAN mode and 2.5 m in multispectral mode delivers very accurate data for remote sensing analysis of crop stress by disease, insect pests or weeds.

(f) ***Hyper Spectral Imaging*:** Hyperspectral imaging is a result of the integration of optical remote sensing and traditional spectroscopic technologies, allowing for differentiation of unique spectral signatures of targets on the surface of the earth (Fig. 4). Unlike the human eye, which just see visible light (blue, green and red); hyperspectral imaging is able to detect wavelengths right into the infrared portion of the electromagnetic spectrum. Hyperspectral sensors are imaging spectrometers that sample the reflected solar region of the electromagnetic spectrum in narrow continuous increments. The continuous increments allows for blue, green, red and near infrared spectra to be recorded instead of single value. One hyperspectral sensor is the NASA jet propulsion laboratory airborne visible/infrared imaging spectrometer (AVIRIS). The AVIRIS sensor is a 224-band system measuring light between 400 and 2500 nm with 10 nm increments. AVIRIS is a commercial hyperspectral aircraft system provides high quality, laboratory like spectroscopy on an image-wide basis, resulting in direct or quantitative surface mapping of minerals, plants, water stress, diseases, insect pests or weeds. Serious economic losses in yield and quality of cultivated plants can be caused by pests and disease (Macleod *et al.*, 2004). Although the broadband multispectral sensors may be helpful in discriminating diseased and healthy crops, the best results for identifying diseases were obtained with hyperspectral information (Moran *et al.*, 1997). Therefore there are indication that the use of hyperspectral remote sensing can be valuable to disease/pest detection and crop damage assessment and also hyperspectral remote sensing increases our ability to accurately map vegetation attributes (Kumar *et al.*, 2001). Apan *et al.* (2005) concluded the study of detection of pests and diseases in vegetable crops using hyperspectral remote sensing as the following: it is feasible to detect the effects of insect pests and disease in field crops using hyperspectral measurements. Different sets of pest and disease symptoms provided different sets of diagnostic spectral regions. The most significant spectral band for tomato disease prediction corresponded to the reflectance red-edge, as well as the visible region and part of near infrared wavelengths. For the eggplants insect infestation, the near infrared region was identified by the regression model to be as equally significant as the red edge in the prediction. However, the inclusion of short wave infrared bands as significant variables has indicated the effect of other contributing factors. It was observed that the use of a portable field spectrometer can provide a means of rapid observation and digital recording of hundreds of plants samples in few hours of scouting through the fields. Combined with global positioning systems (GPS) location data collected simultaneously, field level maps can be created by spatial interpolation among the sampling points. By creating spectral libraries of specific crops comprising a wide range of healthy and diseased crops spectra, such site specific crop data can be used routinely with various spectral matching type algorithms for automated detection of disease spots.

- Site specific crop protection is a system that is designed to strengthen farming management, providing farmers improved means to spot weak productivity zones and react before they become loses, increasing yields for the farmer and reducing annual production costs and adverse environmental impacts. Research into hyperspectral remote sensing is providing important tools for site specific farming by assessing crop health conditions and deriving quantitatively important crop health parameters such as plant canopy vigour, water content, chlorophyll, disease or insect pest incidence or weed infestation. Additionally, hyperspectral sensors can give farmers chance to order spectral imagery of their fields for determining the statues of their lands and whatever is growing on them. For instance the sort of vegetation can be determined through hyperspectral remote sensing.

(g) ***Thermal Imaging*:** If plant disease influences the transpiration of plants, thermal imaging can be applied. The evaluation of thermal image is an approach in plant pathology to detect fungal plant disease (Nilsson, 1995). The temperature of plant leaves and other plant parts depends on degree of transpirations. Due to latent heat necessary for transpiration, the plant temperature decreases with increase in transpiration. Transpiration itself depends on type and state of plant, on humidity and motion of the ambient air. Dry stress of plants reduces the transpiration of plants. Infected leaves of plants are often coated by superficial hyphae or fructification structures of the fungi and in such a way may change the plant transpiration.

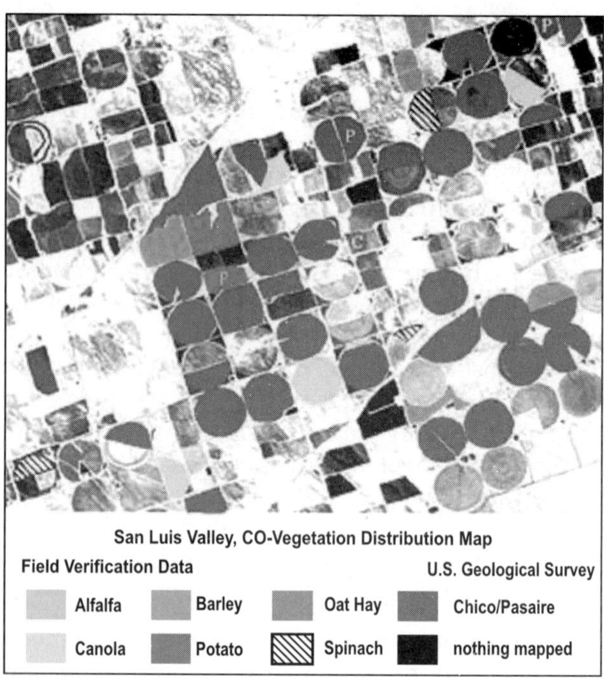

Fig 4: Hyperspectral image made from a single narrow band set acquired by AVIRIS over terrain in San Luis Valley of Colorado, on which of the agricultural fields have been classified to the level of specific crops (Khormi, 2010)

Leaf temperature of plants is the result of external and internal (physiological) factors. The environmental factors solar radiation, air temperature, and relative humidity (RH), and the water status of the shoot tissue determine the temperature of plants via stomatal transpiration. There is a correlation between leaf temperature and water status, as water is the primary source of infrared absorption in plant tissue (Kümmerlen *et al.,* 1999). In addition to water supply and overall metabolic activity regulated by environmental conditions, pathogenic organisms may affect both cuticular and stomatal conductance of plant tissue, resulting in significant modifications in leaf temperature. As leaf temperature may be measured remotely and with high spatial resolution, digital infrared thermography may have the potential for the identification of management zones in disease control. The detection of modifications in plants or canopies associated with low disease severity in the early stages of disease epidemiology is crucial for the targeted, site-specific or on demand application of fungicides in integrated disease control. The sensing of ethylene associated with tissue damage from pathogens and optical methods assessing the reflection and fluorescence characteristics of plants associated with photosynthetic activity (Franke *et al.,* 2005) are some approaches. Thermography allows the quantitative analysis of spatial and dynamic physiological information on the plant status. Infrared thermography is used to study spatial variability of stomatal conductance, to schedule irrigation, for monitoring of ice–nucleation or temperature stress in plants, to screen for mutants with altered stomatal control and for assessment of plant-pathogen interaction by monitoring patterns of surface leaf temperature (Oerke *et al.,* 2006). For remote detection, identification, and quantification of plant diseases and associated pathogens, sensors have to be sensitive to physiological disorders associated with fungal attack and disease resulting from pathogen attack and tissue colonization. In contrast to weeds which can be remotely detected and identified in crops according to their macroscopic shape early after emergence (Gerhards and Christensen, 2003), micro-organisms causing plant diseases may be detected only by their effect on plant tissue; visible symptoms often appear only after latent colonization of the plant tissue. Digital infrared thermography has found to be a useful tool for the presymptomatic detection of cucumber downy mildew caused by *Pseudoperonospora cubensis* (Berk. et Curt.) Rostovzev (Lindenthal *et al*., 2005; Oerke *et al*., 2006). The maximum temperature difference (MTD) within a leaf or a canopy turned out to be suitable for the differentiation of infected and non-infected tissue under controlled conditions. Digital infrared thermography alone seems not to be suitable for disease detection in the field, a prerequisite for a more demand-based use of fungicides or site-specific disease control. This sensor has to be combined with other remote sensing methods offering additional spectral information and systems for the recognition of optical patterns in plant canopies (Oerke *et al*., 2006). Also the use of reference areas or plants may be suitable, especially for the identification of wet and dry canopies.

5. SITE SPECIFIC PLANT DISEASES MANAGEMENT (SSDM)

Detection and identification of plant diseases and planning effective control measures are important to sustain crop production. Studies on use of remote sensing for crop disease assessment started long time ago. For example, in the late 1920s, aerial photography was used in detecting cotton root rot (Taubenhaus, *et al.,* 1929). The use of infrared photographs was first reported in determining the prevalence of certain cereal crop diseases (Colwell, 1956). One of the potential applications of remote sensing technique of agriculture is the detection of plant diseases in an extensive area before the symptoms clearly appear on the plant levels. This has been clearly demonstrated for the first time in remote sensing studies on root wilt disease on coconut plants in Kerala under NASA–ISRO–IARI wilt disease on coconut plants in Kerala under NASA–ISRO–IARI collaborative programme (Dakshinamurti, 1969 and Dakshinamurti *et al.*, 1970). Healthy plants give a higher reflectance in the near infrared region and a lower one in the visible region diseased plants show a higher reflectance in the visible spectrum and a lower one in infrared region. This principle can be used in distinguishing healthy and diseased vegetation and assessing vegetation damage due to diseases.

Johannsen and Bauer (1972) conducted a 'Corn blight watch experiment' to determine the feasibility of crop disease detection by remote sensing over a large geographic area. Accurate estimate of acreage of healthy and blighted corn in the intensive study area were obtained from machine analysis of the multispectral scanner data. Analysis of multispectral scanner data gave a more accurate assessment of the blight situation than that provided by photo interpretation methods. Also remote sensing technology has already employed for detecting crop disease and assessing its impact on productively include using colour infrared (CIR) photography to identify circular areas affected by cotton root rot, *Phymatotrichum omnivorum* (Heald *et al.*, 1972, Henneberry *et al.*, 1979) and to estimate yield losses caused by black root disease in sugarbeets (Schneider and Safir, 1975). In the early 1980s, Toler *et al.* (1981) used aerial colour infrared photography to detect root rot of cotton and wheat stem rust. In these studies, airborne cameras were used to record the reflected electromagnetic energy on analogue films covering broad spectral bands. Since then remote sensing technology has changed significantly. Satellite based imaging sensors, equipped with improved spatial, spectral and radiometric resolutions, offer enhanced capabilities over those of previous systems. Cook *et al.*, (1999) also demonstrated the potential for aerial video imagery to detect *P. omnivorum* in kenef, a crop whose tall growth habit makes it almost impossible to survey from the ground. The thermal infrared (TIR) can provide early, sometimes previsual, detection of disease that interfere with the flow of water from the soil through the plane to the atmosphere. As an example, Pinter *et al.* (1979) found that cotton plants whose root were infected with the soil borne fungus *P. omnivorum* and sugarbeets infected with *Pythium aphanidermatum* and both displayed sunlit leaf temperatures that were 3–5°C warmer than adjacent healthy plants. The TIR was also useful for detecting root diseases in clover under irrigated conditions (Olive *et al.*, 1994). Everitt *et al.* (1999) evaluated colour infrared (CIR) digital imagery as a remote sensing tool for detecting oak wilt disease in live oak. They delineated dead, diseased and healthy live oak by using aerial CIR digital imagery and CIR photography concurrently. Appel *et al.* (1989) used CIR aerial photography to analyze epidemiological parameters for oak wilt to improve options for disease management.

Hyperspectral sensing offers potential to improve the assessment of crop diseases and pests. According to Apan *et al.* (2005) pathogens and pests can induce physiological stresses and physical changes in plants, such as chlorosis or yellowing, necrosis, abnormal growth wilting, stunting, leaf curling etc. These changes can alter reflectance properties of plants. In the visible part of the electromagnetic spectrum (approx. 400 nm to 700 nm), the reflectance of green healthy vegetation is relatively low due to strong absorption by pigments in leaves. If there is a reduction in pigments due to pests or diseases, the reflectance in this spectral region will increase. Vigier *et al.* (2004) found that reflectance in the red wavelengths (*e.g.*, 675–685 nm) contributed the most in the detection of sclerotinia stem rot infection in soybeans. In addition reflection of healthy vegetation is significantly high at about 700 nm to 1300 nm (NIR) (Apan *et al.*, 2005). Therefore, with pests and diseases the reflectance in the NIR region is expected to be lower than the overall reflectance of healthy vegetation. Ausmus and Hilty (1972) in their study of maize dwarf mosaic virus concluded that the NIR wave lengths were useful in reflectance studies of crop disease. On stress in tomatoes induced by late blight disease, it was found that near infrared region was much more valuable than the visible range to detect disease (Zhang *et al.,* 2003). In different spectral region of shortwave infrared (SWIR) range (1300 to 2500 nm), the spectral properties of vegetation are dominated by water absorption bands. Less water on leaves and canopies will increase reflectance in this region. Apan *et al.* (2004) noted the spectral discrimination of healthy and diseased sugarcane. Zhang *et al.* (2003) explores spectral analysis to identify the spectral relationship of late blight of tomato. They found strong relationship between disease infection stages and spectral reflectance and validated using hyperspectral image data with known disease infection locations and stages. On this basis they proposed to apply this approach for image analysis to identify the infected plant from healthy one using hyperspectral remote sensing images. Falkenberg *et al.* (2005) applied remote sensing technique of aerial infrared photography for site specific management of biotic and abiotic stress in cotton. Biotic stress (*Phymatotrichum* root rot) was detected early in the growing season with the Infrared (IR) camera before symptoms could be seen visually. The progress of root rot was monitored until the end of the growing season by comparing the IR images to regular digital aerial images. Abiotic and biotic stress can be differentiated better by the Indigo camera than by the pivot mounted IRTs because of its increased image scanning resolution. The IR camera has a pixel size of 2 × 2 feet while the IRTs have a pixel size of 10 × 10 feet. Disease and water stress can be differentiated from each other using the temperature scale on the colour IR image. The root rot temperature was between 37–42°C while the irrigation stress was between 31–35°C. The canopy temperature range for root rot varies with the stage of the root rot infection on the cotton plant and irrigation regime in which it was detected. Feng *et al*. (2008) presents a method for diagnosis of plant disease and insect pests using narrow band multispectral camera, in which natural colour reproduction of colour patches from 14 band multispectral image was demonstrated which can be used for reference as a colour image database or the examination of temporal colour changes by saving colour data. They also added that the measured and analysis spectral information also can be applied to investigate the disease.

6. SITE SPECIFIC WEED MANAGEMENT (SSWM)

There are potentially economic and environmental benefits to site specific weed management (SSWM). Site-specific weed management is an effective tool to reduce the amount of herbicides applied. In addition to that, weed control efficacy over the total field is very high (Girma *et al.,* 2005).Therefore, it is expected that site-specific weed management will be accepted by practical farmers and supported by government. Herbicide under-dosing and over-dosing are inevitable when weed distribution and density varies across a field. Overseas research has shown that SSWM has the potential to reduce herbicide applications by 10 to 80%. The cost savings are obvious, but additionally weed free crops that are not sprayed may yield 5 to 10% more, when phytotoxic effects of the herbicide are removed.

The aim of SSWM spraying systems is to get the right dose of the right herbicide in the right place. There are number of technical hurdles to be overcome, and this has resulted in an array of research approaches with different level of complexity. SSWM has four main components: (1) weed mapping/sensing (2) treatment decision (3) treatment application and (4) documentation. The system components chosen will be influenced by each individual weed control situation.

1. ***Prior Mapping vs Real Time Detection*:** Mapping prior to spraying is easier, but may involve an extra pass. Real time detection requires sensors and on-board computers to process imagery and control nozzles.
2. ***Weed/crop Biomass vs Weed Species ID*:** Simple reflectance systems can measure total plant biomass—this measures both crop and weed together and can be misleading. Scanning only in the inter rows is more challenging, but more accurate. More complex systems use both reflectance and image shapes to identify plant type. The more advanced research system can identify up to 25 weed species.
3. ***Treatment Decisions*:** The simplest is ON/OFF system to apply another treatment only to patches. More complex systems identify spatial variation in weed species and density and may apply up to

three herbicides at varying rates. These systems use sophisticated weed control expert systems. Most SSWM research has concentrated on treating patches (metres across) using boom section control, but more recent research in Denmark is working on "cells" (11 × 3 cm) or even single plant targets.

4. ***Sensor Types*:** The most readily available and sophisticated sensor is the human eye, but manual mapping prior to spraying is time consuming and real time manual control may not be reliable due to periodic distraction of the operator. Digital imagery can be captured from the ground or remotely placed satellites or aircrafts, but for systems aiming to treat patches smaller than several metres, remote imagery has insufficient spatial resolution. 'WeedSeeker' is currently the only commercialised system linking sensors linked to spray control. This system senses green plant biomass using a ratio of red and near infrared (NIR) reflectance and is mainly used for non-selective weed control in non crop areas. There are other commercially available sensors (CropCircles, GreenSeekers and Yara N-Sensor) that can map biomass using red/NIR. More sophisticated systems are under development in Denmark and Germany, use a combination of red/NIR imagery with image shape analysis to identify weed species.

5. ***Documentation*:** Most systems under development log the "as applied" herbicide application map as a useful record of application.

In many studies, a map–based approach of site specific weed management has successfully been applied (Gerhards and Oebel, 2006). The results of these experiments show that site-specific weed management reduced costs for weed control and resulted in less impact on the environment. However, for a broader acceptance of site-specific weed management in practical agriculture, an online system would be needed, which combines weed detection and herbicide application in one step. For real-time weed control, automatic sensor-based weed sampling techniques need to be developed. Dammer *et al.* (2003) used a reflectance sensor in two wave bands of 650 and 830 nm to measure total plant coverage in the track of the tractor wheel to identify patches with high weed infestation levels. Girma *et al.* (2005) were able to distinguish grass weeds from wheat by measuring reflectance in different wave length. Weis and Gerhards (2007) segmented plants, soil and mulch in digital images using a bi-spectral camera system. Shape features were calculated for each plant and stored in a data-base. Later on, those features were used for automatic plant species classification. Improved application technology for site-specific weed control is needed allowing variable rates and herbicide mixtures in real-time based on weed species observed. Vondricka (2007) developed a direct nozzle injection system, which is able to adapt herbicide dose and –mixture to the actual weed density and species composition within lees than 500 ms. Finally, decision rules for site-specific weed management are needed which determine the correct dose and herbicide mixture for each position in the field. So far, decision support systems give a recommendation for uniform weed control applications across the total field based on the average weed infestation level (Rydahl and Thonke, 1993). None of these models take into account the heterogeneous distribution of weeds within the field. Only few models have been developed for site-specific weed management decisions (Christensen *et al.,* 2003). Christensen *et al.* (2003) consider yield loss functions of weeds, dose response functions of several herbicides and population dynamic aspects for each cell of 8 × 8 m in winter wheat fields. Experiments by Gutjahr *et al.* (2008) showed that herbicide application reduced grain yield in areas with no or low weed infestation. Thus, the effect of herbicides on the crop should be considered in a decision algorithms for site-specific weed control. Selectivity of many herbicides is caused by different kinetics of metabolism in the plant. Herbicides could damage the crop when the uptake and translocation within the crop is increased due to less favourable weather condition for cuticle formation. However, it is very difficult to give a general estimate of the yield effect of herbicide. It differs significantly between active ingredients, herbicide doses and weather conditions before, during and after application, crops and growth stages. Site-specific weed management is feasible and may even have economic benefits when herbicide savings compensate for costs for weed mapping and patch spraying (Schwarz *et al.,* 1999). Since weed species distribution varies within the field, application technologies are needed allowing a variation of active ingredients in real-time. The motivation for site-specific weed management has been to reduce herbicide use, and the focus of research has been patch spraying. However, the distributions of individual species in a field do not usually coincide and the species present vary among patches. Control might be more cost-effective with several herbicides when local weed populations vary within fields. Technology for patch spraying now is sufficient for site-specific weed management with several herbicides applied in a single pass across a field (Gerhards and Oebel, 2006). There have been two approaches to prescribing several herbicides for a single field from a map of the weed population in the field (Gerhards and Oebel, 2006). The simplest approach is to identify two or more groups of species to target with different herbicides in a field and then independently create a patch spraying application map for each herbicide. A common example is the targeting of broadleaves and grasses with different herbicides. The second approach relies on weed management decision models. A field is divided into subunits and the herbicide that maximizes net return is recommended for each subunit (Wilkerson *et al.,* 2004). Prescribing management is easier with herbicides selected for predefined groups of weeds. However, if there is an herbicide that effectively controls species in more than one predefined group, patch spraying might not be more cost-effective than a uniform application. The second approach of maximizing net return for each subunit ensures that when weed control with several herbicides is recommended, it is more cost-effective than a uniform application. A uniform application would be recommended if one herbicide maximizes net return for the field when used for all subunits.

Fig. 5: The experimental/prototype patch sprayer described by Gerhards and Oebel (2006)

***Patch Spraying*:** Site-specific weed management with several herbicides or herbicide combinations in a field will require more complicated and expensive technology, and more management time than patch spraying. Patch spraying is promoted as a strategy to reduce herbicide use. The use of two or more herbicides for patch spraying would be likely to be a less expensive method for a grower to reduce herbicide use over time or across a farm compared to patch spraying with the one herbicide. However, it cannot be assumed that herbicide use will be reduced in every field and the area of the field that would not be sprayed would be a less reliable indicator of the reduction in herbicide use with patch spraying with two or more herbicides compared to patch spraying with one herbicide. If the goal is to reduce herbicide use in every field, herbicide use would have to be calculated when patch spraying with two herbicides is recommended based on maximizing net return from a field. Patch spraying of weeds cannot be recommended enthusiastically to growers based on average net return among fields. The use of several herbicides within a field can increase the net return and address growers' concerns about weeds left in the field. Site-specific weed management might be more acceptable to growers if several herbicides are used within a field in addition to leaving some areas untreated, and if we learn to identify characteristics of weed populations in fields for which site-specific weed management will be most beneficial. Growers will need decision models to help with the complex task of selecting the right combination of herbicides for a field. There might be little benefit in using more than two herbicides within a field when the primary goal is to reduce herbicide use. Increases in herbicide use become more likely with more than two herbicides. For growers who are more concerned about weeds left in the field than about herbicide use, using more than two herbicides could be a better strategy. Greater improvement in weed control is more likely when more than two herbicides are used. Using more than two herbicides for patch spraying further reduces weed escapes, but herbicide use is greater than a uniform application in fields. Growers might be more willing to adopt patch spraying if more than one herbicide is used in a field. The effect of herbicides on the crops depends on the active ingredient, dosage applied, weather conditions during and after application and growing conditions for crops until harvest. So more field experiments during several seasons are needed to clarify, whether, and how, crop herbicide stress would have to be included in a decision model for patch-spraying.

The main hurdles for SSWM are efficient and accurate mapping/ scanning systems and suitable direct injection systems for herbicide. Recent advances suggest that these problems may be overcome. Advanced imaging prototypes can identify 25 weed species real time. Direct injection systems currently suffers from long lag times because of the time taken for the herbicide to travel from the injection point to the nozzle (4 to 30 seconds). German research team developed an effective direct injection nozzle that would allow concentrated herbicide to be injected directly in to nozzles with a lag time of less than a second. There has been excellent technological progress has been made and the closest system to commercial release appears to be a 21 m three tank sprayer (CERBERUS) developed in Germany (Fig. 5). The sprayer has tree parallel independent spray lines supplied by three separate tanks filled with different herbicides. Weed maps are used to switch 7 × 3 m boom sections on and off simultaneously in each of three spray lines (Heap and Trengove, 2008). The control lag time is around 0.5 seconds, and they are currently working on real time weed identification (25 species) and treatment mixture determination. Laser induced leaf fluorescence (Germany) and polarised light reflectance (France) are also being explored for weed identification. A team in Denmark is developing extremely accurate autonomous spray delivery systems. The systems are modelled on ink-jet printers and will initially be used in horticultural crops. One system treats small area (cells) of ca. 11 × 3 cm by switching nozzles on and off, while a second system identifies individual weed seedlings and fires either a laser beam or herbicide micro-droplets (0.2 microlitres) at the growing point.

Remote sensing is a useful tool to manage spatial variability of biotic/ non-biotic factors in agricultural fields. For example, late-season weed infestations were mapped through remote-sensed imagery. A computerized decision method that estimates an economic optimal herbicide dose according to site specific weed composition and density was developed by Christensen *et al.* (2003). It consists in a competition model, a herbicide dose-response model and an algorithm that estimate the economically optimal doses. The software named Sectioning and Assessment of Remote Images® (SARI®) was developed to implement precision agriculture strategies through remote sensing imagery. SARI® has shown being efficient software for sectioning remote images, assessment agri-environmental indicators and implementing weed/pathogen-crop control strategies model in each micro–image/micro-plot. In summary, SARI® software provides geo-referenced, quantitative and visual herbicide prescription application, and this can be transferred to variable-rate application equipment for practical SSM strategies.

Remote images usually cover large areas, from several hundred hectares, if taken from airplanes, to dozens of square kilometres, if taken from satellites. Agriculture operations such as sowing, fertilization and pesticide application, among others, use to be programmed for individual plot of reduced area, normally smaller than 20–30 ha. So, a first step in programming such operations by remote sensing is isolating the image of the plot in which to implement the desired action. In addition, precision agriculture intend to determine the biotic/non-biotic spatial variability of agricultural plots and then to apply at variable rates fertilizers, pesticides or other inputs, fitted to the needs of each small area defined (Blackmore, 1996). Consequently, planning site-specific operations by remote sensing requires sectioning the isolated plot image into small micro-images/micro-plots, usually of a few hundred square meters, and interpreting for each micro-plot the adequate agro-environmental indicator for the desired operation.

Actually, SARI® is effective software sectionalizing plot images and assessing key agro-environmental characteristic of each micro-plot, regardless of the size of the original plot image and of the micro-images. In addition, SARI® software can work with any biotic factors and/or non-biotic factor that can be discriminated in the remote image. Moreover, the parameters characterizing the biotic/non-biotic factor, such as the boundaries digital values and the distance and size of aggregates, can be implemented by SARI® in a very flexible way. Spatial distribution biotic/

non–biotic factors patches studies on remote images and crop competition models implemented through SARI® software are much more cost-effective, this is requires much less ground and office work, than those achieved through conventional sampling ground techniques.

Hanks and Beck (1998) utilized spectral contrasts between green plants and bare soil to trigger real-time spraying of herbicide only on the plants that were present between soybean rows, controlling weeds as effectively as with conventional continuous-spray methods, but reducing herbicide usage and production costs. Machine vision techniques have also been used for identifying weed seedlings based on leaf shapes and for guiding an automatic precision herbicide sprayer (Tian *et al.*, 1999).

The ability to detect accidental herbicide damage to a crop has considerable value to a grower for insurance or litigation purposes. Comparing visual assessment of herbicide injury in cotton with colour-infrared photography, NIR videography, and wideband handheld radiometer approaches, Hickman *et al.* (1991) concluded that remote detection and mapping of moderate herbicide damage was not only possible, but that the application amounts could be estimated. Donald (1998) used video photography to quantify stunting of corn and soybean plants exposed to herbicide damage. Using a laboratory-based multispectral fluorescence imaging system (MFIS), Kim *et al.* (2001) were able to detect changes in soybean leaf fluorescence after they were treated with a herbicide. To improve application efficiency of herbicides, Sudduth and Hummel (1993) developed a portable NIR spectrophotometer for use in estimating soil organic matter as part of the estimation procedure for the amount of herbicide to be applied. Thus, remote sensing can not only offer field-scale assessment of herbicide injury problems but also can help define the optimum rate of herbicide application.

7. SITE SPECIFIC INSECT PEST MANAGEMENT

Agricultural pest management techniques have changed many ways in past few years. One of the most important changes is, shifting towards site specific crop protection. This technique is strengthened by the Global Positioning System (GPS) receivers, microcomputers, geographic information systems (GIS), yield monitors, and enhanced crop simulation models, remote sensing technology (Pinter *et al.*, 2003). The integration of these valuable tools that can enable resource managers to develop maps showing the distribution of insect infestations over large areas. The digital imagery can serve as a permanent geographically located image data base for monitoring future contraction or spread of insect infestations over time (Everitt *et al.*, 2003).

7.1. Approaches in Site Specific Insect Pest Management

7.1.1. *Observing and monitoring insect movement in the field*

The electromagnetic remote-sensing methods used for monitoring the free movement of insects outdoors under natural or semi-natural conditions (Reynold and Riley, 2002).

(i) ***Visual methods*:** At their simplest, observations may consist of watching insects with the naked eye or through binoculars, and scoring movement activities on handheld electronic event recorders or on portable (often 'notebook' or 'palmtop') computers (Wyatt, 1997). For insects that fly in short 'hops', stopping points can be marked by flags, and the ground locations of these later surveyed with an electronic theodolite connected to a data logger (Wiens *et al.,* 1993).

(ii) ***Night vision devises*:** At night, human vision can be augmented with night-vision (image intensification) binoculars and telescopes, used with or without supplementary near infrared illumination. For example, individual *Helicoverpa zea* moth flights near the ground were followed by observers equipped with night–vision goggles and infrared illuminators riding in a 4–wheel drive vehicle. Tests showed that an observer could follow an ascending moth up to heights of 100 m, when viewed against the night sky (Lingren *et al*., 1995).

(iii) ***Videographic techniques*:** The range of detection can be increased by improving the contrast between the insect and the background, in particular by viewing the target against the night sky using some form of artificial illumination. Illuminators in the near infrared region (750–900 nm) appear to be most suitable because they do not perturb insect behaviour (Riley, 1993), although in the case of *Heliothis virescens*, illumination can be extended down to 600 nm without producing any reaction in the moths' orientation to pheromone plumes (Vickers and Baker, 1997).

(iv) ***Thermal infrared imaging*:** Thermal IR imaging technology (sometimes referred to as Forward-Looking Infrared) is designed to detect objects in conditions of obscured visibility (darkness, smoke, dust, haze) by utilizing the long-wave infrared (heat) radiation emitted from the objects rather than the light reflected off them. Termites are the major threat in tropical countries for coconut and other trees. Failure to timely detect termites and not taking a preventive measure often leads to the death of the perennial trees. Termites are also a serious problem in farm buildings and in all wooden structures in domestic buildings. Thermal imaging can be used as a non-destructive and fast method to detect termites in trees and buildings, compared to the traditional methods such as knocking and drilling in wood (James and Rice, 2002).

(v) ***Optical sensors and insect trapping*:** Optical methods can provide a method of recording the time of entry of insects to traps, for example, by the interruption of an infrared beam (Waddington *et al.,* 1996). For larger animals, at least, commercial systems exist which automatically trigger digital cameras mounted inside weatherproof housings, either at regular intervals or when movement is detected by a motion sensor (*e.g.*, 'TrailCam' system, Erdman Video Systems, Miami). An archive of the images can be maintained on a dedicated computer, and/or they can be uploaded to the Internet. In the context of trapping and telemetry, infrared telemetry has been used to transfer data from pheromone traps (these used a piezoelectric detection mechanism) and from meteorological sensors in cotton fields, to a base station computer situated in the farm office (Schouest and Miller, 1994).

(vi) ***Specialist optical electronic devices*:** Among the more specialized insect-monitoring electro-optical devices are Farmery's crossed-beam infrared detectors and Schaefer and Bent's (1984) IRADIT. Schaefer and Bent (1984) overcame this limitation by using a very bright xenon flash lamp, working in the near infrared, and a video camera equipped with a gated image intensifier which provided high-contrast images even of small flying insects against the mid-day sky. This sophisticated device has been used for the calibration of light-traps and suction traps (Schaefer and Bent, 1984; Schaefer *et al.,* 1985) but, perhaps due to its complexity, it has not been taken up by other research groups.

7.1.2. *Radio frequency identification (RFID)*

RFID technology may have potential for monitoring and tracking insects moving over distances of less than a metre up to some tens of metres (for instance, in a warehouse-sized indoor arena, or a similar-sized outdoor cage). The Radio Frequency (RF) heat treatment showed a perfect potential in controlling storage insect pest *Sitophilus oryzae* (L.) and can be a very promising alternative phytosanitary processing in milled rice from conventional chemical fumigation. Exposure of milled rice infested with *Sitophilus oryzae* to RF target temperature and application time, the result showed that the temperature and the application time affected to control *Sitophilus oryzae* (L.). The temperature which is higher than 52°C for 1 minute, application time could eliminate the *Sitophilus oryzae* (L.) completly (100%) (Vassanacharoen *et al.*, 2007). RFID devices currently have an enormous variety of applications and can be used to identify and track almost any object to which an appropriate transponder or tag can be attached. Frequencies used in RFID systems range from 30–500 kHz for short-range, low-cost equipment up to 900 MHz–2.5 GHz for high-performance, longer range, high-cost devices. Systems operating below 135 kHz do not need to be licensed in many countries (Reynold and Riley, 2002).

(i) ***Passive tags*:** Passive transponders have no battery of their own but contain a capacitor which is charged inductively or radiatively by transmissions from the scanner, and they use the stored energy to transmit their unique alphanumeric code on an appropriate frequency. These have a read range of about 7–15 cm at the operating frequency of 125 kHz. This type of tag has apparently been used in an unpublished pilot study of cockroach movement carried out by The Clorox Company (Mark Owens, Biomark, pers. comm.). Similar devices could perhaps be used for recording the transit of large-bodied wasps and bees (*e.g.*, bumble bees) in and out of their nests.

(ii) ***Active tags*:** Active RFID tags contain an internal battery, and are typically read/write devices. They can in principle communicate over substantial ranges, but there is obviously a severe trade-off between the transmission power (and hence range), operational lifetime, and battery size. The size and weight of all commercially available active RFID tags preclude entomological applications, but recently some battery–operated radio transmitters have been miniaturised sufficiently to work as conventional radio tracking tags on large walking insects, and on one large flying insect (the hermit beetle) albeit over relatively short ranges (Hedin and Ranius, 2002).

7.1.3. *X-rays*

X-ray radiography has been used since the 1950s to detect insects in wood, soil and grain (Southwood and Henderson, 2000). A movement-orientated example is provided by the radiography of successive positions of elateridae and scarabeid beetle larvae as they moved around in specially-prepared soil blocks (Villani and Gould, 1986; Villani and Wright, 1988).

7.1.4. *The application of space- and airborne technologies*

Monitoring condition of the habitat of some insects, particularly highly-mobile species adapted to ephemeral habitats, can provide strong indications of the likelihood of migration events (Reynold and Riley, 2002).

(i) ***Satellite remote sensing in the 'optical' spectrum*:** Riedell and Blackmer (1999) predicted the potential usefulness of canopy spectra for identifying outbreaks in actual field situations. They carried out the experiment on the wheat in greenhouse by infesting it with aphids *Diuraphis noxia* or greenbugs *Schizaphis graminum* and measured reflectance properties of individual leaves in an external integrating sphere. They came to conclusion that the leaves from infested plants had lower chlorophyll concentrations and displayed significant changes in reflectance spectra at certain wavelengths (notably 500 to 525, 625 to 635, and 680 to 695 nm) as compared with healthy plant. Forecasters at FAO use the Africa Real Time Environmental Monitoring using Imaging Satellites system (ARTEMIS), a dedicated satellite data acquisition and processing system, to detect areas of rainfall or green vegetation in the desert where Desert locust, *Schistocerca gregaria* outbreaks can be expected to occur (Hielkema, 1990; Cherlet *et al.,* 1991). African armyworm forecasting: the aerial concentration of migrating African armyworm moths, *Spodoptera exempta* by wind convergence in the vicinity of convective rainstorms, followed by moth deposition, egg-laying, and the subsequent development of larvae on the flush of grass produced by the rain, can lead to serious high-density outbreaks in East Africa (Rose *et al.,* 2000). This association between rainstorms and larval outbreaks, particularly following dry periods at the beginning of the armyworm season, has led to the use of satellite imagery to help predict the likely position of new infestations (Tucker, 1997; Tucker and Holt, 1999).

- Elliott *et al.* (2005) has found that data obtained from multi-spectral remote sensing by using the SSTCRIS multi-spectral imaging system was sensitive to variation in the damage caused by the Russian wheat aphid in production of winter wheat fields. Since the damage caused by the aphid is highly correlated with its population density, both attributes can be measured in production of winter wheat fields using multi-spectral remote sensing. With an airborne MS4100 multi-spectral imaging system, Huang *et al.* (2008) has reported the great potential in pest management systems, such as weed control or possibly, detection of insect damage. The multi-spectral image processing produces NIR, red, green, NR, NG, NDVI and NDNG indices or images, which can be used to evaluate biomass and biotypes in agricultural fields.

(ii) ***Airborne digital photography and videography:*** The habitats of migratory insects can of course be surveyed from aircraft using conventional, large format, aerial photography. Airborne surveys often use colour infrared (CIR) digital cameras which are very sensitive to vegetation type, age and condition (*e.g.*, pest-induced stress). Hart and Meyers (1968) used color-infrared (CIR) photography and supporting hyperspectral reflectance data to identify trees in citrus orchards that were infested with brown soft scale insects *Coccus hesperidum*. Geo-referenced airborne videography has also been used to map infestations of *Aleurocanthus woglumi* in orchards (Everitt *et al.,* 1994) and *Bemisia* sp. on cotton (Everitt *et al.,* 1996); both of these whiteflies are detectable because deposits of sooty mould fungus associated with the insects alters plant reflectance on colour-infrared and black-and-white near-infrared imagery. Near-infrared reflectance (NIR) spectroscopy is the best technique to detect the single seed of wheat (Elizabeth, *et al*., 2002). The NIR system used to detect insects in kernels which can scan 1000 kernels per second (Dowell *et al.,* 1999). Perez–Mendoza *et al*. (2003) has given the advantage of NIR system over standard floatation method for detecting insect fragments in wheat flour. They suggested that the standard floatation method is time consuming (about 2 h/sample) and expensive. In contrast, a NIR system is rapid (< 1 min/sample), does not require sample preparation and could easily be automated for a more sophisticated sampling protocol for large flour bulks.

(iii) ***Synthetic aperture radar*****:** Unlike the sensors which use the 'optical' spectrum, synthetic aperture radars (SAR) are 'active' devices which both transmit and receive radiation, in this case in the microwave region. Aircraft–mounted SAR has also been used for entomological surveys. Pope *et al.* (1992) used this technique to assess the flooding status of depressions (known as dambos) which form favourable breeding sites for Culex mosquitoes, vectors of Rift Valley fever (RVF).

(iv) ***Global positioning systems*****:** Outdoor sampling or observational points can now be conveniently and very precisely located by the use of hand-held GPS (global positioning systems) equipment (Reynold and Riley, 2002). By using GPS the accurate location of Australian locust, *Chortoicetes terminifera,* population and habitat can easily track down and automatically transmitted to palmtop computer in the field officer's vehicle. These data are then relayed to a main computer at Australian Plague Locust Commission headquarters several times a day via a high-frequency radio link (Deveson and Hunter, 2000).

7.1.5. *Acoustic detection*

Detection of insects by sound waves is the exception to the otherwise invariable rule that remote sensing involves the propagation of electromagnetic waves between the target and sensing device. Acoustic technique can be utilized for monitoring and also to attract insects like mosquitoes, mole crickets, field crickets and tachnid parasitoids to traps (Reynold and Riley, 2002).

(i) ***Atmospheric sounders*****:** High-frequency sound devices are routinely used to investigate the structure of the lower atmosphere and to determine vertical and horizontal wind components, and these acoustic sounders or 'sodars' (SOund Detection And Ranging) have occasionally been used to obtain complementary meteorological information during insect migration studies (Reynold and Riley, 2002). Hendricks (1980) developed a low-power, 40 kHz sodar for detecting moths approaching close (within 1 m) of a sex pheromone source. The device counted over twice as many *Heliothis irescens* than were actually taken in a pheromone trap.

(ii) ***Hydroacoustics*:** The aquatic larvae of *Chaoborus* spp. (Diptera: Chaoboridae), sometimes called phantom midges, can be a major faunal component of certain lakes. Some larval instars, and the pupae of *Chaoborus* undertake diurnal vertical and transverse 'migrations' between the bottom of the lake and surface waters, and these movements have been studied with the aid of high-frequency (70–200 kHz) echosounding (Eckmann, 1998; Malinen *et al.*, 2001).

(iii) ***'Passive' acoustic detection of insects*:** The insects are detected by the low-intensity incidental sounds (in the range c. 0.5–150 kHz), that they make while moving and feeding in the medium. The feasibility of using acoustic detection depends on factors such as the signal-to-noise ratio, the amount of distortion and attenuation of the sound as it travels through the medium, the distinctiveness of sound patterns from target and non-target organisms, and the fraction of the measurement period during which signals are generated (Mankin *et al.,* 1998).

(iv) ***Acoustic traps*:** Another acoustic technique which can be used to monitor movement is the use of sound to attract insects to traps; taxa include mosquitoes, mole crickets and field crickets and their ormiine tachinid parasitoids, and galleriine moths (wax moths) (Reynold and Riley, 2002). Reviews of the history, operation and limitations of sound-baited traps are given by Service (1993) for mosquitoes and Walker (1996) for a variety of insects. Audio-frequency sound, combined with light, was thought to have potential for control of chironomid midges (Hirabayashi and Nakamoto, 2001).

Site-specific crop protection allows producers to take charge of many aspects of crop protection that previously were assumed to be random acts of chance. It makes producers curious about problems in the field and provides the tools to correct many of them effectively. It provides methods to test different resistant varieties, input rates and new agrochemicals on their farm and analyze the results easily and with confidence. Some growers will benefit from variable-rate application. Some will benefit from changes in management. Others will benefit from GPS-guided machinery or archived production information. The public benefits from inputs being applied at appropriate rates to all areas of the field and limiting exposure of sensitive areas to excess nutrients or chemicals. Site-specific crop protection system need not be prohibitively expensive. Methods have been tested to minimize costs and maximize benefits. Equipment costs have decreased since the inception of site-specific crop protection about 20 years ago so that all growers, regardless of size of operation, can participate if they choose to do so.

REFERENCES

Apan, A. Datt, B. and Kelly, R. (2005). Detection of pests and diseases in vegetable crops using hyperspectral sensing: A Comparison of reflectance data for different sets of symptoms. Proceedings of SSC 2005 spatial intelligence, innovation and praxis: The national biennial conference of the spatial sciences institute, September 2005. Melbourne: *Spatial Sciences Institute*. ISBN 0–9581366–2–9.

Apan, A., Held, A., Phinn, S., and Markley, J. (2004). "Detecting Sugarcane "Orange Rust" Disease Using EO–1 Hyperion Hyperspectral Imagery". *International Journal of Remote Sensing,* **2:** 489–98.

Ausmus, B.S. and Hilty, J.W. (1972) "Reflectance studies of healthy, maize dwarf mosaic virus-infected, and *Helminthosporium maydis*–infected corn leaves". *Remote Sensing of Environment,* **2:** 77–81.

Blackmer, T.M., and Schepers, J.S. (1996). Aerial photography to detect nitrogen stress in corn. *Journal of Plant Physiology,* **148(3–4):** 440–44.

Blackmore, S. (1996). An information system for precision farming. *Brighton Conference on Pests and Diseases*. Brighton, United Kingdom, **3:** 1207–14.

Buick, R.D. (1997). Precision agriculture: An integration of information technologies with farming. *Proc. 50th N.Z. Plant Protection Conf.* **1997:** 176–84.

Casady, W.W. and Palm, H.L. (2002). Precision agriculture: Remote sensing and ground truthing. MU Extension, University of Missouri–Columbia. http://extension. missouri.edu/explorepdf/envqual/eq0 453.pdf

Cherlet, M., Di Gregorio, A. and Hielkema, J.U. (1991). Remote-sensing applications for desert–locust monitoring and forecasting. *Bull. OEPP,* **21:** 633–42.

Christensen, S., Heisel, T., Walter, A.M. and Graglia, E. (2003). A decision algorithm for patch spraying. *Weed Res.*, **43:** 276–84.

Colwell, J.E. 1956. "Determining the prevalence of certain cereal crop diseases by means of aerial photography". *Hilgardia,* **26:** 223–86.

Cook, C.G., Escobar, D.E., Everitt, J.H., Cavazos, I., Robinson, A.F. and Davis, M.R. 1999. Utilizing airborne video imagery in kenaf management and production. *Industrial Crops and Products,* **9:** 205–10.

Dakshinamurti, C. (1969). Remote sensing and earth resource survey. *J. Post. Graduate School* IARI, New Delhi.

Dakshinamurti, C., Krishnamurthy, B. Summanwar, A.S., Shanta, P. and Pisharoty, P.R. (1970). Preliminary investigations on the root wilt disease in coconut plants in Kerala state (India) by remote sensing technique (a NASA–ISRO–IARI–CCRS project). XXIst International Astronautical Conference on the IAF. Germany.

Dammer, K.H., Böttger, H. and Ehlert, D. (2003). Sensor–controlled variable rate application of herbicides and fungicides. *Precision Agriculture,* **4:** 129–34.

Deveson, E.D. and Hunter, D.M. (2000). Decision support for australian plague locust management using wireless transfer of field survey data and automatic internet weather data collection. *In*: Laurini, R. and Tanzi, T. (*Eds*), TeleGeo2000–

Proceedings of the second annual conference on telegeoprocessing, 10–12 May 2000, Nice–Sophia Antipolis. E´ cole des Mines de Paris, Sophia Antipolis, Nice, pp. 103–110.

Donald, W.W. (1998). Estimated soybean (*Glycine max*) yield loss from herbicide damage using ground cover or rated stunting. *Weed Science*. **46:** 454–58.

Dowell, F.E., Throne, J.E., Wang, D. and Baker, J.E. (1999). Identifying stored-grain insects using near-infrared spectroscopy. *Journal of Economic Entomology* **92:** 165–69.

Eckmann, R. (1998). Allocation of echo integrator output to small larval insect (*Chaoborus* sp.) and medium-sized (juvenile fish) targets. *Fisheries Res.* **35:** 107–113.

Elizabeth, B.M., Dowell, F.E., Baker, J.E. and Throne, J.E. (2002). Detecting single wheat kernels containing live or dead insects using near-infrared reXectance spectroscopy, ASAE Paper No. 023067. Chicago, IL: ASAE.

Elliott, N., Mirik, M., Yang, Z., Dvorak, T., Rao, M., Michels, J., Catana, V., Phoofolo, M., Giles, K. and Royer, T. (2005). Airborne multi-spectral remote sensing for Russian wheat aphid infestations 20th Biennial Workshop on Aerial Photography, Videography, and High Resolution Digital Imagery for Resource Assessment October 4–6, Weslaco, Texas.

Everitt, J.H., Escobar, D.E., Summy, K.R., Alaniz, M.A. and Davis, M.R. (1996). Using spatial information technologies for detecting and mapping whitefly and harvester ant infestations in South Texas. *Southwest Entomol.*, **21:** 421–32.

Everitt, J.H., Escobar, D.E., Summy, K.R. and Davis, M.R. (1994). Using airborne video, global positioning system, and geographical information system technologies for detecting and mapping citrus blackfly infestations. *Southwest Entomol.*, **19:** 129–38.

Everitt, J.H., Escobar, D.E., Summy, K.R., Fletcher, R.S. and Davis, M.R. (2003). Applications of airborne remote sensing in integrated pest management. Section F Symposium: *Applications of Remote Sensing in Entomology*. October 29.

Everitt, J.D., Escobar, D., Riggs Appel, W. and Davis, M. (1999). Using airborne digital imagery for detecting oak wilt disease, *Plant Disease,* **83(6):** 502–505.

Falkenberg Nyland, R., Giovanni, Piccinni. and Daniel, I.L. (2005). Remote sensing for site-specific management of biotic and abiotic. Beltwide Cotton Conferences, New Orleans, Louisiana–January 4–7, 2005. pp. 1905–1909.

Feng, J., Liao, N., Wang, G., Luo, Y. and Liang, M. (2008). Application of multispectral systems for the diagnosis of plant diseases. *International Symposium on Photoelectronic Detection and Imaging* 2007: Image Processing, edited by Liwei Zhou, Proc. of SPIE Vol. **6623**, 66230E1–7.

Franke, J., Menz, G., Oerke, E.C. and Rascher, U. (2005). Comparison of multi– and hyperspectral imaging data of leaf rust infected wheat plants. In: Owe M, D'Urso G, eds. Proceedings of SPIE, Vol. 5976. Remote Sensing for Agriculture, Ecosystems, and Hydrology VII, 341–50.

Franz, E., Gebhardt, M.R. and Unklesbay, K.B. (1995). Algorithms for extracting leaf boundary information from digital images of plant foliage, *Transactions of the ASAE*, **36(2):** 625–33.

García–Torres, L., Peña–Barragán, J.M., Caballero–Novella, J.J., López–Granados, F. and Jurado–Expósito, M. (2008). SARI Software, splitting and assessment of remote images, office for the registration of intellectual property, regional department of culture, Seville, Spain, No. 200899900226820, p. 39.

Gerhards, R. and Christensen, S. (2003). Real-time weed detection, decision making and patch spraying in maize (*Zea mays* L.), sugarbeet (*Beta vulgaris* L.), winter wheat (*Triticum aestivum* L.) and winter barley (*Hordeum vulgare* L.), *Weed Research,* **43:** 1–8.

Gerhards, R. and Oebel, H. (2006). Practical experiences with a system for site-specific weed control in arable crops using real-time image analysis and GPS-controlled patch spraying. *Weed Research,* pp. 185–93.

Girma, K., Mosali, J., Raun, W.R., Freemann, K.W., Martin, K.L., Solie, J.B. and Stone, M.L. (2005). Identification of optical spectral signatures for detecting cheat and ryegrass in winter wheat. *Crop Science,* **45(2):** 477–85.

Gutjahr, C., Weis, M., Sökefeld, M., Ritter, C., Möhring, J., Büchse, A., Piepho, H.P. and Gerhards, R. (2008). Erarbeitung von Entscheidungsalgorithmen für die teilflächenspezifische Unkrautbekämpfung. *Journal of Plant Diseases and Protection*, Special Issue, **XXI:** 143–48.

Hanks, J.E. and Beck, J.L. (1998). Sensor-controlled hooded sprayer for row crops, *Weed Technology,* **12(2):** 308–314.

Hanley, C. (1997). Satellite imagery brings focus to the field. *Modern Agriculture* **1(1)***:* 17–19.

Hart, W.G. and Myers, V.I. (1968). Infrared aerial color photography for detection of populations of brown soft scale in citrus groves, *Journal of Economic Entomology,* **61(3):** 617–24.

Heald, C.M., Thames, W.J. and Wiegand, C.L. (1972). Detection of *Rotylenchulus reniformis* infestations by aerial infrared photography. *Journal of Nematology,* **4:** 299–300.

Heap, J. and Trengove, S. (2008). Site specific weed management: Weed mapping and patch spraying. *International Research Review*: *Australian Grains*, pp. 26–28.

Hendricks, D.E. (1980). Low-frequency sodar device that counts flying insects attracted to sex pheromone dispensers. *Environ. Entomol.* **9:** 452–57.

Henneberry, T.J., Hart, W.G., Bariola, L.A., Kittock, D.L., Arle, H.F., Davis, M.R. and Ingle, S.J. (1979). Parameters of cotton cultivation from infrared aerial photography, *Photogrammetric Engineering and Remote Sensing,* **45(8):** 1129–33.

Hickman, M.V., Everitt, J.H., Escobar, D.E. and Richardson, A.J. (1991). Aerial-photography and videography for detecting and mapping Dicamba injury patterns. *Weed Technology,* **5(4):** 700–706.

Hielkema, J.U. (1990). Satellite environmental monitoring for migrant pest forecasting by FAO: the ARTEMIS system. *Phil. Trans. Roy. Soc. Lond. B,* **328:** 705–717.

Hirabayashi, K. and Nakamoto, N. (2001). Field study on acoustic response of chironomid midges (Diptera: Chironomidae) around a hyper-eutrophic lake in Japan. *Ann. Entomol. Soc. Am.*, **94:** 123–28.

Huang, Y., Lan, Y. and Hoffmann, W.C. (2008). Use of airborne multi-spectral imagery for areawide pest management. *Agricultural Engineering International: the CIGR Ejournal. Manuscript* IT 07 010. **X:** 1–14.

James, K. and Rice, D. (2002). Finding termites with thermal imaging. Paper No. ITC 035A. North Billerica, MA: Infrared Training Centre.

Jensen, J.R. (1996). Remote sensing of the environment: An earth resource perspective. 3th Edn., Prentice Hall, USA, pp: 1–28.

Johannsen, C.J. and Bauer, M.E. (1972). "Corn Blight Watch Experiment Results". *LARS Technical Reports.* p. 6.

Khormi, H.M. (2010). A review of hyperspectral application and implication in agriculture community. University of New England, Armidale, Australia. pp. 1–11.

Kim, M.S., McMurtrey III, J.E., Mulchi, C.L., Daughtry, C.S.T., Chappelle, E.W. and Chen, Y.R. (2001). Steady-state multispectral fluorescence imaging system for plant leaves. *Applied Optics,* **40(1):** 157–66.

Kohls, C. (1996). The GPS Challenge. *Precision Farming*. Supplement to *GPS World* Magazine, pp. 29–31.

Krohmann, P., Gerhards, R. and Kuhbauch, W. (2006). Spatial and temporal definitions of weed patches using quantitative image analysis. *J. Agron. and Crop Sci.*, **192:** 72–78.

Ku¨mmerlen, B., Dauwe, S., Schmundt, D. and Schurr, U. (1999). Thermography to measure water relations of plant leaves. *In*: Ja¨ hne B, ed. Handbook of computer vision and applications, Vol. 3. London: Academic Press, pp. 763–81.

Kumar, L., Schmidt, K., Dury, S. and Skidmore, A. (2001. "Imaging spectrometry and vegetation science". *In:* Van der Meer, F.D. and de Jong, S.M. (*Eds*.), Imaging spectrometry: Basic principles and Prospective applications pp. 111–55. Dordrecht: Kluwer Academic Publishers.

Liaghat, S. and Balasundram, S.K. (2010). A review: The role of remote sensing in precision agriculture. *American Journal of Agricultural and Biological Sciences,* **5(1**): 50–55.

Lindenthal, M., Steiner, U., Dehne, H.W. and Oerke, E.C. (2005). Effect of downy mildew development on transpiration of cucumber leaves visualized by digital infrared thermography. *Phytopathology,* **95:** 233–40.

Lingren, P.D., Raulston, J.R., Popham, T.W., Wolf, W.W., Lingren, P.S. and Esquivel, J.F. (1995). Flight behaviour of corn earworm (Lepidoptera: Noctuidae) moths under low wind speed conditions. *Environ. Entomol.,* **24:** 851–60.

López–Granados, F., Jurado–Expósito, M., Peña–Barragán, J.M. and García–Torres, L. (2006). Using remote sensing for identification of late-season grass weed patches in wheat. *Weed Sci.,* **54:** 346–53.

Lundquist, J.E. and Klopfenstein, N.B. (2001). Integrating concepts of landscape ecology with molecular biology of forest pathogens. *Forest Ecology and Management*, **150:** 213–22.

MacLeod, A., Head, J. and Gaunt, A. (2004). "An assessment of the potential economic impact of Thrips palmi on horticulture in England and the significance of a successful eradication campaign". *Crop Protection,* **23(7):** 601–10.

Malinen, T., Horppila, J. and Liljendahl–Nurminen, A. (2001). Langmuir circulations disturb the low-oxygen refuge of phantom midge larvae. *Limnol. Oceanogr*. **46:** 689–92.

Malingreau, J.P. (1980). The wetland rice production system and its monitoring using remote sensing in Indonesia. Ph.D. Dissertation, University of California, Davis.

Mankin, R.W., Crocker, R.L., Flanders, K.L. and Shapiro, J.P. (1998). Acoustic detection and identification of insects in soil. *In*: Kuhl, P.K. and Crum, L.A. (*Eds*), Proceedings of the 16th International Congress of Acoustics and the 135th Annual Meeting of the Acoustical Society of America, pp. 685–86.

Manzer, F.E. and Cooper, G.R. 1967. Aerial photographic methods of potato disease detection. *Maine Agric. Exp. Stn. Bull*. **646**: 1–14.

Medlin, C.R., Shaw, D.R., Gerard, P.D. and Lamastus, U.S. (2000). Using remote sensing to detect weed infestations in *Glycine max*. *Weed Sci.* **48:** 393–98.

Moran, M.S., Inoue, Y. and Barnes, E.M. (1997). "Opportunities and Limitations for Image-based remote sensing in precision crop management". *Remote Sensing of Environment,* **61:** 319–46.

Nilsson, H.E. (1995). Remote-sensing and image-analysis in plant pathology. *Annual Review of Phyotpathology,* **33:** 489–527.

Nutter, F.W. Jr. and Gaunt, R.E. (1996). "Recent developments in methods for assessing disease losses in forage/pasture crops". *In*: Chakraborty, S., Leath, D.T., Skipp, R.A., Pederson, G.A., Barry, R.A., Latch, G.C.M. and Nutter, F.W. Jr (*Eds*), *Pasture and Forage Crop Pathology*, American Society of Agronomy, Crop Science Society of America, and Soil Science Society of America, Madison, W.I., pp. 93–118.

Oerke, E.C., Steiner, U., Dehne, H.W. and Lindenthal, M. (2006). Thermal imaging of cucumber leaves affected by downy mildew and environmental conditions. *Journal of Experimental Botany,* **57(9):** 2121–32.

Oliva, R.N., Steiner, J.J. and Young, W.C. (1994). White clover-seed production: I. Crop water requirements and irrigation timing. *Crop Science,* **34(3):** 762–67.

Perez–Mendoza, J., Throne, J.E., Dowell, F.E. and Baker, J.E. (2003). Detection of insect fragments in wheat flour by near-infrared spectroscopy. *Journal of Stored Products Research,* **39:** 305–312.

Peterson, G.W., Bell, J.C., McSweeney, K., Nielson, G.A., and Robert, P.C. (1995). Geographic Information Systems in Agronomy. *Advances in Agronomy,* **55:** 67– 111.

Pierce, J.F. and Peter, N. (1999). Aspects of precision agriculture. *Advances in Agronomy,* **67:** 1–85.

Ping, W., Xiangnan, L. and Fang, H. (2006). Research on mobile mapping system and its application in precision agriculture. www.gisdevelopment.net

Pinter, P.J.Jr., Hatfield, J.L., Schepers, J.S., Barnes, E.M., Moran, M.S., Daughtry, C.S.T. and Upchurch, D.R. (2003). Remote sensing for crop management. *Photogrammetric Engineering and Remote Sensing,* **69(6):** 647–64.

Pinter, P.J. Jr., Stanghellini, M.E., Reginato, R.J., Idso, S.B., Jenkins, A.D. and Jackson, R.D. (1979). Remote detection of biological stresses in plants with infrared thermometry, *Science,* **205(4406):** 585–86.

Rejmanek, M. and Pitcairn, M.J. (2002). When is eradication of exoticpest plants a realistic goal? pp. 249–53 *In*: Veitch, C.R. and Clout, M.N. *Eds*. Turning the tide: the eradication of invasive species. Gland, Switzerland, and Cambridge, UK: IUCN Species Survival Commission Invasive Species Specialist Group.

Reynolds, D.R. and Riley, J.R. (2002). Remote-sensing, telemetric and computer-based technologies for investigating insect movement: a survey of existing and potential techniques. *Computers and Electronics in Agriculture,* **35:** 271–307.

Ribes–Dasi, M., Sio, J., Planas, S., Almacellas, J. and Avilla Tora, R. (2005). The use of geostatistics and GIS to optimise pest control practices in precision farming systems. Information and technology for sustainable fruit and vegetable production, FRUTIC 05, 12–16 September 2005, Montpellier France.

Riedell, W.E. and Blackmer, T.M. (1999). Leaf reflectance spectra of cereal aphid–damaged wheat. *Crop Science,* **39(6):** 1835–40.

Riley, J.R. (1993). Flying insects in the field. *In*: Wratten, S.D. (*Ed*.), Video techniques in animal ecology and behaviour. Chapman and Hall, London, pp. 1–15.

Ristaino, J.B. and Gumpertz, M. (2000). New frontiers in the study of dispersal and spatial analysis of epidemics caused by species in the genus *Phytophthora*. *Annu. Rev. Phytopathology,* **38:** 541–76.

Robert, M. (2001). Soil carbon sequestration for improved land management, World Soil Research Report, 96, Food and Agriculture Organisation of United Nations, Rome Italy, p. 75.

Rose, D.J.W., Dewhurst, C.F. and Page, W.W. (2000). The african armyworm handbook: The Status, Biology, Ecology, Epidemiology and Management of *Spodoptera exempta* (Lepidoptera: Noctuidae), second ed. Natural Resources Institute, Chatham, UK.

Rydahl, P. and Thonke, K.E. (1993). PC–Plant Protection: optimising chemical weed control. *Bulletin OEPP/EPPO Bulletin,* **23:** 589–594.

Schaefer, G.W. and Bent, G.A. (1984). An infra-red remote sensing system for the active detection and automatic determination of insect flight trajectories (IRADIT). *Bull. Entomol. Res.*, **74:** 261–278.

Schaefer, G.W., Bent, G.A. and Allsopp, K. 1985. Radar and opto-electronic measurements of the effectiveness of rothamsted insect survey suction traps. *Bull. Entomol. Res.* **75:** 701–705.

Schneider, C.L. and Safir, G.R. (1975). Infrared aerial photography estimation of yield potential in sugarbeets exposed to blackroot disease. *Plant Disease Reporter,* **59(8):** 627–31.

Schouest, L.P. and Miller, T.A. (1994). Automated pheromone traps show male pink bollworm (Lepidoptera: Gelechiidae) mating response is dependent on weather conditions. *J. Econ. Entomol.*, **87:** 965–74.

Schwarz, J., Wartenberg, G. and Ackermann, L. (1999). Process-engineering investigations for site-specific weed control. Precision Agriculture 1999 Part 1.

Shanwad, U.K., Patil, V.C. and Gowda, H.H. (2004). Precision farming: Dreams and realities for Indian agriculture. Map India conference proceedings. *www. gisdevelopment.net*

Southwood, T.R.E. and Henderson, P.A. (2000). Ecological methods, third ed. Blackwell Science, Oxford, p. 575.

Sudduth, K.A. and Hummel, J.W. (1993). Portable, near-infrared spectrophotometer for rapid soil analysis. *Transactions of the ASAE,* **36(1):** 185–93.

Taubenhaus, J.J., Ezekiel, W.N. and Neblette, C.B. (1929). "Airplane photography in the study of cotton root rot". *Phytopathology,* **19:** 1025–29.

Taylor, L.R. (1984). Assessing and interpreting the spatial distribution of insect populations. *Annu. Rev. Entomol.*, **29:** 321–57.

Tian, L., Reid, J.F. and Hummel, J.W. (1999). Development of a precision sprayer for site-specific weed management. *Transactions of the ASAE,* **42(4):** 893–900.

Toler, R.W., Smith, B.D. and Harlan, J.C. (1981). "Use of aerial color infrared photography to evaluate crop disease". *Plant Disease,* **65:** 24–31.

Trematerra, P. Gentile, P. and Sciarretta, A. (2004). Spatial análisis of pheromone trap catches of codling moth (*Cydia pomonella*) in two heterogeneous agro-ecosystems, using geoestatistical techniques. *Phytoparasitica,* **32:** 325–41.

Tucker, M.R. (1997). Satellite-derived rainstorm distribution as an aid to forecasting African armyworm outbreak. *Weather,* **52:** 204–12.

Tucker, M.R. and Holt, J. (1999). Decision tools for managing migrant insect pests. *In*: Grant, I.F., Sear, C. (Eds.), Decision tools for sustainable development. Natural Resources Institute, Chatham, UK, pp. 97–128.

Vassanacharoen, P., Pattanapo, W., Lucke, W. and Vearasilp, S. (2007). Control *Sitophilus oryzae* (L.) by radio frequency heat treatment as alternative phytosanitary processing in milled rice. Conference on International Agricultural

Research for Development, October 9–11, 2007, University of Kassel–Witzenhausen and University of Göttingen, pp. 1–4.

Vickers, N.J. and Baker, T.C. (1997). Flight of *Heliothis irescens* males in the field in response to sex pheromone. *Physiol. Entomol.*, **22:** 277–85.

Vigier, B.J., Pattey, E. and Strachan, I.B. (2004). "Narrowband vegetation indexes and detection of disease damage in soybeans". *IEEE Geoscience and Remote Sensing Letter,* **1(4):** 255–59.

Villani, M.G. and Gould, F. (1986). Use of radiographs for movement analysis of the corn wireworm, *Melanotus communis* (Coleoptera: Elateridae). *Environ. Entomol.*, **15:** 462–64.

Villani, M.G. and Wright, R.J. (1988). Use of radiography in behavioural studies of turf–grass scarab grub species (Coleoptera: Scarabaeidae). *Bull. Entomol. Soc. Am.* **34:** 132–44.

Vondricka, J. (2007). Study on the process of direct nozzle injection for real-time Site-Specific Pesticide Application. Dissertation.

Waddington, K.D., Esch, H. and Burns, J.E. (1996). The effects of season, pre-training, and scent on the efficiency of traps for capturing recruited honey bees (Hymenoptera: Apidae). *J. Insect Behav.* **9:** 451–55.

Walker, T.J. (1996). Acoustic methods of monitoring and manipulating insect pests and their natural enemies. *In*: Rosen, D., Bennet, F.D. and Capinera, J.L. (*Eds.*), Pest Management in the Subtropics: Integrated Pest Management—a Florida Perspective. Intercept, Andover, UK, pp. 245–57.

Weis, M. and Gerhards, R. (2007). Feature extraction for the identification of weed species in digital images for the purpose of site-specific weed control. *In*: Stafford, J. (*Ed*) Precision agriculture '07, Vol. **6**, The Netherlands, pp. 537–45. 6th *Europ Conf. Prec. Agri.,* (ECPA): Wageningen Academic Publishers.

Wiens, J.A., Crist, T.O. and Milne, B.T. (1993). On quantifying insect movements. *Environ. Entomol.* **22:** 709–15.

Wiles, L.J. 2009. Beyond patch spraying: site-specific weed management with several herbicides. *Precision Agric.* **10**: 277–90.

Wilkerson, G.G., Price, A.J., Bennett, A.C., Krueger, D.W., Roberson, G.T. and Robinson, B.L. (2004). Evaluating the potential for site-specific herbicide application in soybean. *Weed Technology,* pp. 1101–10.

Wyatt, T.D. (1997). Methods in studying insect behaviour. *In*: Dent, D.R., Walton, M.P. (*Eds.*), Methods in Ecological and Agricultural Entomology. CAB International, Wallingford, UK, pp. 27–56.

Zhang, M., Qin, Z., Liu, X. and Ustin, S. (2003), "Detection of stress in tomatoes induced by late blight disease in California, USA, using hyperspectral remote sensing". *International Journal of Applied Earth Observation and Geoinformation,* **4:** 295–310.

10

Protection of Crop Plants with Newer Chemical Molecules

SUSANTA BANIK[1*] AND JAYDEEP HALDER[2]

ABSTRACT

Chemical method of crop protection has been widespread since its inception. Inspite of popularity and effectiveness of other crop protection methods, chemicals seem to be indispensable in order to sustain the growth rate in agriculture though there are adequate experimental results to prove that many of the early chemicals are harmful to humans and the environment. Demand for new molecules with unique mode of action, lower doses with target specificity are the need of the hour for crop protection specialists around the world. Apart from resistance, resurgence and secondary pest outbreak, target specificity and obvious safety to the non-target organisms are also important reasons why pesticides with novel mode of action have been always looked for. As safety norms get stricter in developed as well as developing nations, R&D of new pesticides become more challenging. In this chapter an attempt has been made to list out these novel groups of pesticides with their chemical names, unique mode of actions, some important target pests and other properties.

Key words: Chemical method, Pesticidal molecules, Newer chemical molecules, Safety norms, Target pests

1. INTRODUCTION

Controlling pests to reduce crop loss is deeply embedded in agriculture activities. It is only the methods of control which vary and have greatly evolved since ancient time to present day agriculture. Several methods of

1 Department of Plant Pathology, School of Agricultural Sciences and Rural Development, Nagaland University, Medziphema - 797 106, Nagaland, India.

2 Economic Botanist-VII, Sugarcane Research Station, Bethuadahari, Nadia, West Bengal, India.

* *Corresponding author*: E-mail: susanta.iari@gmail.com

earlier times are advocated and practiced till today owing to their inherent merits. Present day chemical–dominated agriculture in spite of known hazards is looked at as only savior to feed the 7 billion plus world population. In developing countries chemical crop protection has been advocated as a means to help farmers improve agricultural productivity, contribute to food security and alleviate poverty.

Today's agriculture heavily depends on chemicals as far as crop protection is concerned. Crops require protection mainly against three types of pests—insects, plant pathogens, and weeds. Chemicals offer several advantages in terms of cost effectiveness, rapid controlling effect, increasing yield, selectivity, persistence, systemicity etc. It was estimated that without crop protection chemicals, overall crop yield will be reduced by half and in some crops like cotton by 80% of current level. In some years when weather conditions are conducive, crops like potato and vines may fail entirely because of epiphytotics by fungal pathogens.

The concept of managing agricultural pests organically has been regaining importance. However, crop yield in organic agriculture is typically 30% lower in wheat and barley, 40% lower in potato when compared to chemically managed agriculture.

Chemicals seem to be indispensable to keep the production at high level in this scenario of shrinking agricultural land and burgeoning world population. Modern crop protection chemicals are safer as compared to chemicals of earlier years (arsenic and mercury compounds, DDT etc.). The endeavor to find out safer, newer, more effective plant protection chemicals is always on. Any new product coming to market has to fulfil the following requirements:

- High potency against its target pest *i.e.*, weeds, insects or fungi;
- high selectivity,
- low application rates
- low or no mammalian toxicity
- chemical and metabolic stability
- non-persistence in the environment.

Chemical protection is the dominant method of controlling crop pests. Perhaps it would remain so in the days to come as well in order to sustain the high agricultural output. In earlier times, discovery and/or development of new chemicals was either accidental or as bye product and without much thought on the repercussions in the other ecosystem processes. Publication of Silent Spring by Rachel Carson opened human eyes to the need of risk assessment and safety evaluation of chemicals in use. Subsequent research on chemical crop protection has been tremendous with invention of plethora of new and selective active ingredients. In addition to the potency of those chemicals, the focus was also on the behaviour of those chemicals in the environment, residues in applied crop/crop produce, and potential toxicity to mammals especially human beings.

In spite of development in other plant protection strategies, like biotechnology which can do away with the use of chemicals altogether, nanotechnology is currently under the spotlight with chemical molecule in the centre stage of development.

Application method/mode of delivery of chemicals have been changing in order to increase their efficiency and reduce overall chemical load in the environment.

Newer plant protection chemicals are required in order to protect the crop from a range of pests against which no chemical was developed earlier; to counter the resistance development in the pests to existing chemicals; to manage the pests which turned from secondary to major status due to indiscriminate use of existing chemicals and/or killing of natural enemies; to meet the stringent safety standards with regard to agri-ecosystem and human environment.

Screening of naturally occurring substances for pesticidal value or synthesis of pesticidal compounds and their screening have been greatly helped by development in other areas of science and technology. Many of the processes have been hastened due to application of robotics, automation and data-handling technologies. With small amount of test chemicals it is possible to screen a wide range of crop pests. Computer designing of chemical molecules with wide array of structural variations and their laboratory synthesis for a specific purpose have made drug discovery a lot easier.

New crop protection chemicals like herbicides, insecticides and fungicide which have been developed and commercialized preferably after 1990 have been presented in this chapter in tabular form under Tables 1 to 3.

2. FUNGICIDE GENERATIONS

***First Generation Fungicides*:** (Traditional Protectants–Broad spectrum) (after world war-II till 1966):

- Copper fungicides (CoC–copper oxychloride, Cuprous oxide, Copper hydroxide)
- Organomercurials (PMA–phenyl mercuric acetate, MEMC–methoxy ethyl mercuric chloride)

Table 1: Chemical name, mode of action, target insect pests, and some other properties of newer group of insecticides

Insecticide group	*Chemical name*	*Mode of action (as per IRAC)*	*Target insects*	*Some other properties*
Neonicotinoids	Imidacloprid	Agonist of nicotinic acetylcholine receptor (nAChR)	Whitefly, jassid, aphid, mealybug, thrips, termite, leaf miner, shootfly, sawfly, painted bug, psylla, BPH, WBPH, GLH etc.	Systemic insecticide. Polar molecule and transport well through xylem. Used for seed treatment also.
	Thiamethoxam		Aphid, jassid, whitefly, mealy bug, thrips, leaf miner, BPH, GLH, WBPH, shoot fly, termite, mosquito bug, psylla etc.	Systemic insecticide with stomach and contact action. Used for seed treatment also.
	Acetamiprid		Aphids, jassids, whiteflies etc.	Broad spectrum insecticide with translaminar activity.
	Clothianidin		BPH, jassids, whiteflies etc.	Broad spectrum systemic insecticide.
	Thiacloprid		Aphid, jassid, scale insect, thrips, BPH, mealy bug etc.	Systemic, broad spectrum insecticide.
	Dinotefuran		Leaf miner, beetle, weevil, fruit moth, green leaf hopper, aphid, white fly etc.	Low mammalian, aquatic and avian toxicity. Active against wide range of insects.
Spinosyns	Spinetoram	Allosteric activators of nAChRs	Tobacco caterpillar, army worm, Semilooper, leaf miner etc.	Broad spectrum with translaminar activity.
	Spinosad		DBM, borers, boll worms, aphid, jassid, whitefly etc.	Broad spectrum, contact and stomach poison with some contact action with translaminar activity.

Table 1: (*Contd...*)

Table 1: (*Contd...*)

Insecticide group	***Chemical name***	***Mode of action (as per IRAC)***	***Target insects***	***Some other properties***
Oxadiazines	Indoxacarb	Blockers of voltage-gated sodium channels	*Helicoverpa*, *Spodoptera*, *Plutella xylostella*, boll worms, cut worm, *Lygus* sp., *Empoasca* sp etc.	It is also a pro-insecticide and activated by esterases enzyme inside the insects guts
Phenylpyrazoles	Fipronil	GABA gated chloride channels antagonists	Stem borers, gall midge, leaf folder, GLH, BPH, WBPH, DBM, thrips, whitefly, scales, leaf miner etc.	Broad spectrum insecticide with contact and stomach action. Cause paralysis of the insects leading to death.
Avermectins	Abamectin	Activate chloride channels	Phytophagous mites, leafminers, thrips, psylla *Plutella xylostella* and many lepidopteran insects.	Contact and stomach poison with translaminar activity. Mineral oils or synergist enhances its residual activity (Mizell *et al*., 1986)
	Emamectin benzoate		Boll worms, fruit borer, leaf eating caterpillar, cutworm, DBM, green semilooper etc.	Contact and stomach poison. It is very active against beneficial insects such as honey bees and should not be sprayed during flowering (Fisher *et al*., 1993)
Milbemycins	Milbemectin	Activate chloride channels	Phytophagous mites and leaf miner.	It is an acaricide with contact and stomach action, with limited plant systemic activity (Kodandaram *et al*., 2010)
Pymetrozine	Pymetrozine	Selective homopteran feeding blockers.	Aphid, whitefly, mango hopper, BPH, WBPH etc.	Systemic and translaminar activities and is highly specific against sucking insect pests.

Table 1: (*Contd...*)

Table 1: (*Contd...*)

Insecticide group	*Chemical name*	*Mode of action (as per IRAC)*	*Target insects*	*Some other properties*
Flonicamid	Flonicamid	Selective homopteran feeding blockers.	Aphid, whitefly, plant bugs, plat hopper, leaf hopper, thrips and scale insect etc.	Excellent translaminar and systemic activities. Rapidly inhibit the feeding activity of aphids.
Diafenthiuron	Diafenthiuron	Inhibitors of mitochondrial ATP synthase	Whitefly, thrips, hopper, aphid, mite and DBM etc.	Contact, stomach, acaricidal, IGR and ovicidal activities.
Chlorfenapyr	Chlorfenapyr	Uncoupers of oxidative phosphorylation via disruption of proton gradient	Mite, thrips, DBM, fruit borer and other lepidopteran pests.	Broad spectrum insecticide with stomach and contact activities.
METI (Mitochondrial Electron Transport Inhibitor) acaricides and insecticides	Fenazaquin	Mitochondrial complex-I electron transport inhibitors, Energy metabolism (Good evidence that action at this protein complex is responsible for insecticidal effects).	Mites	Contact and ovicidal activities.
	Fenpyroximate	Same as Fenazaquin	Mites	Contact and ovicidal activities. Also serve as an IGR.
	Pyridaben	Same as Fenazaquin	Mite, whitefly, aphid, jassids and thrips etc.	Rapid knockdown and long residual activity.

Table 1: (*Contd...*)

Table 1: (*Contd...*)

Insecticide group	*Chemical name*	*Mode of action (as per IRAC)*	*Target insects*	*Some other properties*
Buprofezin	Buprofezin	Chitin biosynthesis inhibitor, type-I.	Whitefly, BPH, aphid, jassid, mango hopper, mealy bug, scale insect etc.	Contact and stomach insecticide. Also have ovicidal and Acaricidal activities.
Benzyl phenyl ureas (BPUs)	Diflubenzuron	Chitin biosynthesis inhibitor, type-0.	DBM, Boll worms, leaf folder, leaf eating caterpillar, worn worm, psylla etc.	IGR with stomach and contact activities. Also have ovicidal and Acaricidal properties. Compatible with entomopathogenic fungi like *Beauveria bassiana*.
	Novaluron		DBM, potato tuber moth, psylla, fruit borers, boll worms, tobacco caterpillar, whitefly, leaf miner etc.	IGR with stomach and ovicidal activities.
	Flufenoxuron		Red spider mite, yellow mite, DBM etc.	IGR with stomach and contact activities. Also have ovicidal and Acaricidal ptoperties.
	Lufenuron		Cabbage butterfly, DBM, borers etc.	IGR with stomach and contact activities.
Juvenile Hormone mimics	Hydroprene	Juvenile hormone analogues	Cockroach, beetle, moths etc.	Compatible with many synthetic insecticides like Cypermethrin, Acephate and deltamethrin.
	Methoprene		Mosquito larva	Compatible with Fipronil. Safe to domestic animals there by used as a food additive in cattle feed and preventing fly breeding in the dung piles.

Table 1: (*Contd...*)

Table 1: (*Contd...*)

Insecticide group	***Chemical name***	***Mode of action (as per IRAC)***	***Target insects***	***Some other properties***
	Kinoprene		Aphids, white flies, jassids, fungus gnats, and soft and armoured scales etc.	Safe to natural enemies and pollinators.
	Fenoxycarb		Scale insect, mosquito, cockroach, fire ant, flea etc.	Fenoxycarb is safe to birds but moderately to highly toxic to fish.
Diacylhydrazines	Methoxyfenozide	Ecdysone receptor agonists	Army worm, soybean looper etc.	Safe to pollinators, mite and other beneficial insects.
	Tebufenozide		Leaf eating caterpillars, bud worm etc.	Safe to predatory mites, wasps and other beneficial species.
	Halofenozide		White grub, turf caterpillar, webworm, cutworms, armyworm etc.	Stable to heat, light and water and safe to beneficial fauna
Triazine	Cryomazine	Moulting disruptor for dipteran larvae	Mostly effective against maggots	In veterinary it is used for killing ectoparasites.

Note: "Mode of action" of the insecticide group

Disclaimer: This chapter has been compiled on the basis of available information for the guidance and not for legal purposes.

Abbreviations: BPH: Brown Plant Hopper; WBPH: White Backed Plant Hopper; GLB: Green Leaf Hopper; DBM: Diamond Back Moth; IGR: Insect Growth Regulator

Table 2: Chemical name, nature, target pathogen and some other properties of newer fungicides

Sl. No.	*Chemical name*	*Nature*	*Target pathogen*	*Some other properties*
1.	Boscalid ($C_{18}H_{12}Cl_2N_2O$)	Broad spectrum fungicide for prophylactic use; inhibits spore germination, in some cases reduce mycelial growth and spore development; belongs to chemical group carboxamides	*Alternaria*, *Sclerotinia*, *Botrytis*, *Monilinia*, *Phoma*; Boscalid is the first succinate dehydrogenase inhibitor, introduced into the market in 2003, to control ascomycetes on various fruits and vegetables (Kramer and Schirmer, 2007)	Inhibit succinate–dehydrogenase (complex II) of mitochondrial ETS and disrupt ATP production; application rate 150–500 g a.i/ha for rice
2.	Dimoxystrobin ($C_{19}H_{22}N_2O_3$)	A strobilurin analogue, structurally similar to metominostrobin; an oximino amide compound (others are metominostrobin, orysastrobin)	*Microdochium*, *Fusarium*, *Septoria nodorum*, *S. tritici*, brown rust as a mixture with epoxiconazole	Long lasting (high metabolic stability), xylem systemic; typical biological target is *Fusarium*
3.	Fenoxanil ($C_{15}H_{18}Cl_2N_2O_2$)	Systemic fungicide; moderately water soluble	Rice blast pathogen	For foliar and 'into–water' application in rice
4.	Metrafenone ($C_{19}H_{21}BrO_5$)	Benzophenone fungicide and systemic in nature; first fungicide from benzophenone family; translaminar action; acropetal translocation; have preventive, curative and residual activity; similar to cyflufenamid distribution by vapour phase diffusion	Powdery mildew pathogens in cereals, vegetable, legume, apple, and vines; also against eyespot in wheat and barley	Disturbs the organization or polarization of the actin cytoskeleton (Opalski, 2005), inhibits mycelial growth and penetration, development of appressoria, formation of haustoria and sporulation reduced (Anonymous, 2004)
5.	Orysastrobin ($C_{18}H_{25}N_5O_5$) (Stierl *et al.*,	Novel strobilurin fungicide, systemic with protectant activity; oximino amide	Rice blast (*Pyricularia oryzae*) and sheath blight pathogens (*Rhizoctonia solani*)	High residual effect (metabolically stable)

Table 2: (*Contd...*)

Table 2: (*Contd...*)

Sl. No.	*Chemical name*	*Nature*	*Target pathogen*	*Some other properties*
	2004)	compound; highly water soluble; uptake by root and high xylem mobility		
6.	Pyraclostrobin ($C_{19}H_{18}Cl_1N_3O_4$)	Broad spectrum strobilurin fungicide; high resistance risk; lowest melting point of all the presently commercialized strobilurin; has no pronounced systemic or episystemic mobility; rapid leaf uptake and translaminar movement (Stierl *et al.*, 2002, Köhle *et al.*, 2002, Conrath *et al.*, 2002)	On crops like banana, cereals, citrus, field peas, grape, lentil, pea nut, potato, turf and vegetables	Induces resistance in tobacco plants against *Tobacco mosaic virus* (TMV) and *P. syringae* pv. *tabaci* (Herms *et al.*, 2002)
7.	Triticonazole ($C_{17}H_{20}Cl_1N_3O$) Mugnier *et al.*, 1994	A triazole fungicide, inhibits sterol C_{14} demethylase, has contact and systemic activity (Querou *et al.*, 1997, 1998)	*Septoria*, *Sphacelotheca reiliana*, *Tilletia*, *Ustilago nuda*, *U. tritici*, *U. kolleri*, *U. avenae*, *Erysiphe graminis*, *Puccinia*, *Rhynchosporium*, *Geaumannomyces graminis*; control cereal seed–borne and foliar diseases by seed treatment application	Controls seed-borne and soil-borne diseases, does not delay plant emergence as in case of other fungal SBIs (sterol biosynthesis inhibitors); application rate 150–600 g a.i/ 100 kg seed
8.	Fenamidone ($C_{17}H_{17}N_3OS$), Bartlett *et al.*, 2002	A QoI strobilurin type fungicide, structurally similar to famoxadone; belongs to azolone group; systemic in nature; it has protectant and curative activity; it has translaminar and antisporulant activity	Active against Oomycetes (*Bremia lactucae*, *Phytophthora infestans*, *Peronospora*, *Plasmopara viticola* and *Pythium*), some Ascomycetes (*Mycosphaerella*) and *Alternaria* spp.	It is respiratory inhibitor in mitochondria (complex III inhibitor); developed mainly as combination product to avoid resistance development

Table 2: (*Contd...*)

Table 2: (*Contd...*)

Sl. No.	*Chemical name*	*Nature*	*Target pathogen*	*Some other properties*
9.	Fenhexamid ($C_{14}H_{17}Cl_2NO_2$)	Hydroxyanilide fungicide, protectant in action, limited translocation ability, inhibits germ tube elongation, mycelial growth and appressorium production	Narrow spectrum of activity; very effective against *Botrytis cinerea* and related pathogens like *Monilinia* spp. and *Sclerotinia* spp. (Kuck *et al.*, 1997; Rosslenbroich *et al.*, 1998; Debieu *et al.*, 2001; Leroux *et al.*, 2002a; Rosslenbroich, 1999; Suty *et al.*, 1999; Leroux *et al.*, 2002b)	Exact mode of action is not known; designated as reduced risk pesticide by US EPA; target site is 3–Keto reductase inC_4 demethylation in sterol biosynthesis; application dose 375–1000 g a.i/ha
10.	Flumetover ($C_{19}H_{20}F_3NO_3$)	Preventive and curative activity, it is a mildewicide		Its development has been discontinued
11.	Fluoxastrobin ($C_{21}H_{16}ClFN_4O_5$)	Strobilurin fungicide for foliar application and seed treatment mainly for cereals; curative, leaf systemic (translaminar activity) and broad spectrum; good xylem mobility; not episystemic; first strobilurin molecule marketed for seed treatment	Provides broad spectrum and long-lasting protection from seed and soil-borne pathogens such as snow mould (*Monographella nivalis*) and bunt (*Tilletia caries*); complete control of *Septoria tritici*, *Leptosphaeria nodorum*, *Puccinia* spp., all fungal pathogens of barley	Complex III inhibitor; high metabolic stability (Heinemann *et al.*, 2004; Häuser *et al.*, 2004; Dutzmann *et al.*, 2004)
12.	Fluquinconazole ($C_{16}H_8Cl_2FN_5O$); (Russel *et al.*, 1992)	A quinazoline–based triazole fungicide	Active against ascomycete, basidiomycete and deuteromycete fungi such as *Alternaria*, *Cercospora*, *Colletotrichum*, *Erysiphe*, *Fusarium*, *Helminthosporium oryzae*, *Monilinia*, *Mycosphaerella*, *Phoma*, *Podosphaera*, *Puccinia*, *Rhizoctonia*, *Sclerotinia*, *Sclerotium*, *Septoria*, *Sphaerotheca*, *Tilletia*, *Uncinula*, *Venturia*	Acts by sterol C_{14} demethylase inhibition; also known as take–all fungicide as it was found effective against the pathogen of take–all disease of wheat when applied as seed treatment @ 75 g a.i/100 kg seed (Löchel *et al.*, 1998)

Table 2: (*Contd...*)

Table 2: (*Contd...*)

Sl. No.	*Chemical name*	*Nature*	*Target pathogen*	*Some other properties*
13.	Iprovalicarb ($C_{18}H_{28}N_2O_3$) (Liu *et al.*, 2000)	Carboxylic acid amide (CAA) fungicide belonging to sub-class valinamide carbamate; systemic; affects growth of germ tubes of zoospores, sporangia and mycelia; also affects sporulation; target site–phospholipids biosynthesis and cell wall deposition	Specific for oomycetes such as *Peronospora tabacina*, *Phytophthora* sp., *Plasmopara viticola*, *Pseudoperonospora cubensis*	Application rate 120–240 g a.i/ha; first registered in Indonesia in 1998
14.	Prothioconazole ($C_{14}H_{15}Cl_2N_3OS$)	It is neither an azole or triazole fungicide, it belongs to a new chemical class triazolinthiones; it is a DMI inhibitor and acts by disrupting sterol biosynthesis; systemic in nature; it has protective, curative and eradicant activity; broad spectrum	Important cereal diseases; *Rhizoctonia*, *Fusarium*, *Microdochium*, *Puccinia*, *Erysiphe* on wheat; *Pyrenophora teres* and *Rhynchosporium secalis* on barley; *Sclerotinia*, *Cylindrosporium*, *Botrytis cinerea*, *Alternaria*, *Mycosphaerella*, Fusarium ear blight	Application rate 200 g a.i/ha to control all wheat diseases (Mauler–Machnik *et al.*, 2002; Jautelat *et al.*, 2004; Dutzmann and Suty, 2004; Kuck and Mehl, 2004)
15.	Pyrimethanil ($C_{12}H_{13}N_3$)	Anilinopyrimidine (AP) fungicide; inhibit secretion of hydrolytic enzymes; prevents pathogenesis and fungistatic, systemic; has translaminar and vapour activity	Protectant activity against *Botrytis* and curative activity against *Venturia inaequalis*; also acts against *Ascochyta*, *Alternaria*, *Mycosphaerella* in banana	Target site–methionine biosynthesis, Rainfast when dry (Masner *et al.*, 1994; Leroux *et al.*, 1995, 1999; Miura *et al.*, 1994; Milling and Richardson, 1995)
16.	Spiroxamine ($C_{18}H_{35}NO_2$)	First representative of spiroketalamine compound within the amine fungicide group; systemic foliar fungicide with	Particularly effective against *Erysiphe graminis* on wheat and barley; *Puccinia hordei*, *P. recondita*, *P. striiformis*, *Pyrenophora teres*,	Blocks sterol biosynthesis by inhibition of delta14 reductase, penetrates the leaf and translocated

Table 2: (*Contd...*)

Table 2: *(Contd...)*

Sl. No.	*Chemical name*	*Nature*	*Target pathogen*	*Some other properties*
		protective, curative and eradicant activity; inhibits sterol biosynthesis; target site Δ^{14}–reductase and $\Delta^{8} \rightarrow \Delta^{7}$ isomerase in sterolbiosynthesis	*R. secalis*, *Uncinula necator* (Tiemann *et al.*, 1997; Dutzmann *et al.*, 1996; Krämer *et al.*, 1997; Dutzmann, 1997)	acropetally to the leaf tip; 500–800 g a.i/ha for cereals, 300–400 g a.i/ha for vines
17.	Trifloxystrobin ($C_{20}H_{19}F_3N_2O_4$) (Margot *et al.*, 1998)	Episystemic distribution; lipophylic; have the greatest intrinsic activity; typical biological target is powdery mildews; lack xylem system activity; rapid metabolic degradation	Active against a broad range of ascomycete, basidiomycete, deuteromycete and oomycete pathogens. Main targets are *Alternaria solani*, *Cercospora arachidicola*, *C. beticola*, *Cercosporidium personatum*, *Colletotrichum graminicola*, *Drechslera* sp., *Erysiphe graminis*, *Microsphaera*, *Phomopsis*, *Plasmopara viticola*, *Podosphaera leucotricha*, *Puccinia*, *Pyrenophora teres*, *Pyricularia grisea*, *Pythium* spp., *Rhizoctonia solani*, *Rhynchosporium secalis*, *Septoria nodorum*, *S. tritici*, *Sphaerotheca fuliginea*, *U. necator*, *V. inaequalis*.	Active through mitochondrial respiration inhibition; effective during early stages of pathogen infection cycle
18.	Quinoxyfen ($C_{15}H_8Cl_2FNO$)	Phenoxyquinoline fungicide; target site-G proteins in early cell signalling; protectant fungicide; inhibits appressorium formation;	Active against powdery mildew diseases; primary targets are *Erysiphe* spp. (*Blumeria*), *Leveillula taurica*, *Podosphaera*, *Sphaerotheca*, *Uncinula*	Application rate 150–250 g a.i/ha on cereals, 75–100 g a.i/ha on fruits; its persistence makes it a useful partner for rapid knockdown

Table 2: *(Contd...)*

Table 2: (*Contd...*)

Sl. No.	*Chemical name*	*Nature*	*Target pathogen*	*Some other properties*
		degraded more rapidly in light	(Longhurst *et al.*, 1996; Wheeler *et al.*, 2000, 2003; Arnold *et al.*, 1989; Bartlett *et al.*, 1997)	fungicides such as the azoles and morpholines
19.	Thifluzamide ($C_{13}H_6Br_2F_6N_2O_2S$)	It is a carboxanilide fungicide; inhibits the enzyme succinate dehydrogenase	Primary targets are *Hemileia vastatrix*, *Puccinia arachidis*, *Rhizoctonia* spp., *Sclerotium rolfsii* (Dowagroscience, 2004; Alt *et al.*, 1990)	Application rate 400–600 g a.i/ha on rice, 250–500 g a.i/ on peanuts
20.	Zoxamide ($C_{14}H_{16}Cl_3NO_2$)	It is a benzamide fungicide; cause inhibition of cell division and microtubule disruption; first anti-tubulin fungicide to be commercialized against oomycetes	Primary targets are *Phytophthora infestans*, *Plasmopara viticola*, *Pseudoperonospora cubensis* (Egan *et al.*, 1998; Young and Slawecki, 2001; USEPA, 2001; Young and Vjugina, 2002; Gobert *et al.*, 2000)	Inhibit –tubulin assembly; non-systemic but has penetrant activity; good residual activity and excellent rainfastness
21.	Famoxadone ($C_{22}H_{18}N_2O_4$)	It is a QoI group of fungicides; broad spectrum	Effective in controlling potato blight, late blight on tomato, downy mildew on vines (Barchietto and Genet, 2002; Joshi and Sternberg, 1996; Mercer *et al.*, 1998; Mercer and Latorse, 2003)	
22.	Proquinazid ($C_{14}H_{17}IN_2O_2$)	Belongs to quinazolinone group; inhibits spore germination and appressorium formation; may stimulate expression of host defense genes; preventive action; distribution of a.i. through vapour phase; locally systemic	Against powdery mildew of cereals and grapevine	Application rate 40–50 g a.i/ha (Hanhart, 2006; Selby *et al.*, 2004; Bereznak *et al.*, 1994; Michel, 2004; Agrow World Crop Protection News, 2005)

Table 2: (*Contd...*)

Table 2: (*Contd...*)

Sl. No.	*Chemical name*	*Nature*	*Target pathogen*	*Some other properties*
23.	Harpin (Messenger)	It is harpin protein obtained from *Erwinia amylovora*; acts as SAR inducer; 403 amino acid long, water soluble, heat stable; increase plant vigour	Against a wide variety of bacterial, viral and fungal pathogens; primary targets are *Alternaria citri*, *Botrytis cinerea*, *Diplocarpon rosae*, *Elsinoe fawcetti*, *Erwinia amylovora*, *Globodera solanacearum*, *Guignardia bidwellii*, *Meloidogyne* spp., *Mycosphaerella citri*, *Phytophthora citrophthora*, *Pseudomonas solanacearum*, *Pseudomonas syringae*, *Rhizoctonia solani*, *Sclerotium oryzae*, *Tobacco mosaic virus*, *Venturia inaequalis*, *Xanthomonas campestris*	Application rate 5–28 g a.i/ha
24.	Imibenconazole ($C_{17}H_{13}Cl_3N_4S$)	Triazole foliar fungicide; active through sterol C_{14} demethylase inhibition	Primary targets are *Alternaria mali*, *Cercospora arachidicola*, *Cladosporium carpophilum*, *Diplocarpon rosae*, *Elsinoe ampelina*, *E. fawcetti*, *Exobasidium vexans*, *Gymnosporangium* spp., *Podosphaera leucotricha*, *Puccinia* spp., *Tilletia caries*, *Uncinula necator*, *Venturia* spp. (Ohyama *et al.*, 1988; Ogawa, 1995)	
25.	Tetraconazole ($C_{13}H_{11}Cl_2F_4N_3O$)	Triazole fungicide; active through sterol C_{14} demethylase inhibition; systemic; has curative and protectant activity; has translaminar activity; generally	Primary targets are *Alternaria*, *Cercospora beticola*, *Erysiphe*, *Guignardia bidwellii*, *Gymnosporangium*, *Podosphaera leucotricha*, *Puccinia*, *Ramularia*	Application rate 125 g a.i/ha for cereals, 40–60 g a.i/ha on vegetables (Carzaniga *et al.*, 1991; Gozzo *et al.*, 1995; Huraux and Prove, 1992;

Table 2: (*Contd...*)

Table 2: (*Contd...*)

Sl. No.	*Chemical name*	*Nature*	*Target pathogen*	*Some other properties*
		used as foliar spray, can also be used for cereal seed treatment	*beticola*, *Rhynchosporium*, *Septoria*, *Sphaerotheca*, *U. necator*, *Uromyces*, *U. nuda*, *V. inaequalis*	Uchino and Watanabe, 1999; Khan and Smith, 2005)
26.	Cyazofamid ($C_{13}H_{13}ClN_4O_2S$)	Cyano–imidazoles compound; inhibits mitochondrial respiration at complex III; inhibits ubiquinone–reducing site of cytochrome bc1 rather than ubiquinone–oxidizing site of cytochrome bc1 which is inhibited by strobilurin fungicides	Against oomycete pathogens (Mitani *et al.*, 1998; Mitani *et al.*, 2001)	
27.	Benthiavalicarb ($C_{18}H_{24}FN_3O_3S$)	Systemic fungicides; structurally similar to iprovalicarb; has got curative and some preventive activity	Against oomycete pathogens; (Albert *et al.*, 1991; Jende *et al.*, 1999, 2002; Mehl and Buchenauer, 2002; Thomas *et al.*, 1992; Kuhn *et al.*, 1991)	Long lasting preventive activity; inhibit processes involved in cell wall biosynthesis and assembly
28.	Mepanipyrim ($C_{14}H_{13}N_3$)	Anilinopyrimidine fungicide; target site-methionine biosynthesis; broad spectrum	Particular activity against grey mold fungus *Botrytis cinerea* (Maeno *et al.*, 1990; EPA, 1987; Leroux *et al.*, 1999)	
29.	Metconazole	Triazole fungicide; active through sterol C_{14} demethylase inhibition	Primary targets are *Alternaria*, *Erysiphe*, *Fusarium*, *Guignardia bidwellii*, *Helminthosporium*, *Podosphaera*, *Puccinia*, *Pyrenophora*, *Rhynchosporium secalis*, *Sclerotinia sclerotiorum*, *Septoria*, *Tilletia caries*, *U. necator*, *Ustilago*, *V. inaequalis*	Effective against Fusarium head blight of wheat and barley (Sampson *et al.*, 1992; Laffranque *et al.*, 1994; Gilgenberg–Hartung, 1999; Pirgozliev *et al.*, 2002; Kang *et al.*, 2001)

Table 2: (*Contd...*)

Table 2: (*Contd...*)

Sl. No.	*Chemical name*	*Nature*	*Target pathogen*	*Some other properties*
30.	Ethaboxam ($C_{14}H_{16}N_4OS_2$)	Thiazole carboxamide fungicide; inhibits mitochondrial respiration; reported to act by disruption of microtubule (Uchida *et al.*, 2005)	Highly specific against oomycete pathogens	Application rate 250 g a.i/ha on vines and potato, 150–250 g a.i/ha on vegetables
31.	Penthiopyrad ($C_{16}H_{20}F_3N_3OS$)	Succinate dehydrogenase inhibiting anilide fungicide	Effective against *B. cinerea*, *Rhizoctonia* spp. (Tomlin, 2005; Tomiya *and* Yanase, 2003; Yoshikawa *et al.*, 1996)	
32.	Silthiofam ($C_{13}H_{21}NOSSi$)	Hindered silyl amide, Thiophene–carboxamide derivative; inhibits transport of ATP from mitochondrial matrix to cytosol	Seed treatment fungicide for control of take–all in cereals (*Gaeumannomyces graminis*); highly specific for the above fungus	Application rate 2 litres per tonne of seed (Joseph–Horne *et al.*, 2000; Anonymous, 2006)
33.	Tiadinil ($C_{11}H_{10}ClN_3OS$)	Belongs to thiadiazole carboxamide; target site is unknown; indirect action on fungi, bacteria and viruses; activator of SAR and inducer of defence gene expression	Against rice blast in nursery boxes; also have antibacterial and antiviral effect	Effective when applied to paddy water, less active when applied as a foliar spray (Yasuda *et al.*, 2004)
34.	Cyflufenamid ($C_{20}H_{17}F_5N_2O_2$)	Belongs to phenyl–acetamide group of chemical; a benzamidoxime fungicide; have preventive and curative activity; good residual activity, vapour phase activity and translaminar mobility; poor translocation within the host	Effective against powdery mildew pathogens such as *Erysiphe*, *Sphaerotheca* (Yokota, 2004; Kasahara and Kemikaru, 2005; Ma and Liu, 2005; Kasahara *et al.*, 2006; Agrow World Crop Protection News, 2005; Hanhart, 2006)	Inhibits the infection process by preventing haustorium formation, haustoria development, growth of secondary hypha, and conidia formation

Table 2: (*Contd...*)

Table 2: (*Contd...*)

Sl. No.	*Chemical name*	*Nature*	*Target pathogen*	*Some other properties*
35.	Simeconazole ($C_{14}H_{20}FN_3OSi$)	Triazole fungicide; broad spectrum; for seed treatment; prominent vapour phase activity; systemic; good translaminar movement (Tsuda *et al.*, 2000, 2004; Tsuda and Kato, 2003)	Primarily active against basidiomycete pathogens; primary targets are *Cochliobolus miyabeanus*, *Claviceps virens*, *Gymnosporangium yamadae*, *Monilinia mali*, *Thanatephorus cucumeris*, *Rhizoctonia cerealis*, *Rhizoctonia solani*, *U. nuda*, *V. inaequalis*	Application rate 4–10 g a.i/ 100 kg seed as seed treatment for *U. nuda*; first triazole to control rice sheath blight in submerged paddy condition and for control of *U. nuda* as seed treatment
36.	Diflumetorim ($C_{15}H_{16}ClF_2N_3O$)	Belongs to the class of the amino alkyl pyrimidines, it is an aminopyridazine fungicide; target site–complex I of mitochondrial respiration (Fujii and Takamura, 1998)	Effective against powdery mildew and rust in small–grain cereals and ornamentals	Good protective properties, some curative activity and instantly arrests fungal growth at any stage fromgermination of conidia to formation of conidiophore
37.	Oxpoconazole–fumarate ($C_{19}H_{24}ClN_3O_2 . ½C_4H_4O_4$)	Oxazolidine fungicide; an imidazole derivative	Controls broad spectrum diseases in fruits (including vines) and vegetables; *B. cinerea*; rice blast and sheath blight pathogens (Hayashi, 2003; Morita and Nishimura, 2001; Li *et al.*, 2002)	
38.	Flumorph or flumorlin ($C_{21}H_{22}FNO_4$)	Systemic fungicide; chemical structure closely resembles that of dimethomorph	Against oomycete pathogens (Liu *et al.*, 2000)	
39.	SYP–Z071 or Enestroburin ($C_{22}H_{22}ClNO_4$)	Strobilurin fungicide	Asco–, basidio–, deutero– and oomycete pathogens (Lixin *et al.*, 1998; Zhang *et al.*, 2003)	

Table 2: (*Contd...*)

Table 2: (*Contd...*)

Sl. No.	*Chemical name*	*Nature*	*Target pathogen*	*Some other properties*
40.	Metominostrobin ($C_{16}H_{16}N_2O_3$)	Strobilurin fungicide; root uptake; high xylem mobility; typical biological target–rice diseases; hydrophilic compound; high metabolic stability	*Botrytis cinerea*, *Cochliobolus miyabeanus*, *Erysiphe graminis*, *Gibberella fujikuroi*, *Pseudocercosporella herpotrichoides*, *Pseudoperonospora cubensis*, *Puccinia*, *Pyricularia oryzae*, *Rhizoctonia solani*, *Septoria*, *Spherotheca fuliginea*	Inhibits mitochondrial respiration; 150–180 g a.i/ha on rice, 150–500 g a.i/ha on vegetables, 210–840 g a.i/ha on small grain cereals (Kramer and Schirmer, 2007)
41.	Diclocymet ($C_{15}H_{18}Cl_2N_2O$)	Butyramide fungicide; systemic	Rice blast pathogen (Manabe *et al.*, 2002; Buck and Radatz, 1998; Kramer and Schirmer, 2007)	
42.	Acibenzolar–S-methyl ($C_8H_6N_2OS_2$)	Belongs to benzothiadiazole group; a plant activator, stimulates the natural defence mechanisms in plants; acts as functional analogue of salicylic acid, induces SAR; no direct effect on plant	Oidium on wheat, downy mildew on tobacco, bacterial speck and bacterial spot of tomato, powdery mildew and white rust in spinach and lettuce; also recommended for mango, cereals, banana diseases pathogens	Bion is approved in Brazil for the control of *Xylella fastidiosa* in citrus, *Crinipellis perniciosa* in cocoa and *Alternaria solani*, *P. infestans*, *X. vesicatoria* in tomato (Kramer and Schirmer, 2007)
43.	Cyprodinil ($C_{14}H_{15}N_3$)	Belongs to same chemical group as pyrimethanil and mepanipyrim; for foliar and seed treatment; broad spectrum	Cereals diseases (*Erysiphe* and *Tapesia* sp.), fruit diseases (*Alternaria* and *B. cinerea*) and vegetable diseases (*Botrytis* sp.) (Heye *et al.* 1994; Leroux *et al.*, 1999; Sierotzki *et al.*, 2002; Hilber and Bodmer, 1998)	
44.	Picoxystrobin ($C_{18}H_{16}F_3NO_4$)	Active through mitochondrial respiration inhibition; unique amongst the strobilurins having both xylem systemicity and vapour activity; episystemic distribution	*Puccinia coronata*, *P. hordei*, *P. recondita*, *Rhynchosporium secalis*, *Septoria nodorum*, and *S. tritici*; typical biological target–powdery mildews	More curative than azoxystrobin (Godwin *et al.*, 2000)

Table 3: Common name, nature, target weeds, and some other properties of newer herbicides

Sl. No.	*Common name*	*Nature*	*Target weeds*	*Some other properties*
Cereal sulfonylurea herbicides				
1.	Flupyrsulfuron-methyl-sodium (Andrea and Liang, 1992; Teaney *et al.*, 1995)	Post–emergent cereal herbicide	Grass weeds and select broad leaf weeds; for the control of problem grass weeds, such as *Alopecurus myosuroides* and *Apera spica–venti*, and a wide range of broadleaf weeds *Chenopodium album*, *Lamium purpureum*, *Matricaria* sp., *Polygonum aviculare*, *P. convolvulus*, *Senecio vulgaris* and *Sinapis arvensis*	Application rate 8–10 g a.i/ha
2.	Sulfosulfuron (Ishida *et al.*, 1992; Parrish *et al.*, 1995; Gibson and Kerchove, 1999)	Post–emergent herbicide	Grass (especially *Bromus* species) and broad leaf weeds; *Elymus repens, Apera spica–venti, Agrostis stolonifera*, *Avena fatua* (North America), *Bromus commutatus*, *B. japonicus*, *B. mollis*, *B. rigidus*, *B. secalinus*, *B. sterilis*, *B. tectorum*, *Poa bulbosa* and *Poa trivialis*, *Ambrosia artemisiifolia*, *Amsinckia lycopsoides*, *Atriplex patula*, *Brassica nigra*, *Capsella bursa–pastoris*, *Claytonia per*, *Descurainia pinnata*, *D. sophia*, *Fumaria officinalis*, *Galium aparine*, *Helianthus* sp., *Matricaria chamomilla*, M. *inodora*, *Polygonum*	Application rate 10–35 g a.i./ha; barley and oats are sensitive to this herbicide

Table 3: (*Contd...*)

Table 3: (*Contd...*)

Sl. No.	*Chemical name*	*Nature*	*Target weeds*	*Some other properties*
			aviculare, P. persicaria, Sinapis arvensis, Sisymbrium altissimum, Stellaria media, Thlaspi arvense and *Viola arvensis*	
3.	Iodosulfuron–methyl–sodium (Ort *et al.*, 1992; Hacker *et al.*, 1999; Hacker *et al.*, 2000)	First safened sulfonylurea in the market; for cereals and maize	Broadleaf and grass weeds; *Galium aparine, Matricaria chamomilla, Stellaria media, Raphanus* ssp., *Cirsium arvense, Lamium* ssp.; Grass weeds–*Agrostis gigantea, Apera spica-venti, Lolium multiflorum, L. perenne, L. persicum, L. rigidum, Phalaris brachystachys, P. canariensis, P. paradoxa, Poa annua,* and *P. trivialis*	Application rate 5–10 g a.i./ha + safener Mefenpyr
4.	Mesosulfuron–methyl (Lorenz *et al.*, 1995; Hacker *et al.*, 2001; Atlantis, 2005)	Second safened sulfonylurea herbicide to be commercialized for cereal crops; broad–spectrum, post–emergent	Grass weeds and select broad leaf weeds; *Agrostis* spp., *Alopecurus myosuroides, Apera spica–venti, Avena* spp. *Lolium* spp., *Phalaris brachystachis, P. minor, P. paradoxa, Poa annua, Poa trivialis, Pucciniella* spp. and *Sclerochloa kengiana*	Application rate 6–15 g a.i./ha + safener Mefenpyr; mefenpyr–diethyl, as with iodosulfuron–methyl–sodium, selectively accelerates the degradation of the active ingredient to non-phytotoxic compounds in and cereals but not in weeds.
5.	Tritosulfuron (Mayer *et al.*, 1992; Dombo *et al.*, 2002;	Broad spectrum, post-emergent dicot herbicide for use in cereals, rice, maize and turf	Broadleaf weeds; *Thlaspi arvense, Mercurialis annua, Urtica urens, Cirsium arvense, Veronicahederifolia, Chenopodium* spp., *Sinapis arvensis,*	

Table 3: (*Contd...*)

Table 3: (*Contd...*)

Sl. No.	*Chemical name*	*Nature*	*Target weeds*	*Some other properties*
	Schönhammer *et al.*, 2002; Dressel and Beigel, 2001)		*Capsella bursa–pastoris, Galeopsis tetrahit, Matricaria* spp., *Galium aparine, Polygonum* spp., *Centaurea cyanus, Lamium*spp., *Myosotis arvensis, Stellaria media, Vicia* spp., *Convolvulus arvensis, Sonchus arvensis, Brassica napus*	Application rate 30–50 g a.i./ha; acts mainly through the treated leaves but not via soil; short soil half-life
Rice sulfonylurea herbicides				
6.	Ethoxysulfuron (Kehne *et al.*, 1989; Hacker *et al.*, 1995; Hess and Rose,	Fully selective in all types of seeded and transplanted rice	Annual and perennial broadleaf and sedge weeds; *Cyperus* spp., *Aeschynomene* spp., *Eleocharis* spp., *Sagittaria* spp., *Scirpus* spp., *Amannia* spp., *Lindernia* spp., 1995) *Ludwigia* spp. and *Monochoria vaginalis*	Application rate 6–60 g a.i./ha; differential metabolism in the target crop
7.	Azimsulfuron (Levitt, 1988; Marquez *et al.*, 1995)		Annual and perennial broadleaf and sedge weeds including hard-to-control perennials; *Echinochloa crus–galli, E. hispidula, E. oryzicola* and *E. oryzoides*. Other weeds controlled include *Alisma lanceolatum, A. plantago–aquatica, Butomus umbellatus, Cyperus difformis, Scirpus maritimus, S. mucronatus, S. supinus, Heteranthera limosa, Potamogeton nodosus, Ammannia coccinea, A. robusta, Bergia capensis* and *Lindernia dubia*	Application rate 6–25 g a.i./ha

Table 3: (*Contd...*)

Table 3: (*Contd...*)

Sl. No.	*Chemical name*	*Nature*	*Target weeds*	*Some other properties*
8.	Cyclosulfamuron (Condon *et al.*, 1993; Rodaway *et al.*, 1993)		Annual and perennial broadleaf and sedge weeds; *Cyperus serotinus, C. difformis, Elatine triandra, Eleocharis congesta, E. kuroguwai, Lindernia annua, L. procumbens, Monochoria vaginalis, Rotala indica, Sagittaria pygmaea, S. trifolia* and *Scirpus juncoides;* broad leaf weeds such as *Veronica persica, V. hederifolia, Galium aparine, Matricaria* spp. and *Polygonum convolvulus*	Application rate 10–60 g a.i./ ha; selectivity in the rice paddy is achieved due to various factors, including rapid metabolic degradation of the herbicide in rice shoots, placement of rice seedlings during transplanting and the compound's soil bindingproperties, which retain cyclosulfamuron in the upper soil layer of the paddy
9.	Flucetosulfuron (Kim *et al.*, 2003a, 2003b; Tanaka *et al.*, 2003)		Annual and perennial broadleaf and sedge weeds; excellent control of *Echinochloa crus–galli*; *Alisma* spp., *Ammannia coccinea, Cyperus difformis, Fimbristylis* spp., *Lindernia* spp., *Monochoria vaginalis, Rorippa silvestri, Rotala indica, Scirpus juncoides, S. mucronatus* and *S. maritimus*; at a higher rate of 20–30 g a.i./ha, greater than 90% control of *Aeschymene indica, Butomus umbellatus, Eleocharis kuroguwai, Sagittaria pygmaea, S. trifolia* and *Sparganium erectum* is achieved by flucetosulfuron with a high crop	Application rate 15–60 g a.i./ha

Table 3: (*Contd...*)

Table 3: (*Contd...*)

Sl. No.	*Chemical name*	*Nature*	*Target weeds*	*Some other properties*
			safety margin when applied to soil or foliage in direct–seeded or transplanted rice.	
Maize sulfonylurea herbicide				
10.	Forumsulfuron (Collins *et al*., 2001; Bunting *et al*., 2004, 2005; Meyer, 1992)	Post–emergence sulfonylurea herbicide	Grass and broadleaf weeds; *Echinochloa crus–galli, Setaria* spp., *Agropyron repens, Apera spica–venti, Alopecurus myosuroides, Lolium multiflorum, Panicum dichotomiflorum, Poa annua* and *Sorghum halepense*, and a wide selection of broadleaf weed, species, such as *Abutilon theophrasti, Amaranthus* spp., *Galinsoga parviflora, Lamium purpureum, Solanum nigrum* and *Stellaria media*	Application rate 30–45 g a.i./ha + safener isoxadifen ethyl; the basis of selectivity of forumsulfuron in the presence of the safenerisoxadifen–ethyl is a more rapid rate of metabolic detoxification in maize comparedwith target weeds, in which little or no degradation of the parent sulfonylurea occurs
Sulfonylurea herbicides for other crops				
11.	Oxasulfuron (Brooks *et al*., 1995; Foery, 1997)	Pre- and post-emergence herbicide for soybean crop	Broad leaf weeds; *Abutilon theophrasti, Xanthium strumarium, Amaranthus* spp., *Ambrosia artemisiifolia, A. trifida, Bidens pilosa, Cyperusesculentus, Polygonum pensylvanicum, Sorghum bicolor, Echinochloa crus–galli, Helianthus annuus, Sesbania exaltata* and *Ipomoea* spp.	Application rate 45–90 g a.i./ha; the observed selectivity is due to rapid metabolization in the target crop

Table 3: (*Contd...*)

Table 3: (*Contd...*)

Sl. No.	*Chemical name*	*Nature*	*Target weeds*	*Some other properties*
12.	Trifloxysulfuron sodium (Howard *et al*., 2001; Brown *et al*., 1991)	Post–emergence herbicide for sugarcane, cotton, turf	Sedges and broadleaf weeds; at the lower rates, the following weeds are controlled: *Acanthospermum hispidum*, *Ambrosia artemisiifolia*, *Bidens pilosa*, *Senna obtusifolia*, *Cassia occidentalis*, *Chenopodium album*, *Euphorbia heterophylla*, *Ipomoea* spp., *Melochia corchorifolia*, *Mollugo vertillata*, *Sesbania exalta*, *Trianthema portulacastrum*, *Xanthium strumarium*. With post-directed sprays and higher dosages, additional control is achieved of *Ageratum conyzoides*, *Amaranthus hybridus*, *A. palmeri*, *Cyperus esculentus* and *Tridax procumbens*.	Application rate 5–23 g a.i./ha
Commercial and developmental triazolopyrimidine sulfonamides				
1.	Flumetsulam	Systemic	For use in maize and soybean	Application rate 25–80 g a.i/ha for soybean and maize (Snel *et al*., 1993)
2.	Metosulam		For maize and cereals	Application rate 5–30 g a.i/ha for maize (Snel *et al*., 1993)
3.	Diclosulam, (Sheppard *et al*., 1997; Arnold *et al*., 1998; Bailey *et al*., 1999a, b; Shaw *et al*., 1999)	Diclosulam is rapidly metabolized in soybeans	Diclosulam does not provide control of annual and perennial grass weeds or certain broadleaf weeds such as *Solanum* spp; for use in soybean and peanut	Applications can be made pre-plant surface, pre-plant incorporated and pre-emergence at rates of 17.5–26 g a.i. /ha for the control of numerous broadleaf weed species.

Table 3: (*Contd...*)

Table 3: (*Contd...*)

Sl. No.	*Chemical name*	*Nature*	*Target weeds*	*Some other properties*
4.	Cloransulam–methyl (Hunter *et al.*, 1994; Jachetta *et al.*, 1994; Jachetta *et al.*, 1995; Hunter *et al.*, 1995; Braxton *et al.*, 1996; Choate *et al.*, 1996; Stabler *et al.*, 1996; Nelson and Renner, 1998)	Cloransulam–methyl is rapidly metabolized in soybeans	Annual broadleaf weed and certain perennial sedges in soybeans; Cloransulam–methyl does not provide control of annual and perennial grass weeds or certain broadleaf weeds such as *Solanum* spp.	Applications can be made pre-plant surface, pre-plant incorporated, pre-emergence and post-emergence for the control of broadleaf weed species. Post-emergence applications of cloransulam–methyl at 17.5 g a.i./ha or soil-applied treatments at rates of 35–44 g a.i./ha provide control of a large number of soybean relevant weeds.
5.	Florasulam (Thompson *et al.*, 1999; Lepiece *et al.*, 2000; Daniau and Prove, 2001; Jackson *et al.*, 2000; Owen and deBoer, 1994)	Specialized spectrum of broadleaf weed control; in wheat, florasulam has been shown to undergo rapid metabolism; in contrast, slow metabolism is observed in *Galeopsis tetrahit* L. and *Polygonum papathifolium* with little degradation of florasulam observed in *Galium aparine* L. Differences in the rate of metabolism in wheat compared with broadleaf weeds accounted for the sensitivity of broadleaf weeds	Excellent post–emergence selectivity in turf and smallgrain cereal crops such as wheat, barley, oats, rye and triticale	Florasulam is highly active on economically important species in the Compositae, Caryophyllaceae, Cruciferae, Rubiacea and Leguminosae plant families at a typical use rate of 5 g a.i./ ha; relatively short half-life in soil; only post-emergence applications are used in commercial practice

Table 3: (*Contd...*)

Table 3: (*Contd...*)

Sl. No.	*Chemical name*	*Nature*	*Target weeds*	*Some other properties*
6.	Penoxsulam (Larelle *et al.*, 2003; Mann *et al.*, 2003a, b; 2005a, b; Min and Mann, 2004a, b; Shiraishi, 2005; Wang *et al.*, 2004)		For broadleaf, grass and sedge weed control in rice	Application rate 25–40 g a.i./ha; average half-life under field condition is 6.5 days under flooded water-seeded rice conditions
Pyrimidinyl carboxylates				
1.	Pyrithiobac–sodium	Pre- and post-emergent herbicide	One of the best cotton herbicides against broadleaf weeds such as *Abutilon theophrasti* and *Ipomoea lacunose*, and others like *Xanthium strumarium*, *Sida spinosa*, *Sesbania exaltata*, and *Sorghum halepense*	Application rate 35–105 g a.i./ha (Nezu *et al.*, 1998; Saitoh *et al.*, 1996; Takahashi *et al.*, 1991; Saito *et al.*, 1990)
2.	Bispyribac–sodium (Wada *et al.*, 1996, 1990; Watanable *et al.*, 2003; Yokoyama *et al.*, 1994, 1993; Tachikawa *et al.*, 1997)	Post-emergent herbicide	For controlling a wide range of weeds in direct seeded Indica type rice and for vegetative growth reduction; *Echinochloa* spp., *Brachiaria* spp., *Cyperus* spp., *Scirpus* spp., *Polygonum* spp., *Sagittaria* spp., *Commelina* spp. and *Sesbania exaltata*	Application rate 15–45 g a.i./ha
3.	Pyriminobac–methyl	Selective herbicide for rice crop; pre- to late post-emergent herbicide with excellent crop safety	Outstanding efficacy on *Echinochloa* spp. in rice (Tamaru *et al.*, 1991, 1997a, b, 1999; Hanai *et al.*, 1993; Kawamura *et al.*, 2000; Isozumi *et al.*, 2001)	

Table 3: (*Contd...*)

Table 3: (*Contd...*)

Sl. No.	*Chemical name*	*Nature*	*Target weeds*	*Some other properties*
4.	Pyribenzoxim (Cho *et al*., 1997; Koo *et al*., 1997)	Rice herbicide; post emergent activity	Grass and broadleaf weeds, including *E. crus–galli*, *Alopecurus myosuroides* and *Polygonum hydropiper*	Application rate 30–40 g a.i/ha
5.	Pyriftalid (Luthy *et al*., 2001)	Acetolactate synthase (ALS) inhibiting herbicide; rice herbicide	Grasses, especially *Echnochlora* spp.	Application rate 100–300 g a.i./ha
Sulfonylaminocarbonyl–triazolinones (Müller, 2002)				
1.	Flucarbazone–sodium (Santel *et al*., 1999a, b; Brenchley *et al*., 1999; Kirkland *et al*., 2001)	Inhibitors of the acetolactate synthase (ALS) enzyme, also known as aceto hydroxy acid synthase (AHAS)	Wild oats (*Avena fatua*) and green foxtail (*Setaria viridis*) in spring wheat (*Triticum aestivum*) and durum wheat (*Triticum durum*)	15–20 g a.i/ha in Canada, 21–42 g a.i/ha in USA
2.	Propoxycarba-zone–sodium (Feucht *et al*., 1999; Amann and Wellmann, 2001; Amann, 2002; Godley and Bubb, 2001; Wellmann and Feucht, 2002)	Inhibitors of the acetolactate synthase (ALS) enzyme, also known as aceto hydroxy acid synthase (AHAS)	Brome (*Bromus* spp.), loose silky-bent (*Apera spica–venti*), common couch-grass (*Elymus repens*), blackgrass (*Alopecurus myosuroides*), jointed goatgrass (*Aegilops cylindrica*) and several broadleaf weeds from the mustard family in wheat, rye and triticale crops	Application rate 28–70 g a.i/ha
Phytoene desaturate (PDS) inhibitors (Bleaching herbicides)				
1.	Flurtamone (Ward, 1984)	Pre-plant, pre-emergence and post-emergence	For crops like cereals, peanut, pea, cotton, sunflower	Application rate 250–375 g a.i./ha

Table 3: (*Contd...*)

Table 3: (*Contd...*)

Sl. No.	*Chemical name*	*Nature*	*Target weeds*	*Some other properties*
2.	Picolinafen (Foster *et al.*, 1991)	Post–emergence	For crops like cereals, lupines	Application rate 50–100 g a.i./ha
3.	Beflubutamid (Takematsu *et al.*, 1987)	Pre- and early post-emergence; the main metabolite is butanoic acid which is rapidly degraded in soil	For broadleaf weed control in crops like wheat and barley	Application rate 170–255 g a.i./ha
Hydroxyphenylpyruvate dioxygenase (HPPD) inhibitors: Triketones; all known HPPD inhibitors are chelating agents				
1.	Sulcotrione (Beraud *et al.*, 1993; McDougal, 2005)	First triketone herbicide to be commercialized; post–emergence control; readily absorbedthrough the leaves and also via the roots	Broadleaf weeds and some grass weeds (*Digitaria sanguinalis*, *Echinochloa crus–galli*, and *Panicum miliaceum*) in corn	Application rate 300–450 g a.i./ha for corn
2.	Mesotrione (Mitchell *et al.*, 2001)	Second generation triketone product; pre-emergence and post-emergence control	The important broad–leaved weeds in corn together with some of the annual grass weeds (*Digitaria* and *Echinochloa* sp.), which are important in European corn production; broadleaved weeds controlled include *Xanthium strumarium*, *Abutilon theophrasti*, *Ambrosia trifida*, together with *Chenopodium*, *Amaranthus* and *Polygonum* sp.	Selectivity in corn is given by its ability to rapidly metabolize (detoxify) mesotrione into the inactive metabolites; typical use rates range from 100 to 225 g a.i./ha when applied pre-emergence, and 70 to 150 g a.i./ha for post-emergence applications
3.	Benzobicyclon (Komatsubara, 2005; Sekino *et al.*, 2004)	Very selective in rice and environmentally friendly due to its low water solubility	Broadleaf weeds (especially sulfonyl urea resistant weeds, *e.g.*, *Lind* sp., *Lindernia attenuata*, *Monochoria vaginalis*) and some important	

Table 3: (*Contd...*)

Table 3: (*Contd...*)

Sl. No.	*Chemical name*	*Nature*	*Target weeds*	*Some other properties*
			grasses [*e.g.*, *Scirpus juncoides* (sulfonyl urea resistant), *Echinochloa oryzicola* and other *Echinochloa* sp.] in paddy rice	
Hydroxyphenylpyruvate Dioxygenase (HPPD) Inhibitors: Heterocycles				
1.	Isoxaflutole (Luscombe *et al.*, 1995; Cain *et al.*, 1991)	First HPPD inhibitor in corn for pre-emergence application	For broadleaf weed and grass control in corn and sugar cane	Application rate of 75 g a.i./ha is very low compared with other conventional pre-emergence herbicides for corn (*e.g.*, S-metolachlor 0.8–1.6 kg /ha).
2.	Topramezone (von Deyn *et al.*, 1996; Schönhammer *et al.*, 2005)	Post–emergence; for corn	Major grass and broadleaf weeds in corn crops worldwide	Application rate 50–75 g a.i./ha for Clio
3.	Pyrasulfotole (Schmitt *et al.*, 2001)	For cereals		
Tetrazolinones				
1.	Carbamoyl tetrazolinones/ fentrazamide	Pre- or post-emergence; mainly in transplanted rice;	Grass control in rice; high activity to barnyard grass; excellent safety to rice seedlings; low mobility in soil	Application rate 200–300 g a.i./ha

Table 3: (*Contd...*)

Table 3: (*Contd...*)

Sl. No.	*Chemical name*	*Nature*	*Target weeds*	*Some other properties*
	(Covey *et al*., 1985; Yanagi *et al*., 2002; Goto *et al*., 2002; Yanagi, 2001)			
Acetyl–CoA carboxylase inhibitors				
1.	Pinoxaden (Hofer *et al*., 2006)	Post-emergence	Graminicide; applied flexibly from the two-leaf up to the flag leaf stage of grasses; *Alopecurus myosuroides* (blackgrass), *Apera spica venti* (silky bent grass), *Avena* spp. (wild oats), *Lolium* spp. (ryegrass), *Phalaris* spp. (canary grass), *Setaria* spp. (foxtails) and other monocot weed species commonly found in cereals	30–60 g a.i/ha
2.	Amicarbazone (Richards, 2005)	For corn and sugarcane; inhibitor of photosystem II; belongs to the class carbamoyl triazolinones	Major annual dicot weeds affecting the two crops; velvetleaf (*Abutilon theophrasti*), common Lambsquarters (*Chenopodium album*), pigweed (*Amaranthus* spp.), common cocklebur (*Xanthium strumarium*) and morning–glory species (*Ipomoea* spp.)	500 g a.i./ha maximum rate in corn when applied singly

- Dithiocarbamates (Maneb, Mancozeb, Zineb, Thiram)
- Phthalimides (Captan, Captafol, Folpet)
- Tin compounds (TPTC–triphenyl tin chloride, TPTA–triphenyl tin acetate, TPTH–triphenyl tin hydroxide)
- Chloronitrobenzenes (PCNB–pChloronitrobenzene)
- Crotonates (Dinocap)
- Chlorophenyls (Chlorothalonil)
- Guanidines (Dodine)

Second Generation Fungicides: (Systemic, site-specific) 1966–1976

- Oxathiins (Carboxin, Oxycarboxin): Smuts, rusts
- Benzimidazoles/Thiophanates (Benomyl, Carbendazim, TBZ-2-(4′-thiozolyl) benzimidazole): Powdery mildews, Scab, *Botrytris*, *Cercospora*, *Penicillium*, *Fusarium*, *Colletotrichum* etc.
- Organophosphorus (Iprobenphos, Edifenphos)**:** Rice diseases
- Hydroxypyrimidines (Ethirimol, Dimethirimol, Bupirimate): Powdery mildews
- Dicarboximides (Iprodione, Vinclozolin, Procymidone): *Botrytis*, *Sclerotinia*, *Monilinia*

Third Generation Fungicides: (Systemic, specific) 1977–1990

- Phenylamides (Metalaxyl, Ofurace, Oxadixyl): Oomycetes
- Isoxazoles (Hymexazole): *Pythium*
- Carbamates (Prthiocarb, Propamocarb): Oomycetes
- Cyanoacetamideoximes (Cymoxanil): Oomycetes
- Alkylphosphonates (Fosetyl–Al): Oomycetes
- Cinnamic acid deriv. (Dimethomorph): Oomycetes
- Sterol biosysntheis inhibitors (Triazoles, Imidazoles, Piperazines, Piperidines, Pyrimidines, Morpholines): Ascomycetes, Basidiomycetes, Deuteromycetes
- Phenylcarbamates (Diethofencarb, MDPC–methyl N-(3, 5-dichlorophenyl) carbamate: (*–ve cross resistance to benzimidazoles*) *Botrytis*, *Cercospora*, *Fusarium*, *Venturia*

New Generation Fungicides (Novel modes of action, 1991–2006)

- Strobilurins
- Phenylpyrroles

- Melanin biosynthesis inhibitors (MBIs)
- Spiroketalamines
- Phenoxyquinolines
- Anilinopyrimidines
- Oxazolidinediones
- Valinamides
- Benzamides
- SAR activators

2.1. Strobilurins

***Strobilurins*:** Strobilurin group (or QoI fungicides) contains the most widely used and most important fungicides. They are one of the newest fungicide chemical compounds in the market released in 1996. Commercially produced strobilurin fungicides are derivative of the compound (strobilurin-A) first isolated from a wood–rotting mushroom *Strobilurus tenacellus*. Thus by origin strobilurin-A is an antibiotic. However, present day strobilurins are all synthesized products that are not produced in the fungal organism. Therefore, they can be treated as fungicide. These are also regarded as ecofriendly fungicides because these are effective at low rate, have low mammalian toxicity and low risk to non-targets. They are broad spectrum and effective against most fungal pathogens of all types in most of the crops. They have preventive, curative and anti-sporulant activity. However in some cases phytotoxicity of strobilurin compounds are reported against some genotypes of crops. Fungi developing resistance to strobilurins are reported because of their single-site of action.

Fungus class	***Fungicide group***		
	Strobilurin	***Melanin biosynthesis inhibitor (MBI)***	***Benzimidazole***
Ascomycetes	+	+	+
Basidiomycetes	+	+	+
Deuteromycetes	+	+	+
Oomycetes	+	–	–

'+' effective, '–' not effective

2.2. Mode of Action

Strobilurins have single and specific site of action. They interfere with the energy (ATP) production of fungi by specifically blocking the electron transfer between cytochrome b and c1 at quinol oxidation (Qo) site in complex III of electron transport chain present in mitochondria. Therefore strobilurins are also called as Qo inhibitors (QoI). Strobilurins have systemic and trans–laminar action. Some strobilurins are reported to delay leaf senescence and help plant in water–conservation thus promoting plant growth.

***Newer Insecticides Molecules*:** Search for the new molecules with green chemistry is the need of the hour for crop protection. Insecticides with novel mode of action that maximizes safety and minimizes the environmental impact on non-target organisms would be more appropriate in IPM schedule. Newer insecticidal molecules with target specificity, lower doses and obvious safety to human health and natural enemies make a revolutionary approach in the history of the plant protection. Newer molecules like neonicotinoids, phenylpyrazoles, oxadiazines, pymetrozine, spinosyns and Insect Growth regulators (IGRs) are some of them in the list. Their details like mode of action, target insects, and other toxicological properties are summerized in Table 1.

CONCLUSIONS

Progress in the development of crop protection chemicals has been tremendous since the time of its first use and modern agriculture is synonymous with chemical agriculture. New chemical molecules for agricultural use are being developed taking into consideration the key issues of resistance development in target organism and environmental concerns. Chemicals have found application in all agricultural practices right from sowing to harvesting of crop. Such dependence in chemicals helped to achieve the present level of production of food, feed, fibre and fuel. If environmental and health concerns are adequately addressed, chemical methods would remain very handy and affordable without compromising on the agricultural sustainability.

REFERENCES

Agrow World Crop Protection News (www.agrow.co.uk), Article ID A00905436 (December 15, 2005).

Agrow World Crop Protection News (www.agrow.co.uk), Article ID A00884184 (June 2, 2005), A00901062 (November 3, 2005).

Albert, G., Thomas, A. and Gühne, M. (1991). 3rd International Conference Plant Diseases, ANPP, Paris, pp. 887–94.

Alt, G.H., Pratt, J.K., Phillips, W.G. and Srouji, G.H. (1990). Eur. Pat. Appl. EP 371950, 27 *Chem. Abstr.* **113:** 191337.

Amann, A. (2002). *Pflanz.–Nachrichten*, **55(1):** 87–100.

Amann, A. and Wellmann, A. (2001). *Proc. Brighton Conference–Weeds,* **2:** 469–74.

Andrea, T.A. and Liang, P.H.T. (1992). Preparation of Me 2-[[[[(4, 6-dimethoxy-2-primidinyl)amino]carbonyl]amino]-sulfonyl]-6-trifluoromethyl-3-pyridine-carboxylate and salts as herbicides. *Eur. Pat. Appl.* EP 502740 A1.

Anonymous. (2004). Agrow World Crop Protection News (www.agrow.co.uk), Article ID A00831684.

Anonymous. (2006). European commission, health and consumer protection directorate–general, standing committee on the food chain and animal health. 2006, Review report for the active substance silthiofam. http://europa.eu.int/comm/food/plant/protection/evaluation/newactive/ list1_silthiofam_en.pdf.

Arnold, J.C., Shaw, D.R. and Bennett. (1998). *Proc. S. Weed Sci. Soc.*, **51:** 2.

Arnold, W.R., Coghlan, M.J., Krumkalns, E.V., Jourdan, G.P. and Suhr, R.G. (1989). EP, pp. 326–30.

Atlantis, R. (2005). *Pflanzenschutz–Nachrichten Bayer*, **58(2):** 165–299, ISSN 0340–1723.

Bailey, W.A., Wilcut, J.W., Jordan, D.L., Swann, C.W. and Langston, V.B. (1999a). *Weed Technol.* **13:** 450–56.

Bailey, W.A., Wilcut, J.W., Jordan, D.L., Swann, C.W. and Langston, V.B. (1999b). *Weed Technol.* **13:** 771–76.

Barchietto, T. and Genet, J. L. (2002). *BCPC Conference–Pests Dis.* **2:** 835–40.

Bartlett, D.W., Clough, J.M., Godwin, J.R., Hall, A.A., Hamer, M. and Parr Dobrzanski, B. (2002). *Pest Manag. Sci.*, **58:** 649–62.

Bartlett, D.J., Baloch, R. and Guegen, F. (1997). *Proc. Conf. Int. Maladies Plantes.* pp. 1031–38.

Beraud, M., Claument, J. and Montury, A. (1993). In *Proc Brighton Crop Prot Conf–Weeds*, BCPC, Farnham, Surrey, UK, pp. 51–56.

Bereznak, J.F., Chang, Z.Y., Selby, T.P. and Sternberg, C.G. (1994). (E.I. Du Pont de Nemours and Co., USA) PCT Int. Appl. p. 73. WO9426722 A1; *Chem. Abstr.*, **123:** 169646.

Braxton, L.B., Barrentine, J.L., Dorich, R.A., Geselius, T.C., Grant, D.L., Langston, V.B., Redding, K.D. and Richburg, J.S. (1996). *Proc. S Weed Sci. Soc.* **49:** 170–71.

Brenchley, R., Rasmussen, D.E., Santel, H.J., Sorensen, V.M., Bowden, B.A., Bulman, P.G. M., Feindel, D.E., Gibb, B. and Tomolak, B.M. (1999). *Proc. West. Soc. Weed Sci.,* **52:** 125.

Brooks, R.L., Zoschke, A. and Porpiglia, P.J. (1995). *Proc. BCPC Conference–Weeds,* pp. 79–85.

Brown, H.M., Fuesler, T.P., Ray, T.B. and Strachan, S.D. (1991). *In*: Frehse, H. (*Ed*), Pestic. Chem.: adv. iInt. res., dev., legis., proc. int. congr. pestic. chem., 7th (1991), Meeting Date 1990, 257–66. Publisher: VCH, Weinheim.

Buck, W. and Raddatz, E. (1988). *Eur. Pat. Appl.* EP 262393.

Bunting, J.A., Sprague, C.L. and Riechers, D.E. (2004). *Weed Technol.* **52(5):** 711–17.

Bunting, J.A., Sprague, C.L. and Riechers, D.E. (2005). *Weed Technol.* **19(1):** 160–67.

Cain, P.A., Cramp, S.M., Little, G.M. and Luscombe, B.M. (1991). Rhoˆne–Poulenc Agricultutes Ltd., EP 0527036.

Carzaniga, R., Carelli, A., Farina, G., Arnoldi, A. and Gozzo, F. (1991). *Pestic. Biochem. Physiol.* **40:** 274–83.

Cho, J.H., Ahn, S.C., Koo, S.J., Joe, K.H. and Oh, H.S. (1997). *Proc. Brighton Crop Prot. Conf. Weeds,* **1:** 39–44.

Choate, J.H., Wilcut, J.W. and York, A.C. (1996). *Proc. S Weed Sci. Soc.*, **49:** 193.

Collins, B., Drexler, D., Maerkl, M., Hacker, E., Hagemeister, H., Pallett, K.E. and Effertz, C. (2001). *Proc. BCPC Conference–Weeds,* pp. 35–42.

Condon, M.E., Brady, T.E., Feist, D., Malefyt, T., Marc, P., Quakenbush, L.S., Rodaway, S. J., Shaner, D.L. and Tecle, B. (1993). *Proc. BCPC Conference–Weeds*, 41–46.

Conrath, U., Pieterse, C.M.J. and Mauch–Mani, B. (2002). *Trends Plant Sci.*, **7:** 210–16.

Covey, R.A., Forbes, P.J. and Bell, A.R. (1985). Eur. Pat. EP146279.

Daniau, P. and Prove, P. (2001). *Phytoma,* **534:** 49–51.

Debieu, D., Bach, J., Hugon, M., Malosse, C. and Leroux, P. (2001). *Pest Manag. Sci.*, **57:** 1060–67.

Dombo, P., Brix, H.D., Landes, M. and Nuyken, W. (2002). *Mitt. Biol. Bundesanst. Land–und Forstwirtsch.* **390:** 473–74.

Dow AgroSciences, Thifluzamide, Technical Bulletin, product brochure 2004 (http://www.dowagro.com).

Dressel, J. and Beigel, C. (2001). *Proc. BCPC Conference–Weeds,* pp. 119–26.

Dutzmann, S. (1997). *Pflanz.–Nachrichten Bayer* **50:** 21–28.

Dutzmann, S. and Suty–Heinze, A. (2004). *Pflanz–Nachrichten Bayer,* **57:** 249–64.

Dutzmann, S., Berg, D., Clausen, N.E., Krämer, W., Kuck, K.H., Pontzen, R., Tiemann, R. and Weissmüller, J. (1996). *Proc. Brighton Crop Protection Conf–Pests Dis.* pp. 47–52.

Dutzmann, S., Hayakawa, H., Oshima, A. and Suty–Heinze, A. (2004). *Pflanz–Nachrichten Bayer,* **57:** 415–35.

Egan, A.R., Michelotti, E.L., Young, D.H., Wilson, W.J. and Mattioda, H. (1998). *Brighton Crop Protect. Conf.–Pests Dis.* pp. 335–42.

EPA 224 339 (Publ. date 03.06.1987), (*Inventors* Ito, Sh., Masuda, K., Kusano, Sh., Nagat, T., Kojima, Y., Sawal, N. and Maeno Shin–Ichiro), Kumiai Chemical Industries Co., Ltd. and Ihara Chemical Industry Co., Ltd.

Feucht, D., Müller, K.H., Wellmann, A. and Santel, H.J. (1999). *Proc. Brighton Conference–Weeds,* **1:** 53–58.

Fisher, M.H. (1993). *In:* Duke, S.O., Menn, J.J. and Plimmer, J.R. (*Eds*) ACS Symposium Series No. 524, American Chemical Society, Washington, DC, pp. 169–82.

Foery, W. (1997). PCT Int. Appl. WO 9741112 A1.

Foster, C.J., Gilkersen, T. and Stocker, R. (1991). Shell Int. Res., EP 447 004.

Fujii, K. and Takamura, S. (1998). *Agrochem. Jpn.* **72:** 14–15.

Gibson, G. and de Kerchove, G. (1999). *Proc. BCPC Conference–Weeds,* pp. 87–92.

Gilgenberg–Hartung, A. (1999). *Gesunde Pflanzen,* **51:** 55–57.

Gobert, L., Mattioda, H., Raux, A., Arp, U., Egan, A.R., Michelotti, E.L., Young, D.H. and Wilson, W.J. (2000). Mededelingen–Fac. landbouwkundige toegepaste *biol. wetenschappen* (Universiteit Gent), **65(2b):** 799–806.

Godley, N.P. and Bubb, G.W. (2001). *Proc. Brighton Conference–Weeds,* **2:** 633–38.

Godwin, J.R., Bartlett, D.W., Clough, J.M., Godfrey, C.R.A., Harrison, E.G. and Maund, S. (2000). *In*: proc. brighton crop protect. conf.–pests and diseases, BCPC, Farnham, Surrey, UK, pp. 533–40.

Goto, T., Ito, S., Yanagi, A., Watanabe, Y. and Yasui, K. (2002). *Weed Biol. Manag.*, **2:** 18–24.

Gozzo, F., Carelli, A., Carzaniga, R., Farina, G., Arnoldi, A., Lamb, D. and Kelly, S. (1995). *Pestic. Biochem. Physiol.* **53:** 10–22.

Häuser–Hahn, I., Baur, P. and Schmitt, W. (2004). *Pflanz–Nachrichten Bayer,* **57:** 437–50.

Hacker, E., Bauer, K., Bieringer, H., Kehne, H. and Willms, L. (1995). *Proc. BCPC Conference–Weeds,* pp. 73–78.

Hacker, E., Bieringer, H. and Willms, L. *et al.* (2001). *Proc. BCPC Conference–Weeds,* pp. 43–48.

Hacker, E., Bieringer, H., Willms, L., Ort, O., Koecher, H. and Kehne, H. (1999). *Proc. BCPC Conference–Weeds,* pp. 15–22.

Hacker, E., Bieringer, H., Willms, L., Roesch, W., Koecher, H. and Wolf, R. (2000). *Z. PflKrankh. PflSchutz, Sonderh.* **XVII:** 493–500.

Hanai, R., Kawano, K., Shigematsu, S. and Tamaru, M. (1993). *Proc. Brighton Crop Prot. Conf. Weeds,* **1:** 47–52.

Hanhart, H. (2006). Top agrar 1/2006, pp. 58–61 (http://www.topagrar.com).

Hayashi, K. (2003). Wageningen University Dissertation, no. 3459.

Heinemann, U., Benet–Buchholz, J., Etzel, W. and Schindler, M. (2004). *Pflanz–Nachrichten Bayer,* **57:** 299–318.

Herms, S., Seehaus, K., Koehle, H. and Conrath, U. (2002). *Plant Physiol.,* **130:** 120–27.

Hess, M. and Rose, E. (1995). *Proc. BCPC Conference–Weeds,* pp. 763–68.

Heye, U.J., Speich, J., Siegle, H., Wohlhauser, R. and Hubele, A. (1994). *Proc. Br. Crop Prot. Conf.–Pests Dis.,* **2:** 501.

Hilber, U.W. and Hilber–Bodmer, M. (1998). *Plant Dis.* **82:** 496–500.

Hofer, U., Muehlebach, M., Hole, S. and Zoschke, A. (2006). *J. Plant Diseases Prot.* Special Issue, **XX:** 989–95.

Howard, S., Hudetz, M. and Allard, J.L. (2001). *Proc. BCPC Conference–Weeds,* pp. 29–34.

Hunter, J.J., Langston, V.B., Grant, D.L., McCormick, R.W., Barrentine, J.L. and Braxton, L.B. (1995). *Proc. S. Weed Sci. Soc.,* **48:** 201.

Hunter, J.J., Schultz, M.E., Mann, R.K., Cordes, R.C. and Lassiter, R.B. (1994). *Proc. N. Cent. Weed Sci. Soc.,* **49:** 124.

Huraux, M. and Prove, P. (1992). *Phytoma,* **443:** 69–70.

Ishida, Y., Ohta, K. and Yoshikawa, H. (1992). *Eur. Pat. Appl.* EP 477808 A1.

Isozumi, K., Umezu, K., Miyazaki, T. and Kimura, Y. (2001). Japan Patent 3258043.

Jachetta, J.J., Van Heertum, J.C. and Gerwick, B.C. (1994). *Proc. N. Cent. Weed Sci. Soc.* **49:** 123–24.

Jachetta, J.J., Van Heertum, J.C., Gerwick, B.C. and Barrentine, J.L. (1995). *Proc. S. Weed Sci. Soc.,* **48:** 199.

Jackson, R., Ghosh, D. and Paterson, G. (2000). *Pest Manag. Sci.* **56(12):** 1065–72.

Jautelat, M., Elbe, H.L., Benet–Bucholz, J. and Etzel, W. (2004). *Pflanz–Nachrichten Bayer* **57**: 145–62.

Jende, G., Steiner, U. and Dehne, H.W. (1999). *Pflanz.–Nachrichten Bayer,* **52:** 49–60.

Jende, G., Steiner, U. and Dehne, H.W. (2002). *In*: Dehne, H.W., Gisi, U., Kuck, K.H., Russell, P.E. and Lyr, H. *Eds.*, Modern Fungicides and Antifungal Compounds III, pp. 83–90, Agro Concept, Bonn.

Joseph–Horne, T., Heppner, C., Headrick, J. and Hollomon, D.W. (2000). *Pesticide Biochem. Physiol.*, **67:** 168.

Joshi, M.M. and Sternberg, J.A. (1996). *Proc. Brighton Crop Protect. Conf.–Pests and Diseases*, BCPC, Farnham, Surrey, UK, pp. 21–26.

Kang, Z., Huang, L. and Buchenauer, H. (2001). *Z. Pflanz. Pflanz.*, **108:** 419–32.

Kasahara, I. and Kemikaru. (2005). *Fain Kemikaru (Fine Chemicals),* **34:** 29–37.

Kasahara, I., Ooka, H., Sano, S., Hosokawa, H. and Yamanaka, H. (2006). (Nippon Soda Co., Ltd., Japan) PCT Int. Appl. 1996, 89 pp. WO9619442 A1; *Chem. Abstr.* **125:** 167579.

Kawamura, N., Masuyama, N., Takabe, F., Tamaru, M., Isozumi, K. and Umezu, K. (2000). Japan Patent 3046401.

Kehne, H., Willms, L., Bauer, K., Bieringer, H. and Buerstell, H. (1989). *Eur. Pat. Appl.* EP 342569 A1.

Khan, M.F.R. and Smith, L.J. (2005). *Crop Prot,* **24:** 79–86.

Kim, D.S., Koo, S.J. and Lee, J.N. *et al.* (2003a). *Proc. BCPC Int. Congress–Crop Sci. Technol.*, pp. 87–92.

Kim, D.S., Lee, J.N., Hwang, K.H., Kang, K.G., Kim, T.Y. and Koo, S.J. (2003b). *Proc. BCPC Int. Congress–Crop Sci. Technol.,* pp. 941–46.

Kirkland, K.J., Johnson, E.N. and Stevenson, F.C. (2001). *Weed Techno.*, **15:** 48–55.

Köhle, H., Grossmann, K., Jabs, T., Gerhard, M., Kaiser, W., Glaab, J., Conrath, U., Seehaus, K., Herms, S., In Dehne, H.W., Gisi, U., Kuck, K.H., Russell, P.E. and Lyr, H. (*Eds.*). (2002). Proc. 13th Int. Reinhardsbrunn Symposium, Agro–Concept, Bonn, 2002, pp. 61–74.

Kodandaram, M.H., Rai, A.B. and Halder, J. (2010). *Veg. Sci.,* **37(2):** 109–23.

Komatsubara, K. (2005). *Fain Kemikaru,* **34:** 38–48.

Koo, S.J., Ahn, S.C., Lim, J.S., Chae, S.H., Kim, J.S., Lee, J.H. and Cho, J.H. (1997). *Pestic. Sci.,* **51:** 109–14.

Krämer, W., Weissmüller, J., Gau, W., Etzel, W. and Stelzer, U. (1997). *Pflanz–Nachrichten Bayer,* **50:** 5–14.

Kramer, W. and Schirmer, U. (*Eds*). (2007). Modern Crop Protection Compounds. Wiley–VCH Verlag GmbH and Co. KGaA, Weinheim.

Kuck, K.H. and Mehl, A. (2004). *Pflanz–Nachrichten Bayer,* **57:** 225–35.

Kuck, K.H., Krueger, B.W., Rosslenbroich, H.J. and Brandes, W. (1997). Proceedings of ANPP cinquieme conference international sur les maladies des plantes, Tours 1997, ANPP, Paris, pp. 1055–62.

Kuhn, P.J., Pitt, D., Lee, S.A., Wakley, G. and Sheppard, A.N. (1991). *Mycol. Res.* **95(3):** 333–40.

Larelle, D., Mann, R., Cavanna, S., Bernes, R., Duriatti, A. and Mavrotas, C. (2003). *Proc. Int. Congr. Br. Crop Protect. Conf–Crop Sci. Technol.*, Glasgow, UK, Nov. 10–12, **1:** 75–80.

Lepiece, D., Rijckaert, G. and Thompson, A. (2000). *Proc 52nd Int. Symp. Crop Protec., Gent* **65(2a):** 141–49.

Leroux, P., Fritz, R., Albertini, D.C., Lanen, C., Bach, J., Gredt, M. and Chapeland, F. (2002b). *Pest Manag. Sci.* **58:** 876–88.

Leroux, P., Chapland, F., Desbrosses, D. and Gredt, M. (1999). *Crop Prot.,* **18:** 687–97.

Leroux, P., Colas, V., Fritz, R. and Lanen, C. (1995). *Modern Fungicides and Antifungal Compounds, Eds.* Lyr, H., Russell, P.E. and Sisler, H.D. Intercept, Andover, pp. 61–67.

Leroux, P., Debieu, D., Albertini, C., Arnold, A., Bach, J., Chapeland, F., Fournier, E., Fritz, R., Gredt, M. *et al.* (2002a). *In*: Dehne, H.W., Gisi, U., Kuck, K.H., Russell, P.E. and Lyr, H. (*Eds.*). Modern Fungicides and Antifungal References 649 Compounds III, Agroconcept, Bonn, pp. 29–40.

Levitt, G. (1988). Tetrazole–containing sulfonylureas, their herbicidal compositions, and their use in weed control, U.S. Pat. US 4,746,353 A.

Li, Z., Guan, A. and Liu, C. (2002). *Xiandai Nongyao,* **1(4):** 31–33.

Liu, C.L., Liu, W.C. and Li, Z.C. (2000). *Brighton Crop Protection Conf.*, pp. 549–56.

Lixin, Z., Zongcheng, L., Zhinian, L., Hong, Z., Changling, L., Bin, L., Shaber, S.H. (Rohm and Haas), EP 936,213, priority date February 10, 1998. The Rohm and Haas agrochemical business unit is now merged into Dow Chemical.

Löchel, A.M., Wenz, M., Russell, P.E., Buschhaus, H., Evans, P.H., Cross, S., Puhl, T. and Bardsley, E. (1998). *Proc. Brighton Crop Protection Conf.–Pests Dis.* pp. 89–96.

Longhurst, C., Dixon, K., Mayr, A., Bernhard, U., Prince, K., Sellars, J., Prove, P., Richard, C., Arnold, W., Dreikorn, B. and Carson, C. (1996). *Brighton Crop Prot. Conf.–Pests and Diseases,* **1:** 27–32.

Lorenz, K., Willms, L., Bauer, K. and Bieringer, H. (1995). PCT Int. Appl. WO 9510507 A1.

Luscombe, B.M., Pallett, K., Loubiere, P., Millet, J. C., Melgarejo, J. and Vrabel, T.E. (1995). *Proc. Br. Crop Prot. Conf.–Weeds,* **1:** 35.

Luthy, C., Zondler, H., Papold, T., Seifert, G., Urwyler, B., Heinis, T. and Steinrucken, H.C. (2001). *J. Allen, Pestic. Manag. Sci.,* **57:** 205–24.

Ma, Y.S. and Liu, C.L. (2005). *Nongyao (Chin. J. Pest.),* **44:** 128–29.

Maeno, S., Miura, I., Masuda, K. and Nagata, T. (1990). *Proc. Br. Crop Prot. Conf.–Pests Dis.*, **2:** 425.

Manabe, A., Maeda, K., Enomoto, M., Takano, H., Katoh, T., Yamada, Y. and Oguri, Y. (2002). *J. Pestic. Sci.*, **27:** 257–66.

Mann, R.K., Haack, A.E., Langston, V.B., Lassiter, R.B. and Richburg, J.S. (2005a). *Proc. Weed Sci. Soc. Am.*, **45:** 308.

Mann, R.K., Huang, Y.H., Larelle, D., Mavrotas, C., Min, Y.K., Morell, M., Nonino, H. and Shiraishi, I. (2003a). *Proc. 3rd Int. Temperate Rice Conf.*, Punta References 113 del Este, Uruguay, March 10–13, abstract WD055, p. 68.

Mann, R.K., Lassiter, R.B., Haack, A.E., Langston, V.B., Simpson, D.M., Richburg, J.S., Wright, T.R., Gast, R.E. and Nolting, S.P. (2003b). *Proc. Weed Sci. Soc. Am.*, **43:** 40.

Mann, R.K., Mavrotas, C., Huang, Y.H., Larelle, D., Patil, V., Min, Y.K., Shiraishi, I., Nguyen, L., Nonino, H.L. and Morell, M. (2005b). *Proc. 20th Asian–Pacific Weed Sci. Soc.*, Vietnam, pp. 289–94.

Margot, P., Huggenberger, F., Amrein, J. and Weiss, B. (1998). In *Proc. Brighton Crop Protect. Conf.–Pests and Diseases*, BCPC, Farnham, Surrey, UK, pp. 375–82.

Marquez, T., Joshi, M.M., Pappas Fader, T. and Massasso, W. (1995). *Proc. BCPC Conference–Weeds,* 65–72.

Masner, P., Muster, P. and Schmid, J. (1994). *Pestic. Sci.*, **42:** 163–66.

Mauler–Machnik, A., Rosslenbroich, H.J., Dutzmann, S., Applegate, J. and Jautelat, M. (2002). *Proc. BCPC Conf.–Pests Dis.* pp. 389–94.

Mayer, H., Hamprecht, G., Westphalen, K.O. *et al.* (1992). Herbicidal N–[(1,3,5–triazin–2–yl)–aminocarbonyl]benzenesulfonamides and their preparation Ger. Offen. DE 4038430 (Publ. 04.06.1992), PCT WO92/09608 (Publ. 11.06.1992).

McDougal, P. (2005). Agri Service. *Products Section, 2004 Market*, November, 2005.

Mehl, A. and Buchenauer, H. (2002). *In:* Dehne, H.W., Gisi, U., Kuck, K.H., Russell, P.E. and Lyr, H. *Eds.*, Modern Fungicides and Antifungal Compounds III, Agro Concept, Bonn, pp. 75–82.

Mercer, R.T. and Latorse, M.P. (2003). *Pflanz.–Nachrichten Bayer,* **56:** 465–76.

Mercer, R.T., Lacroix, G., Gouot, J.M. and Latorse, M.P. (1998). *Proc. Brighton Crop Protect. Conf.–Pests and Diseases*, BCPC, Farnham, Surrey, UK, pp. 319–26.

Meyer, W. (1992). Eur. Pat. Appl. EP 496701 A1.

Michel, P. (2004). *Phytoma,* **577:** 47–48.

Milling, R.J. and Richardson, C.J. (1995). *Pestic. Sci.* **45:** 43–48.

Min, Y.K. and Mann, R.K. (2004a). *Korean J. Weed Sci.*, **24(3):** 192–98.

Min, Y.K. and Mann, R.K. (2004b). *Korean J. Weed Sci.*, **24(3):** 199–205.

Mitani, S., Araki, S., Matsuo, N. and Camblin, P. (1998). Proc. Brighton Crop Protect. Conf.–Pests and Diseases, BCPC, Farnham, Surrey, UK, pp. 351–58.

Mitani, S., Araki, S., Takii, Y., Oshima, T., Matsuo, N. and Miyoshi, H. (2001). *Pest. Biochem. Physiol.,* **71:** 107–15.

Mitchell, G., Bartlett, D.W., Fraser, T.E., Hawkes, T.R., Holt, D.C., Townson, J.K. and Wichert, R.A. (2001). *Pest. Manage. Sci.* **57:** 120–28.

Miura, I., Kamakura, T., Maeno, S., Hayashi, S. and Yamaguchi, I. (1994). *Pestic. Biochem. Physiol.*, **48**: 222–28.

Mizell, R.F.III, Schiffhauer, D.E. and Taylor, J.L. (1986). *J. Entomol. Sci.,* **21:** 329–37.

Morita, T. and Nishimura, T. (2001). *Agrochem. Jpn.* **79:** 10–12.

Müller, K.H. (2002). *Pflanz.–Nachrichten*, **55(1):** 15–28.

Mugnier, J., Chazalet, M., Gaulliard, J.M., Anelich, R.Y. and Gouot, J.M. (1994). *Proc. Brighton Crop Protection Conf.–Pests Dis.* 1994, pp. 325–30.

Nelson, K.A. and Renner, K.A. (1998). *Weed Technol.* **12:** 293–99.

Nezu, Y., Wada, N., Yoshida, F., Miyazawa, T., Shimizu, T. and Fujita, T. (1998). *Pestic. Sci.* **52:** 343–53.

Ogawa, Y. (1995). *Agrochem. Jpn.*, **67:** 20–21.

Ohyama, H., Wada, T., Ishikawa, H. and Chiba, K. (1988). *Proc. Brighton Crop Protection Conf.–Pests Dis.*, pp. 519–26.

Opalski, K. (2005). PhD Thesis, University of Giessen, Germany (http://geb.uni–giessen.de/geb/volltexte/2005/2365/).

Ort, O., Bauer, K. and Bieringer, H. (1992). PCT Int. Appl. WO 9213845 A1.

Owen, W.J. and deBoer, G.J. (1994). *In*: Eighth International Congress of Pesticide Chemistry: Options 2000. Ragsdale, N.N., Kearney, P.C. and Plimmer, J.R. *Eds.*, ACS Conference Proceedings Series, American Chemical Society, Washington, DC, 1994, pp. 257–268.

Parrish, S.K., Kaufmann, J.E., Croon, K.A., Ishida, Y., Ohta, K. and Itoh, S. (1995). *Proc. BCPC Conference–Weeds,* pp. 57–63.

Pirgozliev, S.R., Edwards, S.G., Hare, M.C. and Jenkinson, P. (2002). *Eur. J. Plant Pathol.* **108:** 469–78.

Que, J.P., Thienpont, E. and Garford, B. (1994). *Phytoma,* **459:** 49–51.

Querou, R., Euvrard, M. and Gauvrit, C. (1997). *Pestic. Sci.,* **49:** 284–90.

Querou, R., Euvrard, M. and Gauvrit, C. 1998. *Pestic. Sci.*, **53:** 324–32.

Richards, C. (2005). President and CEO of Arysta LifeSciences, Credit Suisse/First Boston Agrochemicals Conference, 15th February 2005, Internet address: http://www.arysta. hu/feltoltes/Egyeb/Erdekes/ arysta2005feb.pdf.

Rodaway, S.J., Tecle, B. and Shaner, D.L. (1993). *Proc. BCPC Conference–Weeds,* pp. 239–46.

Rosslenbroich, H.J. (1999). *Pflanz.-Nachrichten Bayer,* **52:** 131–48.

Rosslenbroich, H.J., Brandes, W., Krueger, B.W., Kuck, K.H., Pontzen, R., Stenzel, K. and Suty, A. (1998). *Proc. Brighton Conf.* 1998, pp. 327–35. British Crop Protection Council, Farnham.

Russel, P.E., Percival, A., Coltman, P.M. and Green, D.E. (1992). *Proc. Brighton Crop Protection Conf.-Pests Dis.* pp. 411–18.

Saito, Y., Wada, N., Kusano, S., Miyazawa, T., Takahashi, S., Toyokawa, Y. and Kajiwara, Y. (1990). US Patent 4932999 (1990).

Saitoh, Y., Wada, N., Kusano, S., Toyokawa, Y. and Miyazawa, T. (1996). Japan Patent 2561524.

Sampson, A.J., Cazenave, A., Laffranque, J.P., Jones, R.G., Kumazawa, S. and Chida, T. (1992). *Proc. Brighton Crop Protection Conf. –Pests Dis.* pp. 419–26.

Santel, H.J., Bowden, B.A., Sorensen, V.M. and Müller, K.H. (1999a). *Proc. Brighton Conference–Weeds,* **1:** 23–28.

Santel, H.J., Bowden, B.A., Sorensen, V.M., Mueller, K.H. and Reynolds, J. (1999b). *Proc. West. Soc. Weed Sci.,* **52:** 124.

Schmitt, M., Van Almsick, A., Preuss, R., Willms, L., Auler, T., Bieringer, H. and Thuerwaechter, F. (2001). Aventis CropScience GmbH, WO 2001074785.

Schönhammer, A., Freitag, J. and Koch, H. (2005). BASF AG, 46. Ö sterreichischen Pflanzenschutztagen, 30.11.–01.12.2005, Stadthalle Wels.

Schönhammer, A., Pörksen, N., Freitag, J. and Nuyken, W. (2002). *Biol. Bundesanst. Land-und Forstwirtsch.* **390:** 241–42.

Sekino, K., Komatsubara, K., Koyanagi, H. and Yamada, Y. (2004). *Shokubutsu no Seicho Chosetsu,* **39:** 23–32.

Selby, T.P., Sternberg, C.G., Bereznak, J.F., Coats, R.A. and Marshall, E.A. (2004). Abstract of papers, 228th ACS national meeting, Philadelphia, PA, August 22–26, AGRO–001.

Shaw, D.R., Bennett, A.C. and Grant, D.L. 1999. *Weed Technol.* **13:** 791–98.

Sheppard, B.R., Braxton, R.L., Barrentine, J.L., Geselius, T.C., Grantm D,L., Langston, V.B., Redding, K.D., Richburg, J.S. and Roby, D.B. (1997). *Proc. S. Weed Sci. Soc.*, **50:** 161.

Shiraishi, I. (2005). *J. Pestic. Sci.,* **30(3):** 265–68.

Sierotzki, H., Wullschleger, J., Alt, M., Bruye‘re, T., Pillonel, C., Parisi, S. and Gisi, U. (2002). Modern fungicides and antifungal compounds III, *Eds.* Dehne, H.W., Gisi, U., Kuck, K.H., Russell, P.E. and Lyr, H. Agro Concept, Bonn, pp. 141–48.

Snel, M., Watson, P., Gray, N.R., Kleschick, W.A. and Carson, C.M. (1993). Mededelingen–Facul. *landbouwkundige Toegepaste Biol. Wetenschappen* (Univ. Gent) **58(3a):** 845–52 (Chem. Abstr. 1994, 120, 291962).

Stabler, G.F., Murdock, E.C., Keeton, A. and Isgett, T.D. (1996). *Proc. S. Weed Sci. Soc.*, **49:** 18.

Stierl, R., Ammermann, E., Lorenz, G., Schelberger, K., In Dehne, H.W., Gisi, U., Kuck, K. H., Russell, P.E. and Lyr, H. (*Eds.*). (2002). Modern fungicides and antifungal compounds III, Proc. 13th int. reinhardsbrunn symposium, Agro Concept, Bonn, 2002, pp. 49–59.

Stierl, R., Grote, T., Bross, M., Washioka, S. and Schöfl, U. (2004). Proceedings of the 15th International Plant Protection Congress, Beijing, China, 2004, p. 166.

Suty, A., Pontzen, R. and Stenzel, K. (1999). *Pflanz.–Nachrichten Bayer,* **52:** 149–61.

Tachikawa, S., Miyazawa, T. and Sadohara, H. (1997). *Proc. 16th Asian–Pacific Weed Sci., Soc. Conf.*, **2A:** 114–117.

Takahashi, S., Shigematsu, S., Morita, A., Nezu, Y., Claus, J. S. and Williams, C.S. (1991). *Proc. Brighton Crop Prot. Conf. Weeds,* **1:** 57–62.

Takematsu, T., Takeuchi, Y., Takenaka, M., Takamura, S. and Matsushita, A. (1987). Ube Industries, EP 239 414.

Tamaru, M., Inoue, J., Hanai, R. and Tachikawa, S. (1997b). *J. Agric. Food Chem.* **45:** 2777–83.

Tamaru, M., Kawamura, N., Sato, M., Tachikawa, S., Yoshida, R. and Takabe, F. (1991). EP Patent 435170.

Tamaru, M., Kawamura, N., Satoh, M., Takabe, F. and Tachikawa, S. (1997a). Japan Patent 2603557.

Tamaru, M., Kawamura, N., Satoh, M., Takabe, F. and Tachikawa, S. (1999). Japan Patent 2977591.

Tanaka, Y., Kajiwara, Y., Noguchi, M., Kajiwara, T. and Tabuchi, T. (2003). PCT Int. Appl. WO 03061388.

Teaney, S.R., Armstrong, L., Bentley, K., Cotterman, D. and Leep, D. *et al.* (1995). *Proc. BCPC Conference–Weeds* 49–56.

Thomas, A., Albert, G. and Schlösser, E. (1992). Med. Fac. Landbouww. Univ. Gent, **57(2a):** 189–97.

Thompson, A.R., McReath, A.M., Carson, C.M., Ehr, R.J. and DeBoer, G.J. (1999). *Brighton Conf. Weeds,* **1:** 73–80.

Tiemann, R., Berg, D., Krämer, W. and Pontzen, R. (1997). *Pflanz.–Nachrichten Bayer,* **50:** 29–48.

Tomiya, K. and Yanase, Y. (2003). *Congress Proceedings–BCPC International Congress: Crop Science and Technology*, Glasgow, **1:** 99–104.

Tomlin, C. (*Ed.*). (2005). The e–Pesticide Manual, version 3.2, British Crop Protection Council, Hampshire.

Tsuda, M. and Kato, S. (2003). *Kitanippon Byogaichu Kenkyukaiho,* **54:** 32–34.

Tsuda, M., Itoh, H. and Kato, S. (2004). *Pest Manage. Sci.,* **60(9):** 875–80.

Tsuda, M., Itoh, H., Wakabayashi, K., Ohkouchi, T., Kato, S., Masuda, K. and Sasaki, M. (2000). *Proc. 2000 Brighton Conf.* pp. 557–62.

U.S. Environmental Protection Agency. (2001), Internet website:/www.epa.gov/ opprd001/factsheets.

Uchida, M., Roberson, R.W., Chun, S.J. and Kim, D.S. (2005). *Pest Manag. Sci.,* **61:** 787–92.

Uchino, H. and Watanabe, H. (1999). *Tensai Kenkyu Kaiho,* **41:** 80–84.

Von Deyn, W., Hill, R.L., Kardorff, U. and Engel, S. *et al.* (1996). Westphalen, BASF AG, WO 96/26206.

Wada, N., Kusano, S. and Toyokawa, Y. (1990). US Patent 4906285.

Wada, N., Kusano, S. and Toyokawa, Y. (1996). Japan Patent 2558516.

Wang, C.L., Lee, M.S., Li, Y.W., Yao, Z.W., Shieh, J.N., Mann, R.K. and Huang, Y.H. (2004). *15th Int. Plant Protect. Congr.* Abstracts. Beijing, China, p. 598.

Ward, C.E. (1984). Chevron Research, DE 3 422 346.

Watanabe, O., Yokoyama, M., Fujita, S., Miyazaki, T. and Wada, N. (2003). *J. Weed Sci., Technol.*, **48:** 24–30.

Wellmann, A. and Feucht, D. (2002). *Pflanz-Nachrichten,* **55(1):** 67–86.

Wheeler, *I.E.*, Hollomon, D.W., Gustafson, G., Mitchell, J.C., Longhurst, C., Zhang, Z. and Gurr, S.J. (2003). *Mol. Plant Pathol.* **4:** 177–86.

Wheeler, I., Hollomon, D.W., Longhurst, C. and Green, E. (2000). *Brighton Crop Prot. Conf.–Pests and Diseases,* **3:** 841–46.

www.irac-online.org

Yanagi, A. (2001). *Pflanz. Nachrichten Bayer,* **54:** 2–12.

Yanagi, A., Watanabe, Y., Narabu, S., Ito, S. and Goto, T. (2002). *J. Pestic. Sci.*, **27:** 199–209.

Yasuda, M., Nakashita, H. and Yoshida, S.J. (2004). *Pestic. Sci.* **29:** 46–49.

Yokota, C. (2004). *Agrochem. Jpn.*, **84:** 12–14.

Yokoyama, M., Watanabe, O., Kawano, K., Shigematsu, S. and Wada, N. (1993). *Proc. Brighton Crop Prot. Conf. Weeds,* **1:** 61–66.

Yokoyama, M., Watanabe, O., Yanagisawa, K., Ogawa, Y., Wada, N. and Shigematsu, S. (1994). *J. Weed Sci. Technol.,* **42(Supplement):** 32–33.

Yoshikawa, Y., Tomiya, K., Katsuta, H., Kawashima, H., Takahashi, O., Inami, S., Yanase, Y., Kishi, J., Shimotori, H. and Tomura, N. (1996). *Eur. Pat. Appl.* EP 737682, 59 pp., *Chem. Abstr.* **126:** 7982.

Young, D.H. and Slawecki, R.A. (2001). *Pestic. Biochem. Physiol.* **69:** 100–111.

Young, D.H. and Vjugina, U. (2002). *Phytopath.,* **92:** S89.

Zhang, L.X. *et al.* (2003). *Proc. BCPC Int. Congr.*, Glasgow, Vol. 1, p. 93.

11

Botanicals in Crop Protection

JAYDEEP HALDER[1*] AND SUSANTA BANIK[2]

ABSTRACT

Non-chemical method is in the centre of activity of all advanced plant protection strategies. Use of plant-derived products to ward off various pests including that of agricultural importance is also not new. Intensive and vigorous reseach activities on plant extracts for managing agricultural pests have generated many useful products which have been successfully incorporated in integrated management of plant pathogens, insect- and non-insect pests for sustainable crop protection. Such plant products offer several advantages in terms of target specificity, biodegradability, environmental effect, and compatibility issues with other plant protection products etc. for which they can now be used as complete or partial replacement of synthetic harmful pesticides. This chapter attemps to highlight the effectiveness of various plants and products derived from them against a wide range of plant pests.

***Key words*:** Botanicals, Neem, Pest management, Compatibility.

1. INTRODUCTION

Plant species have provided enormous benefits to the mankind. For all our needs of food, fuel, fodder, fiber, furniture, life saving drugs we depend directly or indirectly on plants. Plants with their own mechanisms of food production have evolved over millions of years to adopt and survive in this planet. Efforts of adoption and survival of plants have been directed not only to abiotic factors of a niche area but also to defend against the biotic stresses like microorganisms and various herbivores. Presence of a wide array of morphological and physiological modifications including production

1 Economic Botanist-VII, Sugarcane Research Station, Bethuadahari, Nadia, West Bengal, India.

2 Department of Plant Pathology, School of Agricultural Sciences and Rural Development, Nagaland University, Medziphema - 797 106, Nagaland, India.

* *Corresponding author*: E-mail: jaydeep.halder@gmail.com

of secondary metabolites in a plant is a testimony to that. These secondary metabolites have been an interesting area of scientific research since a long time though their utilities were known even in the earliest human civilization *i.e.*, Vedic civilization. Plant extracts or botanicals containing secondary metabolites have been used to ward off various pests and pathogens affecting human beings as well as to protect crop plants for better yield. Well known plant extracts like neem, tobacco, karanja and garlic are only few to name.

2. ANCIENT USE

Vriksha Ayurveda compiled by Surpala describes the use the various plant-based preparations to protect crop against various diseases. This has analogy with Ayurveda the well-known treasure of wisdom on human medicine. Vriksha Ayurveda reflects the ethos of organic way of managing crop pests, making us believe that sustainable agriculture was very much the way of life in ancient time. Use of several plants such as Adhatoda, garlic, Lantana, mahuva, neem, nirgundi, peepli, Pongamia, tulsi etc. has been mentioned and their role in crop protection has been described in the ancient manuscripts (Dodia *et al.*, 2008).

Insecticidal use of plant extracts was known during the period of Roman Empire. The plant pyrethrum was used for delousing during Persian king Xerxes' reign in 400 B.C. Insecticidal use of nicotine obtained from tobacco plant to kill plum beetle was recorded in 17th century. Rotenone was reported to have insecticidal value in the middle of 19th century (Dodia *et al.*, 2008).

Stringent evaluation process to test plants for possessing insecticidal and antimicrobial properties has resulted in discovery of many plants. Purohit and Vyas (2004) reports that there is 2121 number of plant species which can be useful in pest management. On the other hand Grainage *et al.* (1985) provided a greater detail on the use of plants for pest management (Table 1).

Table 1: Number of plants with pesticidal values

Sl. no.	*Effect of plant*	*Estimated no. of plant species*
1.	Insecticidal	1053
2.	Anti–feedant	384
3.	Repellent	297
4.	Growth inhibitor	32
5.	Attractant	27
6.	Rodenticidal	29
7.	Acaricidal	2

Table 1: (*Contd...*)

Table 1: (*Contd...*)

Sl. no.	*Effect of plant*	*Estimated no. of plant species*
8.	Nematicidal	58
9.	Fungicidal	100
10.	Bactericidal	4
11.	Molluscicidal	6
12.	Herbicidal	14
13.	Antiseptic	35
14.	Fish poison	147
15.	Arrow poison	90
16.	Poisonus	69

Chemical analysis of potentially pesticidal plant extracts has elucidated the nature, structural formula and other properties of active principles (Table 2). Almost all the tested botanicals have diverse modes of action (Table 3) against crop pests; however, their effectiveness varies depending on the pest species (Table 4).

Table 2: List of some plants of pesticidal value with their active components (Dodia *et al.*, 2008)

Family	*Common name of plant*	*Scientific name*	*Active principle*
Euphorbiaceae	Castor	*Ricinus communis*	Ricin, ricinine
	Jatropha	*Jatropha curcas*	Curcusone, jatrophol, jatrophin
	Euphorbia	*Euphorbia tirucalli*	Euphorbosterol, euphorbol
Solanaceae	Chilli	*Capsicum annum, C. frutescens*	Capsaicin
	Kantakari, Surrattense nightshade	*Solanum surrattense*	Solasodine, solanolone, solasonine
	Datura	*Datura stramonium*	Hyoscyamine, atropine, scopolamine
	Tobacco	*Nicotiana tabacum, N. rustica*	Nicotine, nornicotine, anabasine
Asteraceae	Marigold	*Tagetes erecta, T. minuta*	Tagetone, allopatuletin
	Parthenium	*Parthenium hysterophorus*	Parthenin
	Chrysanthemum	*Chrysanthemum cinerariifolium*	Pyrethrin I, pyrethrin II, cinerin I, cinerin II, jasmolin I, jasmolin II

Table 2: (*Contd...*)

Table 2: (*Contd...*)

Family	*Common name of plant*	*Scientific name*	*Active principle*
Meliaceae	Neem	*Azadirachta indica*	Azadirachtin, meliantriol, nimbinin, nimbidin, salannin, nimbin
	Dharek, Chinaberry	*Melia azedarach*	Nimbolin A, nimbolin B, meliatoxin A1
Verbenaceae	Lantana	*Lantana camara*	Lantanolic acid, lantic acid
	Clerodendron	*Clerodendrum indicum*	Trans–decalin, clerodin
	Nirgudi, Chaste tree	*Vitex negundo*	Vitexin, negundoside
Zingiberaceae	Ginger	*Zingiber officinale*	Gingerols, alpha–zingiberene, arcurumene
	Turmeric	*Curcuma longa*	Curcumol, curcumin
Rutaceae	Murraya, curry tree	*Murraya koenigii*	Curryanine, curryangine, murraxonin, murrayanone
Lamiaceae	Vilayati Tulsi, American mint	*Hyptis suaveolens*	Hyptolide
	Dronapushpi, Chhota halkusa	*Leucas aspera*	—
	Ocimum	*Ocimum sanctum, O. basilicum*	Juvocimene I, II, ocimin
	Mint	*Mentha spicata, M. arvensis*	Menthole, limonene, dihydrocarvone, menthone
Alliaceae	Onion	*Allium cepa*	Oleic acid, cepocide–D, alpha and beta-tocopherols
	Garlic	*Allium sativum*	Allicin, diallyl sulphide and diallyl disulphide
Apocynaceae	Vinca	*Catharanthus roseus*	Vincristine, vinblastine, vindesine
	Pink oleander	*Nerium oleander*	Oleanderol, oleanderin, oleanderolicacid
	Yellow oldeander	*Thevetia neriifolia*	—
Fabaceae	Vilayati babul, Algaroba, Mesquite	*Prosopis juliflora*	Juliprosopine, julifloridine, juliflorinine, prosopidione
Leguminosae	Karanja, Indian beech	*Pongamia glabra*	Pongamol, pongapin, pongaglabrone, pongallone
Fabaceae	Derris	*Derris chinensis*	Rotenone
	Gliricidia	*Gliricidia sepium*	Sepinol, gliricidol
Simaroubaceae	Ardusa, Tree of heaven	*Ailanthus excelsa*	Ailanthone

Table 3: Mode of action of botanicals (Dodia *et al.*, 2008)

Target organism	***Mode of action***
Insects	Ovipositional deterrent (not allowing the females to deposit eggs), ovicidal (killing of eggs or disrupting embryonic development thus preventing hatching), attractant (to lure and eliminate insects by forming traps), repellents (to revel insects from a source), feeding deterrents or antifeedants, antigonadal agents, growth regulation, insecticidal, insect phytohormones (phytojuvenoids, antijuvenile hormones, phytoecdysteroids), physiological effects, confusants, inhibition of chitin formation, disruption in sexual communication
Fungi	Inhibition of conidial (spore) germination, mycelial growth
Bacteria	Antibacterial
Viruses	Inhibition of multiplication of virus
Weeds	Allelopathy; herbicidal; inhibition of germination, radicle elongation, seedling growth; reduction in dehydrogenase activity, chlorophyll content
Nematodes	Inhibition of egg hatching and emergence of larvae
Mite	Oviposition deterrent, ovicidal and acaricidal, antifeedant, repellent
Rodents	Decrease in sperm count and sperm motility, alteration of mating behaviour

Table 4: Use of botanicals on common insects

	Neem	***Lime***	***Sulphur***	***Pyrethrum/ pyrethrins***	***Rotenone***
Ants		*		*	*
Aphids	*			*	*
Apple maggot				*	*
Armyworm	*			*	
Asparagus beetle				*	*
Blister beetles					*
Cabbage worms, butterflies	*				
Caterpillars, leaf-feeding	*			*	*
Codling moth					
Colorado potato beetle	*			*	*
Corn earworms				*	
Cucumber beetles	*			*	*
Earwigs				*	
Elm leaf beetles	*				
Flea beetles					*

Table 4: (*Contd...*)

Table 4: (*Contd...*)

	Neem	***Lime***	***Sulphur***	***Pyrethrum/ pyrethrins***	***Rotenone***
Geranium budworm				*	
Grubs				*	
Leaf hoppers					*
Leaf miner	*				
Looper caterpillars	*				
Mealy bugs	*				
Psyllids			*		
Root knot nematodes	*				
Scales		*			
Spider mites	*	*	*		*
Squash bugs				*	*
Squash vine borers					*
Stink bugs					*
Thrips	*	*	*		*
Whiteflies (green house)	*				*
Whiteflies (sweet potato)	*				

Source: David Hillock and Pat Bolin, Earth-kind gardening series: Botanical pest controls, *Oklahoma Cooperative Extension Service* HLA–6433. Division of Agricultural Sciences and Natural Resources, Oklahoma State University (http://osufacts.okstate.edu).

Advantages of Botanicals

1. Easily degradable through the action of air, water and sunlight and thereby possessing less residual toxicity problem.
2. Act quickly and stop feeding of insects often causing immediate paralysis
3. Possess low to moderate toxicity to mammals
4. In most cases non-toxic to plants (not phytotoxic)
5. Broad-spectrum
6. May be used shortly before harvesting
7. Farmers can make the local formulations on their own based on their requirement.
8. Resistance development in the target organisms is not as quick as in synthetic pesticides
9. Mostly compatible with insect pathogenic microbes

10. Multiple modes of action like anti-feedant, repellent, anti-ovipositional, insect growth regulation etc. Many of the botanicals have antifungal, antibacterial, antihelminthetic and antiviral properties as well.
11. Well-fit for the sustainable farming
12. Relatively safer to the user during handling, other non–target organisms in the environment.

Disadvantage of Botanicals

1. Easily degradable and therefore required to be applied more often.
2. May not cause death of insects.
3. Though less toxic, may be poisonous to humans or the environment.
4. May not be readily available all throughout the season and everywhere.
5. Broad-spectrum and therefore sometimes toxic to beneficial insects and pathogens as well.
6. Generally have poor water solubility.
7. Non-systemic in nature.
8. Temporal and spatial variation in the performance of a particular botanical product.
9. All botanicals do not have knockdown effect as in synthetic insecticides.

3. TOXICITY OF BOTANICALS

Usually toxicity levels are expressed in terms of LC_{50} or LD_{50} value which refers to the concentration or the dose at which 50% of the test organisms are killed. Lower LD_{50} value indicates higher toxicity and vice-versa. Oral LD_{50} value of some of the insecticides and botanicals are given below for comparison (*Source*: http://osufacts.ukstate.edu).

Insecticide	***Oral*** LD_{50}
Nicotine	55
Sevin	246–283
Aspirin	1,200
Rotenone	1,500
Pyrethrin	1,500
Malathion	2,800
Salt	3,320
Neem	>5,000

In addition to its LD_{50} value, residual toxicity of botanicals is also an important consideration where botanicals have an edge over the conventional pesticides. Several reports suggest that active ingredient of botanicals are degraded rapidly on the applied site of the plant and thereby reduces residual problem. Besides this, most of the plant based products have low mammalian toxicity and therefore safe to use. In mango tree, residue of azadirachtin was beyond detectable level two days after application (0.05%) (Indumathi, 2002).

4. CLASSIFICATION

Based on modes of action botanicals on target organisms botanicals can be grouped into two categories: contact and stomach poisons.

4.1. Contact Poisons

1. ***Pyrethrins*:** Pyrethrins act as fast-acting insecticide with knockdown effect on the target insects by interrupting their nerve impulses. Pyrethrins, chemical components of pyrethrum are derived from the dried flower buds of Chrysanthemum species. Pyrethrum was used in the early 19th century to control body lice during the Napoleonic wars. Pyrethrum is a mixture of four esters: pyrethrins I and II and cinerins I and II. These esters are formulated from two acids (chrysanthemic acid and pyrethric acid) and two alcohols *viz.*, cinerolone and pyrethrolone.

   Pyrethrolone + Chrysanthemic acid = Pyrethrin I

   Pyrethrolone + Pyrethric acid = Pyrethrin II

   Cinerolone + Chrysanthemic acid = Cinerin I

   Cinerolone + Pyrethric acid = Cinerin II

   - To improve the insecticidal value of pyrethrins against which insects can develop resistance quickly, certain synergists like piperonyl butoxide, sesamin, sesamolin etc. are added. Pyrethrins are effective against many sucking and chewing insects as well as flying insects such as gnats. Pyrethrum is nontoxic to most mammals, making it one of the safest insecticides in use (Dubey *et al.*, 2010)

2. ***Rotenone*:** Rotenone is extracted from the roots of *Derris* plants in Asia and cube roots in South America. Rotenone primarily acts in nerve and muscle cells of insects (being particularly effective against leaf-feeding beetles and certain caterpillar pests), and inhibits cellular respiration. It is considered both a contact and stomach poison.

Rotenone is quite toxic to mammals when inhaled and extremely toxic to fish.

4.2. Stomach Poisons

***Neem*:** Neem, the name of a tree *Azadirachta indica*, has been in use for crop protection and other human needs since ancient times. Various plant parts like leaves, seed, and bark are sources of the active ingredient azadirachtin that can be used for controlling plant pathogens, and insect pests. Neem extracts can act as feeding deterrent as well as growth regulator to insects. Neem extracts interfere with the production of insect growth regulators. Neem based products with no residual effect on crop can have systemic action when applied in root region. Azadirachtin is considered nontoxic to mammals, fish and pollinators, having low mammalian toxicity with LD_{50} of >5000 mg/kg for rat (Dubey *et al.*, 2010).

5. COMMERCIAL PRODUCTS

Gahukar (1998) enlisted more than 90 neem-based commercial formulations that are manufactured in India. Some of the randomly selected neem-based products that have been commercialized in India are presented below (Table 5). Biological effects of NSKE and neem oil have been studied on various insect pests and the findings are summarized in Table 6.

Table 5: Commercialized neem-based products

Sl. No.	*Product*	*Name of the manufacturer*
1.	Achook (0.03% Aza)	Bahar Agrochem and Feed Pvt. Ltd., 411, AGH Chambers, Masjid Bunder Road, Mumbai-400009
2.	Azadirachtin (0.03% Aza)	Goyal Pesticides, Sanchi Road, Vidisha-464001
3.	Azadit	Pesticides (India) Ltd., Udaipur
4.	Bioneem (0.03% Aza)	Ajay Biotech (India) Ltd., 100/3, Kalpana Apartment, Dr. Ketkar Marg, Pune-411004
5.	Bioneem (0.03%Aza)	Zuari Industries Ltd., Margoa-403726, Goa
6.	Field Marshal	Laxmi Enterprises, Vadodora-390001
7.	Gronim–T (10% Aza), Gronim–Coat, Gronim–EC	National tree growers Co-operative federation Ltd., Anand-388001
8.	Limonool	Sri Bio-Multi Tech. (Pvt.) Ltd, Bangalore-560032
9.	Margocide CK 20 EC (0.15% Aza), Margocide CK 80 EC (0.03% Aza)	Monofix Agro Products Ltd., Hubli–580030
10.	Margosom	Agri Life, SOM Phytopharma (India) Limited, 154/A5, SVCIE, IDA Bollaram 502 325, Medak Dist. (Hyderabad), A.P.
11.	Multineem (0.03% Aza)	Karnataka Agro Chemicals, 180, Bangalore-560086
12.	Neem Plus	Sio Agro Research labs, Mumbai-400080
13.	Neem Spray	BDK Agro Chemicals, 47–48, Gokul Road, Hubli-580030

Table 5: (*Contd...*)

Table 5: (*Contd...*)

Sl. No.	*Product*	*Name of the manufacturer*
14.	Neemactin (0.15% Aza)	Wockhardt (Biostadt Agrosciences), Veera Desai Road, Andheri (West), Mumbai-440010
15.	NeemAzal F (5% Aza), NeemAzal T.S (1% Aza), Aza 25% (for export)	EID Parry (I) Ltd., 234, Netaji Subhas Chandra Bose Road, Chennai-600001
16.	Neemex (EC)	National Pesticides and Chemicals, C-8, MIDC, Amravati-444606
17.	Neemguard	Gharda Chemicals Ltd., Bandra (W), Mumbai-400050
18.	Neemhit	Skylark Agrochem, Thane, Maharastra
19.	Neemitaf (0.15% Aza)	Rallis India Ltd., Sangam Bridge, Pune-411001
20.	Neemite	Ecomax Agrosystems Ltd., Thane (W)-400604
21.	Neemstar (0.03%)	Universal Pesticides and Chemicals Industry, Trichy Road, Coimbatore-641045
22.	Nimbecidine (0.03% Aza)	T–Stanes and Company Ltd., Cimbatore-641018
23.	Rakshak (0.15% Aza)	Murkumbi Bioagro Pvt. Ltd., BC 105, Havelock Road, Cantonment, Belgaum-590001
24.	Suneem	Sunida Export, Mumbai-400049
25.	Superguard	Shiv Agro Chemicals Industries, Baroda-390010

Table 6: Biological effects of NSKE and neem oil on various insect pests

Pradhan *et al.* (1962)	First to prove the antifeedant property of neem and showed that 0.001 per cent aqueous suspension of crushed neem kernel when sprayed on cabbage plant totally stopped the feeding of desert locust, *Schistocerca gregaria* on treated foliage.
Mane (1968)	Aqueous extract of neem seeds kernel gave considerable protection to various crops against larvae of *Euproctis lunata, S. litura, Utetheisia pulchella.*
Gill and Lewis (1971)	*Pieris* sp larvae fed on foliage treated with neem kernel extract failed to develop to maturity and most of them died while moulting.
Sandhu and Singh (1975)	1 per cent neem kernel extract reduced the damage by *Pieris brassicae.*
Attri (1975)	Neem oil extractive was found to be 40 times less effective as antifeedant than water extract to *Schistocerca gregaria.*
Ladd and Jacobson, (1980)	1 per cent neem kernel extracts as spray on soyabean deterred the feeding of *Popillae japonica.*
Saxena (1982)	Neem kernel extract showed per cent anti-feedance of 95.1 to 98.4 against *Amsacta moorie.*
Joshi *et al.* (1984)	Spray application of 0.5, 0.75 and 0.1 per cent suspension of neem seeds kernel protected tobacco plants from feeding by *S. litura*

Table 6: (*Contd...*)

Table 6: (*Contd...*)

Saxena and Khan (1984)	Phloem feeding by *N. lugens* on rice plants treated with neem oil was significantly less than on untreated plants.
Chari and Muralidharan (1985)	2 per cent kernel suspension completely inhibited the feeding by *Achea janata* larvae.
Singh and Sharma (1986)	Cabbage and cauliflower crops sprayed with various concentrations of neem kernel suspension showed strong repellant and antifeedant property against aphid *Brevicoryne brassicae*.
Prabhakar *et al.* (1986)	Neem kernel extract caused larval mortality, prolonged larval duration and inhibited pupation of *Trichoplusia ni* and *Spodoptera exigua*.
Meisner *et al.* (1986)	Aqueous extract of neem significantly affected the larval weight and pupation of *Ostrinia nubilalis*.
Singh *et al.* (1988)	Neem oil and ethanolic extract at 0.5% exhibited 48.00 and 9.00 per cent mortality of *Lipaphis erysimi* and the aqueous extract casued nil mortality at the same concentration.
Gujar and Mehrotra (1988)	Effective concentration for 50 per cent antifeedant activity of neem oil against *Aulacophora foveicollis* was 0.4 per cent.
Kareem *et al.* (1988)	3 per cent NSKE protected the mung bean crop from *Heliothis* sp., *Etiella* sp. and *Maruca* sp.
Mikolajczak *et al.* (1989)	Neem seeds extracts reduced the larval growth rate and increased the time of pupation of *Spodoptera frugiperda*.
Singh *et al.* (1990)	Neem oil was at par with monocrotophos in reducing the *Scirpophaga incertulas* incidence in rice
Patel *et al.* (1993)	Five per cent suspension sprays of neem seeds found to be effective against 2nd instar larvae of *Amsacta moorie*.
Sarode and Gabhane (1994)	5 per cent NSKE reduced the infestation of okra fruit borer
Dhingra (1996)	Neem oil provided 14 and 12 fold synergism of cypermethrin at 1:1 and 1:5 ratios against *Mylabris pustulata*.
Behera and Satpathy (1996)	NSKE caused 100 per cent mortality of 4th instar larvae of *S. litura* at 10 DAT.
Rao and Dhingra (1997)	Synergistic activity of neem oil in mixed formulation with cypermthrin (1:5) was pronounced against 4 and 9 day old larvae of *S. litura*.
Jeyakumar and Gupta (1999)	NSKE (10%) showed oviposition deterrency and ovicidal effect against *H. armigera*
Dhingra *et al.* (2002)	Neem oil microemulsion reduced the larval and pupal weight gain, pupal development and adult emergence, and increased the larval mortality, larval–pupal intermediates and the pupal deformity in the 3rd instar treated–larvae *H. armigera*. Its relative effectiveness in inhibiting adult emergence compared to the macroemulsion (1.0) was 1.68.
Bajpai and Sehgal (2003)	Neem oil caused high level pupal mortality, abnormal pupae of *S. litura* which could not emerge.
Simmonds *et al.* (2004)	Amongst the important limonoids of neem seeds, isosalanninolide showed antifeedant activity greater than that of nimbin or salannin and comparable to azadirachtin against *S. littoralis S. frugiperda* and *H. armigera* larvae.

6. USE OF BOTANICALS AGAINST PLANT PESTS

Available reports on effectiveness of various botanicals on plant pathogens and insects are summarized below in the form of tables (Tables 7 and 8).

Table 7: Use of botanicals/plant extracts against plant pathogens (fungi, bacteria and viruses)

Target pathogen/ disease	***Crop and/or application (evaluation method)***	***Reference(s)***
Azadirachta indica		
Rhizoctonia solani	Soil incorporation of neem cake	Singh *et al.*, 1980
Chickpea pathogens *R. bataticola*, *F. oxysporum*, *Alternaria cymopsidis*, *A. Niger*	Poison food technique using NSKE	Sharma and Gaur, 2003
Club root of yellow sarson	Neem cake and Nimin	Bhattacharya and Pramanik, 1998
Alternaria blight and white rust of mustard	Two applications of ethanol extract of neem	Godika and Pathak, 2003
Stem rot of mustard leaf extract	Soil application of neem	Prasad *et al.*, 2003
Rice blast	Neem-based compound Repelin	Thangavelu *et al.*, 1995
Rhizoctonia root rot of fenugreek	Soil application of neem cake @ 1 ton per ha	Rajan *et al.*, 1997
Powdery mildew and die-back or fruit rot of chilli	Aqueous extract of neem leaf and neem oil (3%)	Jayalakshmi *et al.*, 1997; Gehlot, 2005
Brinjal stem and branch blight (*Sclerotium rolfsii*) and collar rot disease (*Phomopsis vexans*)	Neem leaf extract and neem cake extract	Jadeja, 2003; Patel *et al.*, 2005a
Damping off of tomato (*Pythium aphanidermatum*)	Aqueous extract of neem leaves	Tripathi, 2005
Basal stem rot of coconut (*Ganoderma lucidum*)	Neem leaf extract 10 per cent	Karunanithi *et al.*, 2003
Mango wilt (*Fusarium solani*)	Soil application of neem cake under pot culture condition	Sharma and Verma, 2003
Bacterial blight of rice	Neem cake extract 5% and NSKE 2%	Eswaramurthy *et al.*, 1993; Maharishi, 1993
Bacterial blight of cluster bean (*Xanthomonas axonopodis*)	Seed treatment with neem extract @25%	Gour and Mali, 2005

Table 7: (*Contd...*)

Table 7: (*Contd...*)

Target pathogen/ disease	***Crop and/or application (evaluation method)***	***Reference(s)***
Bacterial blight of rice (*X. oryzae* pv. *oryzae*)	Neem leaf extract 5%	Madhiazhagan *et al.*, 2002
Cowpea aphid borne mosaic virus	Neem extract	Kannan and Doraiswamy, 1993
Chilli leaf curl and tomato leaf curl	Neem formulations	Maharishi, 1993; Somasekhara *et al.*, 1997
TMV in tomato	Acetone extract of neem leaf by inoculation in local lesion host *N. glutinosa*	Sawant *et al.*, 2005
Pratylenchus thornei in chickpea	Application of neem cake 400 kg per ha	Sebastian and Gupta, 1996
Meloidogyne incognita, *Rotylenchulus reniformis*, *Tylenchorhynchus brassicae* in tomato	Soil application of neem cake	Azam and Tiyagi, 2005
Reniform nematode *R. reniformis* in okra	Aqueous oil cake extract of neem	Bhargava *et al.*, 2002
Melia azedarach		
Rhizoctonia solani (root rot in French bean)	Ethanolic and acetone extract in poison food technique	Sharma and Gupta, 2003
Aspergillus flavus, *Fusarium moniliforme*	Ethanolic extract in poison food technique	Carpinella *et al.*, 1999
Pongamia glabra, Pongamia pinnata		
Fusarium pallidoroseum and *F. moniliforme* (leaf blight of mulberry), *Fusarium oxysporum* (leaf spot in mulberry)	Leaf extract of *P. pinnata*	Gupta *et al.*, 1996
Basal stem rot of coconut	10% extract	Karunanithi *et al.*, 2003
Rice yellow dwarf virus transmission by *Nephotettix virescens*	Neem and pongamia oil formulations	Rajappan *et al.*, 1999
Reniform nematode *R. reniformis* in okra	Aqueous oil cake extract of pongamia	Bhargava *et al.*, 2002
Annona squamosa		
Transmission of rice tungro virus by leaf hopper *N. virescens*	Mixture of seed oil of custard apple and neem	Mariappan and Saxena, 1984

Table 7: (*Contd...*)

Table 7: (*Contd...*)

Target pathogen/ disease	*Crop and/or application (evaluation method)*	*Reference(s)*
***Madhuca indica* (Synonym–*M. longifolia*)**		
Alternaria solani (leaf blight of tomato)	Oil of Madhuca *In Vitro*, pot and field study	Vadivel and Ebenezar, 2006
Cowpea aphid borne mosaic virus	Extract of *M. longifolia*	Kannan and Doraiswamy, 1993
***Ailanthus excelsa* (Ardusa)**		
Seed mycoflora of fennel *Alternaria alternata*, *Aspergillus niger*, *Drechslera australiensis*, *F. solani*, *R. solani*	Ethanolic extract of ardusa leaf *In Vitro* condition	Trivedi, 2004
F. oxysporum wilt pathogen of fenugreek	Leaf extract *In Vitro* condition	Bansal and Gupta, 2000
Eucalyptus globules		
Fusarium pallidoroseum and *F. moniliforme* (leaf blight of mulberry), *Fusarium oxysporum* (leaf spot in mulberry)	Leaf extract of *Eucalyptus* sp.	Gupta *et al.*, 1996
R. solani (black scurf of potato)	Essential oil of *Eucalyptus*	Dhaliwal *et al.*, 2003
Seed mycoflora of fennel *Alternaria alternata, Aspergillus niger, Drechslera australiensis, F. Solani, R. Solani*	*Eucalyptus* oil	Trivedi, 2004
Penicillium italicum	Essential oil of *E. citriodora*	Tripathi, 2005
Prosopis juliflora		
Colletotrichum capsici, Gloeosporium piperatum	Leaf extract of *P. juliflora*	Gomathi and Kannabiran, 2000
Rice blast pathogen *P. oryzae*	Leaf extract of *P. juliflora*	Kamalkannan *et al.*, 2001
Bacterial blight of rice	Leaf extract of *P. juliflora* in pot trials	Madhiazhagan *et al.*, 2002
***Vitex negundo* (Nirgundi)**		
Penicillium expansum (fruit rot of apple)	Pre- and post infection dip with aqueous leaf extract	Singh and Sumbali, 2003
Phytoplasma (causing phyllody in *Sesamum*)	Leaf extract	Srinivasulu and Narayanswamy, 1993

Table 7: (*Contd...*)

Table 7: (*Contd...*)

Target pathogen/ disease	***Crop and/or application (evaluation method)***	***Reference(s)***
Cowpea aphid borne mosaic virus	Plant extract	Kannan and Doraiswamy, 1993
Nerium oleander* (pink oleander), *Nerium indicum		
Colletotrichum capsici, Gloeosporium piperatum (chilli anthracnose)	Leaf extract	Gomathi and Kannabiran, 2000
Penicillium italicum	Various organic solvent extract of *Nerium indicum*	Tripathi, 2005
Clerodendron indicum, Clerodendron inerme		
Rice blast pathogen *Pyricularia oryzae*	*C. inerme* extracts	Kamalkannan *et al.*, 2001
Sclerotium rolfsii	–do–	Tiwari *et al.*, 2002
Penicillium italicum	Various organic solvent extract of *C. inerme*	Tripathi, 2005
Xanthomonas campestris	*Clerodendron setatum* and *C. indicum*	Tiwari *et al.*, 2005
Ring spot virus and TMV	Aqueous leaf extract of *C. inerme*	Patel, 1990
Cowpea aphid borne mosaic virus	Extract of *C. inerme*	Patel, 1999
Papaya ring spot virus	Leaf extract of *C. inerme*	Singh and Awasthi, 2005
***Ipomoea* sp.**		
Xanthomonas campestris pv. *campestris*	Extract of *Ipomoea mauritiana* under *in Vitro* conditions	Tiwari *et al.*, 2005
Murraya koenigii		
Colletotrichum lindemuthianum (anthracnose of French bean)	Aqueous leaf extract	Gupta *et al.*, 2005
Penicillium italicum	Various organic solvent extract of *Murraya* leaf	Tripathi, 2005
Xanthomonas campestris	Leaf extract	Bissa *et al.*, 2005
Jatropha curcas		
Cowpea aphid borne mosaic virus	Pre-inoculation and post-inoculation treatment with extract of *J. curcas* was reported to give 100% inhibitory effect upto 16 days	Patel, 1999

Table 7: (*Contd...*)

Table 7: (*Contd...*)

Target pathogen/ disease	***Crop and/or application (evaluation method)***	***Reference(s)***
Meloidogyne sp. in tomato	Jatropha cake @ 2 tonnes per ha	Patel *et al.*, 2005b
Bougainvillea spectabilis		
Sugarcane mosaic potyvirus in sorghum	Extract of *B. spectabilis*	Molina and Leon, 1991; Molina and Sanchez, 1993
Potato virus X, TMV	–do–	Duarte *et al.*, 1996
Cowpea aphid borne mosaic virus	Pre-inoculation and post-inoculation treatment with extract of *J. curcas* was reported to give 100% inhibitory effect upto 6 days	Patel, 1999
***Lantana* sp.**		
Cercospora cocciniae (leaf spot of Ivy gourd)	10% *Lantana* extract	Davis *et al.* 2003
Phomopsis vexans (stem and branch blight of brinjal)	Leaf extract of *Lantana*	Jadeja, 2003
R. solani (root rot and web blight in French bean)	Acetone extract	Sharma and Gupta, 2003
Seed mycoflora of fennel *Alternaria alternata, Aspergillus niger, Drechslera australiensis, F. Solani, R. solani*	Ethanolic extract	Trivedi, 2004
Pythium aphanidermatum and *F. solani* (storage rot of ginger)	Extract of *Lantana*	Jeeva Ram and Thakore, 2005
Pseudomonas solanacearum	Phenolic compounds from leaf extracts of *L. camera*	Sukul and Chaudhari, 1999
Cowpea aphid borne mosaic virus	Extract of Lantana	Patel, 1999
Meloidogyne incognita	Methanolic extract of Lantana	Anon, 2005
***Calotropis* sp.**		
Alternaria blight and white rust of mustard	Two sprays of ethanolic extract of *Calotropis*	Godika and Pathak, 2003
Mentha piperata		
R. solani (black scurf of potato)	Essential oil both under laboratory and pot culture	Dhaliwal *et al.*, 2003

Table 7: (*Contd...*)

Table 7: (*Contd...*)

Target pathogen/ disease	*Crop and/or application (evaluation method)*	*Reference(s)*
Ocimum sanctum* and *O. basilicum		
Phytophthora meadii	10% leaf extract	Naik and Anahosur, 2001
Colletotrichum capsici (die-back of chilli)	Leaf extract	Gehlot, 2005
Alternaria brassicae (mustard leaf blight)	Methanolic extract	Patni *et al.*, 2005
TMV in tomato	Acetone extract	Sawant *et al.*, 2005
M. incognita	Plant extract	Sharma *et al.*, 2005
Parthenium		
Xanthomonas campestris pv. *vesicatoria* (seedling leaf spot disease)	Leaf extract of *Parthenium*	Agrawal and Sharma, 2005
Alternaria blight and white rust of mustard	Two sprays of ethanolic extract of *Parthenium*	Godika and Pathak, 2003
Allium cepa		
Alternaria, Colletotrichum, Fusarium	Spraying of onion bulb extract at 5%	Stoll, 2000
Alternaria tenuis (fruit rot of grape)	Bulb extract of onion	Khilare *et al.*, 2002
TMV in chilli	Extract	Anon, 2004
***Tagetes erecta* (marigold)**		
Alternaria alternata, Drechslera halodes and Helminthosporium speciferum	Extract	Srivastava and Srivastava, 1998
Colletotrichum capsici, R. solani and *Sclerotinia sclerotiorum*	Leaf extract	Shivpuri *et al.*, 1997
Colletotrichum lindemuthianum (anthracnose of French bean)	Aqueous leaf extract	Gupta *et al.*, 2005
Root knot nematode	Fermented marigold extract	Stoll, 2000
Curcuma longa		
Macrophomina phaseolina (dry root rot/stem blight of cowpea)	Rhizome extract	Bhatnagar *et al.*, 2003

Table 7: (*Contd...*)

Table 7: (*Contd...*)

Target pathogen/ disease	***Crop and/or application (evaluation method)***	***Reference(s)***
Zingiber officinale		
Brown spot of rice, mango anthracnose	Ginger extract	Sridhar *et al.*, 2002
R. solani (sheath blight of rice)	Acetone extract	Verma *et al.*, 2003
Yellow vein mosaic of okra	Ginger extract	Sridhar *et al.*, 2002
Leucas aspera		
Drechslera oryzae	Leaf extract	Ganesan and Krishnaraju, 1995
Alternaria porri	Leaf extract	Choudhuri *et al.*, 2006
Allium sativum		
F. solani and *Erwinia carotovora* causing wilt and bulb rot of garlic	Garlic bulb extract	Alice *et al.*, 1997
Macrophomina phaseolina	–do–	Datar, 1999
Sclerotinia sclerotiorum (stem and fruit rot in tomato)	–do–	Kumar and Anamika, 2003
Mustard white rust, powdery mildew, Alternaria blight of sunflower	–do–	Chattopadhyay *et al.*, 2005
***Datura* sp.**		
Alternaria cyamopsidis (cluster bean leaf spot and blight)	Extract of black datura	Pangavhane and Solanky, 2003
Exobasidium vexans (blister blight of tea)	Ethanolic and aqueous extracts of Datura	Saha and Saha, 2003
Drechslera oryzae	Datura leaf extract	Krishnappa *et al.*, 2005
TMV and Tobacco ring spot virus in tobacco	Leaf extract of *D. metel*	Verma and Mukherjee, 1979
Acorus calamus		
Alternaria solani (tomato leaf blight)	Extract of dried root @ 5%	Vadivel and Ebenezar, 2006

Table 8: Use of botanicals/plant extracts against phytophagous insects

Name of the plant(s)	***Target insect(s)***	***Crop and/or application (Evaluation method)***	***Reference(s)***
Bitter gourd, *Momordica charantia*	Serpentine leaf miner, *Liriomyza sativae*	Showed contact toxicity, ovicidal as well as larval growth regulatory activity.	ShaoQin, 2001
Black pepper, *Piper nigrum*	Lesser grain borer, *Rhizopertha dominica*	3 g pepper when mixed with per 100 g wheat resulted in 100% mortality after 15 days, as did 2.5 g after 25 days.	Tripathi *et al.,* 2000
Dirca palustris	Neonate larvae of *Helicoverpa zea, Lymantria dispar, Orgyia leucostigma,* and *Malacosoma disstria.*	Potent feeding deterrent	Ramsewak *et al.*, 2001
Sophora alopecuroides	Green peach aphid, *Myzus persicae*	Strong aphidicidal activity when evaluated by glass slide dipping method	Chong *et al.*, 2000
Cleome viscose	2nd and 4th instar larvae of *Anopheles stephensi*	Larvicidal activity	Saxena *et al.*, 2000
Vetiver grass, *Vetiveria zizanioides*	Formosan subterranean termite, *Coptotermes formosanus*	Repellent activity	Zhu *et al.*, 2001
Vinca rosea	Neonate and third instar larvae of *Helicoverpa armigera*	Artificial diet incorporation and leaf dip method	Halder *et al.*, 2010 Halder *et al.*, 2009a, b
	Viviparous, apterous adults of *Lipaphis erysimi*	Direct spray and leaf residue methods	Halder *et al.*, 2010
Gnidia glauca	Sixth instar larvae of *Helicoverpa armigera*	Antifeedant activity and larval toxicity	Sundararajan and Kumuthakalavalli, 2001
Chinaberry tree, *Melia azedarach*	Sweetpotato whitefly, *Bemisia tabaci*	Repellent and lethal effect	Hammad *et al.*, 2000
Datura stramonium	Pulse beetle, *Callosobruchus chinensis*	Reduced fecundity	Mishra, 2000

Table 8: (*Contd...*)

Table 8: (*Contd...*)

Name of the plant(s)	*Target insect(s)*	*Crop and/or application (Evaluation method)*	*Reference(s)*
Lantana camera	Orange banded blister beetle, *Zonabris pustulata*	Caused toxicity	Oudhia, 2000
Stellera chamaejasme (Wikstroemia chamaejasme)	Larvae of *Pieris rapae, Ostrinia furnacalis, Plutella xylostella* and *Spodoptera litura*	Caused mortality	Zhou *et al*. 2000
Gnaphalium affine	*Spodoptera litura*	Strong antifeedant activity	Morimoto, 2000
Cedrela salvadorensis	Fall Armyworm, *Spodoptera frugiperda*	Insect Growth regulatory activity	Cespedes *et al*., 2000
Withania somnifera	Khapra beetle, *Trogoderma granarium*	Oviposition detterent	Dwivedi *et al*., 1999
Ageratum conyzoides	Gundhy bug, *Leptocorisa acuta*	Toxic to nymphs and adults	Lu, 1982
Croton sparsiflora	Gundhy bug,	Reduced survival *Leptocorisa acuta*	Narasimhan and Mariappan, 1988
Pongamia glabra	*Chilo partellus*	Reduced survival	Sharma and Bhatnagar, 1990
Annona squamosa	Green leafhopper, *Nephotettix virescens* and Brown Planthopper, *Nilaparvata lugens*	Insecticidal activity	Epino and Saxena, 1982
Acorus calamus	Potato tuber moth, *Phthorimaea operculella*	Larval toxicity	Pandey *et al*., 1982
Callistemon lanceolatus	Neonate and third instar larvae of *Helicoverpa armigera*	Artificial diet incorporation and leaf dip method	Halder *et al*., 2010
	Viviparous, apterous adults of *Lipaphys erysimi*	Direct spray and leaf residue method	Halder *et al*., 2010
Madhuca indica	Green leafhopper, *Nephotettix virescens* and rice leaf folder, *Cnaphalocrosis medinalis*	Reduced survival	Narasimhan and Mariappan, 1988

Table 7: (*Contd...*)

Table 8: (*Contd...*)

Name of the plant(s)	***Target insect(s)***	***Crop and/or application (Evaluation method)***	***Reference(s)***
Rhododendron molle	Rice yellow stem borer, *Scirpophaga incertulas*	Lethal toxicity	Chiu, 1982
Annona squamosa	*Oryzaephilus surinamensis, Tribolium castaneum*	Insect growth regulatory activity	Prakash and Rao, 1997
Solanum xanthocarpum	*Henosepilachna vigintioctopunctata*	Insect growth regulatory activity	Chitra *et al.*, 1991
Eucalyptus camaldulensis	*Helicoverpa armigera*	Antifeedancy and insect growth regulatory activity	Birah *et al.*, 2009
Cannabis sativa	*Spodoptera litura*	Antifeedancy	Sharma *et al.*, 2005
Andrographis paniculata	*Leptocorisa acuta*	Protect the rice panicle	Gupta *et al.*, 1990
Hyptis suavelobens	Cabbage aphid, *Myzus persicae*	Caused mortality	Roy and Pande, 1991
Eclipta alba	*Sitotroga cerealella*	Ovicidal activity	Prakash and Rao, 1997
Dodonaea viscosa	Teak defoliator, *Hyblaea puera*	Insecticidal activity	Deepa and Remadevi, 2007
Petunia parodii	*Schistocerca gregaria* and *Blattella germanica*	Antagonist of (GABA)–mediated chloride ionophore	Isman *et al.*, 1997
Aloe vera (A. barbadensis)	Sissoo defoliator, *Plecoptera reflexa*	Antifeedant activity	Kulkarni *et al.*, 1997
Citrus reticulata	*Leptocorisa acuta*	Lethal toxicity	Gupta *et al.*, 1990
Gynandropsis gynandra	*Henosepilachna vigintioctopunctata*	Lethal toxicity	Chandel *et al.*, 1987
Calotropis procera	*Spodoptera litura*	Antifeedancy	Sharma *et al.*, 2005
Clerodendrum inerme	Castor semilooper, *Achaea janata*	Lethal effect	Basappa and Lingappa, 2005
Lavendula officinalis	Two spotted spider mite, *Tetranychus urticae*	Acaricidal activity	Rangrez and Rather, 2007
Ocimum sanctum	*Amrasca biguttula biguttula* and *Earias* spp.	Insecticidal activity	Adiroubane and Letchoumanane, 1998
Adenocalyma alliacea (Adenocalymma alliaceum)	*Drosophila busckii*	Inhibited egg-laying, hatching and pupation	Sinha, 1998

7. BOTANICALS TO CONTROL WEEDS

There are reports that weed control can be achieved using certain plant extracts. Extracts of some plants contain allelopathic chemicals like phenolic acids, alkaloids, coumarins and quinones which are nothing but secondary metabolites. These plants through allelopathic effect can negatively influence the surrounding flora (*i.e.*, weeds) some of which are listed below (Table 9).

Table 9: Weed management using botanicals

Plant extract	*Weeds targeted*	*Reference(s)*	*Remark(s)*
Parthenin	*Ageratum conyzoides*	Batish *et al*. 1997	Inhibits germination, radicle elongation, and seedling growth of target weed
	Avena fatua and *Bidens pilosa*	Batish *et al*. 2002	Reduced root and shoot length and seedling dry weight of target weed
Propenyl derivative of parthenin	*Cassia tora*	Datta and Saxena, 2001	
Leaf extract of Vana Tulsi (*Ocimum gratissimum*)	Parthenium	Thapar and Singh, 2003	Adverse effect on seed germination, seedling growth, chlorophyll content and protein of target weed
Parthenium	*Hydrilla verticillata*	Pandey *et al*., 1993; Pandey, 1995; Mackay, 1992	Reduction in dehydrogenase activity, membrane integrity, chlorophyll content and plant growth
Ipomoea tricolor	*Amaranthus hypocondricus, Echinocloa crusgalli*	Anaya *et al*., 1990	*I. tricolor* contains tricolorin A, a plant growth inhibitor

8. BOTANICALS TO CONTROL PLANT PARASITIC NEMATODES

Reddy and Khan (1995) observed that okra plot when treated with karanj and neem cake significantly reduce the infestation of root knot nematode, *Meloidogyne incognita* resulting in higher okra fruit yield compared to control. They also noticed that groundnut cake when applied along with carbofuran gave the highest yield. Otiniano *et al*. (2002) from Spain documented that neem oil at 10 l/ha was effective against *Meloidogyne* and *Tylenchorhynchus* spp. affecting *Solanum melongena* var. Cristal grown in greenhouses. In another study, Sharma *et al*. (2001) reported that neem oil cake when mixed with *Verticillium chlamydosporium* and applied at 2 t/ha at the mulberry field infested with root knot nematode (*Meloidogyne incognita*) resulted significantly less number of root knot per plant, egg masses per plant and nematode population in soil as well as high leaf yield as compared to untreated control. To control the nematode infestation in sugarcane in Brazil application of neem oil (3%) reduce the *Pratylenchus* population where as the crushed neem (100 kg) reduced *Meloidogyne* population significantly from the rhizosphere (Rossi *et al*., 2006). Neem The combination of neem cake and carbofuran, in reduced doses, revealed the best response in reducing the nematode multiplication with an increase in plant vigour in cowpea infested with disease–complex caused by *M. incognita* and *Fusarium oxysporum* (Singh and Goswami, 2001). To control the root-knot nematode (*Meloidogyne incognita*) in brinjal Singh *et al*. (2005) demonstrated that Indian mustard cake in combination with press mud resulted in the greatest reduction in *M. incognita* population (37.59%) in addition to the highest plant height (24.02 cm), shoot fresh weight (9.02 g) and root fresh weight (3.95 g). Compatibility of neem leaf extract with microbial was confirmed by the study of Nagesh and Reddy (1995) and they concluded that the combinations of *V. lecanii* and *P. lilacinus* (each at 2×10^4 spores ml^{-1}) with 5% neem leaf extract resulted in significantly higher plant growth and flower yield in *Crossandra undulaefolia* infested by *Meloidogyne incognita*. Root knot nematode in cow pea was controlled by the application of neem leaf extract (15%) reduces the number of nematodes in soil and root which was 90 and 90.3% decrease over control, respectively (Maheswari and Sundarababu, 2001).

9. BOTANICALS TO CONTROL MITES

Phytophagous mites often pose serious problems in crop cultivation. Plant extracts have been found to have controlling effect on such mites of economic importance some of which are listed below (Table 10).

10. BOTANICALS AND THEIR HARMFUL EFFECTS

Use of products of plant origin does not guarantee safety in all cases. There is a variety of plant produced compounds which vary in terms of toxicity to mammals. The acetogenins from *Annona* spp., rotenone from *Derris* sp. block energy production in mitochondria and therefore toxic to mammals (Londershausen *et al*., 1991). Acute exposure of rats to rotenone produces brain lesions consistent with those observed in humans and animals with Parkinson's disease (Betarbet *et al*., 2000). In another study the half-life of rotenone was reported to be 4 days, and at harvest time of olives the levels of residue which concentrated in olive oil was above the tolerance limit

(Cabras *et al.*, 2002). Though technical grade pyrethrum is less toxic (LD_{50} 1500 mg/kg), pure pyrethrins are moderately toxic to mammals (rat oral acute LD_{50} 350 to 500 mg/kg) (Casida and Quistad, 1995).

Table 10: Effect of plant extracts on phytophagous mites

Plant extract	***Mite targeted***	***Reference(s)***	***Remarks***
Lavendula officinalis	Two spotted spider mite, *Tetranychus urticae*	Rangrez and Rather, 2007	Acaricidal activity
Tulsi leaf extract	Tetranychus in tomato, jowar, rice, sugarcane, bhendi, brinjal, papaya etc.	Anon, 2004	Acaricidal
Neem products	*T. cinnabarinus, T. urticae* and *T. macfarlanei*	Dreyer, 1983; Bhanderi, 1991; Meman *et al.*, 1997	Oviposition and feeding deterrence, repellency
	T. urticae	Dimetry *et al.*, 1993	Ovicidal activity
	T. cinnabarinus	Patel, 1996	Ovicidal activity
	T. macferlanei and coconut mite *Aceria guerreronis*	Naik, 2000	Acaricidal
Essential oil of *Mentha piperita*	Adult female mite of *Tetranychus cinnabarinus*	Mansour and Ravid, 1986	Caused repellency, mortality and reduced egg laying

In pure form, the active principles, cevadine–type alkaloids of sabadilla (*Schoenocaulon officinale*), a tropical lily that grows in Central and South America are extremely toxic to mammals (rat oral LD_{50} is 13 mg/kg). Pure nicotine also produced extreme toxicity to mammals (rat oral LD_{50} is 50 mg/kg) and rapid dermal absorption in humans (Isman, 2006). Broad-spectrum insecticides based on essential oils are poisonous for pollinators and natural enemies.

11. COMPATIBILITY OF BOTANICALS/PLANT EXTRACTS

Botanicals alone or for that matter any single strategy, be it microbial bio-pesticides or other biocontrol agent may prove inadequate in managing the crop pests. It is often envisaged that combining botanicals with other natural enemies of crop pests would achieve the target of pest control. Though botanicals have been found to have adverse effect on some beneficials, in most cases they have either positive effects or no adverse effect on natural enemies of insects (Table 11). Pesticidal activities of microbial agents like *Bacillus thuringiensis*, *Beauveria bassiana*, *Paecilomyces* sp., *Metarrhizium anisoplieae* including nuclear polyhedrosis virus (NPV) were enhanced when combined with plant based products

(Table 12). NPV combined with rotenone had improved efficacy against *Galleria mellonella*; with nicotine sulphate or pyrethrum showed increased larval mortality of *S. litura*; with methanolic extracts of *Ocimum sanctum* and *Acorus calamus* showed increased efficacy against *H. armigera* and *S. litura*; and with crude extract of *Prosopsis juliflora* leaves (at 20% concentration) proved 20 times more effective against *H. armigera* larvae (Dodia *et al.*, 2008).

Table 11: Compatibility of botanicals with natural enemies of insects

Botanical product	***Adverse effects on***	***Reference(s)***
Rotenone	Egg parasitoid *Trichogramma* spp.	Iannacone and Lamas, 2003
Commercial neem products *viz.*, Neem Azal, Neem Azal F, Nimbecidine and Neem gold	Ovicidal and larvicidal effect on green lacewing, *Chrysoperla carnea*	Srinivasan and Babu, 2000
The unfortified oils	Reduction in the adult emergence of *T. chilonis* and parasitization, reduction in the population of *Bracon brevicornis*	Saminathan, 2000
Different neem formulations as (i) unfortified (ii) fortified (iii) dust (iv) seed kernel extracts	Reduction in the populations of natural enemies like spiders, coccinellids, chrysopids and syrphids	Saminathan, 2000
Rosmarinus officinalis oil	Toxic to predatory mite *Amblyseius barkeri*	Momen and Amer, 1999
Neem oil	Moderately toxic to braconids, *Chelonus blackburni*	Shelke *et al.*, 1990
Neem products	Toxic effects on *Rhynocoris marginatus*	Jhansilakshmi *et al.*, 1998
Ecotypes of neem	Toxic effect on parasitoid *Bracon brevicornis*	Srivastava *et al.*, 1997
Nicotine sulphate	Very toxic to *Apanteles africanus, Telenomus remus, Dacnusa sibirica* and *Diglyphus isaea*	Chari *et al.*, 1996
Aqueous, ethanolic and hexane extracts of neem seed kernel	Oviposition and feeding of *Trichogramma chilonis*	Raguraman and Singh, 1995
Garlic oil spray	Kills the natural enemies of aphids	Olkowski *et al.*, 1995

Table 11: (*Contd...*)

Table 11: (*Contd...*)

Botanical product	***Adverse effects on***	***Reference(s)***
Botanicals	***Positive/no adverse effect on***	***Reference(s)***
Neem formulations	Braconid *Diadegma semiclausum* (larval parasitoid of *Plutella xylostella*)	Jayaraj and Ignacimuthu, 2005
NSKE	Larval parasitoids *Cotesia flavipes* on *Chilo sachariphagus indicus*	Reddy and Srikanth, 1996
	Cotesis plutella on *Plutella xylostella*	Srivastava *et al.*, 1997
Neem formulations Fortune Aza (0.1%) and Neem Azal (0.3%)	*Trichogramma japonicum* on *Scirpophaga incertulas*	Jhansi Lakshmi *et al.*, 1997
Oils from *Vinea rosea* (1% and 5%)	Lady bird beetle *coccinella septempunctata*	Halder *et al.*, 2010
Oils from *Callistemon lanceolatus* (1% and 5%)	–do–	Halder *et al.*, 2010

Table 12: Effect of botanicals on microbial agents

Microbial agent	***Effect***	***Reference(s)***
Bacillus thuringiensis var. *kurstaki*	Enhanced mortality of *Spodoptera frugiperda* larva in combination with neem product AZT–VR–K	Jayaraj and Ignacimuthu, 2005
B. thuringiensis	Increased efficacy against *P. xylostella* in mixture with neem products	Mau *et al.*, 1995
Bacillus thuringiensis var. *kurstaki*	Increased mortality of *H. virescens* larvae in combination with Neemix	Walter, 1999
Beauveria bassiana	Greater morality of *S. litura* with sub-lethal dose of *B. bassiana* in combination with sub-lethal dose of NSKE	Jayaraj and Ignacimuthu, 2005
Paecilomyces fumosoroseus	Enhanced efficacy of entomogenous fungus *Paecilomyces*	Landa and Bohata, 1999

Table 12: (*Contd...*)

Table 12: (*Contd...*)

Microbial agent	*Effect*	*Reference(s)*
	fumosoroseus strain PFR 97–Apopka against several pests *viz, Trialeurodes vaporariorum, Aphis gossypii* or *Tetranychus urticae* in combination with neem oil or azadirachtin	
B. thuringiensis and *Metarrhizium anisopliae*	Significant reduction in the growth and development, pupation (46.7%) and adult emergence (76.7%) of *S. litura* larvae in combination with fortified formulation of plant products	Saminathan, 2000

12. COMPATIBILITY AMONG BOTANICALS

When botanicals were mixed for the purpose of agricultural pest control they showed synergistic rather than antagonistic effect.

Raman *et al.* (2000) documented that *Annona* oil @ 0.72% when applied on castor the mean larval population of castor semilooper, *Achea janata* was 18.33 and 22.00 per ten plants after 1 and 3 days of spraying but when it is mixed with neem oil @ 1:1 ratio and applied at the same dose, the mean larval population decreased and these were 15.33 and 18.33 per ten plants after 1 and 3 days respectively. One per cent chilly and garlic @1:1 mixture when sprayed over cowpea there was 34.17, 20.76 and 13.88 per cent reduction in aphids, pod borer and jassids population as compared with untreated control (Ukey *et al.*, 1999). In another study Jayakumar *et al.* (2004) observed highest percentage of reduction (45%) in *Aphis craccivora* population with *Hyptis suaveolens* (at 5% concentration) followed by *Melochia corchorifolia* (29.45%) after 7 DAT but when these plant extracts were mixed at the ratios of 3:1, 1:1, 1:3 and 1:4, the per cent reduction was significantly lower as compared with *H. suaveolens* (5%) alone and the mean reduction were 39.38, 40.69, 39.21 and 30.79 per cent respectively.

Santhosh Kumar (1994) tested the efficacy of a mixture of aqueous extract of *Andrographis paniculata* and garlic in chilli fields infested with aphids (*Aphis gossypii*), thrips (*Scirtothrips dorsalis*) and mite

(*Polyphagotarsonemus latus*). There was a significant reduction in the population of these pests in plots treated with the mixture which was comparable with that of fields treated with malathion and monocrotophos. Leaf extracts of *Tabernaemontana divaricata* was blended with that of *Quisqualis indica, Chenopodium album, Annona squamosa, Anethum sowa* and *Tamarindus indicus* at 1:1 ratio (v/v) and assessed for their oviposition deterrent action against the pulse beetle, *Callosobruchus chinensis*. When mixed with *A. squamosa* and *C. album*, the mixtures resulted in 97.15 and 94.70 per cent deterrence, respectively while with *Q. indica*, it gave 75.15% reduction in oviposition over control. The other two combinations resulted in moderate reduction in oviposition varying from 51.83 to 57.44 per cent (Dwivedi and Venugopalan, 2004).

Chowdhury *et al*. (2001) reported that the activity of azadirachtin was considerably enhanced by incorporation of different concentrations of turmeric oil in the mixtures. Among the three different proportions (1:1, 2:1 and 3:1), azadirachtin–turmeric oil (1:1) mixtures recorded pronounced IGR activity (EC_{50} $1.26 \times 10^{-2}/2.16 \times 10^{-2}$ %) and considerable antifeedant activity (EC_{50} $2.90^{-2} \times 10/1.0 \times 10^{-2}$ %) against third and fifth instar larvae of *Spodoptera litura*, respectively. The apparent enhancement in activity of azadirachtin–turmeric oil mixtures were attributed to the possible additive, synergistic and/or stabilizing effect of turmeric oil. Other than azadirachtin, several other allelochemicals particularly salanin, from Indian neem tree *A. indica* have antifeedant and IGR properties against various insect pests. When salanin was mixed with 3-O-acetyl salanol, salanol and combination of both these with salanin and tested against 48 h old neonate *H. armigera* larvae it recorded pronounced IGR activity (EC_{50} values 78.5, 80.5 and 71.9 ppm, respectively) and in case of *S. litura* the EC_{50} values were 77.8, 81.1 and 70.7. Non-aza limnoids having structural similarities and explicitly similar mode of action like feeding deterrence, IGR have no potentiating effect in any combination (Juan *et al*., 2000). Halder *et al*., 2010 reported that 1:3 mixture of methanolic extracts of *V. rosea* and *C. lanceolatus* showed higher antifeedancy, larval growth inhibition, cumulative larval mortality as well as more adult inhibition against neonate larvae of *H. armigera* when applied through artificial diet incorporation method as compared to 1:1 and 3:1 mixtures and also assumed that many botanical insecticides including azadirachtin is having diverse mode of action. At higher concentration (75%) of methanolic extract of *Vinca* in mixture might be blocking the other side of action or inhibiting the ion channels. However, Mineo *et al*. (2000) from Italy reported that azadirachtin when mixed with mineral paraffin oil showing teratological symptoms against the natural parasitoids of *Phyllocnistis citrella* Stainton.

Table 13: Synergistic effect of botanicals in combination in managing agricultural pests

Botanicals in combination	***Target pest***	***Reference(s)***
Seed extracts of neem, *Pongamia pinnata* and *Vitex negundo*	*H. armigera*	Babu *et al.*, 2000
Neem, sweet flag and pongam	*Earias vitella* in okra	Rao *et al.*, 2002
Annona + neem	*H. armigera* in pigeon pea	Das *et al.*, 2000
A. squamosa and neem oil	*N. virescence* and transmission of rice tungro virus	Mariappan and Saxena, 1984a
Aloe+ *Vitex* extract	*H. armigera* in chickpea	Barapatre and Lingappa, 2003
Lantana leaf extract 20%+ neem seed kernel extract 3%	Castor jassids, *Amarasca bigutula bigutula*	Patel and Patel, 1996

13. COMPATIBILITY OF BOTANICALS WITH CHEMICAL PESTICIDES

There have been several reports of replacing chemical pesticides which are inherently harmful to the ecosystem with the botanicals. In many case, mixture of botanicals and chemical pesticides were found to have synergistic effect. Such attempts will reduce the application of synthetic pesticides and thereby the environmental risk. Karanj oil or citronella oil mixed with deltamethrin was found to be more toxic to stored product pest *Tribolium castaneum* (Sridevi and Dhingra, 1997). Highest mortality of *S. litura in vitro* was found when *Annona* seed extract (1%) was combined with monocrotophos 0.05% (Boreddy and Chitra, 2001). Improvement in the efficacy of insecticides in controlling *P. xylostella* was reported in combination with 4% aqueous extract of neem leaves (Facknath, 1993). The combination of indoxacarb and *Pongamia glabra* oil (5%) showed encouraging result in controlling *S. litura* (Loganathan *et al.*, 2003). Bandyopadhyay and Kumar (2004) documented that highest percentage reduction of the mulberry whitefly, *Dialeuropora decempuncta* population (77.35%) was obtained by the combination of neem oil and triazophos (0.1%).

CONCLUSIONS

Plants innate ability to defend the invading pests through production of secondary metabolites has been widely explored. Such exploration drives have yielded many plant based products which supplemented the efforts of crop protection. Advantages of botanicals over synthetic chemicals have placed them in a rank which is highly sought after in present days eco-agriculture and environment friendly crop protection endeavor. In practicality only neem occupies the most prominent and successful position for crop protection. In order to further the agenda of chemical free crop protection, many innovations in terms of new sources and combination products are the need of the hour.

REFERENCES

Adiroubane, D. and Letchoumanane, S. (1998). *Indian Journal of Agricultural Sciences,* **68(3):** 168–70.

Agrawal, K. and Sharma, D.K. (2005). *J. Mycol. Pl. Pathol.,* **35(3):** 586.

Alice, D., Seetharaman, K. and Ebenezar, *E.G.* (1997). In: National symposium on plant protection Towards Sustainability, 22–24 Dec. 1997, Hyderabad, pp. 28.

Anaya. (1990). *In* Compendium: Summer school: Pesticides, chemistry a modern methods of residue analysis, held at GAU, Anand 28 May–17 June. 2002. pp. 46–62.

Anonymous. (2004). *Amaltas,* **2(10):** 32–49.

Anonymous. (2005). *Natural Product Research,* **19(6):** 609–13.

Attri, B.S. (1975). *Indian Journal of Entomology,* **37(3):** 417–19.

Azam M.F. and Tiyagi, S.A. (2005). *In* (Abs.): Second Global Conference. Plant health-global wealth, at Udaipur, India, 25–29 Nov. 2005, p. 296.

Babu, R., Murugan, K., Kavitha, R. and Sivaramakrishnan, S. (2000). *Indian J. Env. Toxicol.* **10:** 42.

Bajpai, N.K. and Sehgal, V.K. (2003). *Indian Journal of Entomology,* **65(1):** 427–33.

Bandyopadhyay, U.K. and Kumar, M.V.S. (2004). *Journal of Entomological Research,* **24(4):** 325–29.

Bansal, R.K. and Gupta, R.K. (2000). *Indian Phytopath.* **53(1):** 107–108.

Barapatre, A.B. and Lingappa, S. (2003). In: National symposium on frontier areas of entom. Res., 5–7 Nov. 2003, New Delhi: pp. 333–36.

Basappa, H. and Lingappa, S. (2005). *Indian Journal of Plant Protection,* **33(2):** 223–25.

Batish. *et al.* (1997). *In* Compendium: Summer school: Pesticides, Chemistry a modern methods of residue analysis, held at GAU, Anand, 28 May–17 June, 2002. 46–62.

Batish. *et al.* (2002). *In* Compendium: Summer school: Pesticides, Chemistry a modern methods of residue analysis, held at GAU, Anand, 28 May–17 June, 2002. pp. 46–62.

Behera, U.K. and Satpathy, C.R. (1996). *Insect Environment,* **11(2):** 43–44.

Betarbet, R., Sherer, T.B., MacKenzie, G., Garcia–Osuna, M., Panov, A.V. and Greenamyre, J.T. (2000). *Nature Neuro Sci.,* **3:** 1301–06.

Bhanderi, G.R. (1991). A thesis submitted to GAU, Navasari. M. Sc. Agri.

Bhargava, S., Verma, M.K. and Dashora, P.K. (2002). *J. Mycol. Pl. Pathol.* **32(3):** 436.

Bhatnagar, Kalpana., Cheema, H.S., Bansal, R.K. and Chitle, K. (2003). *J. Mycol. Pl. Pathol.* **33(3):** 505.

Bhattacharya, I. and Pramanik, M. (1998). *Indian Phytopathology* **51(1):** 87–90.

Birah, A., Raghuraman, M., Kaushik, N. and Gupta, G.P. (2009). *Pesticide Research Journal,* **21(1):** 9–12.

Bissa, S., Songara, D. and Bohra, A. (2005). *J. Mycol. Pl. Pathol.* **35(3):** 579.

Boreddy, Y. and Chitra, K.C. (2001). *Pest Management and Economic Zoology* **9(1):** 79–82.

Cabras, P., Caboni, P., Cabras, M., Angioni, A. and Russo, M. (2002). *J. Agric. Food Chem.*, **50:** 2576–80.

Carpinella, M.C., Herrero, G.G., Alonso, R.A. and Palacios, S.M. (1999). *Fitoterapia* **70:** 296–98.

Casida, J.E. and Quistad, G.B. (1995). Pyrethrum Flowers: Production, Chemistry, Toxico. and Uses, Oxford, UK: Oxford Univ. Press., p. 356.

Cespedes, C.L., Calderon, J.S., Lina, L. and Aranda, E. (2000). *Journal of Agricultural and Food Chemistry,* **48(5):** 1903–1908.

Chandel, B.S., Pandey, U.K. and Kumar A. (1987). *Indian Journal of Entomology,* **49(2):** 294–96.

Chari, M.S. and Muralidharan, C.M. (1985). *Journal of Entomological Research,* **9(3):** 243–45.

Chari, M.S., Sreedhar, U., Rao, R.S.N. and Reddy, S.A.N. (1996). *Tobacco Research* **22(1):** 32–35.

Chattopadhyay, C., Meena, P.D., Godika, S., Yadav, M.S., Meena, R.L. and Bhunia, C.K. 2005. *In* (Abs.): Second Global Conference. Plant health-Global wealth, at Udaipur, India, 25–29 Nov. 2005, pp. 289–90.

Chitra, K.C., Reddy, P.V.R. and Rao, K.P. (1991). *Journal of Applied Zoological Research* **2(1):** 1–9.

Chiu, S.F. (1982). *Plant protection, South China Agricultural College, Cuangzhon, China,* pp. 42

Chong, L.C., Li, W.G., Fang, K.T., Liu, C.C., Wang, G.L. and Kang, T.F. (2000). *Plant Protection,* **26(6):** 20–22.

Choudhuri, C., Isha, M. and Saha, S. (2006). *Indian Phytopathology,* **59(3):** 392.

Chowdhary, H., Walia, S. and Dhingra, S. (2001). *Pesticide Research Journal* **13(2):** 165–72.

Das, N.D., Sankar, G.R.M. and Biswas, K.K. (2000). *Annals of Plant Protection Sciences,* **8**: 2.

Datar, V.V. (1999). *J. Mycol. Pl. Pathol.,* **29:** 251–53.

Datta, S. and Saxena, D.B. (2001). *Pest Management Sci.* **57:** 95–101.

David, Hillock. and Pat, Bolin. Earth kind gardening series: Botanical pest controls, Oklahoma Cooperative Extension Service HLA–6433. Division of Agricultural Sciences and Natural Resources, Oklahoma State University (http://osufacts.okstate.edu).

Davis, Deepta., Beena, S. and Mathew, S.K. (2003). *J. Mycol.Pl.Pathol.* **33(3):** 474.

Deepa, B. and Remadevi, O.K. (2007). In national conference on organic waste utilization and Eco–friendly technologies for crop protection, March, 2007, at Hyderabad, Andhra Pradesh, pp. 178–80.

Dhaliwal, G.S., Arora, R. and Dilwari, V.K. (1996). *In*: Narwal, S.S. and Tauro, P. (*Eds.*) Allelopathy in pest managemnet for sustainable agriculture. Scientific Publishers, Jodhpur, India, pp. 93–119

Dhaliwal, H.J.S., Thind, T.S. and Mohan, C. (2003). *J. Mycol. Pl. Pathol.* **33(3):** 399–402.

Dhingra, S., Hedge, R.S., Vohra, S. and Parmar, B.S. (2002). Bioefficacy of neem oil microemulsions against gram pod borer, *Helicoverpa armigera* (Hub). *Pesticide Research Journal,* **14(2):** 224–48.

Dhingra, S. (1996). *Journal of Entomological Research,* **20(1):** 19–22.

Dimetry, N.Z., Amer, S.A. and Reddy, A.S. (1993). *J. Applied Ent.* **116(3):** 308–312.

Dodia, D.A., Patel, I.S. and Patel, G.M. (2008). Botanical pesticides for pest management. Scientific Publishers (India), Jodhpur, pp. 1–35.

Dreyer, M. (1983). *RAW(A)*, **78(6):** 5580.

Duarte, L.M.L., Noronha, A.B., Alexendere, M.A., Vicente, M. and Chagas, C.M. (1996). *Microbios,* **84:** 13–20.

Dubey, N.K., Shukla, R., Kumar, A., Singh, P. and Prakash, B. (2010). Prospects of botanical pesticides in sustainable agriculture. *Current Science,* **98(4):** 479–80.

Duke, S.O. (1990). *In*: advances in new crops (*Eds* Janick, J. and Simon, J.E.), Timber Press, Portland, OR, pp. 511–17.

Dwivedi, S.C. and Venugopalan, S. (2004). *Pest management and Economic Entomology,* **12(1):** 93–95.

Dwivedi, S.C., Kumar, R. and Kumar, R. (1999). *Journal of Advanced Zoology,* **20(1):** 6–9.

Epino, P.B. and Saxena, R.C. (1982). *Philippines Association of Entomological Convention*, Baguna, Philippines, p. 23.

Eswaramurthy, S., Mariappan, V., Muthusamy, M., Alagiangalingam, M.N. and Subramanian, K.S. (1993). *Neem and Environment,* **2:** 783–85.

Facknath, S. (1993). *In*: World Neem Conference, Bangalore, India. pp. 550–51.

Gahukar, R.T. (1998). *Pestology,* **22(1):** 15–41.

Ganesan, T. and Krishnaraju, J. (1995). *Advances in Plant Science,* **8(1):** 194–96.

Gehlot, P. (2005). *In* (Abs.): Second Global Conference. Plant health-Global wealth, at Udaipur, India, 25–29 Nov. 2005, p. 314.

Gill, J.S. and Lewis, C.T. (1971). *Nature,* **232:** 402–403.

Godika, S. and Pathak, A K. (2003). *J. Mycol. Pl. Pathol.,* **33(3):** 476.

Gomathi, and Kannabiran, B. (2000). *Indian Phytopath.,* **53(3):** 305–308.

Gour H.N. and Mali, B.L. (2005). *In* (Abs.): Second Global Conference. Plant health-Global wealth, at Udaipur, India, 25–29 Nov. 2005, p. 310.

Grainage, M., Ahmed, S., Mitchell, W.C. and Hylen, J.H. (1985). An EWC/UM database; Resources Systems Institute, East West Centre, Honolulu.

Gujar, G.T. and Mehrotra, K.N. (1988). *Phytoparasitica,* **16:** 293–302.

Gupta, S., Kalha, C.S., Vaid, A. and Rizvi, S.E.H. (2005). *J. Mycol. Pl. Pathol.,* **35(3):** 432–36.

Gupta, S.P., Prakash, A. and Rao, J. (1990). *Journal of Applied Zoological Research* **1(2):** 55–58.

Gupta, V.P., Govindaiah, and Datta, R.K. (1996). *Current Science,* **71(5):** 406–409.

Halder, J., Srivastava, C., Dhingra, S. and Dureja, P. (2009a). Annals of Plant Protection Sciences, **17(2):** 288–91.

Halder, J., Srivastava, C., Dhingra, S., Dureja, P. and Tanwar, R.S. (2009b). Annals of Plant Protection Sciences, **17(2):** 496–97.

Halder, J., Srivastava, Chitra, Dhingra, Swaran, Dureja, Prem and Rai, A.B. (2010). *Vegetable Sciences,* **37(2):** 153–55.

Halder, J., Srivastava, C. and Dureja, P. (2010). *Indian Journal of Agricultural Sciences,* **80(9):** 820–23.

Halder, J., Srivastava, C., Dhingra, S. and Dureja, P. (2010). Pesticide Research Journal, **22(2):** 174–76.

Hammad, E.M.A.F., Nemer, N.M., Hawi, Z.K. and Hanna, L.T. (2000). *Annals of Applied Biology,* **137(2):** 79–88.

Hollingworth, R.M., Johnstone, E.M. and Wright, N. (1984). *In:* Pesticide synthesis through rational approaches (*Eds* Magee, P.S., Kohn, G.K. and Menn, J.J.), ACS Symposium, Series No. 255, American Chemical Society, Washington, DC, 1984, pp. 103–25.

Iannacone, J. and Lamas, G. (2003). *Boletin de Sanodod Vegetal Plagas,* **29(1):** 123–42.

Indumathi, N. (2002). M.Sc. (Ag) thesis submitted to ANGRAU, Rajendra Nagar, Hyderabad.

Isman, M.B. (2006). *Annu. Rev. Entomol.* **51:** 45–66.

Isman, M.B., Jeffs, L.B., Elliger, C.A., Miyake, T. and Matsumura, F. (1997). *Pesticide Biochemistry and Physiology,* **58(2):** 103–107.

Jadeja, K.B. (2003). *J. Mycol. Pl. Pathol.,* **33(3):** 446–50.

Jayakumar, M., Raja, N. and Ignachimuthu, S. (2004). *Journal of Entomological Research,* **28(3):** 321–27.

Jayalakshmi, C., Durairaj, P. and Seetharaman, K. (1997). National symposium on plant protection towards sustainability, 22–24 Dec. 1997. Hyderabad, pp. 28–29.

Jayaraj, S. and Ignacimuthu, S. (2005). *In*: Sustainable insect Pest Management, *Eds.* Ignacimuthu, S. and Jayaraj, S. Narosa Publishing House, New Delhi.

Jeeva, Ram. and Thakore, B.B.L. (2005). *In* (Abs.): Second Global Conference. Plant health-Global wealth, at Udaipur, India, 25–29 Nov. 2005, p. 189.

Jeyakumar, P. and Gupta, G.P. (1999). Effect of neem seed kernel extract (NSKE) on *Helicoverpa armigera*, *Pesticide Research Journal,* **11(1):** 32–36.

Jhansi Lakshmi, V., Katti, G.V. and Lmgalah, T. (1997). *Journals of Biological Control,* **11:** 29–32.

Jhansi Lakshmi, V., Katti, G., Krishnaiah, N.V. and Mahesh Kumar, K. (1998). *Journals of Biological Control,* **12:** 119.

Joshi, B.G., Ramprasad, C. and Rao, S.N. (1984). *Phytoparasitica,* **12(1):** 3–12.

Juan, A., Sans, A. and Riba, M. (2000). *Phytoparasitica,* **28(4):** 311–19.

Kamalkannan, A., Sanmugam, V., Surendran, M. and Srinivasan, R. (2001). *Indian Phytopath.* **54(4):** 490–92.

Kannan, N.R. and Doraiswami, S. (1993). *Madras Agric. J.,* **80:** 393–95.

Kareem, A.A., Saxena, R.C. and Palanginan, E.L. (1988). *International Rice Research Newsletter,* **13(6):** 41–42.

Karunanithi, K., Sarala, L., Rabindran, R., Rajaruthinam, S. and Kalamani, T. (2003). *J. Mycol. Pl. Pathol.,* **33(3):** 495.

Khilare, V.C., Kadam, K.S. and Gangawane, L.V. (2002). *J. Mycol. Pl. Pathol.,* **32(3):** 416–17.

Krishnappa, M., Nataraja, S. and Ramesh Raby, H.N. (2005). *In* (Abs.): Second Global Conference. Plant health-Global wealth, at Udaipur, India, 25–29 Nov. 2005, p. 291.

Kulkarni, N., Joshi, K.C., Kalia, S. and Kalia, S. (1997). *Indian Journal of Forestry,* **20(4):** 390–94.

Kumar, N. and Anamika. (2003). Natn. Symp. On biodiversity management for 21st Century, BHU, Varanasi, 28–30 June, 2003, p. 153.

Kumar, R., Mishra, A.K., Dubey, N.K. and Tripathi, Y.B. (2007). *Int. J. Food Microbiol.*, **115:** 159–64.

Ladd, T.L. and Jacobson, M. (1980). *Japanese Beetle Research Laboratory*, OARDC, Ohio.

Landa, Z. and Bohata, A. (1999). Collection of scientific papers, faculty of agriculture in ceske budejovice. *Series for Crop Sciences*, **16(2):** 99–106.

Loganathan, J., Dhingra, S. and Walia S. (2003). National Symposium on Frontier Areas of Entom. Res., 5–7 Nov. 2003, New Delhi, pp. 515–16.

Londerhausen, M., Leicht, W., Lierb, F. and Moesschlev, H. (1991). *Pesticide Science*, **33:** 427–33.

Lu, R.D. (1982). *Insect Knowledge,* **19(4):** 22–25.

Mackay, D.B. (1992). *In*: Viability of seeds *Ed*. Roberts, E.H. Chapman and Hall Ltd., London, pp. 172–208.

Madhiazhagan, K., Ramadoss, N. and Anuradha, R. (2002). *J. Mycol. Pl. Pathol.* **32(1):** 68–69.

Maharishi, R.P. (1993). *Neem and Environment,* **2:** 810–12.

Maheswari, U. and Sundarababu, R. (2001). *Indian Journal of Nematology,* **31(2):** 126–28.

Mane, S.D. (1968). M.Sc thesis, I.A.R.I., New Delhi.

Mansour, F. and Ravid, V. (1986). *Phytoparasitica,* **14(2):** 137–42.

Mariappan, V. and Saxena, R.C. (1984). *J. Ecol. Entomol.*, **77(2):** 519–21.

Mau, R.F.L., Gusukuma–Minuto, L. and Shimabuku, R.S. (1995). Arthropod Management Tests, **20:** 327.

Meisner, J., Melamed, M.J., Tam, S and Ascher, K.R.S. (1986). *Zeitschrift fur Pflanzenkrankheiten und Pflanzenschutz,* **93(6):** 585–89.

Meman, F.M., Reda, A.S. and Amar, S.A.A. (1997). *RAW* (1998), **86(10):** 9496.

Mikolajizak, K.L., Ziloroski, B.W. and Bartell, R.J. (1989). *Journal of Chemical Ecology,* **15(1):** 121–28.

Mineo, N., Mineo, G. and Sinacori, A. (2000). *Bollettino di Zoologia Agraria e di Bachicoltura,* **32(2):** 157–62.

Mishra, H.P. (2000). *Indian Journal of Entomology,* **62(2):** 218–20.

Molina M. and Leon, O. (1991). *Revista de Proteccion Vegetal.,* **6:** 150–55.

Molina, M. and Sanchez, L.M. (1993). *Revista de Proteccion Vegetal.*, **8:** 131–34.

Momen, F.M. and Amer, S.A.A. (1999). *Acta Phytopathologica et Entomologica Hungarica,* **34(4):** 355–61.

Morimoto, M., Kumeda, S. and Komai, K. (2000). *Journal of Agricultural and Food Chemistry,* **48(5):** 1888–91.

Nagesh, M. and Reddy, P.P. (1995). *Journal of Biological Control,* **9(2):** 109–12.

Naik D.G. and Anahosur, K.H. (2001). *Indian Phytopath.,* **54(4):** 514–15.

Naik, D.B. (2000). A thesis submitted to GAU, Navsari (unpublished).

Narasimhan, V. and Mariappan, V. (1988). *International Rice Research Newsletter,* **13(1):** 28–29.

Olkowski, W., Daar, S. and Olkowski, H. (1995). The Gardener's Guide to Commonsense Pest Control. The Taunton Press, USA.

Otiniano, J., Marin, A., Sanchez, M.S. and Arjona, *E.G.* (2002). *IDESIA* **20(1):** 43–50.

Oudhia, P. (2000). *Crop Research,* **20(3):** 558–59.

Pandey, D.K. (1995). *In*: Pesticides, crop protection and environment. *Eds*. Walia, S. and Parmer, B.S. Oxford and IBH Pub. Co. Pvt. Ltd., New Delhi.

Pandey, D.K., Kaurav, L.P. and Bhan, V.M. (1993). *J. Chem. Ecol.,* **19:** 2651–62.

Pandey, U.K., Srivastava, A.K., Chandel, B.S. and Lekha, C. (1982). *Angewandte Zoology,* **69(3):** 267–70.

Pangavhane, C.P. and Solanky, K.U. (2003). *In*: Proceedings. national symposium on crop diversification and natural resource management, 20–22 December 2003. IIPR, Kanpur, p. 255.

Patel, B.N. (1990). *Botanical Pesticides in IPM,* **12:** 404–406.

Patel, B.A., Patel, S.K., Panikar, B., Dabhi, M.V. and Patel, B.N. (2005a). *In*: Second Global Conference. Plant health–Global wealth, at Udaipur, India, 25–29 Nov. 2005, p. 283.

Patel, D.J. (1999). A thesis submitted to GAU, Anand (unpublished).

Patel, J.R., Patel, J.I, Mehta, D.M and Shah, B.R. (1993). *Botanical Pesticides in Integrated Pest Management,* pp. 343–50.

Patel, K.N. (1996). A thesis submitted to GAU, Navsari (unpublished).

Patel, S.J., Solanki, K.U. and Naik, B.M. (2005b). *In* (Abs.): Second Global Conference. Plant health–Global wealth, at Udaipur, India, 25–29 Nov. 2005, p. 304.

Patel, Z.P. and Patel, J.R. (1996). *Indian J. Plant Prot.* **64:** 39–43.

Patni, C.S., Gupta, R., Kolte, S.J. and Awasthi, R.P. (2005). *In* (Abs.): Second Global Conference. Plant health-Global wealth, at Udaipur, India, 25–29 Nov. 2005, p. 304.

Prabhakar, N., Coedriet, D.L., Kishba, A.N. and Meyerdirk, D.E. (1986). *Journal of Economic Entomology,* **79(1):** 39–41.

Pradhan, S., Jothwani, M.G. and Rai, B.K. (1962). Neem seed deterrent to locust. *Indian Farming,* **12:** 7–11.

Prakash, A. and Rao, J. (1997). Botanical pesticides in agriculture. CRC Lewis Publication, Boca Raton, USA, p. 481.

Prasad R., Dixit, R.K. and Saxena, D. (2003). National symposium on biodiversity management for 21st Century, 28–30 June, BHU, Varanasi, p. 113.

Purohit, S.S. and Vyas, S.P. (2004). Medicinal plant cultivation, a scientific approach. Agriobios (India), Jodhpur, pp. 158–70.

Pyke, B., Rice, M., Sabine, B. and Zalucki, M. (1987). In workshop on australian cotton grower, Australia, 7–9 May 1987.

Raguraman, S. and Singh, R.P. (1995). National symposium on organic farming, 27–28 Oct, 1995, Agriculture College and Research Institute, Madurai, Abstract, pp. 130–31.

Rajan, F.S., Peter, G.S., Velamurthu, M.D., Khader, A. and Jeyareajan, R. (1997). *South Indian Horticulture,* **39(4):** 221–23.

Rajappan, K., Malini, C.U., Narasimhan, V. and Abdul Kareem, A. (1999). *Annals Plant Prot. Sci.,* **7:** 220.

Raman, G.V, Rao, M.S. and Sriramnnaraya, G. (2000). *Journal of Entomological Research,* **24(3):** 235–38.

Ramsewak, R.S., Nair, M.G., Murugesan, S., Mattson, W.J., Zasada, J. and Murugesan, S. (2001). *Journal of Agricultural and Food Chemistry*, **49(12):** 5852–56.

Rangrez, M.A. and Rather, A.Q. (2007). In national conference on organic waste utilization and eco-friendly technologies for crop protection, March, 2007, at Hyderabad, Andhra Pradesh, pp. 184–86.

Rao, G.R. and Dhingra, S. (1997). *Journal of Entomological Research,* **21(2):** 153–60.

Rao, N.S., Rajendran, R. and Raguraman, S. 2002. *J. Entomol. Res.* **26:** 34.

Reddy, P.P. and Khan, R.M. (1991). *Current Nematology,* **2(2):** 115–16.

Reddy, P.V.R. and Srikanth, J. (1996). *In*: Compendium: Pesticides chemistry a modern methods of residue analysis. Summer school, held at GAU, Anand, 28 May–17 June, 2002, pp. 46–62.

Regnault–Roger, C. and Philogene, B.J.R. (2008). *Pharm. Biol.*, **46:** 41–52.

Rossi, C.E., Almeida, J.E. M–de., Lima, C.B., Ribeiro, L.D. and Petri, J. (2006). *STAB–Acucar,* **24(4):** 34–36.

Roy, D.C. and Pande, Y.D. (1991). *In*: IVth International symposium on growth, development and control technology of insect pests, p. 37.

Saha, D. and Saha, A. (2003). *J. Mycol. Pl. Pathol.*, **33(3):** 475.

Saminathan, V.R. (2000). Ph.D Thesis submitted to *Dept. of Agril. Ento.*, AC and RI, Madurai.

Sandhu, G.S. and Singh, D. (1975). *Indian Journal of Plant Protection,* **3:** 177–80.

Santhosh Kumar T. (1994). M. Sc. Thesis submitted to Kerala Agricultural University, p. 104.

Sarode, S.V. and Gobhane, A.T. (1994). *Journal of Entomological Research,* **18(4):** 327–30.

Sawant, D.M., Sopkal, R.T. and Latake, S.B. (2005). *In* (Abs.): Second Global Conference. Plant health-Global wealth, at Udaipur, India, 25–29 Nov. 2005, pp. 281–82.

Saxena, B.R., Koli, M.C. and Saxena, R.C. (2000). *Ethnobotany,* **12:** 47–50.

Saxena, R.C. (1982). *Indian Journal of Agricultural Sciences,* **52:** 51–52.

Saxena, R.C. and Khan, Z.R. (1984). Neem oil disrupts *Nephotettix virescens* feeding. *Neem Newsletter,* **1:** 28–29.

Sebastian, S. and Gupta, P. (1996). *ICPN* **3:** 40–41.

ShaoQin, Li., WangXi, D., Dong, Z.Q., Li, S.Q., Deng, W.X. and Zhang, Q.D. (2001). *Journal of Huazhong Agricultural University,* **20(6):** 539–43.

Sharma, Satyawati., Mishra, S., Bharwaj, A. and Vasudevan, P. (2005). *In* (Abs.): Second Global Conference. Plant health–Global wealth, at Udaipur, India, 25–29 Nov. 2005, p. 166.

Sharma, D.D., Chandrashekar, D.S., Gunasekhar, V., Rekha, M. and Sarkar, A. (2001). *Indian Journal of Sericulture,* **40(2):** 151–57.

Sharma, I.N.S., Singh, A.K. and Singh, S.P. (1998). First national symposium allelopathy in agro-ecosystems, Indian society of Allelopathy, HAU, Hisar, Haryana, pp. 157–76.

Sharma, M. and Gupta, S.K. (2003). *J. Mycol. Pl. Pathol.,* **33(3):** 345–61.

Sharma, R. R. and Gaur, R. B. 2003. *J.Mycol. Pl. Pathol.,* **33(3):** 504.

Sharma, R.K., Saxena, K. and Singh, C.P. (2005). *Indian Journal of Applied Entomology,* **19(1):** 45–49.

Sharma, S.K. and Verma, V.S. (2003). *J. Mycol. Pl. Pathol.*, **33(3):** 502.

Sharma, V.K. and Bhatnagar, A. (1990). In national symposium on problems and prospects of botanical pesticides in Integrated Pest Management. p. 19.

Shelke, S.S., Jadhav, L.D. and Salunkhe, G.N. (1990). *Indian J. Entomol.*, **52(4):** 709–11.

Shivpuri, Asha, Sharma, O.P. and Jhamaria, S.L. (1997). *J. Mycol. Pl. Pathol.*, **27(1)**: 29–31.

Simmonds, M.S.J., Jarvis, A.P., Johnson, S., Jones, G.R. and Morgan, E.D. (2004). *Pest Management Science,* **60(5):** 459–64.

Singh, J., Sukhiya, H.S. and Singh, P. (1990). *Proceedings Symposium Botanical Pesticides in IPM*, Rajamundry, pp. 288–90.

Singh, K. and Sharma, U.L. (1986). *Agricultural Research and Development Report*, **3:** 33–35.

Singh, R.P., Devakumar, C. and Dhingra, S. (1988). *Phytoparasitica,* **16(3):** 225–30.

Singh, S. and Awasthi, L.P. (2005). *In*: Second Global Conference. Plant health-Global wealth, at Udaipur, India, 25–29 Nov. 2005, p. 299.

Singh, S. and Goswami, B.K. (2001). *Indian Journal of Nematology,* **31(2):** 122–25.

Singh, S., Singh and Gupta, D.C. (2005). *Environment and Ecology*, **23(2):** 319–22.

Singh, U.P., Singh, H.B. and Singh, R.B. (1980). *Mycologia,* **72:** 1077–93.

Singh, Y.P. and Sumbali, Geeta (2003). *J. Mycol. Pl. Pathol.*, **33(3):** 504.

Sinha, P. (1998). *Annals of Plant Protection Sciences,* **6(1):** 112–15.

Smid, E.J. and Gorris, L.G.M. (1999). *In*: Handbook of food preservation (*Ed.* Rehman, M.S.), Marcel Dekker, New York, pp. 285–308.

Somasekhara, Y.M., Nateshan, H.M. and Maniyappa, V. (1997). *Indian, J. Pl. Prot.* **25(1):** 56–59.

Sridevi, D. and Dhingra, S. (1997). *Tropical Agricultural Research,* **9:** 358.

Sridhar, S., Arumugasamy, S., Saraswathy, H. and Vijayalakshmi, K. (2002). *In*: Organic vegetable gardening. Center for Indian knowledge systems, Chennai, pp. 33–34.

Srinivasan, G. and Babu, P.C.S. (2000). *Pesticide Research Journal,* **12(1):** 123–26.

Srinivasulu, B. and Narayanswamy, P. (1993). *Indian J. Pl. Protection,* **21(1):** 77–79.

Srivastava, A. and Srivastava, M. (1998). *Philippine Journal of Science,* **127(3):** 181–87.

Srivastava, M., Paul, A.V.N., Rengaswamy, S., Kumar, J. and Parmer, B.S. (1997). *J. Appl. Entomol.*, **121:** 51.

Stoll, G. (2000). Natural Protection in the tropics, Margraf verlay, Weiker–Sheim.

Sukul, S. and Chaudhari, S. (1999). *J. Phytol. Res.,* **12(1–2):** 449–55.

Sundararajan, G. and Kumuthakalavalli, R. (2001). *Journal of Environmental Biology,* **22(1):** 11–14.

Thangavelu, R., Padmasree, M., Satyanarayana, K. and Reddy, A.P.K. (1995). *Indian J. Pl. Prot.,* **23:** 93–94.

Thapar, R. and Singh, N.B. (2003). In: National symposium on biodiversity management for 21st Century, at BHU, Varanasi, 28–30 June 2003, p. 165.

Tiwari, R.K.S., Chandravansi, S.S., Ojha, B.M., Thakur, R.S. and Dantre, R.K. (2002). *J. Mycol. Pl. Pathol.*, **32(3):** 418–19.

Tiwari, R.K.S., Singh, A., Ghai, T. and Thakur, B.S. (2005). In (Abs.): Second Global Conference. Plant health-Global wealth, at Udaipur, India, 25–29 Nov. 2005, p. 313.

Tripathi, G.K., Verma, A.K. and Pandey, B.N. (2000). Proceedings of the 10th All india congress of zoology and national symposium on environmental degradation and animal biodiversity: The problems of India and its remedial measures, held on 14–18th October, 1998 at Magadh University, Bodh Gaya, Bihar, India, 2000, pp. 108–109.

Tripathi, S.C. (2005). *In* (Abs.): Second Global Conference. Plant health-Global wealth, at Udaipur, India, 25–29 Nov. 2005, p. 309.

Trivedi, R.S. (2004). A thesis submitted to S.D. Agricultural University, SK Nagar, Gujarat (unpublished).

Ukey, S.P., Sarode, S.V., Naitam, N.R. and Patil, M.J. (1999). *Pestology*, **23(6):** 23–25.

Vadivel, S. and Ebenezar, E.G. (2006). *J. Mycol. Pl. Pathol.*, **36(1):** 79–83.

Varma, J. and Dubey, N.K. (1999). *Curr. Sci.*, **76:** 172–79.

Verma, H.N. and Mukherjee, K. (1979). *Indian Phytopathol.*, **32:** 95–97.

Verma, P.R., Chowdhary, A., Dutta, S., Chowdhary, A.K. and Laha, S.K. (2003). *J. Mycol. Pl. Pathol.*, **33(3):** 378–86.

Walter, F. (1999). (*Eds*) Singh, R.P. and Saxena, R.C. Oxford and IBH Pub Co. Ltd., New Delhi.

Zhou, Z.G., Wei, W.Y., Hong, X.U., Huan, Z.S., Zhang, G.Z., Wang, Y.W., Xu, H.H. and Zhao, S.H. (2000). *Journal of Hunan Agricultural University*, **26(3):** 190–92.

Zhu, B.C.R., Henderson, G., Chen, F., Fei, H.X. and Laine, R.A. (2001). *Journal of Chemical Ecology,* **27(8):** 1617–25.

12

Conservation Biological Control of Insect Pests in Agriculture: Progresses and Prospects

ACHINTYA PRAMANIK[1] AND JAYDEEP HALDER[2]

ABSTRACT

Enhancement in the performance of natural enemies of crop pests with the aim of crop protection has been a matter of concern and gaining more importance these days. The idea of conservation biological control (CBC), perhaps the oldest approach to biological control involves the practices that maintain and enhance the reproduction, survival, and efficacy of natural enemies (predators, parasitoids as well as pathogens) of pests. There is now a clear shift of focus from traditional biological control to CBC. Several interrelated phenomena operating in the crop field ecosystem contribute to the ultimate performance of the applied biocontrol agents which are essentially living beings. An active approach to acknowledge the role of native natural enemies and promotion of their effectiveness through landscape ecology is the need of the hour. CBC requires integration of fundamental insect life history and natural enemy-pest interactions with knowledge of the ecology of the systems which makes CBC research and implementation a challenging and yet a potentially rewarding job for the future.

Key words: Chemical ecology, Conservation biological control, Food sprays, Habitat management, Landscape ecology, Natural enemy, Refugia.

[1] Bidhan Chandra Krishi Sugarcane Research Station, Viswavidyalaya, Mohanpur, Nadia, West Bengal - 741252, India.

[2] Economic Botanist-VII, Sugarcane Research Station, Bethuadahari, Nadia, West Bengal, India.

* *Corresponding author*: E-mail: achintyapramanik@gmail.com

1. INTRODUCTION

Biological control is a form of pest control that uses living organisms to suppress pest densities to lower levels. It is a form of ecologically based pest management that uses one kind of organism (the "natural enemies") to control another (the pest species). Four broad ways can be considered in which people have manipulated natural enemies to enhance their action: natural enemy importation (Classical biological control), augmentation (Augmentative biological control), inundation (Inundation biological control) and conservation (Conservation biological control). Each of these approaches has its own rationale, history, and level of past successful use (Eilenberg, 2006).

Managing the insect pests and diseases by enhancing natural enemies without adversely affecting the main crops is gaining more attraction now-a-days. Conservation biological control (CBC) involves the practices that maintain and enhance the reproduction, survival, and efficacy of natural enemies (predators, parasitoids as well as pathogens) of pests. Natural enemies are important in regulating populations of many agricultural and forest insect pests. Like other animals, insect natural enemies also require food, water, and shelter, and protection from adverse weather conditions. Thus CBC aims at preservation and facilitation of native natural enemies to increase their quantitative, limiting effects on populations of invasive species, both native and exotic with the goal of modifying the environmental factor(s) that may limit the control effectiveness of natural enemies. To achieve these goals of CBC, fundamental knowledge of the biology and requirements of natural enemies is required.

2. DEFINITIONS

CBC is among the earliest documented pest control measures, dating to third century BC in China, where bamboo poles were placed between trees in orchards to facilitate dispersal of predatory ants, thus enhancing predation of citrus pests (Coulson *et al.*, 1982). In recent times CBC has been defined by many authors; significant among those are mentioned below:

(a) CBC is the modification of the environment or existing practices to protect and enhance specific enemies or other organisms to reduce the effect of pests (Eilenberg *et al.*, 2001).

(b) CBC is a pest management approach based on manipulation of agroecosystems to promote pest suppression by naturally occurring predators, parasitoids and pathogens (Barbosa, 1998).

(c) CBC encompasses efforts to conserve or enrich the biological control agents that are already present, through either manipulation of the environment or crop and pest management practices (Lazarovits *et al.*, 2007).

3. CONCEPTS

A pest occurs at high population levels due to insufficient effects of the natural enemies. All the macro and microorganisms controlling invertebrates, weeds and plant diseases, including the antagonistic microorganisms are considered as natural enemies which are responsible for biological regulation. At certain time interval the environment is modified or the practice is changed in order to enhance the natural enemies, which are already present. They increase in population size and their effect results in a lower pest population (Eilenberg, 2006). CBC is thus completely different from the three other biological control strategies, since no organisms are released and organisms, which are already present, are enhanced in order to avoid damage.

Both passive and active conservation are also the part of CBC. An example of passive conservation is the avoidance of such actions that disfavour the natural enemies, for example, spraying with certain broad spectrum chemical pesticides. An active conservation could be the initiation of actions to support the natural enemies actively by establishing for example 'beetle banks' (Landis *et al.*, 2000). Many other examples of habitat manipulation at different levels, from landscape to crop plants are found in the writings of Barbosa (1998) and Pickett and Bugg (1998).

Among the four biological control strategies, CBC seems to be most tightly connected to the main principles of organic farming, which have the protection of the existing natural enemies as one of the main principles (Anonymous, 2002). There is a tight connection also to 'conservation biology' (Letourneau, 1998). CBC can thus be seen as an example of habitat restoration with the specific purpose of supporting natural enemies to control pests (Eilenberg, 2006).

Under following conditions CBC can be proved as the most useful control tactics when:

- There is a potentially effective natural enemy that is rendered ineffective due to adverse biotic or abiotic environmental factor(s).
- Pests are not adequately or cost-effectively controlled by other means.
- Other pest management tactics would disrupt the environment.

4. DIFFERENCE BETWEEN BIOLOGICAL CONTROL AND CBC

Despite these apparent differences, much could be gained by cross-discipline discussion and collaboration between proponents of CBC and the management of biological control system. The encouragement of diverse natural enemy communities through increased habitat diversity and connectivity between crop and non-crop habitats may be the best for both long-term biological pest suppression and as well as sustainable crop production.

Sl. No.	*Particular*	*Biological control*	*CBC*
1.	Fundamental goal	Curbing insecticide use	Manipulation of properties of crop and non-crop vegetation
2.	Focus	Agroecosystems	Specific crop field
3.	Area of interest	Promote the conservation of natural enemy diversity at the landscape level ranging from single cropping system, to complex areas with diverse cropping systems	Promote the conservation of natural enemy diversity within the field, and within a few adjacent fields within a farming unit by increasing taxonomic diversity of plants through mixed cropping, alley cropping, intercropping, and strip cropping etc.
4.	Approach	Natural enemy community as a whole	Enhance the activity of one or a few key natural enemies.

According to the insurance hypothesis, species richness can buffer against spatio–temporal disturbances, thereby ensuring stable levels of pest suppression in changing environments (Tscharntke *et al.*, 2007). Additionally, most pests and natural enemies exploit their habitats at spatial scales much larger than a crop field and its immediate surroundings. However, current CBC practices ignore the effect of large-scale dispersal on pest and enemy species (Landis *et al.*, 2000).

5. BACKGROUND INFORMATION REQUIRED FOR SUCCESSFUL CBC PROGRAMME

Before planning for conservation biological control in a place, the following details are needed to be addressed:

- The local existing predator, parasitoids and parasite guilds
- Identification of the natural enemies sampled
- Identification of potentially effective natural enemies
- Testing the possible effective entomophages in the lab to see if they do consume/parasitize the target host/prey

- Observing the natural enemies in the field to determine which actually attack the target pest in the field, prioritizing entomophages based on their impact on the target pest
- Studying high priority entomophages to determine their habitat requirements and what may be limiting their effectiveness in the crop field
- Determining whether the limiting factor can be moderated. If so, is it potentially viable and cost effective?
- Conducting field experiments for modifying the crop field habitat to test feasibility of CBC

6. TACTICS FOR CBC

Conservation biological control includes identification and remediation of negative human influences that suppress natural enemies, as well as enhancement of systems as habitats for natural enemies. They seem to fall into two strategies:

(a) Reducing direct mortality or interference

(b) Providing supplementary resources

6.1. Reducing Direct Mortality or Interference

The idea behind this strategy is if there is reduction in mortality of natural enemies, there are simply more beneficial insects present and they can exert greater control over the pest population. However, the factors that can kill natural enemies are numerous, so the tactics required to reduce mortality are diverse. Similarly, if there is enhancement of entomophage performance, they will give better control. Direct mortality is related to

(i) Use of insecticides and/or

(ii) Cultural practices

6.1.1 *Use of insecticides*

Many insecticides can have both direct and indirect effects on natural enemies. Direct effects include acute or chronic mortality as a result of direct contact with pesticides. Direct sub-lethal effects, such as decreased adult fecundity, reduced viability of offspring, and changes in feeding habits or other behaviors can also occur. Indirect effects can result from mortality in populations of alternate prey or hosts of the natural enemies. Use of broad-spectrum insecticides that have detrimental effects on natural enemies can lead to rapid resurgence of targeted pest populations. In addition, secondary pest outbreaks can result if naturally occurring biological control of these secondary arthropods is disrupted. For example negative effect of dust on natural enemy can be considered. Dust was found to be disruptive to biological control of scales in citrus (DeBach, 1958) and spider mites in vineyards (Flaherty and Huffaker, 1970). Presumably this involves several phenomena. The more active and exposed natural enemies encounter the dust more frequently and extensively. This could interfere with their searching for prey. This might simply reduce their efficacy or induce emigration. Also, their cuticular wax layer is more likely to be disturbed inducing desiccation. This was the primary impact detected in these studies.

Modifications of pesticide use practices are the most commonly implemented form of conservation biological control, and have long been considered an important component of integrated pest management programs. Pesticide use can be modified to favor natural enemies in a variety of ways (Ruberson *et al.*, 1998; Orr and Suh, 1998):

- Treating only when economic thresholds dictate
- Use of active ingredients and formulations selectively less toxic to natural enemies
- Insecticides with short residual activity
- Use of the lowest effective rates of pesticides
- Temporal and spatial separation of natural enemies and pesticides
- Use of more selective "biorational" insecticides (microbials, soaps and botanicals)
- Timing of insecticide applications when natural enemies are absent or in the life stages that are not susceptible to the insecticide(s)
- Development of pesticide resistant natural enemies

6.1.2 *Cultural practices*

Conservation biological control also includes implementation of agricultural and silvicultural practices compatible with maintenance of natural enemy populations. Monoculture environments are highly advantageous to many herbivorous pests, but are usually very poor environments for natural enemies. Ploughing, mowing, and harvesting operations, burning of crop residues, and poorly timed irrigation practices can cause direct mortality of natural enemies. This disruption can create harsh conditions for natural enemies. Alternate hosts or prey, nectar and pollen sources, free water, and refugia are generally found in greater quantities in habitats with greater diversity of vegetation. Microclimatic conditions are generally more moderate as well. However, increasing reliance on intensive agriculture and forestry practices has tended to decrease habitat heterogeneity as well as genetic diversity of crops. Densities of parasitoids, as well as parasitism rates of pest species, have been found to be higher in mixed species habitats and in agricultural field edges near mixed species habitats. Avoidance or modification of cultural practices that disrupt natural enemy populations is an important strategy for maintaining natural enemy effectiveness.

Examples emphasizing importance of cultural practices on CBC was reported from Australia. It was found that pruning apple twigs infested with the wooly apple aphid, *Eriosoma lanigerum* (Hausman) (Homoptera: Eriosomatidae), killed much of the population of its dominant parasite, *Aphelinus mali* (Haldeman) (Hymenoptera: Aphelinidae). To deal with this, growers began to hold their pruned material over the winter and carry it back into the orchard to increase the early spring population of the wasp (Wilson, 1966).

Van den Bosch *et al.* (1967) found a similar situation in alfalfa, where harvesting was more detrimental to the parasite *Aphidius smithi* (Hymenoptera: Braconidae) than to its host the pea aphid, *Acyrthosiphon pisum*. Strip cutting conserved the parasitoid due to maintaining a better microclimate for the wasp, maintaining host-parasite synchrony and increasing parasite efficiency.

6.2. Providing Supplementary Resources

Incorporating practices that are beneficial to natural enemies requires fundamental knowledge of ecology and life history of natural enemies. A variety of other approaches to conservation of parasitoids and predators has been studied, and are comparatively complex. These include management of soil, water, and crop residue; modification of cropping patterns; manipulation of non-crop vegetation; and direct provision of resources to natural enemies (Van Driesche and Bellows, 1996). In general, these approaches are aimed at enhancing the density of resident natural enemy populations or communities to increase their effectiveness in pest suppression. Approaches to conservation biological control other than pesticide use modification are the focus of much of the current research in this area (Barbosa, 1998). This research is continuing toward developing an understanding of the ecological processes affecting natural enemies at spatial scales ranging from individual fields to entire landscapes (Ferro and McNeil, 1998; Landis and Menalled, 1998). Many of these alternatives have been shown to be effective (Barbosa, 1998), but are rarely implemented (U.S. Congress OTA, 1995), highlighting an extreme gap between research and implementation for this type of biological control (Ehler, 1998). For example the practices of strip harvesting hay alfalfa or intercropping alfalfa and cotton in California act to conserve natural enemy populations in cotton by preventing migration of *Lygus* bugs to cotton and subsequent pesticide applications for this pest. While both practices can be highly effective, both pose operational problems, and are more expensive, and as a result are not widely implemented by growers (Ehler, 1998).

These practices seem to fall into three broad categories: (i) alternate hosts of natural enemies (ii) non host foods/food sprays and (iii) shelter or refugia.

6.2.1 *Alternate hosts/prey*

Alternate hosts or prey have also been supplied to natural enemies. The first study by Flaherty and Huffaker (1970) involved the Pacific mite, *Tetranychus pacificus* McGregor (Tetranychidae) that is a serious pest of grapes in California vineyards. The Willamette mite, *Eotetranychus willamettei* Ewing (Tetranychidae), also occurs in the vineyards, but is a minor pest. The Willamette mite sometimes serves as prey, allowing the predator *Metaseiulus occidentalis* (Nesbitt) (Phytoseidae) to build up in numbers prior to the Pacific mite reaching damaging numbers. Tydeid mites (Tydeidae) are mostly scavengers and pollen feeders. They have been inoculated into vineyards as a means to increase *Metaseiulus* populations.

6.2.2 *Non-host foods*

Many predators and parasitoids require alternate food sources. For instance, certain ladybird beetles (Coccinellidae), which are important aphid predators, feed on plant pollen before switching to aphids. Many adult parasitoids require food in the form of pollen, nectar, or honeydew. Availability of such foods has been found to increase fecundity, longevity, survival, and effectiveness of many species. Availability of plant food sources may also increase the host searching efficiency of parasitoids, since degree of hunger can influence whether the parasitoids spend more time searching for hosts or for food. Starved parasitoids have greater attraction to flower odors over host associated odors. Maintenance of non-crop plants in or around agricultural fields or forest plantations can provide these foods, and also harbor alternate hosts or prey of natural enemies, helping to maintain natural enemy populations when pest populations are low. Provision of artificial supplementary food sources has also been shown to increase longevity and fecundity of some important agricultural and forest parasitoids in the laboratory and field.

Pollen and nectar or food sprays are most commonly involved. Living sources of non-host foods can be other crops or non-crop plants. Water chestnut, *Eleocharis* spp., can be planted in rice field to maintain populations of *Tetrastichus schoenobii* Ferriere (Hymenoptera: Eulophidae), an important parasite of the rice pest *Scirpophaga incertulus* (Walker). Intercropping on a larger scale can accomplish similar goals. In general, it is believed that intercropping reduces the advantages an herbivore gains in extensive monocultures and provides alternate resources for natural enemies, *e.g.*, pollen as a food prior to host availability. Cover crops grown between or below crop plants can contribute to improving biological control. Clover, *Trifolium* spp., grown between cabbage plants increase populations of staphylinids, coccinellids and syrphids, resulting in lower populations of the cabbage root maggot, *Delia brassicae* Weidemann (Diptera: Anthomyiidae). It is also possible in some cases to plant crops sequentially to gain the advantage of maintaining food sources for natural enemies. It has been also found that adding flowering plants to landscapes provides floral resources and refuge that enhance natural enemies.

Food sprays have typically been a carbohydrate source (sugar or honey) or a protein and carbohydrate source (sugar or honey, plus yeast or casein hydrolyzate). In conservation, the food sprays serve primarily as arrestants, retaining the natural enemies in area, hopefully until the pest population begins to increase. Although, food sprays remain popular with organic growers and home gardeners, but their value is sporadic. The timing of the application should coincide with the presence of natural enemies in the field and the target pest must appear as if on cue before the food spray is consumed or a second spray will be needed to continue keeping the predators in the field. Loss due to rain and delayed appearance of the pest along with high costs are major constraints on the utility of food sprays.

6.2.3 *Refugia*

Conservation biological control involves environmental manipulation to enhance the fecundity and longevity of natural enemies, modify their behaviour, and provide shelter from adverse environmental conditions. Natural enemies also require shelter from the elements. Artificial shelters have been found to increase winter survival of various peach orchard predators, allowing them to provide improved control of early season peach pests. Windbreaks and shelterbelts may increase searching efficiency and oviposition of parasitoids and predators adversely affected by high winds.

Refugia can take the form of ephemeral refugia within a growing season; *e.g.*, unmoved strips of alfalfa as previously discussed. However, there have also been studies looking at supplying overwintering/oversummering refugia. Numerous ideas have been suggested and tried, but none is in wide use. Hedgerows, windbreaks and other areas with perennial vegetation can harbor overwintering Chrysopidae and Coccinellidae species that do not migrate long distances. Trees with grass around them are often best. Corners of center pivot fields as good candidates for serving as refugia.

This basic approach has become dominant in modern conservation efforts. It can be successful, but is most effective on small acreages because the natural enemies must disperse from the refugia. Thus, its impact will be less important on large farms than on small farms. Another easy method *i.e.*, banding trees with corrugated cardboard have been found to harbor large numbers of predaceous mites and insects (Tamaki and Halfhill, 1968). In fact, 90% of the residents were entomophagous. Similar strips used during the summer can become pupation sites for Lepidoptera.

Perhaps the most well known type of shelter habitats provided is "beetle banks" which are usually grass covered earth banks located in the middle of a field (Thomas *et al.*, 1991). These banks provide suitable over-wintering sites for predatory beetles in the families Carabidae and Staphylinidae and for spiders (Jonsson *et al.*, 2008).

Some of the practices that enhance natural enemy effectiveness are summarized below:

- Application of sugar-water or protein sprays to attract/maintain natural enemies
- Providing shelters or avoiding destruction of nests of social wasps
- Planting 'banker plants' that harbor alternative (non-pest) prey
- Diversification of crop plantings using intercropping, mixed cropping, relay cropping, etc.
- Altering harvest and/or cultivation practices to maintain 'refuge strips' for natural enemies
- Using cover crops to increase overwintering survival of natural enemies
- Planting flowers/using cultivars that provide pollen and nectar sources

7. SOIL AS A RESERVOIR FOR NATURAL ENEMIES OF PEST INSECTS AND MITES IN THE CONTEXT OF CBC

Research on soil environmental manipulation of insect and mite pathogens and insect parasitic nematodes has mainly focused on four areas: (1) improved transport of the pathogen from the reservoir, usually the soil, to a site where the insect or mite host can come into contact with the pathogen or the nematode, (2) improvement in persistence of the pathogen or the nematode at the site where it contacts the insect or mite host, (3) overall growth of the pathogen or nematode population, (4) activation of latent infections (especially for viruses) (Fuxa, 1998). The success of conservation biological control is very difficult to measure since it is based on a hierarchy of different criteria involving several trophic levels (Klingen and Haukeland, 2006).

Bing and Lewis (1993) suggested, to enhance the natural occurrence of arthropod pathogenic soil fungi, for example, epizootics of *Beauveria bassiana*, in over-wintering larvae of the European corn borer (*Ostrinia nubilalis*) in maize residues, by modifying agronomic practices, such as no-till or reduced tillage systems. Similar suggestions were also made by Hummel *et al*. (2002) who conducted a field experiment on effects of different production practices on soil borne entomopathogens in vegetable systems. One fascinating example involving soil or soil litter is the blowing of NPV (Nuclear Polyhedrosis Virus) contaminated forest litter up into trees for the initiation of a viral epizootic in larvae of *Lymantria dispar* (Fuxa, 1998). According to Fuxa (1998) there has been only one attempt to enhance natural epizootics of nematodes. In that study tillage, weed management, and irrigation were investigated for enhancement of *Heterorhabditis bacteriophora* (*heliothidis*) in *Diabrotica undecimpunctata howardi* infesting maize. No-till and the presence of weeds significantly increased the numbers of nematodes in soil bioassays, but irrigation had no effect (Fuxa, 1998). Lewis *et al*. (1998) discuss a conservation approach to using entomopathogenic nematodes, and emphasize the need to understand the requirements and structure of natural populations before this approach can be recommended for practical use (Klingen and Haukeland, 2006).

8. NATURAL ENEMY DIVERSITY IN THE CONTEXT OF CBC AS A COMPONENT OF HABITAT MANAGEMENT

Although experimental studies have shown that provision of floral resources, food sprays and shelter habitats can increase natural enemy abundance, diversity and fitness, few have found that this leads to decreased pest damage, increased crop yield and greater profit for farmers (Heimpel and Jervis, 2005; Cullen *et al*., 2008; Griffiths *et al*., 2008; Wade *et al*., 2008b). One reason for this is probably a poor understanding of how the added resources affect interactions among natural enemies. It is important to understand the often complex relationship between natural enemy diversity and biological control (Griffiths *et al*., 2008; Straub *et al*., 2008; Tscharntke *et al*., 2007; Jonsson *et al*., 2008). Straub *et al*. (2008) concentrate their review on this relationship and show that the effect of increased natural enemy diversity on biological control may range from negative to positive. Negative effects occur mainly in systems where intra-guild predation among natural enemies is common. Positive effects primarily occur when the feeding niches of the natural enemies complement each other. This implies that CBC measures that increase the diversity of natural enemies are likely to be most successful in food webs comprising species with a significant degree of niche complementarity (Jonsson *et al*., 2008).

In some systems, particular species may be driving the relationship between diversity and biological control (Straub and Snyder, 2006; Cardinale *et al.*, 2006) and then CBC measures targeting these particular species may be most useful. A high diversity of natural enemies may provide an insurance against disturbances, whereby seemingly redundant natural enemy species may become important biological control agents under conditions of global change (Yachi and Loreau, 1999; Tscharntke *et al.,* 2007).

Tscharntke *et al.* (2007) discuss natural enemy diversity at the landscape scale related to biological control and show that complex landscapes with a high proportion of non-crop area have a higher density and diversity of natural enemies and tend to have lower pest pressures compared with simpler landscapes. However, protocols which clearly show how to design a landscape to maximize biological control are still not available.

9. ECOSYSTEM SERVICES OF CBC OTHER THAN BIOLOGICAL CONTROL

Habitat management practices relating to CBC may not only improve the efficacy of natural enemies but can also contribute to a range of other ecosystem services such as nutrient cycling, biofuel production and mitigation of soil erosion. Fiedler *et al.* (2008) explore the potential for CBC practices to increase ecosystem services other than biological control and conclude that, to date, studies relating to CBC have almost solely focused on improving natural enemy efficacy but that by broadening the range of plant species screened and by considering their suitability for other ecosystem services, the interest for adoption of CBC is likely to increase. Several recently developed agroecological partnerships in California and New Zealand that all deploy non-crop plants to improve multiple ecosystem services (including CBC), indicates attractiveness of such an approach to growers (Cullen *et al.,* 2008; Jonsson *et al.*, 2008).

10. CASE STUDY

Conservation Biocontrol is probably the oldest approach to biological control but, in the modern sense, is also an under-explored realm (Lazarovits *et al.,* 2007). Following case studies provide evidence to demonstrate that CBC has much potential but has been little explored.

10.1. Control of Mites in Pome Fruit by Inoculation and CBC

An effective CBC programme has been developed against phytophagous mites in apples orchards across Quebec, North America to satisfy the demand of high-quality, pesticide residue-free fresh apples. The CBC programme is based on the dominance of at least two predacious mite species per season and often another one to two predator species of minor importance (Bostanian and Lasnier, 2007). A key factor for this success is based on a regular comprehensive evaluation and understanding of the toxicology of pesticides that may be used in an orchard. Whenever possible, this information is generated in confidence with the cooperation of the agro-chemical industry. At the appropriate time, the information becomes public and it is relayed in a suitable format to growers. Armed with this information and the cost of treatments, growers, with the help of their pest control advisors, design their own management programmes and assume all risks, ultimately leading to successful pest suppression (Bostanian and Lasnier, 2007).

10.2. Management of Aphid Populations in Cotton through CBC

One successful effort to utilize natural biological control of cotton aphid, by predicting epizootics caused by the fungus, *Neozygites fresenii* has been developed in USA. This project has been proved rewarding on two levels. (i) purely scientific rewards of conducting in-depth, long-term biological and epizootiological studies in the laboratory and field on *N. fresenii*. (ii) providing practical help to cotton growers by making them aware about a fungal entomopathogen *N. fresenii*, which has been proved rewarding in terms of biological research on the epizootiology of an insect pathogen as well as rewarding in practical help for cotton growers to reduce their input costs in terms of reduced pesticide uses (Steinkraus, 2007).

10.3. Management of Pests of Vineyards in New Zealand and Australia

Biological control of insect pests in New Zealand and Australian vineyards is focused largely on the management of leaf rollers, especially the light brown apple moth, *Epiphyas postvittana* (Lepidoptera: Tortricidae), which is widely considered to be the most serious pest of grapevines. Biological control of pests of vineyards *via.*, the enhancement of soil biological activity and vine debris degradation illustrates the power of simple cultural practices to contribute towards such problems.

As the biological mechanisms that operate in the vineyard system are potentially transferable to other horticultural systems and the system seems to be driven by soil moisture and nutrient content, it could easily be combined with the use of flowering plants that help to boost the activity of natural enemies of insect pests. For example, mulching these plants into the below-vine area after pruning would convert the plants that had supported natural enemies earlier in the season into mulch that speeds breakdown of fungal inoculum on pruning over the winter. The use of some fungicides can be integrated into such a system as the fungicide is unlikely to penetrate the mulch and so the system will continue to operate. This technology has the potential to enhance ecosystem services, wine marketing and sustainability in vineyards by using organic waste to help disrupt the life cycle of *B. cinerea* and potentially of other vine pathogens (Gurr *et al.*, 2007).

11. CBC IN THE CONTEXT OF IPM

The goals and approaches of conservation biological control closely match those of IPM. In both, a fundamental understanding of the ecological mechanisms driving pest dynamics is key to success (Huffaker, 1980). Conserving natural enemies often requires modification of production practices that are similar to changes in practices recommended by IPM principles (*e.g.*, increase diversification of crops, reduction in pesticide use, etc.). The use of thresholds to make decisions, common to IPM systems, is closely tied to the impact of natural enemies whose density, composition and impact on pest dynamics (and damage) are dependent on the crop cultivation practices and environmental milieu. The interdependence of farming practices, pest dynamics and the impact of natural enemies often requires farmers to modify practices. As such, farmer education is key to success (O'Neil *et al.*, 2003). Examples of farmer education span a number of extension approaches that include bulletins, field days, grower meetings, electronic media and farmer field schools (FFS).

One example of interdependence of CBC and IPM is illustrated by the BPH (brown plant hopper in paddy) management in south Asia. Initial efforts using resistant varieties were met with success, but resistance was short lived, leading to cycles of varietal development, deployment and eventual failure. BPH problem was further complicated by governmental policies that subsidized the purchase of insecticides causing widespread mis- and over-use which further exasperated BPH (and other pests) management programs. Rice IPM program was relied on conservation biological control and the use of FFS to investigate, implement and extend IPM technologies. Because rice IPM is built upon a foundation of conserving endemic natural enemies (principally predatory bugs, egg parasitoids and spiders), farmers are involved in observing natural enemy attacks on rice pests, and investigating relationships between production practices and natural enemy diversity, dynamics and effectiveness. The success of this approach is evident not only through its implementation in the rice production system, but also in its application to other crop-pest systems in Asia and other parts of the world (Raj and Suresh, 2000).

12. INTERACTIONS BETWEEN HOST PLANT RESISTANCE AND CBC

Scanty old records indicate that people in early societies selected crops for breeding based in part on their resistance to insects. Sericulturists in China selected for mulberry varieties that supported high silkworm production and yielded high quality silk. In the United States, selection of early maturing wheat varieties that were resistant to the Hessian fly, *Mayetiola destructor* (Say) (Diptera: Cecidomyiidae), began in 1788. By 1792, the cultivar "Underhill" was recognized as being resistant to this pest. Apples resistant to the woolly apple aphid, *Eriosoma lanigerum* (Hausman) (Homoptera: Eriosomatidae), were known by 1831.

The interactions between host plant resistance and biological control were soon recognized. Initial observations were of negative interactions, where a plant interfered with biological control agents. However, it is now widely recognized that plant characteristics can positively influence the attractiveness and suitability of a plant to entomophages. Thus, plant breeders could select for traits that favor natural enemies. This would increase "extrinsic resistance."

Leaf trichomes can interfere with both herbivores and entomophages. The balance of these impacts will determine whether hairy or bare leaves are more advantageous in pest management. Cuticular waxes on plants also influence biological control. For example, cabbages and peas with abundant (normal) and reduced (glossy) surface waxes can be considered. In the lab, normal and glossy plants of near isolines support the same number of aphids and other herbivores, *e.g.*, larvae of diamondback moth, *Plutella xylostella* (Linnaeus). This indicates that there is no intrinsic resistance to the pests. Yet, in the field, the glossy lines frequently support fewer aphids. This is due to improved predator (Coccinellidae and Chrysopidae) efficiency on the glossy lines. The predators simply get better traction and better grip. They spend less time scrambling (trying to get footing so they can move) and more time walking so they search more efficiently and can encounter more prey. Improved traction also changes the proportion of time predators spend on various parts of the plant. So, with the glossy plants, the aphids have fewer, less effective refugia. The net effect is often fewer aphids on glossy plants in the field.

Plants may emit volatiles that are attractive to natural enemies including herbivore induced plant volatiles, and in some cases, volatiles emitted by intact plants. Khan *et al*. (2008) reviewed how such volatiles can be exploited in CBC to increase the number of natural enemies attracted to a crop. Exploitation of such volatiles may be particularly effective if they are combined with other CBC techniques that improve natural enemy fitness once they have been attracted to the crop or as an integral part of 'push-pull' strategies (Jonsson *et al*., 2008). The latter approach, which combines bottom up and top down forces to decrease pest damage *via.*, habitat manipulation, has been widely adopted among subsistence farmers in east Africa to control stem borers on sorghum (*Sorghum bicolor* (L.) and maize (*Zea mays* L.) (Cook *et al*., 2007).

13. CBC IN GENETICALLY ENGINEERED CROPS

The new area of genetic engineering of crops for resistance to insects is in its infancy. The long-term agricultural, ecological, scientific and socio-political impacts of this work cannot be predicted with confidence. However, it is a timely topic. Currently available, insect resistant, genetically modified crop plants have had the gene for the *Bacillus thuringiensis* (*Bt*) endotoxin incorporated into their genome. With *Bt* cotton at the centre stage, efforts to develop pest-control systems based on conservation biological control need utmost focus given the reduced insecticide use, changing pest scenario and re-establishing native natural control agents (Biradar and Vennila, 2008).

An attempt to document the taxonomic diversity of arthropods of the current cotton ecosystems inclusive of *Bt* and non-*Bt* has indicated re-establishing native predators from different groups, *viz*., Chrysopidae (*Chrysoperla* sp. (*carnea*–group)) of Neuroptera; Lygaeidae (*Geocoris ochropterus* (Fieber)), and Miridae (*Deraeocoris* sp.) of Hemiptera, Coccinellidae (*Cheilomenes sexmaculata* (F.), *Brumoides suturalis* (F.) and *Scymnus castaneus* Sicaid) of Coleoptera, Syrphidae (*Ischiodon scutellaris* (Fabricius) and *Dideopsis aegrota* (Fabricius)) of Diptera and spiders from different families, *viz.,* Araneidae, Clubiondae, Lycosidae, Oxyopidae and Salticidae of Arachnidae. These diverse groups of predators have greater potential to offer natural control of emerging sucking insects (Biradar and Vennila, 2008). Apart from increasing diversity, report of maximum abundance of coccinellids, chrysopids and syrphids on *Bt* over non-*Bt* cultivars, and IPM over insecticide–treated plots is available (Kulkarni *et al.,* 2004). Transgenic crops through biotechnology, natural control of pests through CBC, enhancement of natural enemy effectiveness through bio-signalling (Ananthakrishnan, 2004, 2006) and need for 'Integrated Biodiversity Management' (Ignacimuthu, 2007) have been the realm of recent research trends in pest management. As the field research cum-application relating to the natural enemy fauna through CBC awaits success, an active approach to acknowledge the role of native natural enemies and promotion of their effectiveness through landscape ecology is the need of the hour, for sucking-pest management in the rapidly expanding area of *Bt* cotton (Biradar and Vennila, 2008).

14. COST–BENEFIT OF CBC

Very few studies have included cost benefit assessments of CBC to date (Cullen *et al*., 2008). Cullen *et al*. (2008) and Grifiths *et al*. (2008) discussed how such economic assessments can be conducted. Cullen *et al*. (2008) also reviewed factors that may influence uptake of CBC amongst growers and consider what policies or strategies might be introduced to increase the incentive to adopt CBC. They concluded that the relative advantage of CBC over conventional control methods, suitability of the protocols for preliminary testing by the growers themselves and the social dynamics of CBC development and extension are key factors influencing adoption.

15. IMPORTANCE OF CBC

CBC has long been a rather neglected form of biological control, but research in this field has increased markedly during the last decade (Zehnder *et al.*, 2007; Wade *et al.*, 2008a).

This approach has received much interest in recent years because:

1. It does not involve introducing exotic organisms into the environment, thus avoiding some major environmental and legal concerns associated with classical biocontrol and
2. When successful, it provides a sustainable pest control solution free of the continuous input of mass-reared natural enemies required by augmentative biocontrol.

16. ADVANTAGES

CBC can contribute to safer and more effective biological control practices but it requires in-depth knowledge of the ecology of natural enemies and the ecological communities which they are part of. CBC also has several other tangible advantages which include the following:

- It is based on concepts which are easy for growers to understand: 'the value of diversity', 'natural enemies need more than their prey/host', 'inter-cropping', 'companion planting' etc.
- It is a practice which individual growers can adopt, in contrast with classical biological control schemes which are usually coordinated at the regional, national or continental scale.
- It usually involves a conspicuous change to the farm landscape (*e.g.*, 'beetle banks' or flower strips) so it can readily be used to support 'green' marketing strategies, including, but not exclusively organic ones.

17. DISADVANTAGE

Some of the disadvantages associated with CBC are as follows:

- Not possible to practice against severe and key pests
- Detail knowledge of agroecosystem analysis and pest bionomics are required

- Not useful when the pest population is approaching economic threshold level (ETL)
- May not yield satisfactory controls when it is utmost required
- May not be compatible with other existing pest management practices
- The grower and buyers must accept the higher price of the produce

18. CBC RESEARCH IN THE FUTURE

Future researchers working on CBC should investigate the following issue as mentioned by Jonsson *et al.* (2008):

- The effect of CBC on decreased pest damage, increased crop yield or quality and improved economic profit for growers.
- There is also a need for a better understanding of the conditions under which certain CBC techniques will reduce pest populations in the field as improved fitness of natural enemies or increased densities and diversity of such species in the crop does not automatically ensure effective pest management.
- Resources that selectively benefit key natural enemies–may be single or multiple species, but not their antagonists or the pest (*e.g.*, selective food plants) (Straub *et al.*, 2008).
- Combining different techniques of biocontrol with each other–'attract and reward' or 'Push-pull' strategy.
- How to optimize landscape management for biological control–by involving farmers in the research process, *e.g.*, agroecological partnerships.

CBC can help improve other ecosystem services than biological control (Gurr *et al.*, 2003) and this could be considered when plants for CBC trials are selected (Fiedler *et al.*, 2008). If multiple ecosystem services can be improved through CBC, this can increase grower profits in several ways and thereby also the likelihood of grower adoption.

19. CONCLUSIONS

Conservative biological control has been most useful in situations where there are potentially effective natural enemies that are rendered ineffective by manipulable biotic or abiotic factors, the pests cannot be satisfactorily (economically and/or environmentally) controlled by other methods. This method requires a sound understanding of the system, insect biology, insect ecology and the constraints operating on the entomophage population. It is currently most popular among organic growers and home gardeners, but has potential in additional commercial settings, especially small acreages. It is necessary that the scientists and practitioners involved in CBC and bio control system need to communicate with each other, collaborate, and take full advantage of their complementary perspectives and knowledge. This should greatly improve our ability to design and manage the landscape, and to implement biological control strategies based on scientific understanding of ecological processes. Not only ecological but also socio-economical considerations may also influence any management recommendation.

Conservation biological control is likely to increase in importance as agricultural and commercial forest systems become more intensively managed and restrictions on the use of conventional insecticides increase. It is an approach that requires integration of fundamental insect life history and natural enemy/pest interactions with knowledge of the ecology of the systems in which these interactions take place. This combination of factors makes conservation biological control research and implementation a challenging, but potentially rewarding, endeavor.

REFERENCES

Ananthakrishnan, T.N. (2004). Biodiversity, biosignalling and biotechnology in insect-plant interactions. *Current Science,* **86:** 376–78.

Ananthakrishnan, T.N. (2006). Emerging technologies in resistance dynamics in insect-plant interactions. *Current Science,* **90:** 477–80.

Anonymous. (2002). IFOAM basic standards for organic production and processing, approved by the IFOAM general assembly, Victoria, Canada, August 2002, Section A–D. pp. 1–72.

Barbosa, P. (*Ed.*). (1998). *Conservation Biological Control.* Academic Press, New York, pp. 1–369.

Bing, L.A., and Lewis, L.C. (1993). Occurrence of the entomopathogen *Beauveria bassiana* (Balsamo) Vuillemin in different tillage regimes and in *Zea mays* L. and virulence towards *Ostrinia nubilalis* (Hubner). *Agriculture, Ecosystems and Environment,* **45:** 147–56.

Biradar, V.K. and Vennila, S. (2008). Pest management for *Bt* cotton: Need for conservation biological control. *Current Science,* **95(3):** 317–18.

Bostanian, N.J. and Lasnier, J. (2007). Control of mites in pome fruit by inoculation and conservation. *In*: Vincent, C., Goettel, M.S. and Lazarovits, G. (*Eds.*). *Biological control: A global perspective*. CABI, Oxfordshire, UK, pp. 374–82.

Cardinale, B.J., Srivastava, D., Duffy, J.E., Wright, J.P., Downing, A.L., Sankaran, M. and Jouseau, C. (2006). Effects of biodiversity on the functioning of trophic groups and ecosystems. *Nature,* **443:** 989–92.

Cook, S.M., Khan, Z.R. and Pickett, J.A. (2007). The use of 'push-pull' strategies in integrated pest management. *Annual Review of Entomology,* **52:** 375–400.

Coulson, J.R., Klaasen, W., Cook, R.J., King, E.G., Chiang, H.C. and Hagen, K.S. (1982). Notes on biological control of pests in China. *In*: US department of agriculture, biological control of pests in China. Washington, DC: USDA–OICD, pp. 1–192.

Cullen, R., Warner, K.D., Jonsson, M. and Wratten, S.D. (2008). Economics and adoption of conservation biological control. *Biological Control,* **45:** 272–80.

DeBach, P. (1958). The role of weather and entomophagous species in the natural control of insect populations. *Journal of Economic Entomology,* **51:** 474–84.

Ehler, L.E. (1998). Conservation biological control: past, present, and future, *In*: Barbosa, P. (*Ed.*). *Conservation Biological Control.* Academic Press, New York, pp. 1–8.

Eilenberg, J. (2006). Concepts and visions of biological control. *In*: Eilenberg, J. and Hokkanen, H.M.T. (*Eds.*). *An ecological and societal approach to biological control.* Springer, The Netherlands, pp. 1–11.

Eilenberg, J., Hajek, A. and Lomer, C. (2001). Suggestions for unifying the terminology in biological control. *BioControl*, **46:** 387–400.

Ferro, D.N. and McNeil. J.N. (1998). Habitat enhancement and conservation of natural enemies of insects. *In*: Barbosa, P. (*Ed.*). *Conservation Biological Control.* Academic Press, New York, pp. 123–32.

Fiedler, A.K., Landis, D.A. and Wratten, S.D. (2008). Maximizing ecosystem services from conservation biological control: the role of habitat management. *Biological Control,* **45:** 254–71.

Flaherty, D.L. and Huffaker C.B. (1970). Biological control of pacific mites and willamette mites: *I. Role of Metaseiulus occidentalis. Hilgardia,* **40:** 267–308.

Fuxa, J.R. (1998). Environmental manipulation for microbial control of insects. *In*: Barbosa, P. (*Ed.*). *Conservation Biological Control.* Academic Press, New York, pp. 255–68.

Griffiths, G.J.K., Holland, J.M., Bailey, A. and Thomas, M.B. (2008). Efficacy and economics of shelter habitats for conservation biological control. *Biological Control,* **45:** 200–209.

Gurr, G.M., Scarratt, S.L., Jacometti, M. and Wratten, S.D. (2007). Management of pests and diseases in New Zealand and Australian vineyards. *In*: Vincent, C., Goettel, M.S. and Lazarovits, G. (*Eds.*). *Biological control: A global perspective.* CABI, Oxfordshire, UK, 392–98.

Gurr, G.M., Wratten, S.D. and Luna, J. (2003). Multi-function agricultural biodiversity: pest management and other benefits. *Basic and Applied Ecology,* **4:** 107–16.

Heimpel, G.E. and Jervis, M.A. (2005). Does floral nectar improve biological control by parasitoids? *In*: Wackers, F.L., Van Rijn, P.C.J. and Bruin, J. (*Eds.*), *Plant-provided Food for Carnivorous Insects: A Protective Mutualism and its Applications.* Cambridge University Press, Cambridge, UK, pp. 267–304.

Huffaker, C.B. (*Ed.*). (1980). *New Technology of Pest Control.* John Wiley and Sons, New York, pp. 1–500.

Hummel, R.L., Walgenbach, J.F., Barbercheck, M.E., Kennedy, G.G., Hoyt, G.D. and Arellano, C. (2002). Effects of production practices on soil-borne entomopathogens in western North Carolina vegetable systems. *Environmental Entomology,* **31:** 84–91.

Ignacimuthu, S. (2007). Insect Pest Management. *Current Science,* **92:** 1336–37.

Jonsson, M., Wratten, S.D., Landis, D.A. and Gurr, G.M. (2008). Recent advances in conservation biological control of arthropods by arthropods. *Biological Control,* **45:** 172–75.

Khan, Z.R., James, D.G., Midega, C.A.O. and Pickett, J.A. (2008). Chemical ecology and conservation biological control. *Biological Control*, **45:** 210–24.

Klingen, I. and Haukeland, S. (2006). The soil as a reservoir for natural enemies of pest insects and mites. *In*: Eilenberg, J. and Hokkanen, H.M.T. (*Eds.*). *An ecological and societal approach to biological control*. Springer, The Netherlands, pp. 145–212.

Kulkarni, K.A., Kambrekar, D.N., Gundannavar, K.P., Devaraj, K. and Udikeri, S.S. (2004). In international symposium on 'Strategies for sustainable cotton production A Global Vision' 3. Crop Protection. University of Agricultural Sciences, Dharwad, pp. 149–51.

Landis, D.A., Wratten, S.D., and Gurr, G.M. (2000). Habitat management to conserve natural enemies of arthropod pests in agriculture. *Annual Review of Entomology*, **45:** 175–201.

Landis, D.A. and Menalled, F.D. (1998). Ecological considerations in the conservation of effective parasitoid communities in agricultural systems, *In*: Barbosa, P. (*Ed.*). *Conservation Biological Control*. Academic Press, New York, pp. 101–21.

Lazarovits, G., Goettel, M.S. and Vincent, C. (2007). Adventures in biocontrol. *In*: Vincent, C., Goettel, M.S. and Lazarovits, G. (*Eds.*). *Biological control: a global perspective*. CABI, Oxfordshire, UK, pp. 1–6.

Letourneau, D.K. (1998). Conservation biology: lessons for conserving natural enemies. *In*: Barbosa, P. (*Ed.*). *Conservation Biological Control*. Academic Press, New York, pp. 9–38.

Lewis, E.E., Campbell, J.F., and Gaugler, R. (1998). A conservation approach to using entomopathogenic nematodes in turf and landscapes. *In*: Barbosa, P. (*Ed.*). *Conservation Biological Control*. Academic Press, New York, pp. 235–54.

O'Neil, R.J., Yaninek, J.S., Landis, D.A. and Orr, D.B. (2003). Biological control and integrated pest management. *In*: *Integrated pest management in the global arena*. Maredia, K.M., Dakouo, D. and Mota–Sanchez, D. (*Eds.*), CABI Publishing, Wallingford, UK, pp. 19–30.

Orr, D.B. and Suh, C.P.C. (1998). Parasitoids and predators. *In*: Rechcigl, J.E. and Rechcigl, N.A. (*Eds.*). *Biological and Biotechnological Control of Insect Pests*. CRC Press LLC, Boca Raton, New York, pp. 2–33.

Pickett, C.H. and Bugg, R.L. (*Eds.*). (1998). *Enhancing biological control: habitat management to promote natural enemies of agricultural pests*. University of California Press, Berkeley, pp. 1–422.

Raj, D.A. and Suresh, C. (2000). Farmers' research on natural crop protection. *In:* Stoll, G. (*Ed.*). *Natural Crop Protection in the Tropics, 2nd edn. AGRECO / CTA,* Margrof Verlag, Weikersheim, Germany, pp. 291–96.

Ruberson, J.R., Nemoto, H. and Hirose. Y. (1998). Pesticides and conservation of natural enemies in pest management, *In*: Barbosa, P. (*Ed.*). *Conservation Biological Control*. Academic Press, New York, pp. 207–20.

Steinkraus, D. (2007). Management of aphid populations in cotton through conservation: delaying insecticide spraying has its benefits. *In*: Vincent, C., Goettel, M.S. and Lazarovits, G. (*Eds.*). *Biological control: a global perspective*. CABI, Oxfordshire, UK, pp. 383–91.

Straub, C.S. and Snyder, W.E. (2006). Species identity dominates the relationship between predator biodiversity and herbivore suppression. *Ecology*, **87:** 277–82.

Straub, C.S., Finke, D.L. and Snyder, W.E. (2008). Are the conservation of natural enemy biodiversity and biological control compatible goals? *Biological Control,* **45:** 225–37.

Tamaki, G. and Halfhill, J.E. (1968). Bands on peach trees as shelters for predators of the green peach aphid. *Journal of Economic Entomology,* **61:** 707–11.

Thomas, M.B., Wratten, S.D. and Sotherton, N.W. (1991). Creation of 'island' habitats in farmland to manipulate populations of beneficial arthropods: predator densities and emigration. *Journal of Applied Ecology,* **28:** 906–17.

Tscharntke, T., Bommarco, R., Clough, Y., Crist, T.O., Kleijn, D. and Rand, T.A. (2007). Conservation biological control and enemy diversity on a landscape scale. *Biological Control,* **43:** 294–309.

U.S. Congress OTA (Office of Technology Assessment). (1995). Biologically based technologies for pest control. OTA–ENV–636. U.S. Congress, Office of Technology Assessment,Washington, D.C.

Van den Bosch, R., Lagace, C.F. and Stern, V.M. (1967). The interrelationship of the aphid, *Acyrthosiphon*, and its parasite *Aphidius smithii*, in a stable environment. *Ecology,* **48:** 993–1000.

Van Driesche, R.G. and Bellows, T.S.Jr. (1996). *Biological control.* Chapman and Hall, New York, pp. 1–539.

Wade, M.R., Gurr, G.M. and Wratten, S.D. (2008a). Ecological restoration of farmland: progress and prospects. *Philosophical Transactions of the Royal Society B Biological Sciences,* **363:** 831–47.

Wade, M.R., Zalucki, M.P., Wratten, S.D. and Robinson, K.A. (2008b). Conservation biological control of arthropods using artificial food sprays: current status and future challenges. *Biological Control,* **45:** 185–99.

Wilson, F. (1966). The conservation and augmentation of natural enemies. *Procedings of FAO Symposium on Integrated Pest Control,* **3:** 21–26.

Yachi, S. and Loreau, M. (1999). Biodiversity and ecosystem productivity in a fluctuating environment: The insurance hypothesis. *Proceedings National Academy of Sciences*, USA, **96:** 1463–68.

Zehnder, G., Gurr, G.M., Kuhne, S., Wade, M.R., Wratten, S.D. and Wyss, E. (2007). Arthropod pest management in organic crops. *Annual Review of Entomology,* **52:** 57–80.

13

Indigenous Methods of Crop Protection: An Indian Scenario

BISWARUP SAHA[1] AND PRANAB DUTTA[2]

ABSTRACT

The researchers have been trying to introduce eco-friendly approaches in crop protection to reduce harmful chemical load in environment. In this connection, indigenous practices offer safe and better alternative methods of pest management. Indigenous knowledge systems and technologies are found to be socially desirable, economically affordable, and sustainable and involve minimum risk to rural farmers and producers, and above all, they are widely believed to conserve resources. There are situations in which modern science is not appropriate, and use of simpler technologies and procedures are required. Thus Indigenous Knowledge provides basis for problem solving strategies for local communities, especially the poor. The focus of the chapter is on documentation of ITK for sustainable crop protection. The chapter highlights the process of collection, documentation and validation of Indigenous Knowledge Systems and suggests steps to promote Indigenous Knowledge.

Key words: Indigenous knowledge, Indigenous technological knowledge, Crop protection, Documentation, Validation.

When a knowledgeable old person dies, a whole library disappears.

— **African Proverb**

1 College of Fisheries, Central Agricultural University, Lembucherra, Tripura West - 799210.

2 Department of Plant Pathology, Assam Agricultural University, Jorhat, Assam - 785013.

Corresponding author: E-mail: biswarup.ext@gmail.com

The use of non-chemical methods for crop protection is already gaining importance in several countries including India. The researchers recently recognized the harmful effects of pesticides and have been trying to introduce eco-friendly approaches to reduce harmful chemical load in environment by using bio-agents and bio-pesticides but these are not easily available and are costly. So it has been difficult for farmers to utilize these tools in pest management. To overcome these problems, indigenous practices offer safe and better alternative methods of pest management. This is the era of sustainability which involves the use of eco-friendly indigenous pest management practices to obtain the rich harvest and to keep the environment safe for ever. Moreover, interest in indigenous knowledge systems has been fueled by the recent worldwide ecological crisis and the realization that its causes lie partly in the overexploitation of natural resources based on inappropriate attitudes and technologies. Scientists now recognize that indigenous people have managed the environments in which they have lived for generations, often without significantly damaging local ecologies. Many feel that indigenous knowledge can thus provide a powerful basis from which alternative ways of managing resources can be developed. A critical assumption of indigenous knowledge approaches, for example, is that local people have a good understanding of the natural resource base because they have lived in the same, or similar, environment for many generations, and have accumulated and passed on knowledge of the natural conditions, soils, vegetation, food and medicinal plants etc.

1. INDIGENOUS KNOWLEDGE—ITS MEANING AND NATURE

The terms indigenous technical knowledge (ITK), "local knowledge" and "Traditional knowledge" have been used in the literature inter-changeably. Traditional knowledge is gathered over a period of time and transferred from generation to generation. It is synonymous to local knowledge. Wang (1988) defined ITK as "sum total knowledge and practices which are based on people's accumulative experiences in dealing with situations and problems in various aspects of life and such practices are special to a particular culture". Pushpangadan *et al.* (2002) defined "ITK is a community based functional knowledge system, developed, preserved and refined by generations of people through continuous interaction, observation and experimentation with their surrounding environment. It is a dynamic system, ever charming, adopting and adjusting to the local situations and has close links with the culture, civilization and religious practices of the communities".

Therefore, we can summarize that ITK is the knowledge, which has been accumulated by the people in a given community over generations by observation, by experimentation and by handling on old peoples' experiences and wisdom in any particular area of human behavior. Indigenous technical knowledge is the local knowledge that people have gained through inheritance from their ancestors. It is a people-derived science and represents people's creativity, innovations and skills. Indigenous technological knowledge pertains to various cultural norms, social roles or physical conditions. Such knowledge is not a static body of wisdom, but instead consists of dynamic insights and techniques, which are changed over time through experimentation and adoption to environmental and socio-economic changes. Traditional knowledge is deep rooted in many countries of the world. This knowledge system is vital for their well being and for sustainable development, besides monitoring their cultural liveliness. This knowledge has backgrounds of hundreds and sometimes thousands of years of adoption, while bearing odds and evens of the time. Generally speaking, such knowledge evolves in the local environment, so that it is specifically adapted to the requirements of local people and conditions. It is also creative and experimental, constantly incorporating outside influences and inside innovations to meet new conditions. It is usually a mistake to think of indigenous knowledge as "old-fashioned", "backwards", "static" or "unchanging".

Indigenous knowledge is the knowledge that people in a given community has developed over time, and continues to develop. It is:

- Based on experience.
- Often tested over centuries of use.
- Adapted to local culture and environment.
- Dynamic and changing.

Indigenous people have developed a close and unique connection with the lands and environment in which they have lived for generations. They have generated unique system of knowledge, innovation and skills relating to the uses and management of different crops they have been growing in their land. Over millennium people have learnt how to adopt in the adverse climatic situation for their existence. Through trial and experience they established how to get food and survive in different environment, what variety of crops needs to be plant, when and how to planted, which crops are poisonous, which can be used as medicine, how to protect their crops from different pests and diseases.

2. WHY IS INDIGENOUS KNOWLEDGE (IK) USEFUL IN PLANT PROTECTION?

IK is the basis for self-sufficiency and self-determination for the following reasons:

1. Indigenous products have assumed greater importance in recent years all over the world due to growing awareness of harmful effect of indiscriminate use of pesticides. Since the chances of development of resistance by pests, towards a cultural control are very low, it is expected to provide a long term solution for the control of insect pests of agricultural crops.

2. Indigenous products offer a potential alternative or substitute to conventional pesticides.

3. IK provides effective alternatives to western know–how. It gives local people and development workers extra options when designing projects. Instead of searching only among western technologies for feasible solutions, they can choose from indigenous knowledge or combine indigenous and western technology.

4. People are familiar with indigenous practices and technologies. They can understand, handle, and maintain them better than introduced western practices and technologies.

5. IK draws on local resources. People are less dependent on outside supplies, which can be costly, scarce and available only irregularly.

6. Indigenous technologies and practices are often cheaper than western ones. They rely on locally available skills and materials and often require little or no cash outlay.

7. Studies on indigenous technologies are urgently wanted because of great deal of knowledge is being rapidly lost.

8. Some indigenous technologies that have been documented provide an opportunity to evaluate their efficacy on different pests.

3. BLENDING IK SYSTEMS WITH OTHER FORMS OF KNOWLEDGE

According to Dean (1999) the main goal of technology blending is to both maintain local knowledge and supplement it; "expanding it and combining it with modern knowledge and technologies". He goes on to argue that it provides self sustaining livelihood for local people. Local knowledge and technology must be amenable to commercial orientation at least to some degree in order to serve as a means to secure a livelihood. It will lead to more equitable sharing of the resulting benefits and increase the standard of living in traditional societies. This will be as a consequence of up-grading traditional production methods, elevating productivity, income and quality of life in the traditional sector.

World Bank (1998) forwarded the following steps to integrate IK with other forms of knowledge:

1. IK has to be recognized, identified and selected from a multitude of other knowledges.
2. IK's validation/affirmation by identifying its significance, relevance, reliability, functionality, effectiveness and transferability. This signifies an ability to support problem solving.
3. Codification/recording/documentation.
4. Storage of IK for retrieval.
5. IK transfer. Such transfers go beyond focusing on human recipients.
6. Dissemination and use of IK. Use of knowledge put it to testing, acceptance and further validation for development.

Rajshekaran (1993) has advocated a framework for incorporating IK into agricultural research and extension with the following features:

1. Strengthening the capacities of regional research and extension organizations;
2. Building upon local people's knowledge that are acquired through various processes such as farmer to farmer communication and farmer experimentation;
3. Identifying the need for extension scientist/social scientist in an interdisciplinary regional research team;
4. Formation of a sustainable technology development consortium to bring farmers, researchers, NGOs and extension workers together well ahead of the process of technology development;
5. Working with the existing organization and management of research and public sector extension;
6. Bringing research–extension–farmer together at all stages is practically difficult considering the existing bureaucracies and spatial as well as academic distances among the personnel belonging to these organizations. Hence the need for utilizing the academic knowledge gamed by some extension personnel (Subject Matter Specialists) during the process of validating farmer experiment.
7. Outlying areas that research and extension organizations need to concentrate on during the process of working with farmers;

8. Understanding that it is impractical to depend entirely on research stations for innovations; considering the inadequate human resource capacity of the regional research system.

Parasar (1994) developed a working model for incorporating indigenous technological knowledge in agricultural research and extension work.

1. Establishing a centre for indigenous knowledge
2. Giving curriculum sanction for indigenous knowledge having a broad based extension approach
3. Framing appropriate policy
4. Seeking support from general academia
5. Involving NGOs
6. Giving wide publicity.

4. COLLECTION, DOCUMENTATION AND VALIDATION OF INDIGENOUS KNOWLEDGE (IK) SYSTEMS

A comprehensive understanding of the traditional knowledge and expertise is essential for sustainable development. Gaining functional understanding of IK system includes collection and assessment (validation) of such IKs and finally its documentation.

4.1. Collection of IK

An investigator, for collection of IKs, has to employ an appropriate method of data collection/data gathering and have to identify appropriate sources of it. Pal (1998) had employed 'snow-ball' sampling to identify the sources of IKs in a society. He suggested that to employ snow-ball sampling technique, selection of the very first respondent is a very crucial one. However, the following characteristics may be considered for the selection of such respondents.

1. Good knowledge about the historical background of food production and resource conservation of the study villages,
2. A minimum of ten years of farming experience, and
3. Not being involved in other stages of the study.

The Mission Mode Project on "Collection, Documentation and Validation of Indigenous Technical Knowledge (ICAR)" has proposed the following sources of IK systems (Table 1).

Table 1: Sources of Indigenous Knowledge

Primary source	*Secondary source*
1. Farmers	1. Extension workers and others
2. Elderly persons of the area	2. Ancient literature
3. Village Mukhia	3. Media (Print and electronic)
4. Retired persons	4. Folksongs, proverbs and festivals
	5. NGOs/Institutions

Different authors have suggested following methods for collection of ITK (Mane and Sutaria, 1993; Quiroz, 1993; Rajshekaran, 1993; Pal and Das Gupta, 1997; Colfer, 1990):

- Observation and interviewing
- Case studies
- Field observation
- Participant observation
- Interpretative research (IR)
- Local taxonomy
- Cognitive method analysis
- Crop history
- Innovator workshop
- Different PRA methods
- Participative technology analysis
- Illustrations and diagrams
- Mapping/area domains of the ITK
- Audio–visual/photos/slides
- Use of schedule/questionnaire etc.

4.2. Documentation of IK

Documentation may be done with the help of the following methods as proposed in the training project report on Methodology of Collection, Documentation and Validation of ITKs (ICAR, 2004):

1. Audio
2. Visual

3. Audio-visual
4. Print media and
5. Computer aided documentation

Assessment of ITK, setting importance and prioritization of it are essential steps for proper documentation. Besides logical assessment in the light of formal scientific theories ITK needs an informal interaction with the inhabitants of the community where the ITK is documented to elucidate relevant information pertaining to ITKs. This interaction would provide an in-depth understanding of the 'emic' perspectives of local farmers.

4.3. Method of Validation of ITK

Following section has been adopted from the training report on Methodology of Collection, Documentation and Validation of ITKs (ICAR, 2004):

Validation is one of the important steps to study ITK effectively. It is logical step to qualify and quantify the effectiveness of Indigenous Technical Knowledge and also testing of rationale behind its use. For validation, generally following methods are used:

1. Quantifying Indigenous Knowledge (QuIK)
2. On-station and on-farm experimentation
3. Controlled studies
4. Case study

5. DOCUMENTED ITKS ON PLANT PROTECTION

Traditional practices of biological pest control include intercropping of diverse plant species to provide habitat to natural enemies of pests as well as alternate host plants for pest. For example, intercropping/mixed cropping of castor with cotton is practiced in Gurudaspur of Punjab, where castor acts as an alternative host for cotton pests (Sardana, 2001).

Some of conventions technologies, used by the farmers, include the followings:

5.1. ITKs for Pest Management

- ***Rice stem borer***

    (a) Refilling of water at tillering stage. Activities of the pest reduce due to non availability of water.

(b) Throwing of peels of rabab tenga, *Citrus grandis* on standing water at tillering stage. It repels various pests from the field due to aromatic and pungent and acidic nature of the citrus peels. This practice is followed in Assam.

(c) Burning of rice stubbles after harvesting. Burning causes destruction of eggs and pests due to which pest infestation in next crop season become less.

(d) Summer ploughing before sowing of the crop. Infestation by pest in the next crop reduced due to destruction of eggs by UV rays of sun light.

(e) Spreading of bark of sojina, *Moringa oleifera* Lamk. (Moringaceae) in crop field. Sojina has high medicinal property.

(f) Incorporation of wild fern, *Sphaerostiphnos unitus* in the field soil. Wild fern are pungent and highly pubescent which can cause severe irritation on the pests body.

(g) Extract of tobacco waste with 250 g of soap solution in 200 litres of water as spray.

(h) Against rice stem borer and leaf folder of rice, putting bamboo perches and/or branches for birds sitting in the nursery as well as in the main field. Birds act as predator. The sap of raw bamboo shoot contains hydrocyanic acid and posses antiseptic and larvicidal properties (CSIR, 1948).

- ***Rice hispa and insects that feed on epidermis and chlorophyll***

(a) Sweeping of thorny branch of ber in the rice field at tillering stage at an interval of 3–4 days (Dutta, 2004). So that, the pest gets injured and disturbed in movement and fly away from the field.

(b) Broadcasting goats' excreta at tillering stage. Due to disagreeable odour pest fly away from the field.

(c) Spraying of tobacco leaf at tillering stage. Tobacco solution is alkaline in nature which repels the pest.

(d) Dusting of dried and grinded Posotia, *Vitex negundu*, L. (Verbenaceae) leaves at tillering stage. Due to disagreeable odour of the dust pest fly away from the field.

(e) Pouring of kerosene oil directly on the standing water of field at tillering stage. Hispa fly away from the field due to intolerable odour of kerosene.

(f) Scattering of Kodesa, *Cleistanthus collinus* leaves and twigs in rice fields to control hispa and mealy bugs is practiced in Gurudaspur of Punjab (Sardana, 2001).

(g) Release of duck in hispa infested paddy field reduced the hispa infestation in Sibsagar district of Assam (Dutta, 2004).

- ***Rice case worm***

(a) Moving of previously soaked rope of kerosene oil and draining out of standing water at tillering stage. Cases of the worm are fallen down on water and drained out from the field.

(b) Throwing of branches of Germany ban, *Eupatorium odoratum*, Wild fern, *Polygonum hydropiper* on standing water at tillering stage ((Dutta, 2004). The unpleasant odour from these botanicals act as repellent to the pest.

(c) Distribution of chopped black colocasia, *Colocasia esculenta* on floating water surface. Poisonous condition created on floating surface may kill the case worm. CSIR (1950) reported that the acidity due to presence of needle shaped crystals or raphides of calcium oxalate in the tissue cause severe irritation on animal skin. Moreover, species like *C. virosa* and *C. gigantea* contain strong acids like hydrocyanic acid.

- ***Rice gundhi bug***

(a) Using light trap at night at milking and panicle initiation stage. It acts as attractant. Pest jump down on fire and gets killed.

(b) Hanging of dead frogs or crabs or rotten fish on erected bamboo stick randomly in different places of crop field at milking or panicle initiation stage. Rotten smell emitted attract the pest thereby crops are escaped from the infestation.

- ***Rice leaf folder***

(a) Spraying of boiled neem leaves and ground seed solution at tillering stage. Bitter taste of solution created on treated parts prevents the attack of pests. Odours of neem solution act as repellent which drives away the pest from the field.

(b) Apply solution of prepared from neem leaf paste in water (10 kg: 2 litres).

(c) To prevent leaf folder attack in paddy and ragi, sand is sprayed on the leaves that are wet with fog, so that sand sticks to the leaves. This sand prevents the larva from attacking the crop and feeding on the leaves. The sand also abbreviates the skin of the larvae and causes desiccation and death of the larvae.

- ***Hibernated insect pest:***Cutting border edges of crop field before transplanting. Hibernated pests are killed and thus reduces the pest population at later stage of the crop.
- ***Rice weevil and grain moth***

  (a) Mixing curry leaves, *Murraya koenigii* and neem leaves with grains during storage. The odour of leaves keeps away most of stored grain pest. Vaidyaratnam (1995) reported that curry leaves has high medicinal properties against several human diseases. Nimbidin is one of the compounds responsible for bitterness of neem and on hydrolysis it produces nimbinic acid (Deb, 1981). He also reported that the odour produced by burning of neem leaves is fatal to the insect.

  (b) Covering grain with a layer of 2 to 3 inch paddy husk. Pest gets difficulty in infesting the grains as they are unable to penetrate the thick layer of paddy husk.

  (c) Storing grains in bamboo made wall plastered with cow-dung and mud (1–2 inch thickness). The structure protects most of the stored grain pest to reach the grains from outside.

- ***Fruit and shoot borer of brinjal:*** Dusting ash on plants and around plants at active vegetative growing stage and flowering. As ash is alkaline in nature the pest can not stay around the plants and can not attack leaves and stems containing ash. Ash enhances resistance against the pests.
- ***Aphid, fruit and shoot borer of vegetables:*** Spraying of tobacco, *Nicotiana tabacum* leaves solution at flowering stage. Nicotin (an alkaloid) and alkalinity of tobacco leaf extract prevents the attack of pests.
- ***Red ant, Dorylus orientalis***

  (a) Hiding or keeping of banana (var. bhimkal, *Musa balbisinia* Colla. (Musaceae)) just beneath the soil surface of potato field at seedling stage. Banana attracts a large number of the red ants and farmers can easily destroy the same (Dutta and Chakravorty, 2009).

  (b) Application of salt solution at seedling stage of vegetables. Salty characteristics repel the ants from the treated areas.

(c) Incorporation of mustard oil cake in to the soil during the land preparation. Decomposition of oil cake releases butyric acid, formic acid, hydrogen sulphide which cause destruction of ants and many fungi, bacteria. Moreover it also increases the soil fertility.

- ***Soil-borne insects of gourds*:** Soil drenching with fish water (water after washing of fishes) at the base of plants. Potassium present in the fish washing confers resistance to the treated crops. Fish water may also attract predators and parasites.
- ***Fruit fly of vegetables***

  (a) Regular smoking by rice straw + chilli powder at fruiting stage. Smokes with unbearable odour prevent the attack of fruit fly.

  (b) Covering of fruit with rice straw at fruiting stage. Rice straw acts as barrier for penetration by fruit fly.
- ***Field cricket of vegetable:*** Smoke by burning dried chilli in front of the holes of cricket pests at seedling stage. Due to the effect of smoke cricket come out from the holes and farmers can easily kill the same.
- ***Aphid***: Ants nest or hives are kept near pest infested pockets of cauliflower and/ or cabbage growing field at seedling stage. Ants feed upon aphids and the aphid infestation is reduced.
- ***Pulse bruchid***: Smearing grains with edible oils preferably mustard oil before storing. Due to slipperiness of oil pest cannot attack the grains.
- ***Mealy bug of fruit tree:*** Banding of rope, made of straw and cow dung around the tree trunk at flowering stage. The rope disturbs the upward movement of the pest.
- ***Stem borer or trunk borer of citrus***: Sealing tightly the trunk hole made by the pest with bamboo plugs after clearing the holes or otherwise plastering the holes with mud at active vegetative stage. Pest can not come out through bore and ultimately die inside the plant.
- Use of neem leaves and kernels extract of custard apple, garlic and green chilli, salt lemon juice, castor cake, tobacco leaves powder for lepidopetran pest control in Punjab (Sardana, 2001).
- A solution prepared from neem leaf paste to control leaf folder in rice; extract of *Ipomoea* leaves in water (1:1) is used to control brown plant hopper in rice. This is being practiced in Gurdaspur of Punjab (Sardana, 2001).

- Flour and sugar are used in Rajasthan to attract ants and termite maunds. The ants act as biological control agents against termite.
- Farmers of Jammu use dried leaves of *neem* tree or its powder after wrapping in paper, placing them at upper, middle and bottom layers of storage structure and closing them air tightly keep grains safe in storage (Berjesh and Kumar, 2011).
- Urine of cow is sprinkled to minimize aphid infestation in vegetable crops.
- Spraying of cow dung solution: Mixed 200 g raw cow dung in 1 lit of water, kept overnight and apply at an interval of 7 days. It reduces attack of many pest and increases crop health also.
- Treating green gram seeds with 1% neem leaf powder to prevent bruchid (*Callosobruchus* spp.) infestation.
- Placing barriers (stick, leaves, tin etc.) around tree trunk to prevent rat gaining access to coconut trees.
- Erecting bird perches randomly in the field increases the chance of predation of larval parasite.
- Use of crushed bark and extract of *Careya arborea* Roxb. all over the paddy field as a prophylactic measures against pests and diseases.
- Use of *Vitex negundo* Linn. twigs in rice fields to protect the crop from diseases and insect pests.
- Drying of coarse millets and pulses is done on specially built muncha/ tray, a specialized structures providing sufficient ventilation for drying of seeds and produce is protected from livestock besides providing shades to farmers during sunny days.
- Smearing of seeds of wheat, pulses with castor oil for long time preservation.
- Leaves of neem and *Ipomoea fistula* Mart. ex Choisy are used for long term preservation of sorghum, rice and grams.
- Ground green neem leaves and wheat straw at 1:2 ratio are commonly used for storage purpose in Uttar Pradesh (Pal, 2001).
- Storing of millets, maize and pulses in elevated structures made of bamboo over the chullas (cooking furnace) (Plate 1 & 2).
- Pulses mixed with cow dung powder and kept in earthen pots increases the shelf life by preventing insect attack.

- Combination of cow dung, ash and neem leaves are used against storage pests of millets and pulses in Punjab and Himachal Pradesh (Sardana, 2001).
- Keeping the seeds of bottle gourd, bitter gourd, pumpkin etc. in cow dung cakes to provide protection from insect and fungal diseases.
- Storage of cabbage and cauliflower in tunnels by farmers of Himachal Pradesh (Sardana, 2001).
- ***Use of ash:*** Many of the vegetable growers of Jammu broadcast ash in vegetables fields to meet the deficiency of potash and enhance the availability of phosphate. Spreading of ash dust in vegetables like pumpkin, gourd, muskmelon and water melons is done to repel the insect pests like red pumpkin beetle (Berjesh and Kumar, 2011). Papaya mealy bug, *Paracoccus marginatus* infest papaya, okra, *Hibiscus*, tomato, brinjal, *Jatropha*, *Amaranthus*, marigold, *Selvia*, *Crysanthemum* etc. and can cause yield loss upto 40–60% in papaya crop and even cause complete death of the crop. Farmers of West Tripura district control the pest by thorough dusting of fresh ash during the morning hours (Gogoi *et al.*, 2010). It results in complete eradication of the pest from the infested area and ultimately regenerates the plant growth and initiate flowering in the infested crops. Ash is more particularly used to store pulses for long time by mixing the seeds with ash. Brinjal and tomato seeds, collected from the fruits, are dried on beds of ash and allowed for drying; the seeds will be rolled back along with the ash bed and will be stored until the next season. This will prevent the attack of pests on seeds. For protecting the vegetable seed nurseries from ants, finely ground ash powder is put as a band and it acts as a repellant for the ants, those coming to eat away the sown seeds. Ash is used in land preparation of Ragi crop. The ash incorporated in the soil will act as repellant and prevents the soil-borne pests those attacks the crop.
- In *Kati bihu*, a festival observed in Assam, India in the evening of *Shankranti* (last day of Ashina), earthen lamps are lit and placed in the crop field. The insects of field migrate towards light and are killed (Dutta *et al.*, 2009).

6.1. For Silk Worm in NE States

- A mixture of well ground rice hull and ash is applied at the base of fruit tree or food plant of muga and then earthing up is done. It controls red ants (*Dorylus orientalis:* Formicidae: Hymenoptera) (Dutta *et al.*, 2009).

- During rearing of Muga silk worm rearer take vegetarian food. This is because waste matter of non vegetarian food like fish, meat, bones if thrown around the field it becomes a breeding place for uzi fly and also become food for ants, frogs, snakes etc. Uzi is a major pest of muga rearing. Similarly ants, snakes etc. are major predators of muga, *Antherea assamensis* (the golden silk producer and pride of North East) especially at 5th instars (Dutta and Chakravorty, 2009).
- Out of the different nocturnal predators owl and bats are major ones. Muga rearers prefer bamboo torches for night watch and ward duty for collection of matured worms. These torches act as a light trap for pest and predators of muga. It also acts as attractant. Pests fall into fire and gets killed.
- Kola kachu, *Colocasia esculenta* or taro or black colocasia are pinned on the tree trunk during muga rearing. This is done because of their believe that some people by their supernatural power can cause mortality to the worms by casting an evil eye to them which may eventually lead to complete crop failure. *Colocasia* plants show acridity due to presence of needle shaped crystals or raphides of calcium oxalate in the tissue and thus can cause severe irritation to the pest and some species *viz.*, *C. gigantean* contain strong acids like hydrocyanic acid which prevent the infection by fungus and bacteria (Dutta *et al.*, 2009).
- Controlled burning of undergrowth in the host plantations prior to rearing is a practice followed by the farmers of North East India, which helps in removal of pests and predators, disease causing microflora etc. (Unni *et al.*, 2001).
- Unni *et al.* (2001) reported that traditional silk worm rearer of North East India disinfect the silk worm rearing equipment by continuous hanging of the same above the kitchen fire. This practice is effective as continuous exposure in high temperature helps in desiccating the pebrine spore and other disease causing microbes from the equipment.

6.2. ITKs for Disease Management

- Against bacterial blight and blast of rice, spraying of cow dung solution (Mixed 200 g raw cow dung in 1 lit of water kept overnight and applied at tillering stage of the crop at an interval of 7 days) is effective (Anonymous, 2009–10).
- ***Reason:*** Cow dung contains bacteriophages which destroy the bacterial causal organisms. Further, due to pungent odour of cow dung cattle do not graze the treated crop.

- Application of solution of neem leaves and ground seeds + soap and/ or (detergent powder) + raw turmeric at tillering and panicle initiation stage control the fungal and bacterial disease of rice.
- ***Reason*:** Prepared solution contain some anti-microbial substances like azadirachtin, nimbidin etc. which destroy the fungal and bacterial flora present on the treated surface.
- Brown spot of rice can be controlled by spraying of ash at panicle initiation stage.
- ***Reason*:** Ash contains hydrocarbon and/or sulfur which may prevent the spreading of infection. Moreover, ash is alkaline in nature so it reduces the fungal infection.
- Late blight of potato can be managed to some extent by dusting of ash at seedling and at active vegetative stage.
- ***Reason*:** As mentioned.
- Blight and fruit rot of tomato can be reduced by spreading of rice straw under tomato plants especially during fruiting stage.
- ***Reason*:** Due to straw the branches of plant and fruit do not touch soil and fruit rot and blight diseases are prevented.
- Against blight diseases of tomato, dusting of ash on the plants at active vegetative growing stage is effective.
- ***Reason*:** Alkalinity of ash reduces the disease incidence.
- Bunchy top of chilies can be minimized by dusting ash, spray sour butter milk, spray cow/goat urine.
- ***Reason*:** Ash enhances host resistance against the organism.
- Damping off of vegetables, chilies can be managed by applying fresh cow dung near the collar region.
- ***Reason*:** Beneficial microorganisms present in the cow dung may destroy the fungal organism causing damping off diseases.
- Against collar rot, root rot of vegetables application of castor cake, neem cake, karanj cake is effective.
- ***Reason*:** Castor cake, neem cake, karanj cake etc. act as organic amendment which on decomposition alter the C: N ratio and make the environment favorable for beneficial microorganisms but unfavorable for pathogenic organisms.

- Against powdery mildew and viral diseases of vegetables, apply milk solution: 1 litre milk in 9 litre of water.

  Against rust of wheat, seed soaking in milk is done before sowing.

- Brown spot of rice can be managed by application of solution of 2 kg fresh leaves of papaya in 3–4 liters of water and keep overnight. After filtration dilute it with 50–60 liters of water and 250 ml soap solution.
- ***Reason***: Papain present in the solution may help in development of resistance against the disease.
- Against root knot disease of vegetables, tube rose, growing of 2–3 rows marigold after every row of vegetables or tube rose.
- ***Reason***: Marigold acts as trap crop for the nematode and thus divert the nematode from the vegetable and tube rose.
- Small seeds of vegetable are treated with sindur/vermillion @1–2 g/kg of seed. Treated seeds are then sown in the field in row. This is specially practiced in Sonatola village of West Tripura for small seeds like TPS (True Potato Seed), chilli, tomato, cabbage, cauliflower etc.
- ***Reason***: Colour of most of crop seeds is same with soil particles. This technique helps in detection of seeds after sowing. Germination percent increased with very less pre and post emergence damping off as compared to the common farmers' practices. It also prevents the attack of ants. The technique is eco-friendly and less cost involved as compared to chemical treatment.
- Storing grains in kitchen is a common practice for grain storage followed in the hill districts of North East India specially, Meghalaya, Manipur for post harvest protection be it for their own consumption and maintain the seed for the next season sowing (Sinha, 2010).
- ***Reason***: Smoke of *Chullah* (cooking place) protects the grains from the attack of insects, fungus and also increases the shelf life.
- Seeds of sorghum are treated with cow dung, urine for 2–3 min and then sun drying before sowing protects the crop from gain smut.
- Seed treatment with cow dung paste is done to control seed-borne diseases.
- Soaking the seed in milk prior to sowing avoid rust diseases. Seed soaking in milk followed by smearing with *ghee* and rolling them in cow dung for many times protect the cereal from many diseases.

- Raising vegetable seedlings such as tomato, chilli etc. at 3–4 ft height in artificial tray made of half split bamboo is practiced by many farmers of Manipur. This is mainly practiced to control damping off of seedlings. Less space is required and easy to manage the soil borne disease. It helps in production of diseased free seedlings (Gogoi *et al.*, 2010).
- Domestic storage technique of ripen tomato at harvesting stage—The technique involves air hanging of stalked tomatoes on ceiling by tying with thread. Ropes with 1–2 cm diameter are tied on two sides of bamboo pool. On the rope, tomatoes are tied with stalks is hanged keeping at least 6 ft. height from the ground level and 2 ft. below the roof with free air circulation. To protect direct light from the roof they kept ceiling of thick cotton cloth at a distance of 1.5 ft. from the roof. This increases the shelf life of tomato upto 3–4 months.

CONCLUSIONS

The immediate need of the present time is to identify and find scientific rationale of these indigenous technical knowledge (ITKs) and integrate them into recommended farming practices for wider adoption. In doing so, such technological options developed by using diversified sources with active participation of farmers will have technological options which have originated from farmers' own knowledge, and therefore will be readily accepted; technological options thus developed will enable farmers to choose according to their conditions and decision making system.

Many ITKs on *in situ* plant protection are not adopted everywhere throughout India because of constraints in adoption and unawareness of the effectiveness of such practices. The present documentation process has definite bearing on the future course of action in framing new projects.

This chapter may be beneficial to the following future activities:

- Similar exercise can be undertaken to document the ITKs from all the agro-ecological regions of the country.
- The potential ITKs may be tested for their suitability and adoption in other agro-ecological regions as a dissemination strategy.
- The documented ITKs may be published/translated in all regional languages for the benefit of the farming communities.
- Validation of the ITKs is a logical step to qualify and quantify the effectiveness of these practices. Suitable modifications of the traditional practices through on-farm research would help in developing appropriate and acceptable technologies for different local environments.

Raising seedling in bamboo tray

Hanging of maize over Chullah

Application of ash in tree base

Application of ash on leaves

Use of split bamboo and rope to protect from the birds attack

Plate 1: Indigenous practices followed by farmers

Hanging tools over Chullah

Seed treatment with vermillion

Bamboo perches in tomato field

Bamboo perches in chilli field

Pinning colocasia in muga food plant

Tomato hanging for better shell life

Plate 2: Indigenous practices followed by farmers

REFERENCES

Anonymous. (2009–10). Annual report of divyodaya krishi vigyan kendra–west tripura (2009–10). Submitted zonal project directorate, Zone–III, ICAR, Barapani.

Berjesh, A. and Kumar, A. (2011). Indigenous technical knowledge (ITK) for farmers. *Agrobios Newsletter,* **IX(11):** 57–58.

Colfer, C.J.P. (1990). Rice production and subtleties of cultural change: An example from borneo. In proceedings of International workshop on sustainability through farmers involvement in technology generation and diffusion. Feb. 9, 1990. *Indian Society of Agronomy*. New Delhi.

CSIR. (1948). The wealth of India. *Raw Materials*. Vol. I. Council of scientific and industrial research. New Delhi, India, p. 254.

CSIR. (1950). The wealth of India. *Raw Materials*. Vol. II. Council of scientific and industrial research. New Delhi, India, p. 427.

Dean, D.L. (1999). Indigenous knowledge. Technology blending and gender implications. gender technology and development, New Delhi. *Safe publications*. **3(2)**.

Deb, D.B. (1981). The flora of tripura, Vol. I (Vegetation and Ophioglossaceae: Staphyleaceae). Today and tomorrow printers and publishers, New Delhi, India, p. 509.

Dutta, Pranab and Chakravorty, R. (2009). Commonly used traditional practices in muga silk culture in North East India. *Insect Environment,* **15(1):** 4–7.

Dutta, Pranab, Das, K., Das, R., Mech, D. and Chakravorty, R. (2009). Pest and disease management in muga culture: Traditional practices. *Indian Silk,* (January): 16–18.

Dutta, Pranab (2004). Shyshya rakhyarthe rashanik paddhatir bikalpa upai, *In*: *Assamia Khabar,* 16th March , Tuesday, Section: Kheti Pathar, p. 6.

Gogoi, A.K., Singha, A.K., Sanjeev, M.V. and Dutta, P. (2010). Technological nad methodological innovation by the framers of North East India, (*Eds*. Gogoi, Singha, Sanjeev. and Dutta.), publ by ZPD, Zone–III, ICAR, Barapani (in press).

ICAR. (2004). Training project report on methodology of collection, Documentation and Validation of ITKs held during July 17–Aug. 06, 2004 at CRIDA, Hyderabad, India.

Mane, P.M. and Sutaria, J.N. (1993). Study and documentation of indigenous knowledge/ traditional agricultural practices of tribal farmers. Paper presented in national seminar on indigenous technology for sustainable agriculture. 23–25 March, 1993. IARI, New Delhi, India.

Pal, M.S. (2001). Indigenous technologies for safe storage. *Intensive agriculture* (March–April), pp. 17–19.

Pal, P.K. (1998). Indegenous knowledge base of agro-technology—A study in a lateritic area of west bengal. Ph. D. Thesis. Visva–Bharati, Sriniketan, West Bengal, India.

Pal, P.K and Das Gupta, D. (1997). Participatory rural appraisal in identifying intellects of local people. *In*: Das Gupta, D. and Mukhopadhayay, S.D. (*Eds*). Social Science on the Farm Front. Naya Prakash.

Parasar, B. (1994). Indigenous agricultural knowledge of tribals in eastern ghat highland zone of Orissa. Ph.D. Thesis. BHU, Varanasi, U.P., India.

Pushpangadan, P., Rajasekharan, S. and George, V. (2002). Indigenous knowledge and benefit sharing—A TBGRI experiment In IK strategies for Kerala. NSE Publication, Thiruvananthapuram.

Quiroz, C. (1993). Methodology for the study of farmers agricultural local indigenous knowledge system in rural development programme. Inservice training programme for research/extension practitioner—an experience from Venezuela. Indigenous knowledge and development monitor. Vol. 1, No. 2.

Rajshekaran, B. (1993). A framework for incorporating indigenous knowledge system into agricultural research, extension and NGOs for sustainable agricultural development. Studies on technologies and social change, No. 21. Technologies and social change programme, IOWA, State University, Ames, IOWA, 5001, USA.

Sardana, V. (2001). Indigenous technical knowledge of framers since ancient time. *Indian Farming*, **29:** 13–14.

Sinha, B. (2010). An appraisal of the traditional post harvest management methods in the North East India plains uplands. *Indian Journal of Traditional Practices,* **9(3):** 536–43.

Unni, B.G., Goswami, M., Kakoti, Y., Bhattacharyee, M., Wan, S.B., Rajkohoa, G., Das, S., Devi, B.R. and Chutia, A.D. (2001). Indigenous knowledge of silk worm cultivation and its utilization in North East India, *Indian Journal of Traditional Practices*, **8(1):** 70–74.

Vaidyaratnam, P.O.S. (1995). Indian medicinal plants, Vol. 4, variers arya vaidya sala. Orient Longman Limited, Madras, India, p. 444.

Wang, G. (1988). Indigenous communication systems in research and development. *Journal of Extension Systems*, **4:** 75–86.

World Bank (1998). Indigenous knowledge for development. A framework for action [Online] http://www.worldbank.org/afr/ik/ikrept.pdf.

14

Epidemiology and Management of Phytoplasma Associated Plant Diseases

M. Azadvar[1], V.K. Baranwal[2] and M. Najafinia[1]

ABSTRACT

Phytoplasmas, formerly termed mycoplasma–like organisms (MLOs), are non-culturable, minute cell-wall-less prokaryotes of the class Mollicutes a group of microorganisms phylogenetically related to low G+C, Gram-positive bacteria. They are known to be pathogenic to more than a thousand plant species. Phytoplasmas are phloem–limited, transmitted by specific phloem–feeding insects, sensitive to tetracycline but resistant to penicillin. Phytoplasma associated diseases are spread worldwide, and in several cases are associated with severe epidemic of very often quarantine importance and can cause devastating losses in crops and natural ecosystems worldwide. Control of phytoplasma diseases involves using clean propagative materials, scouting and removal of infected or symptomatic plants, as well as control of the vector with chemical, physical or biological methods. The methods can be used independently or in conjunction with other methods, based on the disease. Many factors need to be considered to arrive at the appropriate management tactics which we will discus in this chapter.

***Key words*:** Control, Epidemiology, Management, Phytoplasma, Vector.

1. INTRODUCTION

Several plant diseases, believed to be caused by viruses, are associated with phloem restricted mycoplasmas which was first identified in 1967 (Doi *et al.*, 1967). Since then, several hundreds of plant syndromes, which some of them associated with important or quarantine subjected plant diseases, have been reported to be associated with the so-called

1 Department of Plant Protection Research, Agricultural Research Center of Jiroft and Kahnouj, Jiroft, Iran.
2 Plant Virology Unit, Division of Plant Pathology, Indian Agricultural Research Institute, New Delhi, India.
Corresponding author: E-mail: m_azadehvar@yahoo.com

"mycoplasma–like organisms", MLOs (Bertaccini, 2007). Ribosomal DNA sequencing provided evidence that these cell-wall-less prokaryotes constitute a large monophyletic group within the class *Mollicutes*. The Phytoplasma Working Team of the International Research Project for Comparative Mycoplasmology (IRPCM) adopted the trivial name 'phytoplasma' to identify the prokaryotes belonging to this group. The "*Candidatus* Phytoplasma" genus has been proposed and adopted in order to start formal classification of these prokaryotes (IRPCM, 2004).

Phytoplasmas are small sized (±500 nm in diameter) single-celled polymorphic mollicutes and are strictly host-dependent and could survive and multiply only in hypotonic habitats such as plant phloem or insect haemolymph (Bertaccini, 2007; Hogenhout and Seruga Music, 2010; Marcone, 2010). They are known to be pathogenic to more than a thousand plant species (Marcone *et al.*, 1999; Lee *et al.*, 2000; Seemüller *et al.*, 2002; Zhao *et al.*, 2010). Phytoplasmas continuously cycle between plants and insects and, in nature, require both organisms for survival and dispersal. This requires adaptation to a broad range of environments, including the phloem of their plant hosts and the gut lumen, haemolymph, saliva and endocellular niches in various organs of their insect hosts (Hogenhout and Seruga Music, 2010). Up to date satisfaction of Koch postulates has not been achieved, but indirect proof, such as phytoplasma and symptoms eliminating after tetracycline treatments, confirmed that they are associated with many plant diseases worldwide (Musetti, 2010; Seemüller and Harries, 2010).

The first report of association of phytoplasma with plant diseases in India was on sandal spike and brinjal little leaf, being worked out at Indian Agricultural Research Institute at 1969 (Verma *et al.*, 1969). Several plant diseases of phytoplasma etiology have been reported from India which earlier reported to be caused by viruses (Raychaudhuri, 1993). Some of them are of minor economic importance while a few cause serious economic losses. Among the diseases induced by phytoplasmas in India, phytoplasmas in association with almost twenty plant diseases have been characterized and group identified using molecular work in last few years (Azadvar, 2010).

2. SYMPTOMS OF DISEASES CAUSED BY PHYTOPLASMAS

In infected plants, phytoplasmas systemically colonize sieve cells of phloem tissue and typically induce disease symptoms involving impaired amino acid and carbohydrate translocation, disrupted hormonal balance, and rapid senescence. Symptoms of diseased plants may vary with the phytoplasma, host plant, stage of the disease, age of the plant at the time of infection and environmental conditions (McCoy, 1979; McCoy *et al.*, 1989; Lee *et al.*, 2000; Seemüller *et al.*, 2002; Seruga Music *et al.*, 2008; Bertaccini, 2007; Hogenhout *et al.*, 2008). Phytoplasma infected plants exhibit a wide range of specific and non-specific symptoms (Fig. 1) suggesting profound

Fig. 1: *(Contd...)*

Fig. 1: (*Contd...*)

Fig. 1: Characteristic symptoms of phytoplasma diseases on different host plants. (a) little leaf and witches broom of lime, (b) and (c) viresence, phyllody, proliferation, inhibition of flowering and abnormalities on toria, (d) little leaf and stunting on brinjal, (e) abnormal elongation of internodes and phyllody on toria, (f) big bud and flower sterility on tomato and (g) bunchy and non-apical dominance growth on cucumber

disturbances in the normal balance of plant metabolism, including yellowing, chlorosis, or bronzing of foliage, stunting (reduction of internodes and leaf size), virescence (the development of green flowers and the loss of normal pigments), phyllody (the development of floral parts into leafy structures), proliferation of secondary auxiliary buds often resulting in a witche's–broom effect, abnormal elongation of internodes leading to slender shoots, proliferation of secondary roots, big bud, enlarged stipules, inhibition of flowering, flower proliferation and other flower abnormalities, abnormal fruits and seeds, off-season growth and brown discoloration of phloem tissue (Bertaccini, 2007; Hogenhout *et al.*, 2008; Hodgetts *et al.*, 2008; Hogenhout and Music, 2010; Marcone, 2010). The same phytoplasma strain may induce different symptoms in different hosts, and different phytoplasma strains may share a common vector(s) or cause diseases exhibiting similar symptoms. The exact mechanisms involved in symptom development or the genes that control these events are still poorly understood (Marcone, 2010).

3. TRANSMISSION METHODS OF PHYTOPLASMAS

Transmission of phytoplasmas from plant to plant in nature as well as laboratory occurs by the phloem feeding of inoculative vector insects, propagation of infected plant material, graft inoculation and by plant parasitic dodder. In nature, phytoplasmas are mainly spread by phloem–feeding insects of the order Hemiptera, especially phloem feeding leafhoppers (*Cicadellidae*) belonging to the families *Cicadellidae, Cixidae, Psyllidae, Delphacidae,* and *Derbidae* and psyllids (family *Psyllidae*) in a persistent, propagative manner (Marzachì *et al.*, 2004; Weintraub and Beanland, 2006). In their insect vectors, phytoplasmas traverse the intestinal wall, circulate in haemolymph, and multiply in tissues including salivary glands, where phytoplasma cells are incorporated into saliva injected into plants during inoculation (Weintraub and Beanland, 2006). Some factors influence transmission, among which are life stage, gender, presence of associated symbionts, flight behaviour, weed control measures, temperature, phytoplasma strain, source and recipient plant species. Insect vectors can acquire more than one phytoplasma species/strain, either by feeding on multiple–infected source plants or by feeding sequentially on different plants infected by different phytoplasmas. Once a vector becomes infectious, infectivity is retained for life (Firrao *et al.*, 2007). The geographical distribution and impact of phytoplasma diseases depends on the host range of the phytoplasma as well as the feeding behaviour of the insect vector (Bertaccini, 2007; Hogenhout *et al.*, 2008; Seruga Music *et al.*, 2008; Foissac and Wilson, 2010). Insect species that transmit the phytoplasma may or may not be present at the time of symptom appearance.

Transmission of phytoplasmas through propagation material can occur and lead to their long-distance dispersal and introduction into regions where they have not previously been found (Constable, 2010). Recent reports on the detection of phytoplasma in both the seed and seedling progeny of alfalfa (Khan *et al.*, 2002), coconut lethal yellowing (Nipah *et al.*, 2007), canola (Olivier *et al.*, 2006) and toria (*Brassica rapa* subsp. *dicotoma* (Roxb.)) plants (Azadvar, 2010) indicate that seed transmission in certain plant-host phytoplasma pathosystems is possible. Phytoplasmas can be transmitted experimentally by grafting infected plant material and by plant parasitic dodder (*Cuscuta campestris*) plants onto healthy plants.

4. RECENTLY DEVELOPED METHODS IN PHYTOPLASMA DIAGNOSIS

Phytoplasmas have been difficult to detect due to their low concentration especially in woody hosts and their erratic distribution in the sieve tubes of the infected plants (Firrao *et al.*, 2007). Although phytoplasma identification was relied for more than 20 years on microscopic observations (DAPI staining), in last decades electron microscopy, serological, and DNA-based techniques have been developed and have made phytoplasma detection more rapid and practical for diagnostic labs. These techniques provide the ability to distinguish among different phytoplasmas and permit specific identification (Lee *et al.*, 2000; Firrao *et al.*, 2007; Seruga–Music *et al.*, 2008).

In order to achieve a general and reliable system of phytoplasma detection and identification, molecular tools such as PCR/RFLP (polymerase chain reaction/restriction fragment length polymorphism) and nested–PCR mostly on the highly conserved (16SrDNA) ribosomal phytoplasma region were developed and applied. Numerous universal and/or group specific PCR primer combinations have been devised to amplify the 16S rRNA gene of phytoplasmas (Bertaccini, 2007; Zhao *et al.*, 2010). However, diagnostics based on these primers can be problematic, with occasional false positives through amplification of other bacteria that might be present in samples (Lee *et al.*, 2010). PCR RFLP is time consuming, requires post amplification steps and increases the risk of carry–over contamination in routine analysis. To overcome these problems, several approaches have been developed, amongst which universal and group-specific real-time PCR protocols have been proposed since 2004 (Monis and Giglio, 2006; Galetto and Marzachi, 2010). This method has many advantages over the conventional PCR in terms of accuracy, dynamic range, high-throughput capacity, and absence of post–PCR manipulations that prevents carryover contamination (Torres *et al.*, 2005). This method has been developed mainly based on the 16S rRNA gene, but primers based on other genes are also being used (Wei *et al.*, 2004; Hren *et al.*, 2007; Hodgetts *et al.*, 2009; Hodgetts and Dickinson,

2010). TaqMan and SYBR Green chemistries have been up to now applied for detection and for quantification of some groups of phytoplasmas (Torres *et al.*, 2005; Hodgetts *et al.*, 2009; Aldaghi *et al.*, 2009; Azadvar, 2010).

5. EPIDEMIOLOGY OF PHYTOPLASMA DISEASES

Phytoplasmas continuously cycle between plants and insects and, in nature, require both organisms for survival and dispersal. This requires adaptation to a broad range of environments, including the phloem of their plant hosts and the gut lumen, haemolymph, saliva and endocellular niches in various organs of their insect hosts (Hogenhout and Seruga Music, 2010). Phytoplasma ecology is complex and affected by the host range and geographic distribution of both the phytoplasma and the insects that transmit them, and is strongly affected by weather conditions. Epidemiology of phytoplasma associated diseases is closely tied to the insect vector of the phytoplasma. In diseased plants, phytoplasmas spread systemically throughout the plant by passing through phloem sieve plate pores (Marcone, 2010). The host range of both phytoplasmas and insects greatly influences the chances that a phytoplasma and a potential vector will come into contact. For example, *Cacopsylla* spp., which is monophagous on pome or stone fruits, are only vectors of '*Ca.* Phytoplasma mali', '*Ca.* Phytoplasma pyri' and '*Ca.* Phytoplasma prunorum', which infect pome or stone fruits. *Orosius argentatus* Evans, a very polyphagous leafhopper species on herbaceous hosts, transmits different phytoplasma species/strains to several plant species (Weintraub and Beanland, 2006).

6. PHYTOPLASMA DISEASE MANAGEMENT STRATEGIES

The life cycle of phytoplasma accounts for the fact that the diseases they cause are difficult to control. There is no authorized curative treatment for phytoplasma diseases. Management of phytoplasma–associated diseases relies on prevention rather than cure (Jarausch *et al.*, 1999; Constable, 2010). Plants are not really cured and symptoms reappear when treatments cease. The most promising and standard recommendations for preventing the outbreaks and management of phytoplasma diseases include using clean planting material, phytosanitary measures, eliminating the pathogen from the infected plants by antibiotics or by other chemicals, identification or development of resistant/tolerant plant varieties and controlling the vector(s). However, none of these approaches is fully satisfactory, some of them can be employed only under restricted and defined environmental conditions and it may prove quite ineffective under field conditions (Bertaccini, 2007). Management and control center on eliminating sources of the phytoplasma and controlling leafhopper vectors. There are some methods that can be used independently or in conjunction with other methods. Many factors need to be considered to arrive at the appropriate management tactics. Some methods are more effective for certain kinds of crops; for example, roguing or destruction of symptomatic plants is more effective in orchards or perennials than in row crops (Weintraub and Wilson, 2010). We will review the effective practices for managing phytoplasma disease.

6.1. Quarantine and Inspection Measures

Quarantine regulations govern the interstate, and even intrastate, sale of nursery stock, tubers, bulbs, seeds, and other propagative organs, especially of certain crops such as potatoes and fruit trees (Agrios, 2005). Import regulations help to protect plants from the introduction and spread of phytoplasma diseases. A simple method for elimination of phytoplasma and insect eggs from trees is by treatment with heat. Today vines are being treated by dipping in hot water (Mannini, 2007). It is important not to transport phytoplasma–infected plants to new areas where potential vectors may be present. Weintraub *et al.* (2004) reported that phytoplasma infected *Limonium* (Plumbaginaceae) plants were brought into an area where potential vectors were present. Shortly after the seedlings were planted, phytoplasma symptoms were observed in other *Limonium* plantations throughout the area.

6.2. Phytoplasma Free Plant Material

Using planting material from reliable approved sources and produced in areas where phytoplasma diseases are not present is very important for prevention of phytoplasma disease. Propagating the plants from seed or from phytoplasma–free plants is recommended.

6.3. Resistant or Tolerant Plant Varieties

A more promising approach to controlling phytoplasmas is the use of resistant plants. Intra- and inter-specific variation in susceptibility to these diseases has been reported for several decades. Developing plants that either are resistant to the phytoplasma or deter vector feeding would be a first line of defense. Work on resistance in trees to phytoplasmas and/or vectors has been more intensive and successful (Seemüller and Harries, 2010). For the oilseed Brassica lines that were tested, the incidence of phytoplasma disease ranged from 4.1 to 11%, indicating that some varieties showed significant resistance (Azadvar, 2010). It was found that anthracnose–resistant cultivars of *Stylosanthes* are resistant to phytoplasmas (De La Rue *et al.*, 2003) and are recommended for use for commercial seed production and cattle pasture land.

6.4. Removal of Diseased Plants

When a pathogen has been introduced into a new area despite a quarantine, a plant disease epidemic frequently follows. To prevent such an epidemic, all the host plants infected by or suspected of harboring the pathogen may have to be removed and burned. This eliminates the pathogen and prevents greater losses from the spread of the pathogen to additional plants (Agrios, 2005). The removal of diseased plants is most effective in orchards where the infected plants are less, insect vector is almost controlled and trees are non-contiguous. This activity should be done in all the near area. Infected plants can be rogued–removed entirely–or ratooned–only symptomatic shoots removed. A classic example of roguing to successfully control X-disease is the removal of the phytoplasma reservoir host, chokecherry (*Prunus virginiana* L.), up to 150 m from peach (*Prunus persica* (L.) (Parker *et al.*, 1963). Ratooning is effective in reducing the phytoplasma–bearing material, but it might not be an effective management practice (Esker *et al.*, 2006). Production of phytoplasma–free strawberries is maintained through inspection and roguing of plants with lethal yellows symptoms. Since mother plants produce runners, allowing for movement of phytoplasmas to offspring, this roguing limits the number of diseased plants (Greber and Gowanlock, 1979). Uyemoto *et al.* (1998) developed a new technique of roguing combined with insecticide spraying. To prevent the infected leafhoppers leaving a symptomatic tree as it is being cut down, they sprayed sweet cherry (*Prunus avium*) with Diazinon first. They found a significant reduction in disease spread around trees thus removed (Weintraub and Wilson, 2010).

6.5. Weed Control

Weeds host both vector development and phytoplasma and manipulating habitat to control vectors has elucidated the importance of weeds in the ecology/epidemiology of phytoplasmas. Control of weed plants around cucumber greenhouses at the sowing time could decrease the incidence of cucumber phyllody disease significantly (Azadvar, unpublished data). It is showed that use of the systemic glyphosate in the winter on weeds reduced vector populations. Vegetation composition surrounding a field/orchard/ vineyard has a profound effect on the presence and dispersal of phytoplasma vectors (Weintraub and Beanland, 2006). Vector capture in bidirectional Malaise traps placed in ecotonal regions between vineyards and forests showed that the primary direction of movement was from the wild to the cultivated vegetation. Weedy species can greatly influence vector infection levels and distribution, but plant density also affects vector distribution.

6.6. Vector Control

Identification of a vector is crucial in determining management strategies of phytoplasma diseases. The first and foremost method of managing vectors is to ensure that they are not transported to new areas. Management of vectors in economically important plants differs for annual versus perennial crops, as losses in annual crops are typically seasonal, while losses in perennial plant systems accumulate over many years (Weintraub and Wilson, 2010). The following methods are useful in vector control.

6.6.1. *Chemical Control*

Until recently, management of phytoplasma–caused plant diseases focused on spraying insecticides, with or without regard to specifically managing the vector(s). The use of insecticides might help control vector populations and thus reduce intra-crop transmission. Conventional insecticides, even when frequently used will not control the appearance of disease because pathogen transmission occurs faster than insecticides can act and there is often a constant influx of new vectors from surrounding habitats (Wally *et al.*, 2004). Saracco *et al.* (2008) have developed a two-pronged strategy vis-à-vis the use of insecticides: to protect plants from infectious migratory vectors use systemics such as neonicotinoids, to suppress vectors use organophosphate insecticides. Pilkington *et al.* (2004) have demonstrated that total vector population control can be achieved and disease incidence reduced by just treating crop borders with systemic insecticides. This tactic targets only the vector population, without extensive insecticide application. It is impossible to eliminate all vectors from environments.

6.6.2. *Barrier sprays*

Insecticides kill by a variety of means, both physiologically and physically. Kaolin is a non-abrasive, fine-grained aluminosilicate mineral applied as a particle film. Kaolin can act in a couple of ways: killing an insect by suffocating it or coating the plant and obstructing feeding and oviposition sites. Initial work by Puterka *et al.* (2003) demonstrated that kaolin protected grape plants from feeding and oviposition by leafhoppers, by physically coating the plant with a mineral film. Tubajika *et al.* (2007) showed that grapevines treated with kaolin were less likely to become infected with phytoplasma, and fewer leafhoppers were found in treated fields. The efficacy of kaolin is greatly hindered by water; in dry areas it may be very effective in controlling pests, but in areas with overhead irrigation or heavy rain it is washed away (Weintraub and Wilson, 2010).

6.6.3. *Biological control (Parasitoids/predators)*

A range of predators such as spiders attack both adults and nymphs of leafhoppers and planthoppers, especially in grassland ecosystems, while Miridae (Hemiptera) may be significant egg predators. There are also specialist parasitoids associated with Auchenorrhyncha (Waloff and Jervis,

1987). Pipunculidae (Diptera) are almost exclusively parasitoids of Auchenorrhyncha. They lay eggs into the adult and nymphal stages and the larvae develop internally, swelling the abdomen in the later stages of growth. Dryinidae (Hymenoptera) deposit eggs into adult and nymphal stages, while Mymaridae (Hymenoptera) attack the egg stage of leafhoppers and planthoppers. The various species of parasitoids associated with a particular species of Auchenorrhycha may reduce the natural population below an economically important threshold and thus they are not recorded as pests. Some introductions of parasitoids have been made to attempt to control introduced species. However, parasitoids are all susceptible to reductions in numbers due to pesticide usage, and increases in populations of leafhoppers and planthoppers may result (Heinrichs, 1994; Weintraub and Wilson, 2010).

6.6.4. *Mulching*

One means of manipulating the habitat is the application of various organic and synthetic mulches. Synthetic mulches, such as plastic sheeting, can physically prevent the movement of vectors into the soil and lay eggs at or just below soil surface. The use of reflective mulches works in another manner, rather than physical control, the vector is repelled from the plant. Summers and Stapleton (2002) achieved better control of the maize leafhopper vector, *Dalbulus maidis* (DeLong and Wolcott), and higher maize yield with plastic reflective mulch than with insecticide treatments. Similar results were achieved with control of the aster yellows phytoplasma vector *Macrosteles quadrilineatus* (Forbes) on carrots using aluminium foil mulch. However the efficacy of the mulch decreases as plant canopy increases (Setiawan and Ragsdale, 1987; Weintraub and Wilson, 2010).

6.6.5. *Physical control covering (barriers and net houses)*

The most reliable means of controlling phytoplasma vectors is by covering the crop with insect-exclusion screening (IES); however, its applicability is so severely limited due to the logistics of large-scale agriculture in major crops that its use cannot even be contemplated. Elder *et al.* (2002) and Walsh *et al.* (2006) demonstrated that vectors of papaya phytoplasma diseases in Australia could be 100% controlled by covering the trees with IES. When the effectiveness of IES was compared with systemic insecticide (Imidacloprid) treatments and a non-treated control, covered trees had significantly fewer phytoplasma symptoms. Insect-exclusion screening is the only method to maintain phytoplasma free vineyards (Mannini, 2007). Production of clonal or mother plants must be covered continuously with IES to maintain the plant free from leafhopper vectors (Weintraub and Wilson, 2010).

7. ANTIBIOTIC THERAPY

Tetracycline antibiotics, which inhibit protein synthesis by binding to the 30S ribosomal subunit, are effective against phytoplasmas. Cherry and peach trees infected by X-disease phytoplasma were treated with tetracycline antibiotics by high-pressure injection or by gravity infusion, and disease severity was significantly reduced, providing economic control (Lee *et al.*, 1987). Symptom remission is usually temporary as tetracycline only persists in plants for 1–4 months (McCoy, 1982; Kaminska and Silwa, 2003). Although phytoplasma disease symptoms may be delayed or alleviated by treatment with certain classes of antibiotics, this approach is not usually practical because use of antibiotics is very expensive, not allowed in several countries, and not always effective for long-time (Weintraub and Wilson, 2010).

8. INDUCED RESISTANCE

Research has focused recently on means of inducing recovery in plants. Stressing plants by severe pruning, pollarding, uprooting or partial uprooting can promote subsequent years of symptom-free plants (Romanazzi and Murolo, 2008). In addition to agrotechnical methods, applications of various chemicals known to induce recovery from various plant pathogens are being tested. Curkovic Perica (2008) has shown that the application of the auxins, indole-3-acetic acid (IAA) and indole-3-butyric acid (IBA) to phytoplasma–infected periwinkle shoots induced recovery. Although there were no symptoms, phytoplasma DNA could still be detected in shoots infected with '*Candidatus* Phytoplasma solani'. Field trials in grapevines are currently under way with a number of commercial products based on chitosan, Fosetyl-Al, glutathione and oligosaccharides (Weintraub and Wilson, 2010).

CONCLUSIONS

Epidemiology of phytoplasma associated diseases is complex and closely tied to the insect vector of the phytoplasma, host range and geographic distribution of both the phytoplasma and the insect(s) that transmit it, and is strongly affected by weather conditions. There is no authorized curative treatment for phytoplasma diseases and thus control of phytoplasma-associated diseases relies on prevention rather than cure. Identification of a vector is crucial in determining management strategies, as their control can prevent spread of disease. The most promising strategies for management of phytoplasma diseases are: eliminating sources of the phytoplasma, using resistant plant varieties, propagate from phytoplasma–free plant materials, control the insect vector in the crop and nearby weeds and eliminate the phytoplasma-weed hosts.

FUTURE ASPECTS

The current promised approaches for controlling the phytoplasma associated diseases are not usually practical. Using genetic/molecular manipulation techniques to produce the truly resistant plants or plants that express some chemical which allows them to be tolerant of phytoplasma infections is the most accepted strategy to control phytoplasma diseases in the world.

REFERENCES

Agrios, G.N. (2005). Plant Pathology, 5th edition. Elsevier Academic Press, San Diego, USA.

Aldaghi, M., Massart, S., Dutrecq, O., Bertaccini, A., Jijakli, M.H. and Lepoivre, P. (2009). A simple and rapid protocol of crude DNA extraction from apple trees for PCR and real-time PCR detection of '*Candidatus* Phytoplasma mali'. *Journal of Virological Methods*, **156:** 96–101.

Azadvar M. (2010). Biological and molecular characterization of phytoplasma associated with *Brassica rapa* in India. PhD Thesis, Indian Agricultural Research Institute, New Delhi.

Bertaccini, A. (2007). Phytoplasmas, diversity, taxonomy, and epidemiology. *Frontiers of Bioscience,* **12:** 673–89.

Constable, F.E. (2010). Phytoplasma epidemiology: grapevines as a model. *In*: Weintraub, P.G., and Jones, P. (*Eds*). Phytoplasmas genomes, plant hosts and vectors. CAB International, UK, pp. 188–212.

Curkovic Perica, M. (2008). Auxin–treatment induces recover of phytoplasma-infected periwinkle. *Journal of Applied Microbiology,* **105:** 1826–34.

De La Rue, S.J., Hopkinson, R., Foster, S. and Gibb, K. (2003). Phytoplasma host range and symptom expression in pasture legume *Stylosanthes*. *Field Crop Research,* **84:** 327–34.

Doi, Y., Teranaka, M., Yora, K. and Asuyama, H. (1967). Mycoplasma–or PLT group-like microorganisms found in the phloem elements of plants infected with mulberry dwarf, potato witches' broom, aster yellows, or paulownia witches' broom. *Annals of the Phytopathological Society of Japan,* **33:** 259–66.

Elder, R.J., Milne, J.R., Reid, D.J., Guthrie, J.N. and Persley, D.M. (2002). Temporal incidence of three phytoplasma–associated diseases of Carica papaya and their potential hemipteran vectors in central and south–east Queensland. *Australasian Plant Pathology,* **31:** 165–76.

Esker, P.D., Gibb, K.S., Padovan, A., Dixon, P.M. and Nutter, F.W.Jr. (2006). Use of survival analysis to determine the postincubation time-to-death of papaya due to yellow crinkle disease in Australia. *Plant Disease,* **90:** 102–107.

Firrao, G., Garcia–Chapa, M. and Marzachì, C. (2007). Phytoplasmas: Genetics, diagnosis and relationships with the plant and insect host. *Frontiers in Bioscience,* **12:** 1353–75.

Foissac, X. and Wilson, M.R. (2010). Current and possible future distributions of phytoplasma diseases and their vector. *In*: Weintraub, P.G., and Jones, P. (*Eds*). Phytoplasmas Genomes, Plant Hosts and Vectors. CAB International, UK, pp. 309–24.

Galetto, L., and Marzachi, C. (2010). Real-time PCR diagnosis and quantification of phytoplasmas. *In*: Weintraub, P.G., and Jones, P. (*Eds*). Phytoplasmas Genomes, Plant Hosts and Vectors. CAB International, UK, pp. 1–18.

Greber, R.S. and Gowanlock, D.H. (1979). Rickettsia–like and mycoplasma–like organisms associated with two yellow-type diseases of strawberries in Queensland. *Australian Journal of Agricultural Research,* **30:** 1101–1109.

Heinrichs, E.A. (1994). Impact of insecticides on the resistance and resurgence of rice planthoppers. *In*: Denno, R.F. and Perfect, T.J. (*Eds*) The Planthoppers: their Ecology and Management. Chapman and Hall, London, pp. 571–98.

Hodgetts, J. and Dickinson, M. (2010). Phytoplasma phylogeny and detection based on genes other than 16S rRNA. *In*: Weintraub, P.G., and Jones, P. (*Eds*). Phytoplasmas Genomes, Plant Hosts and Vectors. CAB International, UK, pp. 93–113.

Hodgetts, J., Boonham, N., Mumford, R., and Dickinson, M. (2009). Panel of 23S rRNA geneased real-time PCR assays for improved universal and group-specific detection of phytoplasmas. *Applied and Environmental Microbiology,* **75:** 2945–50.

Hodgetts, J., Boonham, N., Mumford, R., Harrison, N., and Dickinson, M. (2008). Phytoplasma phylogenetics based on analysis of secA and 23S rRNA gene sequences for improved resolution of candidate species of '*Candidatus* Phytoplasma'. *International Journal of Systematic and Evolutionary Microbiology,* **58:** 1826–37.

Hogenhout, S.A. and Seruga Music, M. (2010). Phytoplasma genomics, from sequencing to comparative and functional genomics–what have we learnt? *In*: Weintraub, P.G. and Jones, P. (*Eds*). Phytoplasmas Genomes, Plant Hosts and Vectors. CAB International, UK, pp. 19–36.

Hogenhout, S.A., Oshima, K.E., Ammar, D., Kakizawa, S., Kingdom, H.N., and Namba, S. (2008). Phytoplasmas: bacteria that manipulate plants and insects. *Molecular Plant Pathology,* **9**: 403–23.

Hren, M., Boben, J., Rotter, A., Kralj, P., Gruden, K. and Ravnikar, M. (2007). Real-time PCR detection systems for Flavescence dorée and Bois noir phytoplasmas in grapevine: comparison with conventional PCR detection and application in diagnostics. *Plant Pathology,* **56**: 785–96.

IRPCM Phytoplasma/Spiroplasma Working Team–Phytoplasma Taxonomy Group. (2004). 'Candidatus Phytoplasma', a taxon for the wall-less, non-helical prokaryotes that colonize plant phloem and insects. *International Journal of Systematic and Evolutionary Microbiology,* **54**: 1243–55.

Jarausch, W., Eyquard, J., Mazy, K., Lansac, M. and Dosba, F. (1999). High level of resistance of sweet cherry (*Prunus avium* L.) towards European stone fruit yellows phytoplasmas. *Advances in Horticultural Science,* **13:** 108–12.

Kaminska, M. and Silwa, H. (2003). Effect of antibiotics on the symptoms of stunting disease of *Magnolia liliiflora* plants. *Journal of Phytopathology,* **151:** 59–63.

Khan, A.J., Botti, S., Paltrinieri, S., Al–Sobhi, A.M. and Bertaccini, A.F. (2002). Phytoplasmas in alfalfa seedlings: infected or contaminated seedling? 14th International Organization of Mycoplasmalogy Conference, Viena, Austria, p. 148.

Lee, I.M., Davis, R.E. and Gundersen–Rindal, D.E. (2000). Phytoplasma: phytopathogenetic mollicutes. *Annual Review of Microbiology,* **54:** 221–55.

Lee, I.M., Zhao, Y. and Davis, R.E. (2010). Prospects to multiple gene–based systems for differentiation and classification of phytoplasmas. *In*: Weintraub, P.G., and

Jones, P. (eds). Phytoplasmas Genomes, Plant Hosts and Vectors. CAB International, UK, pp. 51–63.

Lee, R.F., Nyland, G. and Lowe, S.K. (1987). Chemotherapy of cherry buckskin and peach yellow leafroll diseases: an evaluation of two tetracycline formulations and methods of application. *Plant Disease,* **71:** 119–21.

Mannini, F. (2007). Hot water treatment and field coverage of mother plant vineyards to prevent propagation material from phytoplasma infections. *Bulletin of Insectology,* **60:** 311–12.

Marcone, C. (2010). Movement of phytoplasmas and the development of disease in the plant. *In*: Weintraub, P.G. and Jones, P. (*Eds*). Phytoplasmas Genomes, Plant Hosts and Vectors. CAB International, UK, pp. 114–31.

Marcone, C., Neimark, H., Ragozzino, A., Laur, U. and Seemüller, E. (1999). Chromosome sizes of phytoplasmas composing major phylogenetic groups and subgroups. *Journal of Phytopathology,* **89:** 805–10.

Marzachì, C., Milne, R.G. and Bosco, D. (2004). Phytoplasma–plant–vector relationships. *In*: Pandalai, S.G. and Gayathri, A. (*Eds.*) Recent Research Development in Plant Pathology, Vol. 3. Research Signpost, Trivandrum, Kerala, India, pp. 211–41.

McCoy, R.E., Caudwell, A., Chang, C.J., Chen, T.A., Chiykowski, L.N., Cousin, M.T., Dale, J.L., de Leeuw, G.T.N., Golino, D.A., Hackett, K.J., Kirkpatrick, B.C., Marwitz, R., and Petzold, H. (1989). Plant diseases associated with mycoplasma-like organisms. *In*: Whitcomb, R.F. and Tully, J.G. (*Eds*) The Mycoplasmas, Vol. 5. Academic Press, New York, pp. 545–60.

McCoy, R.E. (1979). Mycoplasmas and yellows diseases. *In*: Whitcomb, R.F. and Tully, J.G. (*Eds*) The Mycoplasmas, Vol. III. Academic Press, New York, pp. 229–64.

McCoy, R.E. (1982). Antibiotic treatment for control of tree diseases associated with mycoplasma–like organism. *Reviews of Infectious Diseases,* **4:** 157–61.

Monis, P.T. and Giglio, S. (2006). Nucleic acid amplification–based techniques for pathogen detection and identification. *Infection Genetics and Evolution,* **6:** 2–12.

Musetti, R. (2010). Biochemical changes in plants infected by phytoplasmas. *In*: Weintraub, P.G., and Jones, P. (*Eds.*). Phytoplasmas Genomes, Plant Hosts and Vectors. CAB International, UK, pp. 132–46.

Nipah, J.O., Jones, P., Hodgetts, J. and Dickinson, M. (2007). Detection of phytoplasma DNA in embryos from coconut palm in Ghana, and kernels from maize in Peru. *Bulletin of Insectology,* **60:** 385–86.

Olivier, C.Y., Séguin–Swartz, G. and Hegedus, D. (2006). First Report of "Candidatus Phytoplasma asteris" related strains in *Brassica rapa* in Saskatchewan, Canada. *Plant Disease,* **90:** 832.

Parker, K.G., Palmiter, D.H., Gilmer, R.M. and Hickey, K.D. (1963). X-disease of peach and cherry trees and its control. *Cornell Extension Bulletin,* **1100:** 1–12.

Pilkington, L.J., Gurr, G.M., Fletcher, M.J., Elliott, E., Nikandrow, A. and Nicol, H.I. (2004). Reducing the immigration of suspected leafhopper vectors and severity of Australian lucerne yellows disease. *Australian Journal of Experimental Agriculture,* **44:** 983–92.

Puterka, G.J., Reinke, M., Luvisi, D., Ciomperlik, M.A., Bartels, D., Wendel, L. and Glenn, D.M. (2003). Particle film, Surround WP, effects on glassy-winged sharpshooter behavior and its utility as a barrier to sharpshooter infestations in grapes. Plant Health Progress. Online, doi:10.1094/PHP–2003–0321–RS.

Raychaudhuri, S.P. (1993). Over-view of plant mycoplasma diseases in India. In: Raychaudhuri, S.P. (*Ed.*). Management of plant diseases caused by fastidious prokaryotes, Proceedings of the 4th regional workshop on plant mycoplasma, University of Quinsland, Australia.

Romanazzi, G. and Murolo, S. (2008). Partial uprooting and pulling to induce recovery in bois noir-infected grapevines. *Journal of Phytopathology,* **156:** 747–50.

Saracco, P., Marzach, C. and Bosco, D. (2008). Activity of some insecticides in preventing transmission of chrysanthemum yellows phytoplasma ('*Candidatus* Phytoplasma asteris') by the leafhopper *Macrosteles quadripunctulatus* Kirschbaum. *Crop Protection,* **27:** 130–36.

Seemüller, E. and Harries, H. (2010). Plant Resistance. *In*: Weintraub, P.G., and Jones, P. (*Eds*). Phytoplasmas Genomes, Plant Hosts and Vectors. CAB International, UK, pp. 147–69.

Seemüller, E., Garnier, M. and Schneider, B. (2002). Mycoplasmas of plants and insects. *In*: Razin, S. and Herrmann, R. (*Eds*) Molecular Biology and Pathology of Mycoplasmas. Academic/Plenum Kluwer, London, pp. 91–116.

Šeruga–Musie, M., Krajacie, M. and Škorie, D. (2008). The use of SSCP analysis in the assessment of phytoplasma gene variability. *Journal of Microbiological Methods,* **73:** 69–72.

Setiawan, D.F. and Ragsdale, D.W. (1987). Use of aluminum–foil and oat–straw mulches for controlling aster leafhopper, *Macrosteles fascifrons* (Homoptera: Cicadellidae), and aster yellows in carrots. *Great Lakes Entomology,* **20:** 102–109.

Summers, C.G. and Stapleton, J.J. (2002). Management of corn leafhopper (Homoptera: Cicadellidae) and corn stunt disease in sweet corn using reflective mulch. *Journal of Economic Entomology,* **95:** 325–30.

Torres, E., Bertolini, E., Cambra, M., Monton, C., Martin, M.P. (2005). Real-time PCR for simultaneous and quantitative detection of quarantine phytoplasmas from apple proliferation (16SrX) group. *Molecular and Cellular Probes,* **19:** 334–40.

Tubajika, K.M., Civerolo, E.L., Puterka, G.J., Hashim, J.M. and Luvisi, D.A. (2007). The effects of kaolin, harpin, and imidacloprid on development of Pierce's disease in grape. *Crop Protection*, **26:** 92–99.

Uyemoto, J.K., Bethell, R.E., Kirkpatrick, B.C., Munkvold, G.P., Marois, J.J. and Brown, K.W. (1998). Eradication as a control measure for X-disease in California cherry orchards. *Acta Horticulturae,* **472:** 715–21.

Verma, A., Chenula, V.V., Raychaudhuri, S.P., Prakash, N. and Rao, P.S. (1969). Mycoplasma like bodies in tissues infected with sandal spike and brinjal little leaf. *Indian Phytopathology,* **22:** 289–91.

Wally, O., Daayf, F., Khadhair, A.H., Adam, L., Elliott, B., Shinners–Carnelley, T., Iranpour, M., Northover, P. and Keyworth, S. (2004). Incidence and molecular detection of yellows-type disease in carrots, associated with leafhoppers in southern Manitoba, Canada. *Canadian Journal of Plant Pathology,* **26:** 498–505.

Waloff, N. and Jervis, M.A. (1987). Communities of parasitoids associated with leafhoppers and planthoppers in Europe. *Advances in Ecological Research,* **17:** 281–402.

Walsh, K.B., Guthrie, J.N. and White, D.T. (2006). Control of phytoplasma diseases of papaya in Australia using netting. *Australasian Plant Pathology,* **25:** 49–54.

Wei, W., Kakizawa, S., Suzuki, S., Jung, H.Y., Nishigawa, H., Miyata, S., Oshima, K., Ugaki, M., Hibi, T. and Namba, S. (2004). *In planta* dynamic analysis of onion yellows phytoplasma using localized inoculation by insect transmission. *Phytopathology,* **94:** 244–50.

Weintraub, P.G. and Beanland, L. (2006). Insect vectors of phytoplasmas. *Annual Review of Entomology,* **51:** 91–111.

Weintraub, P.G. and Wilson, M.R. (2010). Control of phytoplasma diseases and vectors. *In*: Weintraub, P.G. and Jones, P. (*Eds*). Phytoplasmas Genomes, Plant hosts and vectors. CAB International, UK, pp. 233–49.

Weintraub, P.G., Pivonia, S., Rosner, A. and Gera, A. (2004). A new disease in *Limonium latifolium* hybrids. II. Investigating insect vectors. *HortScience,* **39:** 1060–61.

Zhao, Y., Wei, W., Davis, R.E. and Lee, I.M. (2010). Recent advances in 16S rRNA gene–based phytoplasma differentiation, classification and taxonomy. *In*: Weintraub, P.G. and Jones, P. (*Eds*). Phytoplasmas Genomes, Plant Hosts and Vectors.CABInternational,UK, pp. 4–92.

15

Sustainable Agriculture and Challenges in Crop Protection

D.M.J.B. SENANAYAKE[1]

ABSTRACT

Human population is projected to grow at 80 million per annum and large scale crop production programmes are required to meet the demand for food for this growing human population in the world. Crop losses due to pests, diseases and weeds, emergence of new and secondary pests, resurgence of pests, pathogens and weedicide–resistant weeds account for substantial economic losses in agriculture in many countries. In this scenario crop protection to safeguard crop yields in terms of quantity and quality becomes more and more important. In present days, most of the crops are cultivated as monocultures in vast areas where pest, disease and weed controls are essential. Among crops, the total global losses due to pests vary from place to place and crop to crop and are in billions of US dollars. Use of agricultural chemicals in combating pests, diseases and weeds has been the main methods of controlling them during the last few decades. Chemical pesticides have failed to eradicate pests, diseases and weeds and have become one of the major causes for environmental pollution along with an enormous amount of health hazards to the humankind and other living organisms. To address these issues related to present day chemical agriculture alternative agriculture systems have been developed and practiced by man throughout the world. Sustainable agriculture consists of three major components, health of consumers and animals, environment safety and economy of the agriculture. Organic agriculture can also be considered as a part of sustainable agriculture. Crop protection in a sustainable agriculture system is easier when compared to that of an organic agriculture system. Depending only on the cultural practices has also not given promising results. Non

[1] Field Crops Research and Development Institute, Mahailluppalama, Sri Lanka.
Corresponding author: E-mail: jsenanayake@gmail.com

availability of resistant varieties or effective natural enemies for many pests, diseases and weeds are some of the major problems faced in crop protection in sustainable systems where need of chemical agriculture arises. Careful selection of pesticides is important to reduce the negative impact on beneficial organisms, human and animals and environment. In the present plant protection context, a sustainable system may be more appropriate than an organic system. If sustainable crop protection assures economically profitable crop yield margins it will help in continuation of a sustainable system.

Key words: Sustainable agriculture, Crop protection, Agrochemicals, Pests, Diseases, Weeds.

---

1. INTRODUCTION

Human population is projected to grow at 80 million per annum, increasing by 35% to 7.7 billion by 2020, then by about 75% before leveling off at about 10 billion (Tilman, 1999; United Nations, 1996; Pinstrup–Andersen, 2000). Large scale crop production programmes are required to meet the demand for food for this growing human population in the world. The Green Revolutions that took place during the period between 1940s and 1970s can be considered as the major attempt made in the recent past to meet the food demand and food security in the world. Improved crop management systems based upon genetically improved (high-yielding) crops, enhanced soil fertility through chemical fertilization, pest control through synthetic pesticides, and irrigation were hallmarks of the Green Revolution. The combined effect of these factors allowed world food production to double in the past 35 years (Oerke and Dehne, 2004). However, a few common problems related to agriculture can be observed throughout the world. Most of the soils with high productivity potential are already under cultivation and lands suitable for agricultural production are becoming limited. The availability of water for consumption and crop cultivation is also becoming inadequate, and in some regions land resources are depleted and the cultivated area is shrinking (Nelson–Smith, 1995; Oerke and Dehne, 2004). Crop losses due to pests, diseases and weeds, emergence of new and secondary pests, resurgence of pests, pathogens and weedicide resistant weeds also account for substantial economic losses in agriculture in many countries. In this scenario crop protection to safeguard crop yields in quantity wise and quality wise becomes more and more important.

Since the day when the agriculture was started crop losses due to biotic agents; pathogens, insect pests, weeds and abiotic agents such as moisture, plant nutrients, sunlight, wind, etc. have been experienced by man. These losses are billions of rupees worth and the amount of loss vary with the type of crop and the location of cultivation of the crop or the way of transportation and storage of the crop produce. Of the agents that account for crop losses pest, diseases and weeds play a pivotal role. Man has taken many attempts to control pests, diseases and weeds in agriculturally important crops since the date of crop cultivation started. These attempts, better described as techniques, range from simple cultural practices to present day use of agricultural chemicals and genetic engineering solutions.

In present days, most of the crops are cultivated as monocultures in vast land areas where pest, disease and weed control is essential. Use of agricultural chemicals in combating pests, diseases and weeds has been the main methods of controlling them during last few decades. It has become one of the major causes for environmental pollution along with an enormous amount of health hazards to the humankind and other living organisms. Among crops, the total global potential loss due to pests varies from about 50% in wheat to more than 80% in cotton production. The responses are estimated as losses of 26–29% for soybean, wheat and cotton, and 31, 37 and 40% for maize, rice and potatoes, respectively. Weeds produce the highest potential loss (34%), with animal pests and pathogens being less important (losses of 18 and 16%) (Oerke, 2006). One of the best examples of crop losses due to a disease with a big impact on the human population is rice brown spot devastation that caused great famine in Bengal, India causing death in two million people with starvation in 1942. Crop losses may vary from place to place the annual loss caused to agricultural production in India was estimated as much as ₹ 7000 crores in 1988 (Jayaraj, 1988).

In the present agriculture, a huge amount of pesticides are used against insects, pathogens and weeds. In India chemical pesticide usage has increased from 2.2 g (a.i.)/ha in 1950 to the current level of 650 g (a.i.)/ha (Dubey *et al.*, 2003). With the modern agriculture where synthetic fertilizers and agrochemical are used man has experienced several negative impacts to human and animal health, ecological balances and environment. As a result, various other agriculture systems such as organic agriculture and sustainable crop production have emerged and are being practiced by farmers in many parts of the world in order to address those negative impacts associated with the modern synthetic agriculture. The demand for organic foods is in an ever increasing trend in the present day world. These new environmental friendly agriculture systems are gaining wide popularity throughout the world, attracting an increasing number of consumers for the produce.

2. SUSTAINABLE AGRICULTURE

Sustainable agriculture consists of three major components, (1) Health of consumers and animals; in a sustainable agriculture system the food consumed by humans and animals are healthy, (2) Environment safety; the environmental pollution is kept minimum in sustainable agriculture, (3) Economy of the agriculture; in a sustainable system agriculture is economically profitable and brings about a better socio-economic situation. There are number of definitions available for sustainable agriculture and a few are indicated below for the usefulness of the reader.

The word "sustain" from the Latin sustinere (sus–, from below and tenere, to hold), to keep in existence or maintain, implies long–term support or permanence. As it pertains to agriculture, sustainable describes farming systems that are "capable of maintaining their productivity and usefulness to society indefinitely. Such systems must be resource-conserving, socially supportive, commercially competitive, and environmentally sound" (Duesterhaus, 1990).

Sustainable agriculture is a philosophy based on human goals and on understanding the long-term impact of our activities on the environment and on other species. Use of this philosophy guides our application of prior experience and the latest scientific advances to create integrated, reduce environmental degradation, maintain agricultural productivity, promote economic viability in both the short and long term, and maintain stable rural communities and quality of life (Francis and Youngberg, 1990).

A sustainable agriculture system does not mean a return to either the low yields or poor farmers. Rather, sustainability builds on current agricultural achievements, adopting a sophisticated approach that can maintain high yields and farm profits without undermining the resources on which agriculture depends (*Source*: http://www.ucsusa.org).

A sustainable agriculture is one that, over the long term, enhances the environmental quality and the resource base on which agriculture depends; provides for basic human food and fiber needs; is economically viable; and enhances the quality of life for farmers and society as a whole (Francis and Youngberg, 1990).

Douglass (1984) described three major themes in the sustainable agriculture literature: sustainability as food sufficiency, as stewardship of the earth, and as community. According to him, these themes seem to be promoted by three different constituencies. Those who argue for sustainability as food sufficiency tend to see the sole or primary function of agriculture as providing abundant food for growing populations.

3. NATURAL RESOURCES IN SUSTAINABLE FARMING

Maintaining a sustainable agriculture system is not a responsibility of a single individual or an entity. It is an effort and the result of a group consisting of farmers, consumers, researchers, extension officers, policymakers, politicians, traders, labourers and other personnel who are directly and indirectly involved in agriculture. A number of socio-economic factors are involved in success of a sustainable farming system. In such a system priorities should be given to water, soil, air, natural flora and fauna. Balanced and sustainable use of these resources helps in balancing the ecosystem where natural enemies of pests and pathogens survive. Thus, they also become important in pest and disease management in a sustainable system. Water, soil, air and natural flora and fauna are considered as the most important natural resources in such a system. Here, importance of these resources with respect to pest, diseases and weeds is briefly discussed.

3.1. Water

Water is considered as the principle resource for all the living beings and agriculture. Water is becoming limited and water pollution due to industrial wastes and agricultural practices is a common problem throughout the world. Pollution affects the quality of water too. Apart from the scarcity of water, salinity affects the agriculture in many irrigated farm lands in many places in the world, especially in dry areas. A vigorous crop growth cannot be expected and that may account for easy pest and disease infestations in crops in areas where water scarcity or excess water remain or salinity develops. On the other hand, weeds that are tolerant to water scarcity, flooding and salinity can grow well in those places. Sea Barley–grass (*Hordeum marinum* (syn. *Critesion marinum*)), which is found in Australia and some parts of Europe is an example for a salinity tolerant grass.

3.2. Soil

Decreasing soil fertility is a major problem faced by farmers. Soil erosion is one of the major factors that affects the soil fertility at a higher degree. Certain land use patterns cause very high soil erosions. When a crop like tobacco, which exposes the land to a considerable extent, is cultivated in a bare hilly land it may cause considerable soil erosion. Removal of surface soil along with nutrients due to run off causes less vigorous crop growth and in turn it would support for an easy infestation of pests and diseases.

3.3. Air

Air can be polluted due to some agricultural practices such as burning of crop residues or aerial application of agrochemicals and crops can also be affected due to polluted air. High global aerial temperature is known to be caused by green house effect. High N content in the air may cause for acid rains which can devastate crops. Air currents have a big impact on the spread of diseases. The spread of the stem rust fungus that overwinters in Mexico and Texas in the USA is always favored by moist southern air currents (http://www.apsnet.org). It is reported that the southern leaf blight of corn spread from Mississippi to the Midwest by air currents of a tropical storm in the Gulf of Mexico during 1970s (Rosenzweig and Iglesias, 2007).

3.4. Natural Flora and Fauna

It is understood that the balance of the ecosystem is a key component of a sustainable agriculture system. Disturbing the natural habitats of wildlife may result in moving of wild animals to crop lands making considerable crop losses. It may cause imbalance in the population of natural enemies of pests and pathogens.

4. PLANT PROTECTION IN SUSTAINABLE AGRICULTURE SYSTEMS

As per the above definitions, use of natural pest and disease control practices must be an essential component in crop plant protection under a sustainable agricultural system. Use of agriculture chemicals that causes health and environmental hazards should be minimum in such systems. Thus, agricultural practices of a sustainable agriculture system would become close to that of an organic agriculture system.

5. ORGANIC AGRICULTURE IN SUSTAINABLE AGRICULTURE

Organic agriculture can also be considered as a part of sustainable agriculture. Use of methods that maintain the long term productivity of the farming systems and practicing such methods as the guide of the land can be defined as organic agriculture. Biodiversity, ecological balance, sustainability, use of natural fertilizers, pest management methods and soil integrity are the principle components of organic agriculture (http://www.sustainabletable.org).

There are key differences observed between an organic agriculture system and a sustainable agriculture system. The major differences between organic and sustainable agriculture are listed in Table 1.

Table 1: The main differences between organic and sustainable agriculture

Organic agriculture	***Sustainable agriculture***
Must be certified every year and approved by a relevant authority. *i.e.*, USDA/USA, INDOCERT, Kerala, India	No such certification is required
An actual certification system is there	A more philosophical way
A considerable amount of money is required for organic certification	Since there is no such certification required no such cost is incurred
Animals are kept indoor most of the time of their life	Animals are kept out door
Antibiotics are not allowed to feed to animals	Antibiotics can be used
Organic farming may be in a much organized way and may be in large scale	Sustainable farming is mostly in small scale basis
Use of synthetic agrochemicals are not allowed	Judicious integration of pest resistant varieties and agrochemicals are allowed. Practices such as the reduction in number of agrochemical sprays can be done. More environmental friendly chemicals can be included in a sustainable system. Integrated pest management (IPM) is compatible in the system

6. DIFFERENT PLANT PROTECTION PRACTICES IN SUSTAINABLE AGRICULTURE AND CHALLENGES ASSOCIATED WITH THEM

6.1. Plant Disease Management

Different pathogens, fungi, bacteria, viruses, mycoplasma, phytoplasma, rickettsia etc. cause diseases in crops in different parts of the world and different pathogens dominate in different locations. In India, diseases/ pathogens that are of more importance in cropping systems can be shown as follows. Rice: rice blast (*Magnaporthe grisea),* rice bacterial blight (*Xanthomonas oryzae* pv. *oryzae*); wheat: wheat rusts (*Puccinia graminis tritici*), Karnal bunt (*Tilletia indica*); maize: Turcicum leaf blight (*Exserohilum turcicum* (Pass.), rust (*Puccinia polysora*), stem rot (*Fusarium moniliforme*); legumes: root rot (*Fusarium* spp.,) Ascochyta blight (*Ascochyta* spp.,), sterility mosaic virus, yellow mosaic virus, and Phytophthora blight (*Phytophthora drechsleri*); vegetables: damping off (*Sclerotinia* spp, *Sclerotium* spp, wilts (*Fusarium* spp., *Pythium* spp.,) and powdery mildew (*Erysiphe* spp.,); mango: powdery mildew (*Oidium mangiferae* Berthet) and grapes: downy mildew (*Plasmopara viticola*) and powdery mildew (*Uncinula necator*); cotton: angular leaf spots (*Xanthomonas campestris* pv. *malvacearum*) and wilts (*Fusarium oxysporum* f.sp. *vasinfectum*); sugar cane: red rot (*Colletotrichum falcatum* Went) and smut (*Sporisorium scitamineum*); apple: scab (*Venturia inaequalis*) (Dubey *et al.*, 2003).

The plant disease protection practices in a sustainable agriculture system are almost closely related to the practices found in organic agriculture. Therefore, such practices are discussed here in a view point of a sustainable agriculture system.

6.1.1. *Healthy planting material and removal of infected crop*

Practices done by a grower or farmer to prevent or control plant diseases by changing the plant environment can be described as cultural practices. In cultural practices main aim is to reduce the quantity of initial inoculum that is available to the plant. This is carried out by changing the plant's environment; changing the temperature, light, water, altitude etc. Reducing the level of inoculum is basically done by using non-infected planting materials. Use of clean planting material is important since some pathogens can spread to the next crop generation through seeds or planting materials. Transmission through seeds may be as internal seed borne diseases *i.e.*, bacterial grain rot of rice (*Pseudomonas glumae*) or external seed borne diseases *i.e.*, ergot of bajra (*Cleviceps fusiformis*).

Removal of infected plants or plant parts and crop residues of diseased plants help in reducing the initial inoculum potential. Removal of disease crop residues along with nearby soil is commonly recommended by pathologists for controlling soil-borne wilting diseases. However, this is effective when the disease is at very early spreading stage. In India the average land size for a farmer is 1.8 ha (Source: http://www.new–ag.info) and it is the sole income source of the family. Therefore, if this removal of infected plants has to be done in a larger scale a farmer may not be willing to do that. Burning of infected plants or parts or transporting them away from the fields is also being done by farmers. Burning is considered as an effective method of controlling the second crop infestation by certain diseases such as sheath blight in rice (causal organisms: *Thanatephorus cucumeris* (A.B. Farnak) Donk (anamorph: *Rhizoctonia solani)* through stubble and straw.

6.1.2. *Elimination of alternate and alternative hosts*

Most of the plant pathogens have alternative hosts that harbour pathogens during cropping seasons as well in off seasons. One of the best examples in this regard is rust fungi (Family: *Pucciniaceae*), which have alternate host. The sexual stage of the stem–rust fungus of wheat develops on common barberry (*Berberis vulgaris* L.). Common weeds may act as hosts for many pathogens apart from their competition with a crop for nutrients and space. In a sustainable system, there is a less scope for the use of a weedicide for the elimination of alternate hosts though weedicides are not excluded. Manual removal of weeds or other unwanted plants is a labour intensive task. If the crop is not very economical, *i.e.*, as an organic product, manual weeding may not give a profitable return to the farmer. However, it is considered as a more appropriate method to remove alternative hosts in small scale farming.

6.1.3. *Removal of vectors*

Inoculum of pathogens can spread through mechanical means. Contagious viruses such as *Tobacco mosaic virus* or *Potato virus X* spread easily through contaminated hands or tools. Winds and rain splash may also help in inoculum dissemination. Some insects transport inoculum of pathogens from its sources to the new courts of infections. Removal or repulsion of an insect vector of a particular disease through conventional methods may not be effective. Some diseases can be transferred to a susceptible plant from an infected plant even with one vector. *Chilli leaf curl virus* (a begomovirus spp. of Family *Geminiviridae*) is a good example for this type of transmission.

6.1.4. *Crop rotation*

Monocultures which are frequently practiced in present day agriculture helps in establishing diseases and their pathogens. Use of crop rotation, in which a crop that is not harboured or infected by a particular pathogen is cultivated after a crop infection, can be used to control a plant disease. The follow up crop may be a resistant cultivar or another crop that is not attacked by the same pathogen. The time period for the growth of the non-host plant may be higher than the lifespan of the pathogen. Therefore, the pathogen does not find a host to complete its life cycle. Use of mixed varieties or crops therefore, helps to reduce disease incidence and pathogen. Pathogens which are not host-specific and not mobile can be controlled by the choice and sequence of crops grown in the rotation. Soil-borne pathogens, *Sclerotinia sclerotiorum* and *Rhizoctonia solani* are examples for this. A non-solanaceous crop such as maize (*Zea mays* L.) can be grown in a field where solanaceous crops frequently experience wilting due to *Ralstonia solanacearum*. This is one of the most successful methods that can be adapted by farmers to control some soil-borne diseases. Host-specific pathogens with high mobility such as *Phytophthora infestans* in potato cannot be controlled by crop rotation (Kohl, 2006). However, farmers' tendency to go for another crop depends on the income that can be obtained from another crop and the pests and diseases associated with the new crop.

6.1.5. *Changing of the microenvironment of the crop*

Most of the pathogens like fungi prefer high humid nature for their growth. Increasing plant spacing and reducing the plant density mainly help in two ways to reduce a pathogen population. First, it allows proper sunlight among plants for proper and vigorous growth, making plants stronger against a pathogen's attack. Secondly, it reduces the moisture in soil around the plants and in canopy making the moisture level non-favourable for the colonization and growth of pathogens. Soil moisture level can also be effectively used for the control of pathogens. Reducing the amount of water in soil helps in controlling the spread of a pathogen like *Ralstonia solanacearum* which causes bacterial wilt in many solanaceous crops. Reduction of club root disease [*Plasmodiophora brassicae*] by the fungal endophyte *Heteroconium chaetospira* has been observed at high pH (6.3) and lower soil moisture content (Narisawa *et al.*, 2005) in Chinese cabbage.

6.1.6. *Ploughing or tillage practices*

Due to ploughing or tillage practices pathogens, living in the soil surface and on crop residues, are transferred to deeper layers of soil where the environment is unfavourable for their growth and development. Thus, an aerobic pathogen may face an anaerobic condition and vice versa due to tillage, which is not conducive for their living. However, some pathogens may spread by this method as the masses of pathogen colonies tend to disperse with tillage.

6.1.7. *Intercropping or barrier crops*

In an intercropping system for plant diseases control susceptible and resistant or tolerant crops are grown as alternative rows. The resistant or tolerant crop acts as a physical barrier to the spread of a pathogen between susceptible crop rows. Once the farmer is in the opinion to maximize the income from a monoculture with a high value crop making the farmer understanding for such change in order to control a disease may not be successful. On the other hand, it would be tedious for the farmer to use different agriculture practices when more crops are found in the same location. In certain occasions, barrier crops are used to avoid the contact between a susceptible crop and vector insects that are involved in virus transmission. The barrier crop is a crop species that is not favoured by a particular pest. Barrier cropping lessens the incidence and/or hinders the spread of aphid-borne non-persistent viruses (Fereres, 2000). Cucumber mosaic virus incidence in chilli can be reduced with barrier crops such as maize, sorghum, sunflower (Anandam and Doraiswamy, 2002). Barrier cropping can be successfully used to significantly mitigate the severity of yield loss caused by non-persistently transmitted aphid-borne viruses. But this may not be effective under conditions of polycyclic disease spread or when trying to protect a perennial host crop (Hooks and Fereres, 2006).

6.1.8. *Mulching and organic amendments*

Mulching is mainly done for moisture retention in the soil around a crop, weed control and providing organic manures to soil. With mulching incidence of some soil-borne diseases can be increased. High moisture retention around the crown area due to mulching may be a favourable factor for the growth of some pathogens. On the other hand, use of already infected plant materials for mulching such as rice straw infected with sheath blight (*Thanatephorus cucumeris* (A.B. Farnak) Donk) or rice blast (Causal organism: *Magnaporthe grisea*) may cause the same disease in a new rice crop if it is added as a mulching material or organic manure in rice. However, mulching can also support the crop for a vigorous growth by providing nutrients, required moisture and temperature etc., thus, increasing the defending ability of plants to the attacks of pathogens.

During the decomposition of organic material, many plant pathogens and their resting bodies are also decomposed. Therefore, use of well decomposed organic materials is beneficial to the plant and in controlling pathogens as well. However, it is impracticable to find enough quantity of mulching material with said objectives for large scale cultivations.

6.1.9. *Flooding*

Flooding can be used to reduce the inoculum pathogens. Disease causing organisms and their spores, conidia etc., can be destroyed due to flooding. On the other hand certain diseases such as sheath blight (*Thanatephorus cucumeris* (A.B. Farnak) Donk) in rice can be increased due to the flooding since sclerotia of sheath blight can float on water, spreading the disease to new plants. Water-induced soil transport increases dissemination of soil-borne pathogens to noninfected areas and an example is the outbreak of soybean sudden death syndrome (causal organism: *Fusarium solani* f.sp. *glycine*) in the north central U.S. states in 1993 (Rosenzweig *et al.*, 2001). Continuous soil saturation causes long-term problems related to rot development and increase damage by diseases. Crazy top (Causal organism: *Sclerophthora macrospora*) and common smut in maize (Causal organism: *Ustilago maydis*) disease incidences are found to be increasing due to soil water saturation.

6.1.10. *Rain and irrigation*

Excess water is mostly helpful for disease dissemination. Presence of a thin water layer or moisture on a leaf surface is an essential factor for the germination of spores of some fungal pathogens. With rain and application of water on to the plant pathogens and their propagules may splash with water, disseminating the pathogen to nearby plants. Application of water to the root zone, droplet–wise is more appropriate in plant disease management. Aerial moisture is also important in disease management. Humid summers cause epidemics of gray leaf spot of maize (Causal organism: *Cercospora zeae–maydis*).

6.1.11. *Drought*

Drought conditions can be helpful in controlling the spread of some diseases. Many foliar pathogens infect leaves when they are wet. In a drought condition such pathogens may not be able to enter plants. Fungal and bacterial pathogens that cause leaf spots in plants are not favoured by drought. Diseases which spread through rain splashes are observed to be reducing during droughts. Diseases which are found be less severe in drought conditions include downy mildew of sunflower (Causal organism: *Plasmopara halstedii*), Phytophthora root rot of soybeans and damping off of many plants caused by *Pythium* species. Drought also helps in increasing the incidence of some diseases such as powdery mildews of grapes. Water stress reduces plant vigour and alters carbon-to-nitrogen ratios, lowering plant resistance to pathogens, nematodes and insects. Fungal pathogens can attack stems and roots of crops easily when they are weakened. Outbreak of soybean cyst nematode in the north central USA in 1990 was correlated to drought conditions prevailing during that time. Drought also promotes insect outbreaks *e.g.*, summer locust outbreak in Mexico in 1999 was correlated to the drought condition prevailing in that period (Rosenzweig and Iglesias, 2007).

6.1.12. *Susceptible crops and trap crops*

With the continuous cultivation of resistant crop for a particular disease or a pathogen there is a possibility for creation of a more resistant pathogen strain. To avoid this and create more competition among the pathogen strains a susceptible crop can be cultivated during the off season or with the crop. The removal and destruction of the susceptible crop before the completion of the pathogen's life cycle is necessary. However, there may be risk with a trap crop since a trap crop that harbours vectors such as aphids may help in spreading aphid-borne viruses to a protected crop.

6.1.13. *Fertilizers*

A balanced supply of nutrition to a crop is necessary for an ideal plant growth and it is helpful for increasing the plant's ability to withstand the attack of pathogens. In a sustainable agriculture system use of synthetic fertilizers are not barred. However, organic fertilizers are preferred to chemical fertilizers within the system. Nitrogen fertilization can increase emission of gases that have critical roles in tropospheric and stratospheric chemistry and air pollution (Cicerone *et al.*, 1988; Hall *et al.*, 1996). Nitrogen fertilizers help to build more vegetative growth with succulent tissues in a crop to which most of the pathogens can attack more easily. As a result, in some plant diseases such as rice blast (Causal organism: *Magnaporthe grisea*) and bacterial leaf blight (Causal organism: *Xanthomonas oryzae* pv. *oryzae*) farmers are advised to reduce the dose of N in the form of urea as one of the measures of reducing the disease incidence. In those cases ammonium sulphate are more preferred as a source of N. Phosphorus fertilizers sometimes encourage some pathogens and sometimes inhibit the pathogens growth. Potassium fertilizers are generally considered as an inhibitor of disease development. However, potassium fertilizers can stimulate or inhibit the penetration, establishment, survival and multiplication of a pathogen. Potassium nutrient strengthens plant resistance to diseases, by regulating various plant physiological metabolism pathways, mainly phenolic metabolism, carbon and nitrogen metabolism, and active oxygen metabolism (Xiao–yan *et al.*, 2006). In rice brown spot (*Cochilobolus miyabeanus*) it is advised to farmers to apply more potassium fertilizers (muriate of potash) to get rid of the disease. Apart from N, P, K, micronutrients also play an important role in plant disease management.

Silicon is also known to reduce plant diseases, especially in rice. An adequate supply of calcium produces cell walls more resistant to penetration by facultative pathogens. High levels of calcium can raise the pH of the soil, disadvantaging any pathogens that favour acidic soils. However, soils high in calcium can promote the development of some diseases.

6.1.14. *Biocontrol agents*

For controlling plant diseases agrochemicals are widely used by farmers throughout the world. In the present day agriculture a considerable attention is given to the research on biocontrol agents and use of them at commercial level for pest and disease control. Use of agrochemicals may not be effective much in case of soil-borne pathogens such as, *Sclerotium rolfsii*, *Fusarium oxysporum*, *Rhizoctonia solani*, *Macrophomina phaseolina*, *Sclerotinia sclerotiorum*, *Pythium* and *Phytophthora* spp. Chemical control measures are found to be unsuccessful against these pathogens, mainly due to the following three reasons. First, these pathogens have a wide range of adaptability to different soil environments. Secondly, it is not practicable to use fungicides to control these fungi as uniform and proper distribution of a fungicide in soil cannot be attained during the application. Thirdly, the farmer has to use a large quantity of fungicides when it is applied to soil and it may not be practicable and economical at the field level. Therefore, biocontrol agents become more important in managing soil-borne diseases. Biocontrol agents fit nicely within a sustainable agriculture system. Most of the biocontrol agents show hyperparasitism/mycoparasitism and/or the production of antibiotics. In mycoparasitism, biocontrol agents acquire nutrients from the host. On the other hand the growth of the host is hindered by antibiotics produced by biocontrol agents. A few major drawbacks are associated with the use of biocontrol agents in plant disease or pathogen management. Non-availability of effective biocontrol agents for each and every important pathogen is one of the issues faced by farmers who are willing to use biocontrol agents in crop disease management. Identification, testing of effectiveness and release of a biocontrol agent against a plant pathogen involve a substantial amount of money, labour, time and protocols. A biocontrol agent that is effective against a pathogen in one environment may not show the same effectiveness in another environment for the same or a mostly related pathogen. Biocontrol agents are mostly organism-specific and do not show a broad spectrum of action against plant pathogens. Market availability of preparations containing more than one biocontrol agents can address this issue to a certain extent. Availability of biocontrol agents in the market in most of the developing countries is less when compared to the availability of agrochemicals.

6.1.15. *Commercial biocontrol products available in India and uses*

In India number of commercial products are available in the market while a number of products are in the pipeline for releasing (Table 2).

Table 2: Some of the commercial biocontrol agents available in India

Name of the biocontrol agent	***Pathogen targeted***	***Disease/host***	***Name of the commercial product***	***Manufacturer / Distributor***
Aspergillus niger AN27	*Fusarium* spp., *Pythium*, *Macrophomina*, *Rhizoctonia*, *Sclerotinia*, *Rotylenchulus*	Soil borne diseases of vegetable, guava	Kalisena SD, Kalisena SL	Cadila Pharmaceutical Ltd., Ahmadabad
Trichoderma harzianum	*Pythium*, *Rhizoctonia*, *Sclerotinia*, *Sclerotium*, *Fusarium*, *Verticillium* and other soil fungi	Damping off, root rot	Pant biocontrol agent–1	GBPUA and T, Pantnagar, India

Table 2: (*Contd...*)

Table 2: (*Contd...*)

Name of the biocontrol agent	*Pathogen targeted*	*Disease/host*	*Name of the commercial product*	*Manufacturer/ Distributor*
Trichoderma viride	Soil-borne pathogens	Damping off, collar rot and other soil borne diseases in vegetables and pulses	Ecofit	Hoechest and schering AgrEVO Ltd., Mumbai, India
T. viride	*Rhizoctonia solani*, *Macrophomina phaseolina*		Antagon–TV (India)	Green Tech Agro Products, Coimbatore
	Rhizoctonia solani, *Macrophomina phaseolina*		Sun–Derma	Cun Agro–chemiclas, Chennai, India
	Fusarium spp., *Rhizoctonia* spp.,		Bas Derma	Basarass Biocontrol Res, India
T. viride / T. harzianum	*Fusarium* spp., *Rhizoctonia* spp.,		Niprot	Pest Control India Private Ltd., India
	Fusarium spp., *Rhizoctonia* spp.,		Tri–control	Jeypee Biotechs, India
T. viride / T. harzianum			Bioderma	Biotech International Ltd., India
Pseudomonas fluorescence	Numerous	Several diseases	Pant biocontrol agent–2	GBPUA and T, Pantnagar, India
			FB–Sudo	Pramukh Agri Clinic, Gujarat, India

(*Source*: Singh, 2001; Sharma and Sharma, 2007)

6.1.16. *Botanicals and biopesticides*

In some countries botanical biopesticides are used to control pest and diseases in crops. These contain active ingredients derived from various parts of plants and may also be termed plant extract or natural biochemical products. Products based on garlic, neem, plant oils such as thyme, and clove are examples for botanicals and biopesticides.

6.1.17. *Resistant varieties*

Use of resistant varieties and cultivars against plant diseases is practiced by researchers and farmers throughout the world and the practice fits well in a sustainable system. Use of highly resistant varieties against plant pathogens could lead to the development of resistant races and strains in a shorter time period. This is better known as the breaking down of vertical resistance. There are number of reports available in relation to the breaking down of varietal resistance against pathogens in many crops. Breaking down of rice blast resistance in varieties in many Asian countries have been reported. Breaking down of wheat rust resistance in major wheat cultivating areas has been observed (McDonald and Linde, 2002). Stem-rot, caused by *Magnaporthe salvinii* (Catt.) Krause and Webster, is a major rice disease in Punjab and breaking down of the resistance in resistant Basmati–370 was reported (Mayee and Rattan, 1976). Scientists are making efforts to minimize the resistant break down in crops through various measures such as gene pyramiding and multiple gene resistance (horizontal resistance). Developing moderately resistant varieties for plant diseases or cultivation of a susceptible variety along with a resistant variety or during the off season is therefore, useful in reducing the buildup of resistant pathogenic strains or races in a sustainable system. In case of trans-genetically developed varieties for pest resistance it is compulsory to have a refugee crop in the field along with the resistant variety to minimize the development of resistant pests.

6.1.18. *Agrochemicals*

Different kinds of agrochemicals are used by farmers to control plant pathogens, especially fungal pathogens. These chemicals mainly include fungicides and then bactericides. Most of the available fungicides act directly on essential fungal functions such as respiration, sterol biosynthesis or cell division (Cremlyn, 1990). As in insecticides and weedicides, many of the agrochemicals used in pathogen control or plant disease management are known to be accompanied with health hazards and environmental pollution. Agrochemicals can be used in a sustainable system to manage and control plant diseases with much care. In case of a sudden outbreak of a plant disease it is inevitable to use agrochemicals. Use of agrochemical with the least harm to the non-target organisms, humans, animals or environment can be included in a sustainable system once the other cultural or biological methods are found to be inefficient. Practices such as selecting the correct and exactly required agrochemical, minimizing the quantity applied and increasing the time interval between two application etc. can help in maintaining the sustainability of an agriculture system.

6.1.19. *Integrated disease management*

Integrated disease management is the use of above mentioned practices with combinations in a more appropriate way. The basic ideas of an integrated disease management include avoidance, exclusion and eradication of pathogens, protection of crop from pathogens, use of resistance and chemical therapy as per the requirement. Knowledge on the biology, ecology and epidemiology of a pathogen and biology of crops are essential in carrying out an integrated disease management strategy.

6.2. Insect Pest Management

Insect pests in agriculture can be defined as any form of animal life that is harmful or potentially harmful to agriculturally important crops and their products. Insect pest management in agriculturally important crops also ways back to the time when man started agriculture. Pests that are associated with crops can be categorized in to 4 groups basically. (a) major pests that harm agriculturally important crops or their products giving much economical losses or (b) minor pests of which damages to crops or crop products causing low level economic losses. (c) Pests that associate with crops occasionally and occur only under defined environmental conditions. (d) Potential pests; pests that do not cause losses to crops or stored agriculture products in normal environmental conditions but that can cause economical losses under certain ecological conditions. Pest may associate with crops naturally or they may be invasive species. During the 20th century many major pest outbreaks in agriculturally important crops have been experienced by man. Monocultring of crop species at large scale, limited genetic base of improved varieties, succulent nature of crops with high fertilizer applications, transportations of infested agriculture materials from one place to another, high application of agriculture chemicals have been account for various insect pest occurring and crop damages in fields and during storage.

It is essential to control and manage insect pests in crops for better yields and economic gains. Different pest management strategies in the context of sustainable agriculture are briefly discussed here. Pest control and management practices fitting in a sustainable agriculture system can be categorized broadly as: (a) biological, (b) cultural (c) chemical and (d) integrated pest control methods.

6.2.1. *Cultural pest control methods*

This can be considered as the oldest method of pest management. It is an aspect of preventive nature. Knowledge on the crop and insect biology and ecology is essential in this regard.

6.2.2. *Clean insect free planting materials*

It is necessary to take enough precautions to avoid planting materials contaminated with insect eggs or other stages of the life cycle of an insect since it can give a pest outbreak in a new crop otherwise. The storage pest, beetles of Family *bruchidae* that attack seeds of legumes spread through seeds. The adults lay eggs on the seed surface and the larvae bore inside the cotyledons. Seed contamination with insect eggs may not be visible to the naked eye. On the other hand when a stage of a life cycle of an insect is inside the seeds a farmer may not be able to distinguish it from other healthy seeds and discard them.

6.2.3. *Date of planting/sowing*

Some pests are observed to be abundant in certain periods of time in a year. If the farmer plants his crop in such a way that the peak pest abundance is avoided the damage to the crop by pests can be minimized. This method is suitable for annual crops and cannot be applied to perennial crops. In a survey done in pear orchard in Egypt, it was found that mealybugs were the most important and major insect pests that attack pear trees in the first season in 2005, whereas in the second season in 2006 *Cacopsylla pyricola* was the major pest. Temperatures and relative humidity showed a significant effect on the population of such pests (El–Naeim *et al.*, 2008). In such instances date of planting would not show any significance with the pest attack.

6.2.4. *Crop isolation*

Once the crop is grown isolated from the other similar crop fields spread of some insects can be controlled. As an example, chilli crop in a new field that is located to a chilli field already infested with whitefly (*Bemisia tabaci*) or thrips (*Scirtothrips dorsalis* Hood) may get a fast pest infestation of same pests. Strawberry plants will get the infestation of root weevil (*Otiorhynchus ovatus*), strawberry bud weevil (*Anthonomus signatus* Say) from an already infested plant in the same way. When farmers are interested in continuous cultivation of same crops for economic gain or due to some other reasons insect pest perpetuation cannot be avoided.

6.2.5. *Crop spacing*

With proper spacing plants become healthier and vigorous and are able to withstand insect pest attacks. When plants are closer it is easy for insects to spread from one plant to another. However, different behaviours of insect populations with spacing have been observed. In rice, a field with a 10 × 10 cm hill spacing has greater chances of suffering hopperburn than a field with a 20 × 20 hill spacing. Dense planting increases populations of planthoppers, leafhoppers, leaffolders, gall midges, black bugs, and caseworms while whorl maggots, root aphids, root weevils and leaf beetles may become less abundant (*www.knowledgebank.irri.org*).

6.2.6. *Intercropping/mixed cropping*

Intercropping is the agronomic practice of growing two or more crops in the same field at the same time (Andrews and Kassam, 1976). Crops may be planted without regard to rows (mixed intercropping), in alternating rows, or with different crops alternating within the same row (Smith and McSorley, 2000). Intercropping helps in reducing pest incidences. The damage of stem borer to maize and sorghum can be minimized when molasses grass was intercropped (Khan *et al.*, 2000). Non-availability of suitable intercrops may be a problem in this regard. Different management practices for different crops can also deter a farmer going for intercropping or mixed cropping in order to control pest incidences.

6.2.7. *Alternate host control*

Non-crop habitats such as field margins, fallows, hedgerows and woods, which are relatively undisturbed, hold a substantial proportion of the biodiversity in an agricultural system (Bianchi *et al.*, 2006). A number of agricultural pest species are associated with these habitats, such as aphids, herbivorous flies and beetles (Denys and Tscharntke, 2002). Non-crop habitats also support a diversity of natural enemies such as staphylinids (Maudsley *et al.*, 2002), spiders (Schmidt and Tscharntke, 2005), coccinellids (Honek, 1989), chrysopids (Sengonca *et al.*, 2002), predatory mites (Rieux *et al.*, 1999). The role of these natural enemies is very important and essential in a sustainable agriculture system. Insect pests may habitat in a number of host plants including other crop plants. Kelly's citrus thrips (*Pezothrips kellyanus* Bagnall) (Thysanoptera: Thripidae) is a serious insect pest affecting citrus production in New Zealand and *Brassica napa* and *Lycopersicon esculentum* have been found to be alternative hosts for Kelly's citrus thrips (Froud *et al.*, 2001). *Erodium moschatum* and *Brassica rapa* sp. Silvestris have been found to be alternative hosts for *Thrips tabaci* which acts as the vector for *Tomato spotted wilt virus* (Wilson, 1998).

6.2.8. *Fertilization*

Fertilization, especially N in urea form helps in increasing succulent tissues in plants and this nature is attractive to insect pests. The fertilizers and manures change the interaction among the pests and plant organisms so that some species are favoured whereas others are affected adversely. Therefore, application of fertilizers or manures may increase or decrease pest populations in crop. Prasanna and Kumar in 2011 showed that the application of higher dose of FYM will increase the population density of insects. Devasthali and Vishwakarma (1995) found that the application of NPK (20:60:20 kg/ha) significantly increased the population of white fly, Jassids and tobacco caterpillar on soybean crop. On contrary, Shri Ram *et al.* (1990) reported that increasing concentration of phosphorus and potassium alone and in combined form was more effective in reducing the feeding activity of defoliator in cowpea especially from 40 days age. Different aphid species may respond fertilizers in different ways. It has been reported that the aphid, *Brevicoryne brassicae* become more abundant on organically fertilized plants, while the aphid, *Myzus persicae* have higher populations on synthetically fertilized plants. The diamondback moth (*Plutella xylostella*) was more abundant on synthetically fertilized plants and preferred to oviposit on these plants (Staley, 2010).

6.2.9. *Removal of crops/hand picking of insects*

Removal of infested crops is a way of reducing the insect pest population and different stages of a life cycle of an insect pest in a crop cultivation. However, this would become practicable in high value organic crop productions or in small scale farming. Farmers generally show a less acceptance to this method of controlling insects. It was reported that farmers show the least adoption for the technique of hand picking of infected plant parts and eggs in gram cultivations in Udaipur District of Rajasthan, India when compared to the other techniques such as light trapping (Dadheech *et al.*, 2009).

6.2.10. *Mulching and cover crops*

Mulching in a crop field is mainly done for soil and moisture conservation and weed control. Use of cover crops and mulches, are environmentally safe methods for managing some specific insect pests (Schmidt *et al.*, 2007; Teasdale *et al.*, 2004). Organic mulches may be derived from hay, straw, crop residues, or other plant material that is readily available. Mulching is an effective way to provide shelter for predatory insects and to control weeds (Brown and Tworkoski, 2004). It has been reported that the population density and the damage of Lesser cornstalk borer (LCB) *Elasmopalpus lignosellus* (Zeller), which is a serious pest of bean (*Phaseolus vulgaris* L.) and many other crops in Florida, USA can be reduced with mulching with sunhemp (*Crotalaria juncea* L.) (Gill *et al.*, 2010). Cover crops may also help in increasing beneficial insect populations. For example, cover crops such as phacelia (*Phacelia tanacteflolia*) and buckwheat (*Fagopyrum esculentum*) provide nectar for hoverflies and bees, which are beneficial insects (Pontin *et al.*, 2006). Whitefly (*Bemisia argentifolii*) colonization was found to be hindered in zucchini crops when wheat straw and reflective mulches were used (Summers *et al.*, 2004). Artificial material such as black polyethylene can also be used for mulching. In experiments, it has been shown that straw mulch applied to the center of onion beds at the early to mid-bulb growth stage reduced the abundance of thrips as much as 33% when compared to non-treated plots of transplanted onions

(Howard *et al.*, 2009). Finding enough quantity of suitable materials for mulching and application of mulches may involve a considerable amount of money and labour.

6.2.11. *Time of harvesting*

Time of harvesting depends on the time of cultivation. Some insects become abundant in certain periods of the year. In such cases, if the crop is harvested before the high insect pest population the losses due to such insect pests can be reduced. Time of harvesting is also important in minimizing storage damages to crop. Baidoo *et al.* (2010) found that time of harvest on the level of damage caused by *Callosobruchus maculatus* on stored cowpea (*Vigna unguiculata*) can be reduced with prompt harvest of matured cowpea pods.

6.2.12. *Traps*

Sticky traps and pheromone traps are often used for insect monitoring and can be used in organic systems to attract pests away from crops and disrupt mating (Shrivastava *et al.*, 2010). Yellow sticky traps visually stimulate insect pests such as greenhouse whiteflies (*Trialeurodes vaporariorum*) and tarnished plant bugs (*Lygus lineolaris*). When paired with traps, pheromone and scent lures utilizes as both chemical and visual attractants (Foster and Harris, 1997). Squash vine borer (*Melittia cucurbitae*) is a destructive pest that is difficult to manage in organic systems. Mass trapping using pheromones can lure males away from females that would otherwise mate and lay eggs (Jackson *et al.*, 2005; Shrivastava *et al.*, 2010).

6.2.13. *Trap crops*

Trap crops are composed of one or more plant species that are grown to attract insects in order to protect the cash crop from the pest (Hokkanen, 1991). During an important period of a crop cycle a trap crop may attract and retain insects leaving the main crop with less insect pest incidences. Trap crops also help in building up of natural enemies of insect pests. The trap crops are grown as strips, rows, mixed with the main crop or in borders. Marigold can be used as a border crop in cotton (*Gossypium*) to minimize the damage of thrips (MMSU, 2003). Cowpea can be used as a row intercrop to trap *Heliothis* spp in cotton (CIKS, 2000). There is a possibility of spreading insect pests from a trap crop to a new crop. This matter has been mentioned under disease portion of this chapter.

6.2.14. *Plant extracts and botanicals*

Various plant extracts are used in insect pest management in different crops. Natural plant materials or plant extracts can successfully control major stored product insects (Talukder *et al.*,1998). Over 2000 species of plants representing 170 families are said to have some insecticide properties (Feinstein, 1952). Neem (*Azadirachta indica*), garlic (*Allium sativum*), castor (*Ricinus communis*), pine (Pinus spp.) extracts etc. can be mentioned as examples in this regard. Leaf extracts of custard apple (*Annona squamosa*) was found to be effective in controlling pests in sunflower (Ahmed *et al.*, 2010). Preparation of plant extracts for the application of a larger field may be cumbersome. Plant extracts or botanicals that are suitable to control insect pests may not be commercially available in many areas.

6.2.15. *Biological control*

Biological control can be defined as the use of natural enemies to reduce the damage caused by a pest population. These natural enemies may be fungi, bacteria, virus, nematodes or entomophagous insect that relies on predation, parasitism, herbivory, or other natural mechanisms. A very large number of studies, information and materials are available on biological control of insect pests.

6.2.16. *Agrochemicals*

Chemical pest control methods are widely used by farmer in crop production throughout the world. Most of the farmers prefer control or management of crop pest problems through agrochemical applications despite health hazards and environmental issues to other methods since agrochemicals give sensible or visible results in shorter time periods most of the times. Insecticides are classified in different ways and classification according to different agrochemical families such as organochlorines, organophosphates, and carbamates is useful in understanding the nature of action of insecticides. In a sustainable system chemical pest control methods can be applied with much care. Target specific chemicals that are also less hazardous to humans, animals and the environment can be used in minimum quantities with border application intervals.

6.2.17. *Resistant varieties*

Use of resistant varieties against insect pests is a very successful way of dealing with insect damages in crop cultivation. Resistant varieties for various pests are available in many crops throughout the world. There are two major problems associated with crop resistance; unavailability of resistant varieties for some pests and breaking down of the resistance of a crop variety to a particular insect with time. Resistance–breaking of rice varieties carrying Bph1 and bph2 in Korea due to wild brown planthopper (BPH), (*Nilaparvata lugens*) can be mentioned as an example in this regard (Seo *et al.*, 2009).

6.3. Integrated Pest Management

As per the FAO definition the Integrated Pest Management (IPM) means the careful consideration of all available pest control techniques and subsequent integration of appropriate measures that discourage the development of pest populations and keep pesticides and other interventions to levels that are economically justified and reduce or minimize risks to human health and the environment. IPM emphasizes the growth of a healthy crop with the least possible disruption to agro-ecosystems and encourages natural pest control mechanisms. (http://www.fao.org). In an IPM approach all the above described pest management strategies are used in alternative or selective ways. IPM has been practiced by farmers in many parts of the world. Development of more healthy and environmental friendly agriculture products are the expected outcome of an IPM programme while minimizing the damages to natural flora, fauna and ecological balances of living organisms and the environment. Judicious use of pesticides is a major step in IPM by which farmers reduce unnecessary pesticide use. There are success stories in IPM in various countries. Several effective IPM modules which are cost effective, sustainable and eco-friendly in different ecosystems have been developed for various crops *viz.*, rice, pigeonpea, chickpea, French beans, soybean, lentil, green pea, mustard, tomato, brinjal, capsicum, cauliflower, cabbage, chilli and coconut etc. in India (Wahab, 2009).

Adoption of IPM by farmers varies from place to place and it depends on the type of practice too. Farmers prefer some IPM techniques to the others. Farmers' adoption to IPM depends on many factors, such as their technical skill and socioeconomic conditions as well as psychological and cultural factors, etc. In a study to examine the adoption of IPM practices on cotton in Punjab and on paddy in Haryana in India it was observed about the relationship between farm-size and adoption of IPM practices. In the case of paddy, a negative relationship has been observed, while the cotton has shown a positive relationship (Sing *et al.*, 2008).

6.4. Weed Management

Any unwanted plant in a crop field is considered as a weed. Weeds are spread by wind, water, animals, and man. Controlling of weeds becomes more difficult when natural agents are involved in spreading.

Weeds are important in agriculture since:

1. They reduce crop yields. Weeds compete with crops for moisture, light and nutrients. Therefore, the crop yield loss depends on the number of weed plants growing in a particular crop field. A heavy infestation

of weeds in a crop field may cause a complete crop failure and an establishment of a perennial weed can make a land unsuitable for further cultivation.

2. Some weeds act as alternate alternative hosts for pest and diseases of crops. Some weeds belonging to the family Graminae act as alternative host for blast fungus of rice. Onion thrips (*Thrips tabaci*) can survive on weeds around an onion field. Weeds growing in or near potato fields harbour aphids which are known to be carrying many plant viruses including potato virus Y.
3. Some weeds may be poisonous or cause allergies etc. in humans. Castor bean (castor oil plant) (*Ricinus communis*) of *Euphorbiaceae* family is one of such poisonous plants which is not suitable for consumption.

Weeds may be annual species, which complete their life cycle from seed to seed in less than 12 months or biennial species, which complete their life cycle in two years or perennial species that live for more than two years. Farmer should possess the knowledge on proper identification of weeds and the potential damage that can occur due to weeds. Knowledge on the emergence of weeds and sequential way of the emergence of different weeds from time to time in a field is important in planning effective weed control programs. Each weed species has one or more periods of high emergence; while the initial emergence date may vary from year to year. If a weed is not threatening it is not essential to remove it in a sustainable system. The cost for control, method of control and especially, the environmental impact of the control method must be considered in a sustainable system. Controlling a weed may involve more than one strategy. Here, weed management strategies in the context of sustainable agriculture are briefly discussed.

6.4.1. *Prevention*

As in other pest management systems the prevention of entering of a new weed to an area and establishment is the first step in weed management. Once a weed is introduced to a cropping field it is hard to remove and eradicate. Seeds of some weeds such as Dodder (*Cuscuta* spp.,) can remain viable in soil for more than 70 years. Prevention of weeds becomes difficult when natural agents involve in the dispersion of weed seeds. However, some practices such as the use of clean seeds and clean farm machineries help in reducing invasive weeds. In developing countries seed and planting materials requirement is mostly met through farmers own stocks or from the neighborhood. Most of these seeds are unprocessed and contain other inert materials including weed seeds. Such contaminated seeds can keep on spreading weeds in crop fields.

6.4.2. *Removal*

Mechanical removal of weeds by uprooting or cutting weeds prior to the production of seeds is a useful method in weed control. Hand removal of weeds may not be economical except in high value crops. One example is *Cyperus rotundus* (purple nut sedge) that is found in many farmlands in the south Asia. It is also known as the world's worst weed and consists of underground tubers (Holm *et al.*, 1977). As a result, a complete removal of all the plant parts becomes difficult and this situation helps *Cyperus rotundus* to remain in a field for a longer time period. Perennial weeds which are having deeper root system cannot be uprooted completely. In such cases the satisfactory removal of weeds from a field becomes practically impossible.

6.4.3. *Tillage*

This is a very common practice used in crop cultivation and a comparatively easy and environmental friendly way of controlling weeds. Most of the weeds, that are already germinated and surviving can be damaged physically and buried inside the soil by tillage. With deep tillage some weed seeds that are already buried in the soil can come to the surface of the soil and may germinate causing new weed outbreaks. Tillage cannot be used to manage weeds with large physical structures. Tillage can damage underground portions of weed plants. Physical damages to weeds during tillage make them vulnerable to the attacks of soil micro flora and thus they are deteriorated and decomposed.

6.4.4. *Mowing*

Mowing of weeds with a mechanical mower or using other equipment is another way of controlling weeds. This method is more suitable in managing weeds in flower and horticulture gardens. Mowing of weeds before their seed setting gives better results. Mowing, as possibly as close to the ground is important. Because of continuous removal of areal parts of a weed that engage in photosynthesis the weed becomes deprived of food resources and tends to die.

6.4.5. *Grazing by animals*

Animals like cattle, sheep and goats are used by farmers for grazing. Action of grazing by animals is similar to mechanical mowing. However, a few problems are associated with grazing of animals. Grazing animals prefer certain weeds and dislike some other weeds, based on their palatability. Therefore, a selection of weeds is inevitable and the weed control through grazing may not be uniform. Grazing is not a practical method of controlling weeds in an already cultivated land since it is difficult to manage grazing animals in such fields. Certain weed seeds may remain intact and viable in the gut of grazing animals and weeds can spread through their dung from one place to another. Quackgrass (*Elymus repens* (L.) and pigweed (*Amaranthus palmeri*) are examples for such weeds. They can remain viable after passing through cattle's digestive tract. If fresh cowdung is used as a fertilizer there is a possibility of spreading of such weeds. Use of well decomposed animal dung may help in preventing this situation. Maintaining a temperature of 130°F or more for 3 to 4 days favours the destruction of weed seeds in composting. It also helps to destroy plant pathogens associated with animal dung.

6.4.6. *Burning*

Farmers, especially in dry areas, prefer removal of weeds with burning to other methods. Though this seems to be an easy method for controlling weeds in a field before the cultivation it is not recommended as a suitable method in a sustainable system. Some weed seeds that are heat resistant can remain viable to give the next generation. Burning removes much valuable organic materials from the field. It may affect beneficial microorganisms in shallow soil layer.

6.4.7. *Mulching*

Mulching is another way of controlling weeds, especially between plants or rows of plants in a field. Due to mulching weeds would be deprived of photosynthesis and ultimately die. Straw, saw dust, black polyethylene are commonly used for mulching. Apart from weed control, use of mulching material helps in reducing some plant pathogens and pests while conserving soil moisture.

6.4.8. *Crop rotation*

Crop rotation can also be used to control weeds. Some weeds are more preferably associated with some crops. Mutual benefits between the crop and weed due to the root exudates that favour the weed or due to one species favouring competition for space, light, water and nutrients etc. account for such preferences. Weeds commonly associated with wheat crop such as *Anagallis arvensis* (Krishananeel), *Argemone mexicana* (Satyanashi), *Asphodelus tenuifolius* (Piazi) are a few examples of such associations. Crop rotation breaks continuous association of a particular crop and a weed and it helps in controlling weeds. Plant qualities like height, ground covering ability etc., play a role in the effectiveness of crop rotation for weed control.

6.4.9. *Intercropping*

Intercropping is another way of controlling weeds. A crop that does not compete with the major crop for space, light or nutrients or that does not harbour pest and diseases can be selected for this purpose. Incidence of some weeds such as guinea grass (*Panicum maximum*) in a field can be reduced with the inter cultivation of sweet potato (*Ipomoea batatas*).

6.4.10. *Crop density*

Increased crop plant density may help in reducing the weed population. But it may support spreading some pest and diseases since it increases humidity in the micro-environment where a condition favouring many fungal pathogens is created. The decrease in the distance between two plants helps in increasing diseases and pest spread easily through contact.

6.4.11. *Submergence*

In some crops such as rice, the field is kept inundated for one to two weeks after the emergence of rice seedlings or after the transplanting. This is useful in preventing the emergence of many weeds. This is a suitable method for crops such as rice and in places where water is sufficiently available. In case of upland crops this method cannot be used.

6.4.12. *Use of biocontrol agents*

Use of biocontrol agents for controlling weeds may fit in a sustainable system more than any other methods. These biocontrol agents may be natural enemies of weeds such as fungi, bacteria, nematodes or insects. Though it seems that the application of biocontrol agents is much promising, identification of effective natural enemies and preparation for commercial release takes a much longer time. Many studies related to the efficacy, safety and environmental impact of biocontrol agents are needed to be carried out in this regard and it involves a substantial amount of capital and labour. There are two basic issues related to weed biocontrol agents. Firstly, it is hard to find effective biocontrol agents for each and every weed. Secondly, the outcome of an application of a biocontrol agent may not be visible immediately. These facts may deter a farmer in using a biocontrol agent over a chemical weedicide for controlling weeds in his/her field. Weed biocontrol agents would be more appropriate for an uncultivated land since a balance between the biocontrol agents and weed population is required for the perpetuation of the biocontrol agents. Almost weed free state is expected and in a crop field therefore, a biocontrol agent may not find enough number of plants for its survival. In such cases, it is necessary to apply biocontrol agents to fields in periodic intervals.

Biological control is more effective in controlling invasive weed species for which natural enemies are not found. Alligator weed, an invasive species spread in water ways in the United States was successfully controlled with the alligator weed flea beetle (*Agasicles hygrophila*). In the United States, the fungal pathogens *Phytophthora palmivora* and *Colletotrichum gloeosporioides* f.sp. *aeschynomene* have been formulated and used successfully in an inundative approach for noxious weed management (Templeton *et al*., 1979). *Phytophthora palmivora* has proved effective in managing populations of the exotic plant strangler vine (*Morrenia ordorata* Lindl.) in citrus groves (TeBeest, 1993) and *C. gloeosporioides* f.sp. *aeschynomene* has been used to manage populations of northern jointvetch (*Aeschynomene virginica* L.) in rice fields (Tempeleton *et al*., 1979, TeBeest 1993).

6.4.13. *Chemical weed control*

In a sustainable agriculture system, chemical weed control methods are not excluded. Weedicides are basically categorized as selective and non selective. 2,4–Dichlorophenoxyacetic acid (2,4–D) and MCPA or 2-methyl-4–chlorophenoxyacetic acid, which are used to control broad leaf weeds are the most common examples for selective weedicides. Glyphosate (*N*-(phosphonomethyl) glycine) weedicide is one of most commonly used non-selective weedicides. It is essential to understand the nature of action of a weedicide for effective use in a sustainable system. Weedicides are classified as contact and systemic. A contact weedicide does not translocate inside the target weed whereas a systemic weedicide translocates. Weedicides are also characterized as pre-emergence, which are applied to the soil prior to the emergence of weeds and post-emergent, which are applied to the weeds after they germinate. Weedicides that do not persist in soil or environment or do not involve in heavy metal contamination or other forms of environmental pollutions and health hazards to humans, animals and microorganisms must be preferred in a sustainable agriculture system. Volatilization, leaching, plant uptake, action of microorganisms and ultra violet sunlight help in fast removal of weedicides from soil and subsequent degradation.

It is generally known that the soils with higher amount of organic matter and clay require more quantity of a chemical weedicide when compared to a sandy soil with less organic matter. This is due to the absorbance and retention ability of clay and organic matter particles. Time of application of weedicide, growth stage of the weed and crop, and climatic factors also affect the efficacy of a chemical weedicide. If the weed and crop both are not in a suitable age the efficacy of a weedicide may reduce. Application of weedicides in the morning would be more effective. Generally, warm humid climates are found be effective in controlling weeds through chemical weedicides. Most relevant issues related to the chemical weedicides are the water pollution, crop injury, air pollution, development of resistant weeds and negative impact on non-target organisms. Farmers prefer weedicides in their farmlands due to the high efficacy associated with chemical weedicides and faster visible results. Weedicides have to be applied where other methods do not work.

6.4.14. *Integrated weed management*

An integrated weed management system fits best in a sustainable agriculture system. In integrated weed management, all of the above methods are used in a mixed way or in an alternative way. None of the above methods alone provide acceptable levels of weed control. However, if such methods are used in a systemic way an adequate weed control can be achieved. The success of integrated weed management depends on the fitting of different control strategies to the specific weed problem in a field. (Swanton *et al.*, 2008; Killham, 2010).

6.5. Nematode Management

Nematodes are small, worm–like, multicellular animals and they are found in the oceans, in freshwater habitats, and in soils and they are common in soils all over the world (Dropkin and Vitor, 1980). Plant parasitic nematodes cause significant economic losses in a wide variety of crops. Root-knot nematodes, *Meloidogyne* spp., are the main plant-parasitic nematodes that limit yield of vegetable crops worldwide (Talavera *et al.*, 2009). Nematodes live either in association with plant roots and/or with aerial parts of plants. Plant-parasitic nematodes are found in abundance in the rooting zones of crops, mostly in the top 30 cm of soil; nematodes may also penetrate to deeper soil layers when the conditions are unfavourable (Prot, 1980). Parasitic nematodes may reduce crop yield through direct cell destruction, vectoring of viruses, or indirectly by facilitating invasion of fungi and bacteria through their feeding and movement through roots (Westerdahl *et al.*, 1998). Overall losses caused by plant parasitic nematodes are estimated to exceed US $100 billion/year worldwide including 10–20% yield reduction in several cash crops (Hooks *et al.*, 2010). The free-living stages of some species survive the unfavourable season through physiological or structural adaptation (Evans, 1987; Gaur and Perry, 1991), or survive for several seasons as eggs, as in the case of the potato-cyst nematode (Talavera *et al.*, 2009). As mentioned earlier, some nematodes involve in virus transmission. *i.e.*, *Grapevine fanleaf virus* (nematode vector *Xiphinema index*). Some nematodes have the potential to be used as biocontrol agents. Entomopathogenic nematodes belonging to genera *Steinernema* and *Heterorhabditis*, and slug-parasitic nematodes, *Phasmarhabditis* with its symbiotic bacteria *Moraxella* have been considered as promising biocontrol agents for the management of crop insect pests and slugs (Askary, 2010). Damage of the nematodes to a crop depends on the population size of nematodes. The nematode populations are influenced by reproductive rate of nematodes, host factors and many environmental factors. Nematode control measures can be categorized to two basic groups; (1) Reducing the nematode damage to a crop or protecting a crop from nematode damages, and (2) Reducing the nematode population in an infected soil. In sustainable agriculture system, a number of methods can be used to manage nematode damages and such methods are briefly discussed below.

6.5.1. *Resistant varieties*

As in other disease, insect and weed management use of nematode resistant varieties is one of the best ways of managing nematode problems in crop fields. There are number of resistant varieties available in crops such as tomato, eggplant and sweet potato. Tomato is one of the crops in which genetic resistance has specially been effective against root-knot nematodes. The most important source of nematode resistance in tomato is conferred by the *Mi* family of genes from the wild tomato and it is effective against *Meloidogyne* species (Hadisoeganda and Sasser, 1982). Host resistance is the most practical alternative to the use of nematicides (Da Conceicao *et al.*, 2005). However, breaking down of resistance against nematodes has also been reported in crops such as tomato and eggplant.

6.5.2. *Crop rotation*

In this aspect, crops those are not suitable for the infection, growth or reproduction of nematodes can be used to replace nematode susceptible crops in a sequence of crop cultivation for minimizing damages. Crop rotation is found to be more effective in root knot nematodes. Some crops, such as pea, lentil and some cultivars of canola (*Brassica napus* L.), and barley (*Hordeum vulgare*) may lessen the nematode reproduction (Taylor *et al.*, 2000). Farmers' preferences, yielding potential, availability of suitable crops for rotation and other pest and disease associated with such crops etc., may affect the efficiency of a crop rotation for nematode management. The root lesion nematode (*Pratylenchus neglectus*) has a wide host range and such wide host ranges of some nematode species and the unavailability of resistant varieties limit the use of crop rotation in several production systems (Hooks *et al.*, 2010).

6.5.3. *Cover crops/non-preferable hosts*

Use of poor or non-host cover crops may give an effective control of soil nematodes. Some plant species, *i.e.*, marigold (*Tagetes* spp.) and sudangrass (*Sorghum bicolor* subsp. drummondii) can be used effectively to reduce soil nematode populations. Leguminous cover crops, such as hairy indigo (*Indigofera hirsuta*) and American jointvetch (*Aeschynomene americana*) have been found to be effective in managing soil populations of sting or root-knot nematodes. Sorghum (*Sorghum bicolour*) is also a popular cover crop that reduces root knot nematodes. Sorghum may be a favourable host for sting nematodes.

6.5.4. *Removal of alternative hosts*

Many plant species can act as alternative hosts for nematodes and therefore, removal of such plants is important in a good nematode management system. Some weeds, *i.e.*, nutsedges (*Cyperus esculentus* L.) can act as alternative hosts for nematodes. Farmers' knowledge on the correct identification of alternative hosts and plants that help in reducing the nematode incidence is important in this regard.

6.5.5. *Flooding*

Flooding helps to control nematodes, especially root knot nematodes (*Meloidogyne spp.*) Changes in aerobic situation of the nematode habitat in soil to anaerobic situation help in reducing the activity of soil nematodes (Duncan, 1991). Sotomayor *et al.* (1999) showed that the number of galls found in tomato (*Lycopersicon esculentum* cv. Rutgers) roots can be reduced with intermittent and continuous flooding. They also reported that flooding can reduce endogenous soil populations of ring (*Criconemella* spp.) and stubby–root (*Paratrichodorus* spp.). Some of the problems associated with flooding are its unsuitability for highlands, requirement of a considerable water quantity and possibility of spreading of soil pathogens.

6.5.6. *Organic amendments*

Soil organic amendment helps in controlling nematodes, basically in two ways: (1) Some chemicals released during the decomposition of organic materials may be toxic to nematodes, (2) nematode antagonistic microorganisms can be activated with some organic manure such as chicken litter (Reigel *et al.*, 1996). Addition of sunhemp (*Crotalaria juncea*) to soil reduces nematodes associated with maize (*Zea mays* L.). *Crotalaria* spp. are used as cover crops in crop rotations, and occasionally incorporated into the soil, due to their nematode–suppressive effect and nitrogen–fixing ability (Wang *et al.*, 2010). Use of other Asteraceae plants as soil amendments for nematode control has been reported in a few studies (Oka, 2010). Some leguminous plants are known to contain lectins that disrupt nematode behaviours, such as host–finding (Marban–Mendoza *et al.*, 1987). These soil amendments are required to add from time to time. Application of organic manures to a larger field area may not be practicable and such applications may increase the cost of production of agriculture produce. Inconsistent control efficiency, which is highly influenced by amendment and soil type, is another problem associated with the control of nematodes through organic amendments (Oka, 2010).

6.5.7. *Fallowing*

Fallowing is another way of controlling nematode populations in fields. Clean fallowing during off seasons reduces the availability of food sources for nematodes. Controlling of weeds and alternative host is essential during the fallowing. The rice stem nematode, *Ditylenchus angustus* is a serious disease of deepwater rice (*Oryza sativa*) and is able to survive from one crop to the next by remaining coiled in rice stubble and debris and in soil. Infested ratoons are also sources of inoculum (Ou, 1985; Latiff *et al.*, 2011). Therefore, the removal of rice crop debris, stubble and ratoon crop can reduce *D. angustus* incidence. Two major problems are found with bare fallowing: (1) It reduces the organic matter content of soil which in turn can produce a big negative impact on the soil fertility and on soil beneficial microbial populations as well, (2) Bare fallowing may account for a considerable wind and water erosion of soil.

6.5.8. *Soil solarization*

Soil solarization is one of the simple easy methods that can be used to control nematodes in a sustainable system and has been found to be effective in number of crops. In soil solarization, soil radiant heat acts as a lethal agent against nematodes and other microorganisms. Generally a clear polyethylene is used to trap solar heat in adequately moistened soils. The time period of solarization may vary from a few weeks to several months. This method has been used successfully against plant-parasitic nematodes and soil-borne pathogens in several crops around the world, especially in hot climates (Katan, 1981; Stapleton and Devay, 1983). This is a suitable method for the areas where enough sunlight is received.

6.5.9. *Biological control*

As described in insect pests, disease and weed management, controlling of nematodes through biological agents is also a well suitable method in a sustainable system. There are number of examples available for biocontrol agents against nematodes. Non-parasitic micro-organisms that degrade soil amendments and release nematicidal compounds, such as the bacterium which degrades chitin to produce ammonia are likely to kill most nematodes in soil (Spiegel *et al.,* 1991). In the presence of *Rhizobium,* nematode multiplication can reduce (Siddiqui and Mahmood, 1995). *Bacillus thuringiensis* toxins have been shown to be effective in controlling species of *Meloidogyne* (Devidas and Rehberger, 1992). Use of *Trichoderma harzianum* can significantly suppress root knot diseases in maize (Windham *et al.*,1989). Decline in nematode populations in suppressive soils have been found with two parasitic fungi, *Nematophthora gynophila* and *Verticillium chlamydosporium,* which attack the developing female on the root surface in suppressive soils (Kerry *et al.*, 1987). The major challenges associated with nematode biocontrol agents are: (1) As in many other biocontrol systems, it is hard to find effective biocontrol agents for many nematode spp. (2) Efficiency of biocontrol agents may be inconsistent in different environments, (3) Several applications may be needed for better results and it may be cost ineffective.

6.5.10. *Nematicides*

Nematicides are the chemicals that are used to kill nematodes. Methyl bromide is one of the highly used chemicals to control soil nematodes. Several fumigants and nematicides have been withdrawn from the market in the last few decades due to concerns about the environment and human health (Rich *et al.*, 2004). Currently, the most widely used nematicides are cadusafos, carbofuran, ethoprop, fenamiphos, fosthiazate, oxamyl and terbufos, which are organophosphate or carbamate based pesticides (Rich *et al.*, 2004; Wu *et al.*, 2011). Continuous use of nematicides can develop nematicide resistance. Meher *et al.* (2009) showed that the nematicidal treatments resulted in the development of virulent *M. incognita* populations and their invasion potential increased significantly after continuous exposure to low concentrations of the nematicides.

6.5.11. *Integrated management*

Use of above described methods in a mixed way will give a better control and management of nematodes. In a sustainable system, it is easy to include chemical control measures also as per the requirement.

7. BIOTECHNOLOGY IN SUSTAINABLE CROP PROTECTION

Application of biotechnology in agriculture is still considered as a controversial issue in some countries especially in case of development of crop varieties through genetic engineering. Different views regarding the biotechnology in plant protection in sustainable agriculture systems are highlighted by different authors. The major contribution of biotechnology will be in reducing the use of agricultural chemicals, while sustaining the productivity needed to feed the world's population with more productive and resistant crops, through faster and more precise diagnosis of plant diseases (http://www.agnet.org). Disease diagnosis in plants has been revolutionized through the development of DNA probes as well as antisera and monoclonal antibody (MAb) kits. Researches and plant protectionists have been able to identify pest and diseases in crop plants more rapidly and accurately with these tools replacing the need to culture pathogens for identification. In turn, it will avoid the use of unnecessary agrochemicals. Another related technology is the production of biocontrol agents with the potential to replace chemical pesticides. Although wild organisms may be used, as in the case of some *Bacillus thuringiensis* (Bt)–based insecticides, bacteria may also be manipulated in order to achieve a more potent formulation (http://www.agnet.org).

8. SUSTAINABLE CROP PROTECTION AGAINST PRESENT DAY CHEMICAL AGRICULTURE

Crop protection in a sustainable agriculture system is easier when compared to organic agriculture. In organic agriculture integrated pest management is not possible since the use of synthetic agro-chemicals is not permitted. On the other hand certification for organic agriculture products is a cumbersome and expensive process which is not essential in a sustainable agriculture system. Use of biopesticides is still at a low level in many countries. Non availability of information regarding such biopesticides, non availability of them in the village level markets, non willingness of farmers for implementing integrated pest management methods, low market prices for the crop produce etc., are real challenges for developing sustainable crop protection system in agriculture.

As discussed previously in this chapter, use of agrochemicals in crop protection is larger in the world despite the increasing organic and sustainable agriculture practices. Many reasons are found be accounting for this situation. In the present day intensive agriculture, farmers want to get rid of weeds, pathogens and insect pests from their farmlands for quantity-wise and quality wise better yields. Resistance development such as resurgence of the same pest or outbreak of secondary pests, health hazards and environmental pollution including the negative impact on beneficial organisms and soil organisms are the most threatening issues related to the present-day chemical agriculture. Repetitive use, large scale applications, unnecessary use of pesticides contribute to the development of resistance in pests which in turn makes a demand for more application of pesticides. There are number of factors why farmers cannot exclude agrochemicals and four major reasons are indicated below.

1. Non-availability of resistant varieties or cultivars for all of the major pest and diseases. There are no known naturally resistant varieties for many plant viruses such as *Papaya ring spot virus*. Only moderate levels of multigene resistance have been identified in papaya germplasm and used in conventional breeding programs with limited

success in case of *Papaya ring spot virus* (Zee, 1985). Another example is *Chilli leaf curl virus* where varietal resistance is not reported or ineffective in many chilli growing areas in the world. This situation is common for many other diseases and pests too.

2. Non-availability of effective natural enemies, predators or parasites against some pests, pathogens and weeds is another situation where farmers are compelled to apply agrochemicals.
3. Cultural methods may not show results in the short run. Farmers who engage in intensive large scale crop production are not willing to wait for a long time to solve pest and weed problems in their crops.
4. Lack of awareness programmes on cultural practices/IPM and farmers' preferences in some cultural practices over others and the adoption rate of such technologies etc. have been contributed to less efficiency of such practices.

Careful selection of pesticides is important to reduce the negative impact on beneficial organisms, human and animals and environment. Therefore, a product; (a) that is effective in controlling the pest; (b) that is highly specific to the pest and does not significantly affect beneficial organisms; (c) that has a low human and animal toxicity should be selected when use of a pesticide or a weedicide is inevitable. It is also necessary to pay attention to the quantity of pesticide going to be applied and the method of application (http://web.worldbank.org).

Chemical pesticides have failed to eradicate pests, diseases and weeds. Use of pesticides in order to eradicate pests, pathogens and weeds can destabilize ecosystems since pests, pathogens and weeds are also parts of the ecosystem. Modern pest management analyses the life cycle and ecology of the pest and its interactions with the crop and then uses agro-ecosystem management to manipulate the pest's life-cycle in order to minimize impact on the crop. External pest control inputs are considered if agro-ecosystem management does not succeed in suppressing pest populations and pest damage threatens to exceed economic thresholds (http://web.worldbank.org).

Sustainable agriculture uses new knowledge on crops and pests to reduce the use of environmentally hazardous and costly inputs like pesticides and fertilizers. It is important to take measures to retain the profit margin in crops in a sustainable system though some level of yield reductions are experienced. If the lower yields are the result of lower costs for fertilizers and pesticides the profits on the farm might actually go up. However, a reduction in global crop yields can be another controversial issue since the world food production at the moment is not enough to feed the whole population satisfactorily.

REFERENCES

Ahmed, K.N., Pramanik, S.H.A., Khatun, M., Nargis, A. and Hasan, M.R. (2010). Efficacy of plant extracts in the suppression of insect pests and their effect on the yield of sunflower crop under different climatic conditions. *J. Plant Prot. Sci.,* **2(1):** 53–58.

Anandam, R.J. and Doraiswamy, S. (2002). Role of barrier crops in reducing the incidence of mosaic disease in chilli. *J. Plant Dis. Prot.*, **109:** 109–12.

Andrews, D.J. and Kassam, A. (1976). The importance of multiple cropping in increasing world food supplies. *In*: Papendick, R.I., Sanchez, A. and Triplett, G.B. *Eds*., Multiple Cropping. ASA Special Publication 27. American Society of Agronomy, Madison, WI, pp. 1–10.

Askary, T.H. (2010). Nematodes as biocontrol agents. *Sustainable Agric. Rev*., **3:** 347–78.

Baidoo, P.K., Mochiah, M.B., and Akyaw, M.O. (2010). The effect of time of harvest on the damage caused by the cowpea weevil *Callosobruchus maculatus* (Fab.) (Coleoptera: Bruchidae) *J. stored Products and Postharvest Res.,* **1(3):** 24–28.

Bianchil, F.J.J.A., Booij, C.J.H. and Tscharntke, T. (2006). Sustainable pest regulation in agricultural landscapes: a review on landscape composition, biodiversity and natural pest control 2006. *P. Roy. Soc. B–Biol. Sci.*, **273:** 1715–27.

Brown, M.W. and Tworkoski, T. (2004). Pest management benefits of compost mulch in apple orchards. *Agric. Ecosyst. Environ.*, **103:** 465–72.

Cicerone, R.J., and Oreml, R.S. (1988). Biogeochemical aspects of atmospheric methane. *Global Biogeochem. Cycl.*, **2:** 299–327.

CIKS. (2000). *Bollworm control in cotton.* Center for Indian Knowledge Systems. Chennai, India. *Pesticide Post.,* Vol. 8, No. 6.

Cremlyn, R.J. (1990). Agrochemicals; preparation and mode of action. John Wiley and Sons Ltd.

Da Conceicao, I.L.M.P., Dos Santos, M.C.V., De Olievera Abrantes, I.M. and De Almeida, M.S.N. (2005). Using RAPD markers to analyze genetic diversity in Portuguese potato cyst nematode populations. *J. Nematol.,* **5:** 137–45.

Dadheech, B.S., Dangi, K.L. and Jain, H.K. (2009). Adoption of pod borer management in gram cultivation among farmers in Udaipur District of Rajasthan *J. Global Comm.,* **2(2):** 407–14.

Denys, C. and Tscharntke, T. (2002). Plant insect communities and predator prey ratios in field margin strips, adjacent crop fields, and fallows. *Oecologia,* **130:** 315–24.

Devasthali, S. and Vishwakarma, O.P. (1995). Incidence of major pests of soybean as influenced by *Rhizobium*, FYM, Cycocel and fertilizer. *Indian Agric.,* **39(4):** 265–74.

Devidas, P. and Rehberger, L.A. (1992). The effect of exotoxin (Thurigiensin) from *Bacillus thuringiensis* on *Meloidogyne incognita* and *Caenorhabditis elegans. Plant Soil.*, **145:** 115–20.

Douglass, G.K., *Eds*. (1984). Agricultural sustainability in a changing world order. Westview Press, Boulder, CO, USA, p. 282.

Dropkin, V.H. (1980). Introduction to plant nematology. John Wiley and Sons, New York, p. 293.

Dubey, O.P., Ramunujam, B. and Rajendra, R.J. (2003). Present scenario of crop diseases in India and need for their eco-friendly and sustainable management. Proceeding of the group meeting on antagonistic organisms in Plant disease management held at Project Directorate of Biological control, Bangalore on 10th–11th July 2003, pp. 1–12.

Duesterhaus, R. (1990). Sustainability's promise. *J. of Soil and Water Conservation,* **45(1):** 4.

Duncan, L.W. (1991). Current options for nematode management. *Annu. Rev. Phtopathol.,* **29:** 469–90.

El–Naeim, M.A., Osman, M. and Mahmoud, M.F. (2008). Seasonal abundance patterns of insects and mites on pear trees during the blooming and fruiting seasons at Ismailia Governorate, Egypt *Tunisian J. Plant Prot.,* **3(1):** 47–58.

Evans, A.A.F. (1987). Diapause in nematodes as a survival strategy. *Vistas on Nematology,* pp. 180–87.

Feinstein, L. (1952). Insecticides from plants. *In*: Insects: The Year Book of Agriculture, USDA, Washington, D.C., pp. 222–29.

Fereres, A. (2000). Barrier crops as a cultural control measure of non-persistently transmitted aphid–borne viruses. *Virus Res.,* **71:** 221–31.

Foster, S.P. and Harris, M.O. (1997). Behavioral manipulation methods for insect pest-management. *Annu. Rev. Entomol.,* **42:** 123–46.

Francis, C. and Youngberg, G. (1990). Sustainable agriculture: An overview. Sustainable agriculture in temperate zones, *In***:** Francis, C.A., Flora, C.B. and King, L.D. *Eds.,* John Wiley and Sons, New York, pp. 1–23.

Froud, K.J., Stevens, P.S. and Steven, D. (2001). Survey of alternative host plants for Kelly's citrus thrips (*Pezothrips kellyanus*) in citrus growing regions. *N. Z. Plant Protect.,* **54:** 15–20.

Gaur, H.S. and Perry, R.N. (1991). The role of the moulted cuticles in the desiccation survival of adults of *Rotylenchulus reniformis. Rev. Nematol.,* **14:** 491–96.

Gill, H.K., McSorley, R., Goyal, G., and Webb. S.E., (2010). Mulch as a potential management strategy for lesser cornstalk borer, *Elasmopalpus lignosellus* (Insecta: Lepidoptera: Pyralidae), in bush bean (*Phaseolus vulgaris*). *Florida Entomologist,* **93(2):** 183–90.

Hadisoeganda, W.W. and Sasser, J.N. (1982). Resistance in tomato, bean, southern pea, and garden pea cultivars to root-knot nematodes based on host susceptibility. *Plant Dis.,* **66:** 145–50.

Hall, S.J., Matson, P.A. and Roth, P. (1996). NO emission from soil: implications for air quality modelling in agricultural regions. *Annu. Rev. Energy Environ.,* **21:** 311–46.

Hokkanen H. 1991. Trap cropping in pest management. *Annu. Rev. Entomol.,* **36**: 119–38.

Holm, L.G., Plucknett, D.L., Pancho, J.V. and Herberger. J.P. (1977). The world's worst weeds: distribution and biology. University Press of Hawaii, Honolulu, p. 609.

Honek, A. (1989). Overwintering and annual changes of abundance of *Coccinella septempunctata* in Czechoslovakia (Coleoptera, coccinellidae). *Acta Entomol. Bohemoslov.* **86:** 179–92.

Hooks, C.R., Wang, K.H., Ploeg, A. and McSorley, R. (2010). Using marigold (*Tagetes* spp.) as a cover crop to protect crops from plant–parasitic nematodes. *Appl. Soil Ecol.,* **3:** 307–20.

Hooks, C.R.R. and Fereres, A. (2006). Protecting crops from non-persistently aphid–transmitted viruses: A review on the use of barrier plants as a management tool. *Virus Res.,* **120:** 1–16.

Howard, F., Schwartz, D.H.G., Scott, M., Fichtner, R.H., Whitney, S.C., Mahaffey, L., Camper, M., Otto, K. and McMillan, M. (2009). Straw mulch and reduced risk pesticide impacts on thrips and *Iris Yellow Spot Virus* on western–grown onions. *Southwest. Entomol.,* **34(1):** 13–29.

http://web.worldbank.org accessed on 21st September 2011.

http://www.agnet.org as accessed on 22nd September 2011.

http://www.apsnet.org as accessed on 22nd September 2011.

http://www.knowledgebank.irri.org *as accessed on* 18th September 2011.

http://www.new–ag.info as accessed on 05th June 2011.

http://www.sustainabletable.org *as accessed on* 05th June 2011.

http://www.ucsusa.org as accessed on 05th June 2011.

Jackson, D.M., Canhilal, R., and Carner, G.R. (2005). Trap monitoring squash vine borers in cucurbits. *J. Agric. Urban. Entomol.,* **22:** 27–39.

Jayaraj, S. (1998). Integrated approach vital. The Hindu survey of Indian agriculture, pp. 155–60.

Katan, J. (1981). Solar heating (solarization) of soil for control of soil borne pests. *Ann. Rev. Phytopathol.,* **19:** 211–36.

Kerry, B.R. (1987). Biological control. *In:* Brown, R.H. and Kerry, B.R. *Eds., Biological control of nematodes: prospects and opportunities Principles and practice of nematode control in crops,* Academic Press Sydney, Australia, pp. 233–63.

Khan, Z.R., Pickett, J.A., Van den Berg, J., Wadhams, L.J. and Woodcock, C.M. (2000). Exploiting chemical ecology and species diversity: stem borer and striga control for maize and sorghum in Africa. *Pest Manag. Sci.,* **56:** 957–62.

Killham, K. (2010). Integrated soil management *moving* towards globally sustainable agriculture. *J. Agr. Sci.,* pp. 1–8.

Kohl, J. (2006). Integrated disease management. International conferences 2006: Joint organic congress, Theme 5: Protecting plant health. Wageningen University and research centre.

Latif, M.A., Ullah, M.W., Rafii1, M.Y. and Tajul, M.I. (2011). Management of ufra disease of rice caused by *Ditylenchus angustus* with nematicides and resistance. *Afr. J. Microbiol. Res.,* **5(13):** 1660–67.

Marban–Mendoza, N., Jeyaprakash, A. and Jansson, H.B. (1987). Control of root-knot nematodes on tomato by lectins. *J. Nematol.,* **19:** 331–35.

Maudsley, M., Seeley, B. and Lewis, O. (2002) Spatial distribution patterns of predatory arthropods within an English hedgerow in early winter in relation to habitat variables. *Agric. Ecosyst. Environ.,* **89:** 77–89.

Mayee, C.D. and Rattan. B.K. (1976). A factor in the break-down of resistance of 'Basmati-370' rice to stem-rot disease. *J. of Agr. Sci.,* **87:** 459–60.

McDonald, B.A. and Linde, C. (2002). Pathogen population genetics, evolutionary potential and durable resistance. *Annu. Rev. Phytopathol.,* **40:** 349–79.

Meher, H.C., Gajbhiye, V.T., Chawla, G. and Singh, G. (2009). Virulence development and genetic polymorphism in *Meloidogyne incognita* (Kofoid and White) Chitwood after prolonged exposure to sublethal concentrations of nematicides and continuous growing of resistant tomato cultivars. *Pest Manag. Sci.,* **65(11):** 1201–1207.

MMSU. (2003). *7 new botanical extracts vs. garlic pests.* Mariano Marcos State University. Batac, Ilocos Norte, Philippines.

Narisawa, K., Shimura, M, Usuki, F. Fukuhara, S. and Hashiba, T. (2005). Effects of pathogen density, soil moisture, and soil pH on biological control of clubroot in chinese cabbage by *Heteroconium chaetospira*. *Plant Dis.*, **89(3):** 285–90.

Nelson–Smith, D., (1995). Food or famine politics, economics and science in the world's food supply. Brighton Crop Protection Conference. *Weeds*, **1(1):** 3–15.

Oerke, E.C. (2006). Crop losses to pests. *J. Agr. Sci.*, **144:** 31–43.

Oerke, E.C. and Dehne, H.W. (2004). Safeguarding production losses in major crops and the role of crop protection *Crop Prot.*, **23:** 275–85.

Oka, Y. (2010). Mechanisms of nematode suppression by organic soil amendments—A review. *Appl. Soil Ecol.*, **44(2):** 101–15.

Ou, S.H. (1985). Rice Diseases. 2nd Ed. Commonwealth Mycological Institute, Kew, England, 380p.

Pinstrup–Andersen, P. (2000). The future world food situation and the role of plant diseases.

Pontin, M.D., Wade, M.R., Kehrli, P. and Wratten, S.D. (1998). Attractiveness of single and multiple species flower patches to beneficial insects in agroecosystems. *Ann. Appl. Biol.*, **148:** 39–47.

Prasanna, P.M., and Kumar, N.G. (2011). Effect of FYM and chemical fertilizers on the abundance and diversity of insect pests, soil chemicals and growth and yield parameters of soybean (*Glycine max* L.) *Res. J. Agric. Sci.*, **2(2):** 357–63.

Prot, J.C. (1980). Migration of plant–parasitic nematodes towards plant roots. *Revue Nematol.*, **3:** 305–18.

Reigel, C., Fernandez, F.A. and Noe, J.P. (1996). *Meloidogyne incognita* infested soil amended with chicken litter. *J. nematol.*, **28:** 369–78.

Rich, J.R., Dunn, R. and Noling, J. (2004). Nematicides: past and present uses, *In*: Chen, Z.X., Chen, S.Y. and Dickson, D.W. *Eds.*, *Nematology: Advances and Perspectives, vol. 2. Nematode Management and Utilization*, CABI Publishing, Oxfordshire, pp. 1041–82.

Rieux, R., Simon, S. and Defrance, H. (1999). Role of hedgerows and ground cover management on arthropod populations in pear orchards. *Agric. Ecosyst. Environ.*, **73:** 119–27.

Rosenzweig, C., Iglesias, A., Yang, X.B., Epstein, P.R. and Chivian, E. (2001). Climate change and extreme weather events; Implications for food production, plant diseases, and pests. Global change and human health, **2(2):** 90–104.

Rosenzweig, C. and Iglesias, A. (2007). Encyclopaedia of pest management, Vol. II, *In*: D. Pimental, *Eds.*, CRC Press, pp. 87–89.

Schmidt, M.H. and Tscharntke, T. (2005). The role of perennial habitats for Central European farmland spiders. *Agric. Ecosyst. Environ.*, **105:** 235–42.

Schmidt, N.P., O'Neal, M.E. and Singer, J.W. (2007). Alfalfa living mulch advances biological control of soybean aphid. *Environ. Entomol.*, **36:** 416–24.

Sengonca, C., Kranz, J. and Blaeser, P. (2002). Attractiveness of three weed species to polyphagous predators and their influence on aphid populations in adjacent lettuce cultivations. *J. Pest Sci.*, **75:** 161–65.

Seo, B.Y., Jung, J.K., Choi, B.R., Park, H.M. and Lee, B.H. (2009). Resistance breaking ability and feeding behavior of the brown planthopper, *Nilaparvata lugens*. *In*: Heong, K.L. and Hardy, B. *Eds.*, Planthoppers: new threats to the sustainability of intensive rice production systems in Asia. Los Baños (Philippines): International Rice Research Institute, pp. 303–14.

Sharma, S., and Sharma, A.K. (2007). Biocontrol: an eco-friendly disease management approach. *In*; Trivedi, P.C. and Kant, U. *Eds.*, Biocontrol of Plant Diseases, Avishkar Publishers, India, p. 438.

Shri Ram, Patil, B.D. and Purohit, M.L. (1990). Role of fertilizer (potassium and phosphorus) and soil insecticides in the pest management of fodder cowpea, *Vigna unguiculata* (L) (Walp). *Indian J. Entomol.*, **52(4):** 627–36.

Shrivastava, G., Rogers, M., Wszelaki, A., Panthee, D.R. and Chen, F. (2010). Plant volatiles based insect pest management in organic farming, *Crit. Rev. Plant Sci.*, **29(2):** 123–33.

Siddiqui, Z.A. and Mahmood, I. (1995). Some observations on the management of the wilt disease complex on pigeon pea by treatment with vascular arbuscular fungus and biocontrol agent for nematodes. *Bioremediation Technology*, **54:** 227–23.

Singh, A., Vasisht, A.K., Kumar, R. and Das, D.K. (2008). Adoption of Integrated Pest Management Practices in Paddy and Cotton: A Case Study in Haryana and Punjab. *Agric. Econ. Res. Rev.*, **21:** 221–26.

Singh, U.S, Mishra, D.S., Prasad, R.D. and Chaube, H.S. (2001). Biological control of plant pathogens in India: Historical perspective, present status and future prospects *In***:** Singh, S.P. *Ed.*, Hundred years of biological control in India. ICAR, New Delhi.

Smith, H.A. and McSorley, R. (2000). Intercropping and pest management: A Review of Major Concepts. *American Entomologist*, **46(3):** 154–61.

Sotomayor, D., Allen, Jr.L.H., Chen, Z., Dickson, D.W. and Hewlett, T. (1999). Anaerobic soil management practices and solarization for nematode control in Florida. *Nematropica*, **29(2):** 153–70.

Spiegel, Y., Cohn, E., Galper, S., Sharon, E. and Chet, I. (1991). Graduation of a newly isolated bacterium, *Pseudomonas chitinolytica* sp. nov., for controlling the root-knot nematode *Meloidogyne javanica*. *Biocon. Sci. Technol.*, **1:** 115–25.

Staley, J.T., Stewart–Jones, A., Pope, T.W., Wright, D.J., Leather, S.R., Hadley, P., Rossiter, J.T., van Emden, H.F. and Poppy, G.M. (2010). Varying responses of insect herbivores to altered plant chemistry under organic and conventional treatments. *P. Roy. Soc. B–Biol. Sci.*, **277(1682):** 779–86.

Stapleton, J.J., and Devay, J.E. (1983). Response of phytoparasitic and free-living nematodes to soil solarization and 1,3–dichloropene in California. *Phytopathol.*, **73:** 1429–36.

Summers, C.G., Mitchell, J.P. and Stapleton, J.J. (2004). Management of aphid borne viruses and *Bemisia argentifolii* (Homoptera: Aleyrodidae) in zucchini squash by using UV reflective plastic and wheat straw mulches. *Environ. Entomol.*, **33:** 1447–57.

Swanton, C.J., Mahoney, K.J., Chandler, K. and Gulden, R.H. (2008). Integrated weed management: knowledge-based weed management systems. *Weed Sci.*, **56:** 168–72.

Talavera, M., Verdejo–Lucas, S., Ornat, C., Torres, J., Vela, M.D., Macias, F.J., Cortada, L., Arias, D.J., Valero, J. and Sorribas, F.J. (2009). Crop rotations with *Mi* gene resistant and susceptible tomato cultivars for management of root-knot nematodes in plastic houses. *Crop Prot.*, **28(8):** 662–67.

Talukder, D., Malek, M.A., Khanam, L.A.M. and Dey, K.C. (1998). Toxicity of some indigenous plant seed oil against *Tribolium confusum* (Duval) (Coleoptera: Tenebrionidae) adults. *Pakistan J. Zoology,* **30:** 331–34.

Taylor, S.P., Hollaway, G.J. and Hunt, C.H. (2000). Effect of field crops on population densities of *Pratylenchus neglectus* and *P. thornei* in southeastern Australia. *J. Nematol.,* **32:** 591–99.

Teasdale, J.R., Abdul–Baki, A.A., Mill, D.J., and Thorpe, K.W. (2004). Enhanced pest management with cover crop mulches. *Acta Hort.,* **638:** 135–40.

TeBeest D.O. (1993). Biological control of weeds with fungal plant pathogens. *In:* Jones, D.G. *Ed.*, Exploitation of microorganisms. Chapman and Hall, London, pp. 1–17.

Tempeleton, G.E., TeBeest, D.O. and Smith, R.J. (1979). Biological weed control with mycoherbicides. *Ann. Rev. Phytopath.,* **17:** 301–10.

Tilman, D. (1999). Global environmental impacts of agricultural expansion: the need for sustainable and efficient practices. *Proc. Natl. Acad. Sci.,* **96:** 5995–6000.

United Nations, (1996). World population prospects: The 1996 revisions. United Nations, New York.

Wahab, S. (2009). Biotechnological approaches in the management of plant pests, diseases and weeds for Sustainable Agriculture. *J. Biopesticides.,* **2(2):** 115–34.

Wang, K.H., McSorley, R. and Gallaher, R.N. (2003. Effect of *Crotalaria juncea* amendment on nematode communities in soil with different agricultural histories. *J. Nematol.,* **35:** 294–301.

Westerdahl, B.B., Caswell–Chen, E.P. and Bugg, R.L. (1998). Nematodes. *In:* Engels, C. *Ed., Cover Cropping in vineyards: A grower's handbook,* University of California Division of *Agriculture and Natural Resources. Pub.,* **3338:** 113–25.

Wilson, C.R. (1998). Incidence of weed reservoirs and vectors of tomato spotted wilt tospovirus on southern Tasmanian lettuce farms. *Plant Pathol.,* **47:** 171–76.

Windham, G.L., Windham, M.T. and Williams, W.P. (1989). Effect of *Trichoderma* species on maize growth and *Meloidogyne arenaria* reproduction. *Pl. Dis.,* **73:** 493–95.

Windham, G.L., Windham, M.T. and Williams, W.P. (1989). Effect of *Trichoderma* species on maize growth and *Meloidogyne arenaria* reproduction. *Pl. Dis.,* **73:** 493–95.

Wu, H., Wang, C.J., Bian, X.W., Zeng, Lin., K.C., Wu, B., Zhang, G.A. and Zhang, X. (2011). Nematicidal efficacy of isothiocyanates against root-knot nematode *Meloidogyne javanica* in cucumber. *Crop Prot.,* **30(1):** 33–37.

Xiao–yan L., Ping, H.E. and Ji–yun, Jin. (2006). Advances in effect of potassium nutrition on plant disease resistance and its mechanism. *Plant Nutrition and Fertilizer Sci.,* 03.

Zee F.T.P. (1985). Breeding for papaya ringspot virus tolerance in solo papaya, *Carica papaya* L. Doctoral dissertation, University of Hawaii, Honolulu.

Subject Index

D

E

F

G

H

I

J

K

L

M

N

O

P

Q

R

S

T